Toxicology of the Nose and Upper Airways

TARGET ORGAN TOXICOLOGY SERIES

Series Editors

A. Wallace Hayes, John A. Thomas, and Donald E. Gardner

Toxicology of the Nose and Upper Airways. *John B. Morris and Dennis J. Shusterman, editors, 2010*
Toxicology of the Skin. *Nancy A. Monteiro-Riviere, editor, 2010*
Neurotoxicology, Third Edition. *G. Jean Harry and Hugh A. Tilson, editors, 2010*
Endocrine Toxicology, Third Edition. *J. Charles Eldridge and James T. Stevens, editors, 2010*
Adrenal Toxicology. *Philip W. Harvey, David J. Everett, and Christopher J. Springall, editors, 2008*
Cardiovascular Toxicology, Fourth Edition. *Daniel Acosta, Jr., editor, 2008*
Toxicology of the Gastrointestinal Tract. *Shayne C. Gad, editor, 2007*
Immunotoxicology and Immunopharmacology, Third Edition. *Robert Luebke, Robert House, and Ian Kimber, editors, 2007*
Toxicology of the Lung, Fourth Edition. *Donald E. Gardner, editor, 2006*
Toxicology of the Pancreas. *Parviz M. Pour, editor, 2005*
Toxicology of the Kidney, Third Edition. *Joan B. Tarloff and Lawrence H. Lash, editors, 2004*
Ovarian Toxicology. *Patricia B. Hoyer, editor, 2004*
Cardiovascular Toxicology, Third Edition. *Daniel Acosta, Jr., editor, 2001*
Nutritional Toxicology, Second Edition. *Frank N. Kotsonis and Maureen A. Mackey, editors, 2001*
Toxicology of Skin. *Howard I. Maibach, editor, 2000*
Neurotoxicology, Second Edition. *Hugh A. Tilson and G. Jean Harry, editors, 1999*
Toxicant–Receptor Interactions: Modulation of Signal Transductions and Gene Expression. *Michael S. Denison and William G. Helferich, editors, 1998*
Toxicology of the Liver, Second Edition. *Gabriel L. Plaa and William R. Hewitt, editors, 1997*
Free Radical Toxicology. *Kendall B. Wallace, editor, 1997*
Endocrine Toxicology, Second Edition. *Raphael J. Witorsch, editor, 1995*
Carcinogenesis. *Michael P. Waalkes and Jerrold M. Ward, editors, 1994*
Developmental Toxicology, Second Edition. *Carole A. Kimmel and Judy Buelke-Sam, editors, 1994*
Nutritional Toxicology. *Frank N. Kotsonis, Maureen A. Mackey, and Jerry J. Hjelle, editors, 1994*
Ophthalmic Toxicology. *George C. Y. Chiou, editor, 1992*
Toxicology of the Blood and Bone Marrow. *Richard D. Irons, editor, 1985*
Toxicology of the Eye, Ear, and Other Special Senses. *A. Wallace Hayes, editor, 1985*
Cutaneous Toxicity. *Victor A. Drill and Paul Lazar, editors, 1984*

Toxicology of the Nose and Upper Airways

Edited by

John B. Morris
University of Connecticut
Storrs, Connecticut, U.S.A.

Dennis J. Shusterman
University of California
San Francisco, California, U.S.A.

CRC Press
Taylor & Francis Group
Boca Raton London New York

CRC Press is an imprint of the
Taylor & Francis Group, an **informa** business

CRC Press
Taylor & Francis Group
6000 Broken Sound Parkway NW, Suite 300
Boca Raton, FL 33487-2742

First issued in paperback 2019

© 2010 by Taylor & Francis Group, LLC
CRC Press is an imprint of Taylor & Francis Group, an Informa business

No claim to original U.S. Government works

ISBN-13: 978-1-4200-8187-9 (hbk)
ISBN-13: 978-0-367-38450-0 (pbk)

A CIP record for this book is available from the British Library.

Library of Congress Cataloging-in-Publication Data available on application

**Visit the Taylor & Francis Web site at
http://www.taylorandfrancis.com**

**and the CRC Press Web site at
http://www.crcpress.com**

Preface

In recent years there has been a growing recognition of the importance of the upper airway in normal and abnormal respiratory function. Realization of the role of the upper airway, not only as a sentinel of exposure, but also as an integral part of respiratory processes, has brought this area of study to the forefront of inhalation toxicology. At the same time, a major shift in study methods has occurred. Application of molecular biologic methods, recognition of neurogenic inflammatory processes, and utilization of genetic knockout animals are but a few of the recent advances in research tools, yielding qualitatively new types of data. This volume constitutes our attempt to summarize and update the body of knowledge pertaining to upper airway toxicology and to link data obtained in both human and experimental animal studies. In doing so, we hope to have created an indispensable reference text for toxicologists, sensory scientists, and clinicians.

As scientists actively involved in this area, we have observed that the same phenomena are often studied in separate settings from different perspectives, with each discipline unaware of the models, study tools, and insights of the others. To correct this situation and to achieve the broadest possible perspective, we have sought contributions from internationally recognized leaders in the fields of experimental toxicology, respiratory medicine, otolaryngology, allergy, and sensory science. For completeness' sake, a tandem structure was utilized in many places (with separate chapters on human and experimental animal data). Finally, discussions were included of non-cancer risk assessment, quantitative structure-activity relationships, and the interaction of allergy and chemical irritation. Thus, our aim is to provide, in a single volume, an integration of the basic science, clinical science and regulatory aspects of issues relating to the nose and upper airways.

Notwithstanding this broad perspective, experimental toxicology remains a major focus of this volume. Most experimental toxicology is conducted, explicitly or implicitly, to address human health concerns as filtered through regulatory requirements. In that context the emphasis is often quantitative risk assessment. Mechanistic issues, to the extent studied, usually concern either dose-response relationships or interspecies extrapolation. The editors of this volume have had the rare opportunity to step outside of this paradigm and study, in tandem, the effect of selected pollutants on the upper airways of both humans and experimental animals, emphasizing mechanistic issues in the process. We hope that our enthusiasm in studying the upper airway is evident in our choice of topics and authors, and that the reader will find this volume a coherent and valuable information resource, providing new insights across disciplinary lines.

John B. Morris
Dennis J. Shusterman

Contents

Preface v
Contributors xi

PART III FUNCTIONAL AND PATHOLOGIC RESPONSES AND THEIR
         MEASUREMENT

Contributors

M. C. Ávila-Casado   Department of Cellular and Tissular Biology, School of Medicine, National University of Mexico, Mexico City, Mexico

Michael H. Abraham   Department of Chemistry, University College London, London, U.K.

Bahman Asgharian   Applied Research Associates, Raleigh, North Carolina, U.S.A.

James N. Baraniuk   Division of Rheumatology, Immunology and Allergy, Georgetown University Medical Center, Washington, D.C., U.S.A.

Fuad M. Baroody   Pritzker School of Medicine, University of Chicago, Chicago, Illinois, U.S.A.

William S. Cain   Chemosensory Perception Laboratory, Department of Surgery (Otolaryngology), University of California, San Diego, La Jolla, California, U.S.A.

Stephan A. Carey   Department of Small Animal Clinical Sciences, College of Veterinary Medicine, Michigan State University, East Lansing, Michigan, U.S.A.

J. Enrique Cometto-Muñiz   Chemosensory Perception Laboratory, Department of Surgery (Otolaryngology), University of California, San Diego, La Jolla, California, U.S.A.

Jonathan Corren   Allergy Research Foundation, Los Angeles, California, U.S.A.

Pamela Dalton   Monell Chemical Senses Center, Philadelphia, Pennsylvania, U.S.A.

David C. Dorman   College of Veterinary Medicine, North Carolina State University, Raleigh, North Carolina, U.S.A.

C. I. Falcón-Rodríguez   Department of Cellular and Tissular Biology, School of Medicine, National University of Mexico, Mexico City, Mexico

T. I. Fortoul   Department of Cellular and Tissular Biology, School of Medicine, National University of Mexico, Mexico City, Mexico

Melanie L. Foster   College of Veterinary Medicine, North Carolina State University, Raleigh, North Carolina, U.S.A.

Mary Beth Genter Department of Environmental Health, University of Cincinnati, Cincinnati, Ohio, U.S.A.

Javier Gil-Lostes Department of Chemistry, University College London, London, U.K.

Jack R. Harkema Department of Pathobiology and Diagnostic Investigation, College of Veterinary Medicine, Michigan State University, East Lansing, Michigan, U.S.A.

Thomas Hummel Smell & Taste Clinic, Department of Otorhinolaryngology, University of Dresden Medical School, Dresden, Germany

Julia S. Kimbell Otolaryngology/Head and Neck Surgery, University of North Carolina, Research Triangle Park, North Carolina, U.S.A.

Jane Q. Koenig Department of Environmental and Occupational Health Sciences, University of Washington, Seattle, Washington, U.S.A.

C. Frieke Kuper Department of Toxicology and Applied Pharmacology, Business Unit Quality and Safety, TNO Quality of Life, Zeist, The Netherlands

N. López-Valdez Department of Cellular and Tissular Biology, School of Medicine, National University of Mexico, Mexico City, Mexico

Samantha Jean Merck Division of Rheumatology, Immunology and Allergy, Georgetown University Medical Center, Washington, D.C., U.S.A.

L. F. Montaño Department of Cellular and Tissular Biology, School of Medicine, National University of Mexico, Mexico City, Mexico

John B. Morris Department of Pharmaceutical Sciences, School of Pharmacy, University of Connecticut, Storrs, Connecticut, U.S.A.

Owen R. Moss POK Research, Apex, North Carolina, U.S.A.

Murugan Ravindran Division of Rheumatology, Immunology and Allergy, Georgetown University Medical Center, Washington, D.C., U.S.A.

Paige M. Richards Department of Biology, Wake Forest University, Winston-Salem, North Carolina, U.S.A.

Karen Riveles Office of Environmental Health Hazard Assessment, California Environmental Protection Agency, Oakland, California, U.S.A.

V. Rodríguez-Lara Department of Cellular and Tissular Biology, School of Medicine, National University of Mexico, Mexico City, Mexico

Ricardo Sánchez-Moreno Department of Chemistry, University College London, London, U.K.

Andrew G. Salmon Office of Environmental Health Hazard Assessment, California Environmental Protection Agency, Oakland, California, U.S.A.

C. J. Saunders Neuroscience Program, University of Colorado Denver, Anschutz Medical Campus, Aurora, Colorado, U.S.A.

Suzaynn Schick   Division of Occupational and Environmental Medicine, University of California, San Francisco, California, U.S.A.

Dennis J. Shusterman   Division of Occupational and Environmental Medicine, University of California, San Francisco, California, U.S.A.

Wayne L. Silver   Department of Biology, Wake Forest University, Winston-Salem, North Carolina, U.S.A.

Piet J. Slootweg   Department of Pathology, Radboud University Medical Center, HB Nijmegen, The Netherlands

Kathryn Sowerwine   Division of Rheumatology, Immunology and Allergy, Georgetown University Medical Center, Washington, D.C., U.S.A.

Ricardo Tan   Allergy Research Foundation, Los Angeles, California, U.S.A.

Thomas E. Taylor-Clark*   Division of Allergy & Clinical Immunology, Johns Hopkins School of Medicine, Baltimore, Maryland, U.S.A.

Karla D. Thrall   Pacific Northwest National Laboratory, Richland, Washington, U.S.A.

Bradley J. Undem   Division of Allergy & Clinical Immunology, Johns Hopkins School of Medicine, Baltimore, Maryland, U.S.A.

Christoph van Thriel   IfADo—Leibniz Research Centre for Working Environment and Human Factors, Dortmund, Germany

James G. Wagner   Department of Pathobiology and Diagnostic Investigation, College of Veterinary Medicine, Michigan State University, East Lansing, Michigan, U.S.A.

Bruce S. Winder   Office of Environmental Health Hazard Assessment, California Environmental Protection Agency, Oakland, California, U.S.A.

Ruud A. Woutersen   Department of Toxicology and Applied Pharmacology, Business Unit Quality and Safety, TNO Quality of Life, Zeist, The Netherlands

---

Current affiliation: Department of Molecular Pharmacology and Physiology, University of South Florida, Tampa, Florida, U.S.A.

1 Comparative Anatomy of Nasal Airways: Relevance to Inhalation Toxicology and Human Health

Jack R. Harkema and James G. Wagner

Department of Pathobiology and Diagnostic Investigation, College of Veterinary Medicine, Michigan State University, East Lansing, Michigan, U.S.A.

Stephan A. Carey

Department of Small Animal Clinical Sciences, College of Veterinary Medicine, Michigan State University, East Lansing, Michigan, U.S.A.

INTRODUCTION TO NASAL STRUCTURE AND FUNCTION

The nose is a structurally and functionally complex organ in the upper respiratory tract of mammalian species. It is the primary site of entry for inhaled air in the respiratory system and therefore has many important and diverse functions. The nose not only serves as the principal organ for the sense of smell (olfaction), but it also functions to efficiently filter, warm, and humidify the inhaled air (air conditioning) before it enters the more delicate distal airways and alveolar parenchyma in the lung (1). The nasal passages have been described as an efficient "scrubbing tower" for the respiratory tract because they effectively absorb water-soluble and reactive gases and vapors, trap inhaled particles, and metabolize airborne xenobiotics (2). With its role as an "air conditioner" and a "defender" of the lower respiratory tract, the nose may also be vulnerable to acute or chronic injury caused by exposure to airborne toxic or infectious agents. Many diseases afflict nasal airways and associated paranasal sinuses, including allergic rhinitis and chronic sinusitis. The majority of these conditions are a consequence of viral or bacterial infections, allergic reactions, or aging. However, exposure of humans to toxic agents may also cause or exacerbate certain nasal diseases. In recent years, there has been a marked increase in the study of nasal toxicology and in assessing the human risk of nasal injury from inhaled toxicants (3,4). When using animal toxicology studies to estimate the risks of nasal toxicants to human health, it is important to have a good working knowledge of comparative nasal structure and function. Comparative aspects of the mammalian nose that have special relevance to inhalation toxicology are highlighted in this brief review.

COMPARATIVE GROSS ANATOMY OF THE NASAL AIRWAYS AND PARANASAL SINUSES

The nasal airway is divided into two air passages by the nasal septum. Each nasal passage extends from the nostrils to the nasopharynx. The nasopharynx is defined as the airway posterior to the termination of the nasal septum and proximal to the termination of the soft palate. Inhaled air flows through the nostril openings, or nares, into the vestibule, which is a slight dilatation just inside the nares and before the main chamber of the nose. Unlike the more distal main nasal chamber that is surrounded by bone, the nasal vestibule is surrounded primarily by more flexible cartilage. The luminal surface is lined by a squamous epithelium similar to that of external skin. In humans, unlike laboratory animals, the nasal vestibule also contains varying numbers of hairs near the nares. After passing through the nasal vestibule, inhaled air courses through the narrowest part of the entire respiratory tract, the nasal valve (ostium internum), into the main nasal chamber. A lateral wall, septal wall, roof, and floor define each nasal passage of the main chamber. The lumen of the main chamber is lined by well-vascularized and innervated mucous membranes that are covered by a continuous layer of mucus. The nasal mucous layer is moved distally by underlying cilia to the oropharynx where it is swallowed into the esophagus.

Turbinates, bony structures lined by the well-vascularized mucosal tissue, project into the airway lumen from the lateral walls into the main chamber of the nose. Nasal turbinates increase the inner surface area of the nose, which is important in the filtering, humidification, and warming of the inspired air. Though the turbinated, main chamber of the human nose is only about 5 to 8 cm long, the surface area is approximately 150 to 200 cm^2, about four times that of the human trachea (5). These turbinates divide the main nasal chamber into distinct intranasal airways (e.g., dorsal, lateral, middle, and ventral meatus).

There are some general similarities in the nasal passages of mammalian species, but there are also striking interspecies differences in nasal architecture [Fig. 1(A)]. From a comparative viewpoint, humans have relatively simple noses with breathing as the primary function (microsmatic), while other mammals have more complex noses with olfaction as the primary function (macrosmatic). In addition, the nasal and oral cavities of humans (and some nonhuman primates) are arranged in a manner to allow for both nasal and oral breathing. Most laboratory rodents used in inhalation toxicology studies (e.g., rats, mice, hamsters, guinea pigs) are obligate nose breathers because of the close apposition of the epiglottis to the soft palate. Interspecies variability in nasal gross anatomy has been emphasized in previous reviews (4,6) and demonstrated in early studies using silicone rubber casts of the nasal airways (7). Variation in the shape of nasal turbinates contributes to the marked differences in airflow patterns among mammalian species.

The human nose has a superior, middle, and inferior turbinate in the main chamber of each nasal passage. These structures are relatively simple in shape compared to turbinates in most nonprimate laboratory animal species (e.g., rat, mouse, dog) that have complex folding and branching patterns [Fig. 1(A)]. In laboratory rodents, evolutionary pressures concerned chiefly with olfactory function and dentition have defined the shape of the turbinates and the type and distribution of the cells lining the turbinates. In the proximal nasal airway, the complex maxilloturbinates of small laboratory rodents and rabbits provide far better protection of the lower respiratory tract, by better filtration, absorption, and disposal of airborne particles and gases, than do the simple middle and inferior turbinates of the human nose. The highly complex shape of the ethmoturbinates, lined predominantly by olfactory neuroepithelium, in the distal half of the nasal cavity of laboratory rodents, rabbits, and dogs is especially designed

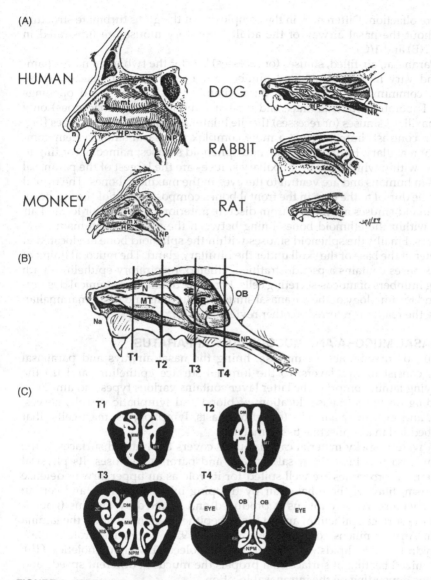

FIGURE 1 (**A**) Diagrammatic representation of the exposed mucosal surface of the lateral wall and turbinates in the nasal airway of the human, monkey, dog, rabbit, and rat. The nasal septum has been removed to expose the nasal passage. *Abbreviations*: HP, hard palate; n, naris; NP, nasopharynx; et, ethmoturbinates; nt, nasoturbinate; mx, maxilloturbinate; mt, middle turbinate; it, inferior turbinate; st, superior turbinate. (**B**) Illustration of the lateral wall and turbinates in the nasal passage of a mouse. The septum has been removed to expose the nasal passage. Vertical lines indicate the location of the anterior faces of four tissue blocks routinely sampled for light microscopic examination (T1–T4). (**C**) Anterior faces of selected tissue blocks from the proximal (T1) to the distal (T4) nasal airway. *Abbreviations*: N, nasoturbinate; MT, maxilloturbinate; 1E–6E, six ethmoturbinates; Na, naris; NP, nasopharynx; HP, hard palate; OB, olfactory bulb of the brain; S, septum; V, ventral meatus; MM, middle meatus; L, lateral meatus; DM, dorsomedial meatus; arrow in T2, nasopalatine duct; MS, maxillary sinus; NPM, nasopharyngeal meatus.

for acute olfaction. Differences in the complexity of the gross turbinate structure throughout the nasal airway of the adult laboratory mouse are illustrated in Figure 1(B) and 1(C).

Paranasal, air-filled, sinuses (or recesses) border the two main nasal chambers and vary in size, shape, and number among mammalian species. These sinuses communicate to the main nasal chamber by varying sized openings (ostia). Laboratory rodents (mice and rats) and monkeys (e.g., macaques) only have maxillary sinuses (or recesses) that lie bilaterally to the nasal passages [Fig. 1(C)]. In contrast, humans have a more complex paranasal sinus system composed of maxillary, frontal, ethmoid, and sphenoid sinuses, named according to the bone within which they lie. Maxillary sinuses are the largest of the paranasal sinuses in humans and are ventral to the eyes in the maxillary bones. The frontal sinuses lie dorsal to the eyes in the frontal bones composing part of the forehead. The ethmoid sinuses are formed from discrete anterior and posterior located air spaces within the ethmoid bones lining between the eyes and the main nasal chambers. Finally, the sphenoid sinuses within the sphenoid bone are located at the center of the base of the skull under the pituitary gland. The mucosal lining of all the sinuses contains a pseudostratified, ciliated, respiratory epithelium with varying numbers of mucus-secreting cells. For more detailed anatomical descriptions and terminology of the paranasal sinuses in humans and other mammalian species, the reader is referred to other reviews (6,8,9).

THE NASAL MUCOSA AND MUCOCILIARY APPARATUS
The mucous membranes, or mucosa, lining the nasal airways and paranasal sinuses consist of two layers: (*i*) the luminal surface epithelium and (*ii*) the underlying lamina propria. The latter layer contains various types and amounts, depending on the intranasal location, of blood and lymphatic vessels, nerves, glands, and mesenchymal cells (e.g., fibroblasts, lymphocytes, mast cells) that are embedded in a connective tissue matrix.

A watery, sticky material called mucus covers the luminal surfaces of the nasal mucosa that lines the nasal airways and paranasal sinuses. Its physical and chemical properties are well suited for its role as an upper airway defense mechanism, filtering the inhaled air by trapping inhaled particles and certain gases or vapors. Airway mucus is produced and secreted by mucous (goblet) cells in the surface epithelium and in the subepithelial glands within the lamina propria. Airway mucus contains approximately 95% water, 1% protein, 0.9% carbohydrate, 0.8% lipids, and other small molecular weight moieties (10). Synchronized beating of surface cilia propels the mucus at different speeds and directions depending on the intranasal location.

Mucus covering the olfactory mucosa moves very slowly, with a turnover time of probably several days. In contrast, the mucus covering the transitional and respiratory epithelium is driven along rapidly (1–30 mm/min) by synchronized beating of the surface cilia with an estimated turnover time of about 10 minutes in the rat (11). The mucus with the entrapped materials ultimately is propelled by the beating cilia to the naso- and oropharynx, and then is swallowed into the esophagus and cleared through the digestive tract. The nasal mucociliary apparatus exhibits a range of responses to inhaled xenobiotic agents and can be a sensitive indicator of toxicity (12). Since this upper airway apparatus is one of the first lines of defense against inhaled pathogens, dusts, and irritant gases, toxicant-induced compromises in its defense capabilities could lead to increased nasal infections and increased susceptibility to lower respiratory tract diseases.

The amount of intraepithelial mucosubstances (i.e., stored mucous product within mucus-secreting cells present in the surface epithelum) in the nasal airways of macaque monkeys (e.g., *Macaca radiata*) has been estimated using histochemical and morphometric techniques (13–15). Like the anterior–posterior gradient increase of mucous cells in the human nasal airway (16), there is an anterior–posterior gradient increase in the amount of intraepithelial mucosubstances in the nasal cavity of these monkeys. Scant amounts of both neutral and acidic mucosubstances are also present in the anterior nasal airway, while the respiratory epithelium covering the maxilloturbinates and nasopharynx has copious amounts of this stored secretory product.

Similar estimates of intraepithelial mucosubstances in the anterior nasal airways and nasopharynx have been made for the F344 rat (17). In contrast to the monkey, the rat has considerably more intraepithelial mucosubstances in the anterior septal respiratory epithelium than that it does in the more distal respiratory epithelium lining the nasopharynx. Like monkeys, however, laboratory rats normally contain very little mucosubstances in the transitional epithelium lining the lateral wall of their proximal nasal passages.

Since mucus is a protective substance for upper airway epithelium, intranasal regional differences in intraepithelial mucosubstances may be useful in predicting sites of certain toxicant-induced nasal injury. For example, mucus is known to be a strong antioxidant agent (18), and inhalation of ambient concentrations of ozone, a strong oxidant in urban smog, has been reported to injure regions of both the monkey and rat nasal airways that contain very little intraepithelial mucosubstances, and spare adjacent regions that contain abundant stored secretory product (14,19–21). More studies designed to examine the protective effects of mucus, and other endogenous antioxidants, in upper airways are needed to fully understand the pathogenesis of oxidant-induced injury, or other toxicant-induced injury, to the nasal mucosa.

NASAL BLOOD VESSELS AND BLOOD FLOW

The subepithelial connective tissue (lamina propria) of the nasal mucosa has a rich and complex network of blood vessels, with each of the epithelial regions receiving blood from a separate arterial supply (22). The vascular system in the nose is composed of resistance and capacitance vessels. Resistance vessels are small arteries, arterioles, and arteriovenous anatomoses. A rich microvascular circulation lies just beneath the surface epithelium of the nasal mucosa. Blood flow to the mucosa is regulated by constriction and dilation of these vessels (23,24). Interestingly, the direction of blood flow in the nose runs toward the naris and countercurrent to inspired airflow. This helps to quickly and efficiently warm the incoming air. Extravasation of plasma from the subepithelial microcirculation occurs in the nasal airways in response to noxious stimuli and may contribute both to mucosal defense (e.g., extravasated immunoglobulins to neutralize allergens) and promotion of nasal airway inflammation (e.g., extravasated proinflammatory mediators) (25,26).

A unique feature of the vasculature of the nose is the large venous sinusoids (i.e., capacitance vessels, venous erectile tissue, or swell bodies) that lie deeper in the lamina propria of the nasal mucosa (24). In humans and laboratory animals, these blood vessels are well developed in specific sites of the anterior or proximal aspects of the nasal passages. Capacitance vessels have dense adrenergic innervation, and the congestion and constriction of these vascular structures are regulated by the sympathetic nerve supply to the nose (27). Congestion of blood in these vessels increases the thickness of the mucosal lamina propria, resulting in a narrowing of the nasal airways, an increase in airway resistance, and changes in intranasal airflow patterns.

Though the rodent nose receives less than 1% of the cardiac output, the vascular uptake of nonreactive gases has been shown to be strongly dependent on nasal blood perfusion rates (28). Therefore, nasal blood flow may be important in removing certain toxic materials from the nose and protecting the respiratory tract from toxicant-induced injury. In contrast, the vascular system may also deliver noninhaled systemic xenobiotics or their metabolites to the nasal mucosa. A wide range of chemicals have been administered to rodents by noninhalation routes that result in subsequent nasal damage (e.g., nitrosamines, acetaminophen) (29–31).

NASAL INNERVATION AND NASAL REFLEXES

At the entrance of the respiratory tract, the nose is in the ideal position to detect toxic airborne particles and gases that could potentially injure the lower tracheobronchial airways and alveolar parenchyma in the lungs. Olfactory and trigeminal nerves innervate the nasal mucosa and provide a sensitive sensory detection system for odors and noxious stimuli, respectively. Olfactory sensory nerves extend from the olfactory epithelium (OE) to the olfactory bulb of the brain without any synaptic junctions between nose and brain. Detecting odorants is the chief function of these chemoreceptor cells (see more detailed description below under section "Olfactory Epithelium").

The trigeminal nerves provide the sense of touch, pain, hot, cold, itch, and the sensation of nasal airflow. These nasal nerve endings detect irritating inhaled chemicals, such as ammonia and sulfur dioxide, and a range of organic substances, such as methanol, acetone, and pyridine. Stimuli from inhaled chemical or physical irritants may initiate respiratory and cardiovascular reflexes via the trigeminal nerves, resulting in apnea and bradycardia. Concentration-dependent reductions in respiratory rate in rodents have been demonstrated after exposure to a number of sensory irritants (32,33).

The nasal mucosa is innervated by both sympathetic and parasympathetic nerve fibers (34). Parasympathetic fibers supply nasal mucosal glands and regulate their secretion. Sympathetic fibers innervate the blood vessels in the lamina propria of the mucosa. Stimulation of these fibers causes nasal vasoconstriction, reduction in blood flow, decongestion of capacitance vessels, and subsequent decrease in nasal airway resistance. In humans, the airway caliber of the left and right nasal cavities normally alternates every 50 minutes to four hours according to an endogenous circadian rhythm that causes vasodilatation in one nasal passage and concurrent vasoconstriction in the other (35). Though the regulation of the nasal cycle in humans and animals is poorly understood, this is likely to be under autonomic control and altered by various inhaled irritants. A more detailed review of the neural regulation of the nasal mucosa is found in a recent article by Baraniuk (36).

CELL POPULATIONS OF THE NASAL SURFACE EPITHELIUM

Besides differences in the gross architecture of the nose among mammalian species, there are also species differences in the surface epithelial cell populations lining the nasal passages. These differences among species are found in the distribution of nasal epithelial populations and in the types of nasal cells within these populations. There are, however, four distinct nasal epithelial populations in most animal species. These include the squamous epithelium, which is primarily restricted to the nasal vestibule; ciliated, pseudostratified, cuboidal/columnar epithelium, or respiratory epithelium, located in the main chamber and nasopharynx; nonciliated cuboidal/columnar epithelium, or often termed transitional epithelium, lying between the squamous epithelium and the respiratory epithelium in the proximal or anterior aspect of the main chamber; and OE, located in the dorsal or dorsoposterior aspect of the nasal cavity. Figure 2(A) illustrates the general distribution of these distinct epithelial cell populations in the nasal cavity of the laboratory rat and monkey. Figure 2(B) illustrates the histologic features of the different nasal epithelial populations in the laboratory rat. The reader is referred to other reports for a more thorough and detailed description of the intranasal distribution of airway surface epithelia in laboratory rodents and nonhuman primates (37,38).

Olfactory Epithelium

The major difference in nasal epithelium among animal species is the percentage of the nasal airway that is covered by OE. For example, the OE covers a much greater percentage of the nasal cavity in rodents, which have an acute sense of smell, as compared to monkeys or humans, whose sense of smell is not as well developed. Gross et al. (39) morphometrically determined that approximately 50% of the nasal cavity surface area in F344 rats is lined by this sensory neuroepithelium. OE of humans is limited to an area of about 500 mm^2, which is only 3% of the total surface area of the nasal cavity (22). Mice, rabbits, and dogs are much closer to rats than humans or monkeys in respect to the relative amount of OE within their nasal passages.

Three epithelial cell types compose the OE. These are the olfactory sensory neuron (OSN), the supporting (sustentacular) cell, and the basal cell [Fig. 2(B1)]. The OSNs are bipolar neuronal cells interposed between the sustentacular cells (40). The dendritic portions of these neurons extend above the epithelial surface and terminate into a bulbous olfactory knob from which protrude on average 10 to 15 immotile cilia (41). These cilia, about 50 μm in length and 0.1 to 0.3 μm in diameter, are enmeshed with each other and with microvilli in the surface fluid, and provide an extensive surface area for reception of odorants. It has been estimated that the ciliary membranes increase the receptive surface of the OSN by 25 to 40 fold (42).

It should be emphasized that the ciliary membranes of the OSN contain the odorant receptors (ORs) responsible for the chemical interaction with and initial detection of inhaled odors. ORs are G protein-coupled, seven transmembrane membrane proteins that are encoded by the largest gene families known to exist in a given animal genome (43). Odorant genes were discovered by Linda Buck and Richard Axel who were awarded the 2005 Nobel Prize in Physiology or Medicine for their landmark work in the cellular and molecular biology of olfaction (44). Their elegant and novel work was one of the first applications

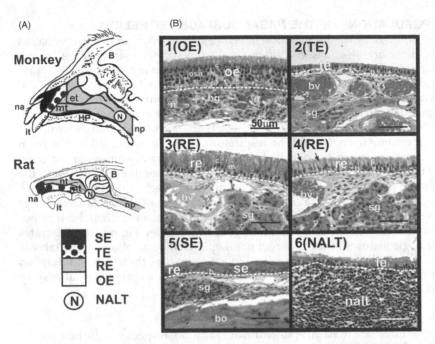

FIGURE 2 (**A**) Distribution of the surface epithelia lining the nasal lateral wall of the monkey and rat. Four distinct epithelial cell populations line both mammalian species: SE, squamous epithelium; TE, transitional epithelium; RE, respiratory epithelium; OE, olfactory epithelium. However, considerably more OE lines the intranasal surface of the rat compared to the monkey. *Abbreviations*: NALT, nasal-associated lymphoid tissue; et, ethmoturbinate; mt, maxilloturbinate; nt, nasoturbinate; na, naris; it, incisor tooth; B, brain. (**B**) Light photomicrographs of the different types of surface epithelia that line the rat nasal airways. Tissue sections are stained with hematoxylin and eosin for routine cellular and acellular morphology, and alcian blue (AB; pH 2.4) to identify epithelial cells with acidic mucosubstances (i.e., mucous cells). (**1**) Olfactory epithelium (oe) lining the dorsal septum from tissue section T3 and containing a prominent apical row of nuclei and cytoplasm of sustentacular cells (s), several middle layers of nuclei in the cell bodies of olfactory sensory neurons (OSN), and basal cells (b) lining the basal lamina. Arrow points to the luminal surface that contains numerous cilia from OSNs and microvilli projecting from the apical surface of sustentacular cells. Bowman's glands (bg), with large amounts of intracellular AB-stained mucosubstances, and nerve bundles (n) are present in the lamina propria. Dotted lines identify the basal lamina separating the surface epithelium from the underlying lamina propria. (**2**) Transitional epithelium (te) lining the proximal lateral meatus in section T1 and composed of nonciliated, cuboidal to low columnar apical cells, and basal cells (b). Blood vessels (bv) and subepithelial glands (sg) are present in the lamina propria. (**3**) Respiratory epithelium (re) lining the proximal septum in section T1 containing columnar, ciliated cells (c), mucous (goblet) cells (m) with large amounts of AB-stained mucosubstances, and basal cells (b). (**4**) Respiratory epithelium lining the midseptum from the middle aspect of the nasal airway in section T2 and containing ciliated cells (c), basal cells (b), and narrow, nonciliated serous cells (arrows), interspersed among the ciliated cells, which normally contain no or scant amounts of acidic mucosubstances. (**5**) Stratified squamous epithelium (se) lining the floor of the ventral meatus in the proximal nasal airway (T1). There is a sharp transition from respiratory epithelium (re) containing numerous AB-staining mucous cells to se. A thin lamina propria containing some subepithelial glands (sg) and blood vessels (bv) lies between the surface epithelium and bone (bo). (**6**) Lymphoepithelium (le) containing both ciliated and nonciliated cells, along with some intraepithelial lymphocytic cells, overlie the nasal-associated lymphoid tissue (NALT) located in the floor of the nasal airway at the opening of the nasopharyngeal meatus (T3).

of degenerate polymerase chain reaction. It is now estimated that there are 500 to 1000 OR genes in the rat and mouse (45), and ~1000 sequences in humans, residing in multiple clusters spread throughout the genome, with more than half being pseudogenes (43).

A single OR gene is expressed in a minute subset of OSN with the current belief that each OSN expresses only a single OR (one receptor–one neuron rule). Interestingly, rat and mouse OR genes are expressed in OSNs within one of four, even-sized, distinct topographical zones in the OE lining the nasal cavity (46,47). OSNs expressing a given OR are distributed in a random, punctate, manner within a zone. Within the nasal cavity of a mouse, there are approximately two million OSNs.

The axon of the OSN originates from the base of the cell and passes through the basal lamina to join axons from other OSNs forming nonmyelinated nerve fascicles, or bundles, in the lamina propria. These olfactory nerves perforate the bony cribriform plate, which separates the nasal cavity from the brain, and form the outer olfactory nerve layer of the olfactory bulb. Axons of OSNs that express the same OR gene converge with extreme precision on ~2000 signal-processing modules called "glomeruli" that reside in distinct locations within the olfactory bulb (48). Glomeruli are relatively large spherical neuropils (100–200 μm in diameter) in which the axons of OSNs form synaptic connections on the dendrites of mitral and tufted cells, the output neurons of the olfactory bulb (49). Transmission of olfactory information is further sent through the axons of the mitral and tufted cells to the olfactory cortex.

Because the OE is in direct contact with the environment, inhaled xenobiotic agents, such as airborne chemical toxicants or infectious microbial agents, may induce cell injury and death of OSNs. Unlike other neurons in the body, the OSNs are able to regenerate when there is neuronal cell loss and there is continual neurogenesis in this nasal epithelium to maintain its olfactory function. Initial studies suggested that OSNs have a steady 28- to 30-day turnover rate in the rat (50,51). Others have shown that many OSNs are more long-lived despite continuous neurogenesis of the OE (52,53). The constant turnover of OSNs is due to the capacity of progenitor cells in the basal cell layer of the OE to proliferate and differentiate into mature OSNs (54,55).

The rate of basal cell proliferation is markedly increased with experimental induction of OSN injury and death whether that be through surgical bulbectomy (56) or axonomy (57), or intranasal exposure to some chemical toxicants such as zinc sulfate (58,59). The unique ability of OSNs to regenerate makes the OE an excellent model tissue to study the underlying cellular and molecular mechanisms of neurogenesis and axon regeneration (51). Though the process of neurogenesis and regeneration of the OE is still not fully defined, recent studies of olfactory mucosal injury and repair suggest that inflammatory signaling pathways may play a key role in the regulation of OSN regeneration (60,61).

The OE also contains two types of basal cells—horizontal (HBC) and globose (GBC). HBCs are thin cells located along the basal lamina and share many of the same morphological and histochemical features as the basal cells of nasal respiratory epithelium (e.g., contain keratins). In contrast, GBCs are morphologically more round or oval and are located above the HBCs. These cells have a more electron-lucent cytoplasm than the HBCs and are not immunohistochemically reactive for keratin. Some of the GBCs are the progenitor cells for OSNs, while the HBCs give rise to GBCs (62). Multipotent basal cells within the OE or in Bowman's gland ducts are the likely progenitors for sustentacular cells that are described below (63,64).

Sustentacular (or supporting) cells are columnar epithelial cells that span the entire thickness of the OE from the airway surface to the basal lamina. The distinct oval nuclei of the sustentacular cells are aligned in a single row along the apical aspect of the OE [Fig. 2(B2)]. The supranuclear portion of the cell is broad, while the portion of the cell below the nucleus tapers to a thin foot-like process that attaches to the basal lamina. These supporting cells surround the OSNs making multiple contacts with OSNs through fine cellular extensions (65). The apical surfaces of sustentacular cells are lined by numerous long microvilli that intermingle with the thin cilia of the OSNs along the surface of the airway lumen. The supranuclear cytoplasm of sustentacular cells has abundant smooth endoplasmic reticulum (SER) and xenobiotic-metabolizing enzymes (e.g., cytochrome P-450, flavin-containing monooxygenases, N-acetyltransferases). The metabolism in these cells may be important in detoxification of inhaled xenobiotics and in the function of smell (66–68). Sustentacular cells are also thought to contribute to the regulation of the ionic composition of the overlying mucous layer that undoubtedly affects the chemical interactions between odors and their ORs. The microvilli of these cells contain amiloride-sensitive sodium channels (69), while the lateral surfaces contain a water channel, aquaporin type 3 (70). Mammalian sustentacular cells do not contain mucin glycoproteins characteristic of the columnar mucus-secreting epithelial cells in nasal respiratory epithelium [e.g., mucous (goblet) cells]. The production and secretion of mucus covering the luminal surface of the OE is restricted to the subepithelial Bowman's glands.

Besides the principal epithelial cells of the OE that include the sustentacular cells, OSNs, and basal cells, there are at least five other morphologically distinct but much less abundant epithelial cells in the OE that have been reported in the literature. Collectively these cells have been termed as microvillous cells because of their distinct luminal surfaces that are lined by numerous microvilli (71). Though these apically located and widely scattered cells have specific morphological or immunohistochemical features that distinguish them from sustentacular cells (another cell with a distinct microvillar apical surface), the exact function of these microvillous cells has not yet been determined.

Bowman's glands, located in the underlying lamina propria and interspersed among the olfactory nerve bundles, are simple tubular-type glands composed of small compact acini [Fig. 2(B1)]. Ducts from these glands transverse the basal lamina at regular intervals and extend through the OE to the luminal surface. Bowman's glands contain copious amounts of neutral and acidic mucosubstances that contribute to the mucous layer covering the luminal surface of the OE. Like the sustentacular cells, both the acinar and duct cells of Bowman's glands also contain many xenobiotic-metabolizing enzymes.

Squamous Epithelium

The nasal vestibule is completely lined by a lightly keratinized, stratified squamous epithelium. It is composed of basal cells along the basal lamina and several layers of squamous cells, which become progressively flatter toward the luminal surface of the airway [Fig. 2(B5)]. Only 3.5% of the entire nasal cavity of the F344 rat is lined by squamous epithelium. This region of the nasal mucosa probably functions like the epidermis in the skin, to protect the underlying tissues from potentially harmful atmospheric agents.

Transitional Epithelium

Distal to the stratified squamous epithelium and proximal to the ciliated respiratory epithelium is a narrow zone of nonciliated, microvilli-covered surface epithelium, which has been referred to as nasal, nonciliated, respiratory epithelium or nasal transitional epithelium [Fig. 2(B2)]. Common, distinctive features of this nasal epithelium in all laboratory animal species and humans include (i) anatomical location in the proximal aspect of the nasal cavity between the squamous epithelium and the respiratory epithelium; (ii) the presence of nonciliated cuboidal or columnar surface cells and basal cells; (iii) a scarcity of mucous (goblet) cells and a paucity of intraepithelial mucosubstances; and (iv) an abrupt morphological border with squamous epithelium, but a less abrupt border with respiratory epithelium.

In rodents, this surface epithelium is thin (i.e., one to two cells thick), pseudostratified, and composed of three distinct cell types (basal, cuboidal, and columnar) (72). In contrast, transitional epithelium in monkeys is thick (i.e., four to five cells thick), stratified, and composed of at least five different cell types (15). The luminal surfaces of transitional epithelial cells lining the nasal airway possess numerous microvilli. Luminal, nonciliated cells in the transitional epithelium of rodents have no secretory granules but do have abundant SER in their apices (15). SER is an important intracellular site for xenobiotic-metabolizing enzymes, including cytochromes P-450. The prominent presence of SER in these cells, like the sustentacular cells in the OE, suggests that they may have roles in the metabolism of certain inhaled xenobiotics.

Respiratory Epithelium

The majority of the nonolfactory nasal epithelium of laboratory animals and humans is ciliated respiratory epithelium [Fig. 2(B3/4)]. Approximately 75% and 65% of the nasal cavity in the adult and infant rhesus monkey, respectively, is lined by respiratory epithelium (73,74), compared to only 46% of the nasal cavity in the adult laboratory rat (39). Although this pseudostratified nasal epithelium is similar to ciliated epithelium lining other proximal airways (i.e., trachea and bronchi), it also has unique features. Nasal respiratory epithelium in the rat is composed of six morphologically distinct cell types: mucous, ciliated, nonciliated columnar, cuboidal, brush, and basal (72). These cells are unevenly distributed along the rat mucosal surface. Using scanning electron microscopy, Popp and Martin demonstrated a proximal-to-distal increase in ciliated cells along the lateral walls of the rat. In the nasal septum of the rat, ciliated cells are evenly distributed from proximal to distal sites.

Like the respiratory epithelium of other mammals, the nasal respiratory epithelium in macaque monkeys is primarily composed of ciliated cells, mucous cells, and basal cells (15). Unlike that of the rat, this nasal epithelium of the monkey also contains small mucous granule cells and cells with intracytoplasmic lumina. Brush cells that have been reported in rodents are not found in the nasal epithelium of macaque monkeys.

The mucous cell is also unevenly distributed in the respiratory epithelium of the nasal cavity. In the normal rat, mucous cells are predominantly located in respiratory epithelium lining the proximal septum and the nasopharynx (17). Serous cells are the primary secretory cells in the remainder of the respiratory epithelium in rodents. Interestingly, secretory cells in the respiratory epithelium of both rats (75) and mice (76) have abundant SER. This suggests that these cells, like the nonciliated cell in the transitional epithelium, may have metabolic capacities for certain xenobiotic agents. Research in the area of xenobiotic metabolism in nasal respiratory epithelium, like the OE, has demonstrated the presence of many enzymes previously described in other tissues (77–79). In particular, carboxylesterase, aldehyde dehydrogenase, cytochrome P-450, epoxide hydrolase, and glutathione S-transferases have been localized by histochemical techniques. The distribution of these enzymes appears to be cell type specific, and the presence of the enzyme may predispose particular cell types to enhanced susceptibility or resistance to chemical-induced injury.

Lymphoepithelium and Nasal-Associated Lymphoid Tissue

In addition to the four principal nasal epithelia already described, there is another specialized epithelium, that is, lymphoepithelium (LE), in animal nasal airway that covers discrete focal aggregates of nasal-associated lymphoid tissue (NALT) in the underlying lamina propria [Fig. 2(B6)]. In rodents, NALT with associated LE is restricted to the ventral aspects of the lateral walls at the opening of the nasopharyngeal duct (80–82). The overlying LE is composed of cuboidal ciliated cells, a few mucous cells, and numerous noncilitated, cuboidal cells with luminal micovilli (so-called membranous or M cells) similar to those in the gut- and bronchus-associated lymphoid tissues (GALT and BALT, in the intestinal and lower respiratory tracts, respectively). M cells are thought to be involved in the uptake and translocation of inhaled antigen from the nasal lumen to the underlying lymphoid structures.

NALT, with its specialized lymphoepithelium, has also been described in the nasopharyngeal airways of the monkey, but these lymphoid structures LE are more numerous and are located on both the lateral and septal walls of the proximal nasopharynx (15). The correlate of NALT in humans is Waldeyer's ring, the orophayngeal lymphoid tissues composed of the adenoid, and the bilateral tubule, palatine, and lingual tonsils (83).

The location of NALT at the entrance of the nasopharyngeal duct is a very strategic position, as most of the nasal secretions and inhaled air, both presumably laden with antigenic material, pass over this area. Though the function of NALT and its place in the general mucosal-associated lymphoid system are not fully understood, these mucosal lymphoid tissues may have an important function in regional immune defense of the upper airways. NALT has been studied primarily in rat and mouse models (80,81,82,84–87). Immunohistochemical characterization of rat NALT has demonstrated that B and T cells are distributed in distinct areas with a high CD4-to-CD8 T-cell ratio and a predominance of B over T cells (87). Initial studies in mice suggest that the NALT is distinct from that found in rats and, if examined solely on immune cell content and subset ratios, more closely resembles the spleen and not the Peyer's patches located in the intestinal mucosa (85). However, the capability of NALT to elicit specific IgA responses locally suggests that this structure might represent a unique mucosal lymphoid tissue that is capable of expressing both mucosal and systemic immune responses.

Though it is clear that NALT plays a key role in nasal mucosal immunity, the toxicity to NALT by inhaled toxicants has unfortunately not been the focus of specific investigation. It has been recently recommended that more research efforts be made in this area and that the histopathological examination of NALT be routinely included in standard guideline-driven inhalation toxicity studies (82).

VOMERONASAL ORGAN

The vomeronasal organ (or Jacobson's organ) is a paired tubular diverticulum located in the vomer bone in the ventral portion of the proximal nasal septum of most mammals. It is a chemosensory structure that contributes to the sense of smell, like the OE, in macrosmotic species (e.g., laboratory rodents, dogs, rabbits). In laboratory rodents, the lateral wall of this organ is lined by tall columnar, respiratory-like, epithelium (nonchemosensory), while the medial wall is lined by a sensory neuroepithelium (chemosensory) similar in morphology to the OE lining the main nasal chamber. Vomeronasal sensory neurons project from the vomeronasal organ to the accessory olfactory bulb of the brain. The lumen of the vomeronasal organ communicates anteriorly with the nasopalatine duct. Therefore, the vomeronasal chemosensory system may detect pheromones and other chemicals through oral or nasal cavities.

The presence and functionality of the vomeronasal organ in primate species is variable (88). The vomeronasal organ has been identified in New World monkeys, prosimians, chimpanzees, and even humans. New world monkeys and prosimians have well-developed vomeronasal organs with sensory epithelium. However, the vomeronasal organs of chimpanzees and humans are nonchemosensory homologs consisting of bilateral septal tubes lined only by nonsensory ciliated epithelium. Macaques have no structures that resemble the vomeronasal organs of either prosimians or humans.

SUMMARY

In this chapter, we have briefly reviewed some of the important anatomical features of the nasal airways in both humans and laboratory animals. We have illustrated several similarities and differences of nasal structure among laboratory animals, and between these animals and humans. In this comparative overview, we have tried to emphasize the complexity and diversity found not only in the nasal organ itself, but also in the different animal species commonly used in nasal toxicological research. In general, nonprimate laboratory animals (i.e., rodents, rabbits, and dogs) have much more complex turbinate structures than do primates (i.e., both humans and monkeys). Primate nasal airways are lined by relatively less olfactory mucosa than that of other animal species. The surface epithelium lining the nasal airways also varies significantly in (*i*) the types of cells present in various intranasal locations within the same animal species, and (*ii*) the types of cells in different species at relatively similar anatomical locations within the nose. The comparative diversity of the nasal airways at the gross and cellular levels undoubtedly translates into differences in normal nasal function and in responses to toxic agents. These species differences (and similarities) must be recognized in order to (*i*) choose appropriate animal models for nasal toxicology studies and (*ii*) appropriately extrapolate data from animal toxicology studies in estimating the human risk of nasal toxicity. Future studies will expand our knowledge of comparative nasal biology by identifying key pheno-typic, genotypic, and proteomic expression differences (and similarities) in nasal tissues from various animal species (and strains) and human populations.

REFERENCES

1. Cole P. The Respiratory Role of the Upper Airways. St. Louis, MO: Mosby Year Book, 1993.
2. Brain JD. The uptake of inhaled gases by the nose. Ann Otol Rhinol Laryngol 1970; 79:529–539.
3. Miller FJ. Nasal Toxicology and Dosimetry of Inhaled Xenobiotics: Implications for Human Health. Washington, D.C.: Taylor & Francis, 1995.
4. Harkema JR, Carey SA, Wagner JG. The nose revisited: a brief review of the comparative structure, function, and toxicologic pathology of the nasal epithelium. Toxicol Pathol 2006; 34:252–269.
5. Guilmette RA, Wicks JD, Wolff RK. Morphometry of human nasal airways in vivo using magnetic resonance imaging. J Aerosol Med 1989; 2:365–377.
6. Negus VE. Comparative Anatomy and Physiology of the Nose and Paranasal Sinuses. Edinburgh, UK: E&S Livingstone, 1958.
7. Schreider JP, Raabe OG. Anatomy of the nasal–pharyngeal airway of experimental animals. Anat Rec 1981; 200:195–205.
8. Gross EA, Morgan KT. Architecture of nasal passages and larynx. In: Parent RA, ed. Comparative Biology of the Normal Lung: Treatise on Pulmonary Toxicology. Boca Raton, FL: CRC Press, 2000:7–25.
9. Wright ED, Bolger WE, Kennedy DW. Anatomic terminology and nomenclature of the paranasal sinuses and quantification for staging sinusitis. In: Sih T, Clement PAR, eds. Pediatric Nasal and Sinus Disorders: Lung Biology in Health and Disease, vol. 199. Boca Raton, FL: Taylor & Francis Group, 2005:187–207.
10. Henkin RI, Doherty A, Martin BM. The role of nasal mucus in upper airway function. In: Sih TS, Clement PAR, eds. Pediatric Nasal and Sinus Disorders: Lung Biology in Health and Disease, Vol 199. Boca Raton, FL: Taylor & Francis Group, 2005:19–58.
11. Morgan KT, Jiang XZ, Patterson DL, et al. The nasal mucociliary apparatus. Correlation of structure and function in the rat. Am Rev Respir Dis 1984; 130:275–281.
12. Morgan KT, Patterson DL, Gross EA. Responses of the nasal mucociliary apparatus of F-344 rats to formaldehyde gas. Toxicol Appl Pharmacol 1986; 82:1–13.
13. Harkema JR, Plopper CG, Hyde DM, et al. Regional differences in quantities of histochemically detectable mucosubstances in nasal, paranasal, and nasopharyngeal epithelium of the bonnet monkey. J Histochem Cytochem 1987; 35:279–286.
14. Harkema JR, Plopper CG, Hyde DM, et al. Effects of an ambient level of ozone on primate nasal epithelial mucosubstances. Quantitative histochemistry. Am J Pathol 1987; 127:90–96.
15. Harkema JR, Plopper CG, Hyde DM, et al. Nonolfactory surface epithelium of the nasal cavity of the bonnet monkey: a morphologic and morphometric study of the transitional and respiratory epithelium. Am J Anat 1987; 180:266–279.
16. Morgensen C, Tos M. Density of goblet cells in the normal adult human nasal turbinates. Anat Anz 1977; 142:322–330.
17. Harkema JR, Hotchkiss JA, Henderson RF. Effects of 0.12 and 0.80 ppm ozone on rat nasal and nasopharyngeal epithelial mucosubstances: quantitative histochemistry. Toxicol Pathol 1989; 17:525–535.
18. Cross CE, Halliwell B, Allen A. Antioxidant protection: a function of tracheobronchial and gastrointestinal mucus. Lancet 1982; 1:1328–1330.
19. Harkema JR, Hotchkiss JA, Barr EB, et al. Long-lasting effects of chronic ozone exposure on rat nasal epithelium. Am J Respir Cell Mol Biol 1999; 20:517–529.
20. Hotchkiss JA, Kimbell JS, Herrera LK, et al. Regional differences in ozone-induced nasal epithelial cell proliferation in F344 rats: comparison with computational mass flux predictions of ozone dosimetry. Inhal Toxicol 1994; 6:440–443.

21. Johnson NF, Hotchkiss JA, Harkema JR, et al. Proliferative responses of rat nasal epithelia to ozone. Toxicol Appl Pharmacol 1990; 103:143–155.
22. Sorokin SP. The respiratory system. In: Weiss L, ed. Cell and Tissue Biology: A Textbook of Histology. Baltimore, MD: Urban & Schwarzenberg, 1988;751–814.
23. Cauna N. Blood and nerve supply of the nasal lining. In: DF Proctor, Anderson I, eds. The Nose, Upper Airway Physiology and the Atmospheric Environment. Amsterdam, The Netherlands: Elsevier Biomedical Press, 1982:45–69.
24. Widdicombe J. Physiologic control: anatomy and physiology of the airway circulation. Am Rev Respir Dis 1992; 146:S3–S7.
25. Grieff L, Andersson M, Erjefält JS, et al. Airway microvascular extravasation and luminal entry of plasma. Clin Physiol Funct Imaging 2003; 23:301–306.
26. Persson CGA, Erjefält JS, Grieff L, et al. Plasma-derived proteins in airway defence, disease, and repair of epithelial injury. Eur Respir J 1998; 11:958–970.
27. Olsson P, Bende M. Sympathetic neurogenic control of blood flow in human nasal mucosa. Acta Otolaryngol Stockh 1986, 102, 482–487.
28. Morris JB, Hassett DN, Blanchard KT. A physiologically based pharmacokinetic model for nasal uptake and metabolism of nonreactive vapors. Toxicol Appl Pharmacol 1993; 123:120–129.
29. Brittebo EB, Eriksson C, Feil V, et al. Toxicity of 2,6-dichlorothiobenzamide (chlorthiamid) and 2,6-dichlorobenzamide in the olfactory nasal mucosa of mice. Fundam Appl Toxicol 1991; 17:92–102.
30. Genter MB, Llorens J, O'Callaghan JP, et al. Olfactory toxicity of beta,beta'-iminodipropionitrile in the rat. J Pharmacol Exp Ther 1992; 263:1432–1439.
31. Jeffrey AM, Iatropoulos MJ, Williams GM. Nasal cytotoxic and carcinogenic activities of systemically distributed organic chemicals. Toxicol Pathol 2006; 34:827–852.
32. Alarie Y. Bioassay for evaluating the potency of airborne sensory irritants and predicting acceptable levels of exposure in man. Food Cosmet Toxicol 1981; 19:623–626.
33. Buckley LA, Jiang XZ, James RA, et al. Respiratory tract lesions induced by sensory irritants at the RD50 concentration. Toxicol Appl Pharmacol 1984; 74:417–429.
34. Eccles R. Neurological and pharmacological considerations. In: Proctor DF, Anderson IB, eds. The Nose: Upper Airway Physiology and the Atmospheric Environment. Amsterdam, The Netherlands: Elsevier Biomedical Press, 1982:191–214.
35. Hasegawa M, Kerns EB. The human nasal cycle. Mayo Clin Proc 1977; 52:28.
36. Baraniuk JN. Neural regulation of mucosal function. Pulm Pharmacol Ther 2008; 21:442–448.
37. Kepler GM, Joyner DR, Fleishman A, et al. Method for obtaining accurate geometrical coordinates of nasal airways for computer dosimetry modeling and lesion mapping. Inhal Toxicol 1995; 7:1207–1224.
38. Mery S, Gross EA, Joyner DR, et al. Nasal diagrams: a tool for recording the distribution of nasal lesions in rats and mice. Toxicol Pathol 1994; 22:353–372.
39. Gross EA, Swenberg JA, Fields S, et al. Comparative morphometry of the nasal cavity in rats and mice. J Anat 1982; 135:83–88.
40. Farbman AI. The cellular basis of olfaction. Endeavour 1994; 18:2–8.
41. Menco BPM. The ultrastructure of olfactory and nasal respiratory epithelium surfaces. In: Reznik G, Stinson SF, eds. Nasal Tumors in Animals and Man. Anatomy, Physiology and Epidemiology, vol 1. Boca Raton, FL: CRC Press Inc., 1983:45–102.
42. Menco BP. Qualitative and quantitative freeze-fracture studies on olfactory and nasal respiratory epithelial surfaces of frog, ox, rat, and dog. III. Tight-junctions. Cell Tissue Res 1980; 211:361–373.
43. Mombaerts P. The human repertoire of odorant receptor genes and pseudogenes. Annu Rev Genomics Hum Genet 2001; 2:493–510.
44. Buck L, Axel R. A novel multigene family may encode odorant receptors: a molecular basis for odor recognition. Cell 1991; 65(1):175–187.
45. Buck LB. The olfactory multigene family. Curr Opin Neurobiol 1992; 2:282–288.
46. Ressler KJ, Sullivan SL, Buck LB. A molecular dissection of spatial patterning in the olfactory system. Curr Opin Neurobiol 1994; 4:588–596.

47. Vassar R, Ngai J, Axel R. Spatial segregation of odorant receptor expression in the mammalian olfactory epithelium. Cell 1993; 74:309–318.
48. Mori K, Nagao H, Yoshihara Y. The olfactory bulb: coding and processing of odor molecule information. Science 1999; 286:711–715.
49. Nagayama S, Takahashi YK, Yoshihara Y, et al. Mitral and tufted cells differ in the decoding manner of odor maps in the rat olfactory bulb. J Neurophysiol 1994; 91:2532–2540.
50. Graziadei PPC, Monti-Graziadei GA. Continuous nerve cell renewal in the olfactory system. In: Jacobson M, ed. Handbook of Sensory Physiology. New York, NY: Springer-Verlag, 1977:55–82.
51. Graziadei PPC, Monti-Graziadei GA. The olfatory system: a model for the study of neurogenesis and axon regeneration in mammals. In: Cotman CW, ed. Neuronal Plasticity. New York, NY: Raven Press, 1978:131–153.
52. Hinds JW, Hinds PL, McNelly NA. An autoradiographic study of the mouse olfactory epithelium: evidence for long-lived receptors. Anat Rec 1984; 210:375–383.
53. Mackay-Sim A, Kittel P. Cell dynamics in the adult mouse olfactory epithelium: a quantitative autoradiographic study. J Neurosci 1991; 11:979–984.
54. Jang W, Youngentob SL, Schwob JE. Globose basal cells are required for reconstitution of olfactory epithelium after methyl bromide lesion. J Comp Neurol 2003; 460: 123–140.
55. Schwob JE. Restoring olfaction: a view from the olfactory epithelium. Chem Senses 2005; 30(suppl 1):i131–i132.
56. Kastner A, Moyse E, Bauer S, et al. Unusual regulation of cyclin D1 and cyclin-dependent kinases cdk2 and cdk4 during in vivo mitotic stimulation of olfactory neuron progenitors in adult mouse. J Neurochem 2000; 74:2343–2349.
57. Suzuki Y, Takeda M. Basal cells in the mouse olfactory epithelium after axotomy: immunohistochemical and electron-microscopic studies. Cell Tissue Res 1991; 266:239–245.
58. Margolis FL, Roberts N, Ferriero D, et al. Denervation in the primary olfactory pathway of mice: biochemical and morphological effects. Brain Res 1974; 81:469–483.
59. McBride K, Slotnick B, Margolis FL. Does intranasal application of zinc sulfate produce anosmia in the mouse? An olfactometric and anatomical study. Chem Senses 2003; 28:659–670.
60. Bauer S, Rasika S, Han J, et al. Leukemia inhibitory factor is a key signal for injury-induced neurogenesis in the adult mouse olfactory epithelium. J Neurosci 2003; 23:1792–1803.
61. Getchell TV, Shah DS, Partin JV, et al. Leukemia inhibitory factor mRNA expression is upregulated in macrophages and olfactory receptor neurons after target ablation. J Neurosci Res 2002; 67:246–254.
62. Goldstein BJ, Schwob JE. Analysis of the globose basal cell compartment in rat olfactory epithelium using GBC-1, a new monoclonal antibody against globose basal cells. J Neurosci 1996; 16:4005–4016.
63. Huard JM, Youngentob SL, Goldstein BJ, et al. Adult olfactory epithelium contains multipotent progenitors that give rise to neurons and non-neural cells. J Comp Neurol 1998; 400:469–486.
64. Weiler E, Farbman AI. Supporting cell proliferation in the olfactory epithelium decreases postnatally. Glia 1998; 22:315–328.
65. Morrison EE, Costanzo RM. Morphology of the human olfactory epithelium. J Comp Neurol 1990; 297:1–13.
66. Ding X, Dahl AR. Olfactory mucosa: composition, enzymatic localization, and metabolism. In: Doty RL, ed. Handbook of Olfaction and Gustation. New York, NY: Marcel Dekker, Inc., 2003:51–73.
67. Genter MB. Update on olfactory mucosal metabolic enzymes: age-related changes and N-acetyltransferase activities. J Biochem Mol Toxicol 2004; 18:239–244.
68. Ling G, Gu J, Genter MB, et al. Regulation of cytochrome P450 gene expression in the olfactory mucosa. Chem Biol Interact 2004; 147:247–258.

69. Menco BP, Birrell GB, Fuller CM, et al. Ultrastructural localization of amiloride-sensitive sodium channels and $Na^+,K^{(+)}$-ATPase in the rat's olfactory epithelial surface. Chem Senses 1998; 23:137–149.

70. Verkman AS. Physiological importance of aquaporins: lessons from knockout mice. Curr Opin Nephrol Hypertens 2000; 9:517–522.

71. Menco BP, Morrison EE. Morphology of the mammalian olfactory epithelium: form, fine structure, function and pathology. In: Doty RL, ed. Handbook of Olfaction and Gustation. New York, NY: Marcel Dekker, Inc., 2003:17–49.

72. Monteiro-Riviere NA, Popp JA. Ultrastructural characterization of the nasal respiratory epithelium in the rat. Am J Anat 1984; 169:31–43.

73. Harkema JR, Plopper CG. The respiratory system and its use in research. In: Wolfe-Coote S, ed. The Laboratory Primate. Amsterdam, The Netherlands: Elsevier Academic Press, 2005:503–526.

74. Carey SA, Minard KR, Trease LL, et al. Three-dimensional mapping of ozone-induced injury in the nasal airways of monkeys using magnetic resonance imaging and morphometric techniques. Toxicol Pathol 2007; 35:27–40.

75. Yamamoto T, Masuda H. Some observations on the fine structure of the goblet cells in the nasal respiratory epithelium of the rat, with special reference to the well-developed agranular endoplasmic reticulum. Okajimas Folia Anat Jpn 1982; 58:583–594.

76. Matulionis DH, Parks HF. Ultrastructural morphology of the normal nasal respiratory epithelium of the mouse. Anat Rec 1973; 176:64–83.

77. Bogdanffy MS. Biotransformation enzymes in the rodent nasal mucosa: the value of a histochemical approach. Environ Health Perspect 1990; 85:177–186.

78. Bogdanffy MS, Randall HW, Morgan KT. Biochemical quantitation and histochemical localization of carboxylesterase in the nasal passages of the Fischer-344 rat and B6C3F1 mouse. Toxicol Appl Pharmacol 1987; 88:183–194.

79. Keller DA, Heck HD, Randall HW, et al. Histochemical localization of formaldehyde dehydrogenase in the rat. Toxicol Appl Pharmacol 1990; 106:311–326.

80. Kuper CF. Histopathology of mucosa-associated lymphoid tissue. Toxicol Pathol 2006; 34:609–615.

81. Kuper CF, Hameleers DM, Bruijntjes JP, et al. Lymphoid and non-lymphoid cells in nasal-associated lymphoid tissue (NALT) in the rat. An immuno- and enzyme-histochemical study. Cell Tissue Res 1990; 259:371–377.

82. Kuper CF, Arts JH, Feron VJ. Toxicity to nasal-associated lymphoid tissue. Toxicol Lett 2003; 140/141:281–285.

83. Brandtzaeg P. Immune function of human nasal mucosa and tonsils in health and disease. In: Bienenstock J, ed. Immunology of the Lung and Upper Respiratory Tract. New York, NY: McGraw-Hill, 1984:28–95.

84. Asanuma H, Inaba Y, Aizawa C, et al. Characterization of mouse nasal lymphocytes isolated by enzymatic extraction with collagenase. J Immunol Methods 1995; 187:41–51.

85. Heritage PL, Underdown BJ, Arsenault AL, et al. Comparison of murine nasal-associated lymphoid tissue and Peyer's patches. Am J Respir Crit Care Med 1997; 156:1256–1262.

86. Ichimiya I, Kawauchi H, Fujiyoshi T, et al. Distribution of immunocompetent cells in normal nasal mucosa: comparisons among germ-free, specific pathogen-free, and conventional mice. Ann Otol Rhinol Laryngol 1991; 100:638–642.

87. Koornstra PJ, Duijvestijn AM, Vlek LF, et al. Immunohistochemistry of nasopharyngeal (Waldeyer's ring equivalent) lymphoid tissue in the rat. Acta Otolaryngol 1993; 113:660–667.

88. Smith TD, Siegel MI, Bonar CJ, et al. The existence of the vomeronasal organ in postnatal chimpanzees and evidence for its homology with that of humans. J Anat 2001; 198:77–82.

Functional Anatomy of the Upper Airway in Humans

Fuad M. Baroody

Pritzker School of Medicine, University of Chicago, Chicago, Illinois, U.S.A.

INTRODUCTION

When discussing the different influences that our environment and its toxins have on the nose and its function, it is essential to have a clear understanding of the anatomy and physiology of the nasal cavity. This chapter is designed to provide such an understanding of the structure and function of both the nasal cavity and the paranasal sinuses, with the aim of facilitating the appreciation of impacts of different environmental influences discussed in subsequent chapters of this textbook.

NASAL ANATOMY

External Nasal Framework

The external bony framework of the nose consists of two oblong, paired nasal bones located on either side of the midline that merge to form a pyramid (Fig. 1). Lateral to each nasal bone is the frontal process of the maxilla, which contributes to the base of the nasal pyramid. The piriform aperture is the bony opening that leads to the external nose.

The cartilaginous framework of the nose consists of the paired upper lateral, the lower lateral, and the sesamoid cartilages (Fig. 1). The upper lateral cartilages are attached to the undersurface of the nasal bones and frontal processes superiorly and their inferior ends lie under the upper margin of the lower lateral cartilages. Medially, they blend with the cartilaginous septum. Each lower lateral cartilage consists of a medial crus, which extends along the free caudal edge of the cartilaginous septum, and a lateral crus, which provides the framework of the nasal ala, the entrance to the nose (Fig. 1). Laterally, between the upper and lower lateral cartilages, are one or more sesamoid cartilages and fibroadipose tissue.

Nasal Septum

The nasal septum divides the nasal cavity into two sides and is composed of cartilage and bone. The bone receives contributions from the vomer, perpendicular plate of the ethmoid, maxillary crest, palatine bone, and the anterior spine of the maxillary bone. The main supporting framework of the septum is the septal cartilage, which forms the most anterior part of the septum and articulates posteriorly with the vomer and the perpendicular plate of the ethmoid bone. Inferiorly, the cartilage rests in the crest of the maxilla, whereas anteriorly it has a free border when it approaches the membranous septum. The latter separates the medial

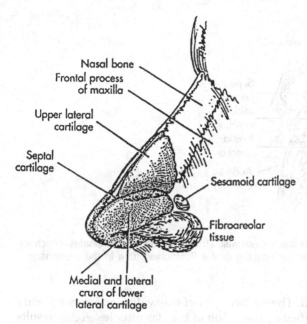

Nasal bone
Frontal process
of maxilla

Upper lateral
cartilage

Septal
cartilage

Sesamoid cartilage

Fibroareolar
tissue

Medial and lateral
crura of lower
lateral cartilage

FIGURE 1   External nasal framework. *Source*: From Ref. 1.

crura of the lower lateral cartilages from the septal cartilage. In a study of cadaveric specimens, Van Loosen and colleagues showed that the cartilaginous septum increases rapidly in size during the first years of life, with the total area remaining constant after the age of two years (2). In contrast, endochondral ossification of the cartilaginous septum resulting in the formation of the perpendicular plate of the ethmoid bone starts after the first 6 months of life and continues until the age of 36 years. The continuous, albeit slow, growth of the nasal septum until the third decade might explain the frequently encountered septal deviations in adults. In addition to reduction of nasal airflow, some septal deviations obstruct the middle meatal areas and can lead to impairment of drainage from the sinuses with resultant sinusitis. Severe anterior deviations can also prevent the introduction of intranasal medications to the rest of the nasal cavity and therefore interfere with the medical treatment of rhinitis (3). It is important to examine the nose in a patient with complaints of nasal congestion to rule out such deviations. It is also important to realize that not all deviations lead to symptoms and that surgery should be reserved for those deviations that are thought to contribute to the patient's symptomatology.

Nasal Vestibule/Nasal Valve

The nasal vestibule, located immediately posterior to the external nasal opening, is lined with stratified squamous epithelium and numerous hairs (or vibrissae) that filter out large particulate matter. The vestibule funnels air toward the nasal valve, which is a slit-shaped passage formed by the junction of the upper lateral cartilages, the nasal septum, and the inferior turbinate. The nasal valve accounts for approximately 50% of the total resistance to respiratory airflow from the

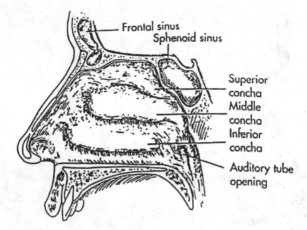

FIGURE 2 A sagittal section of the lateral nasal wall. This shows the three turbinates (conchae), frontal and sphenoid sinuses, and the opening of the Eustachian tube in the nasopharynx. *Source*: From Ref. 5.

anterior nostril to the alveoli. The surface area of this valve, and consequently resistance to airflow, is modified by the action of the alar muscles. Aging results in loss of strength of the nasal cartilages with secondary weakening of nasal tip support and the nasal valve with resultant airflow compromise (3).

Lateral Nasal Wall
The lateral nasal wall commonly has three turbinates, or conchae, the inferior, middle, and superior (Fig. 2). The turbinates are elongated laminae of bone attached along their superior borders to the lateral nasal wall. Their unattached inferior portions curve inwards toward the lateral nasal wall, resulting in a convex surface that faces the nasal septum medially. They not only increase the mucosal surface of the nasal cavity to about 100 to 200 cm^2 but also regulate airflow by alteration of their vascular content and, hence, thickness through the state of their capacitance vessels (4). The large surface area of the turbinates and the nasal septum allows intimate contact between respired air and the mucosal surfaces, thus facilitating humidification, filtration, and temperature regulation of inspired air. Under and lateral to each of the turbinates are horizontal passages or meati. The inferior meatus receives the opening of the nasolacrimal duct, whereas the middle meatus receives drainage originating from the frontal, anterior ethmoid, and maxillary sinuses (Fig. 3). The sphenoid and posterior ethmoid sinuses drain into the sphenoethmoid recess, located below and posterior to the superior turbinate.

PARANASAL SINUS ANATOMY
The paranasal sinuses are four pairs of cavities that are named after the skull bones in which they are located: frontal, ethmoid (anterior and posterior), maxillary, and sphenoid. All sinuses contain air and are lined by a thin layer of respiratory mucosa composed of ciliated, pseudostratified columnar, epithelial cells with goblet mucous cells interspersed among the columnar cells.

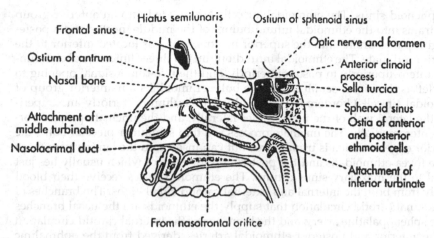

From nasofrontal orifice

FIGURE 3 A detailed view of the lateral nasal wall. Parts of the inferior and middle turbinates have been removed. Visualized are the various openings into the inferior, middle, and superior meati. *Source*: From Ref. 6.

Frontal Sinuses

At birth, the frontal sinuses are indistinguishable from the anterior ethmoid cells and they grow slowly after birth so that they are barely seen anatomically at one year of age. After the fourth year, the frontal sinuses begin to enlarge and can usually be demonstrated radiographically in children over six years of age. Their size continues to increase into the late teens. The frontal sinuses are usually pyramidal structures in the vertical part of the frontal bone. They open via the frontal recess into the anterior part of the middle meatus, or directly into the anterior part of the infundibulum. The natural ostium is located directly posterior to the anterior attachment of the middle turbinate to the lateral nasal wall. The frontal sinuses are supplied by the supraorbital and supratrochlear arteries, branches of the ophthalmic artery, which in turn is a branch of the internal carotid artery. Venous drainage is via the superior ophthalmic vein into the cavernous sinus. The sensory innervation of the mucosa is supplied via the supraorbital and supratrochlear branches of the frontal nerve, derived from the ophthalmic division of the trigeminal nerve.

Ethmoid Sinuses

At birth, the ethmoid and maxillary sinuses are the only sinuses that are large enough to be clinically significant as a cause of rhinosinusitis. By the age of 12 years, the ethmoid air cells have almost reached their adult size and form a pyramid with the base located posteriorly. The lateral wall of the sinus is the lamina papyracea, which also serves as the paper-thin medial wall of the orbit. The medial wall of the sinus functions as the lateral nasal wall. The superior boundary of the ethmoid sinus is formed by the horizontal plate of the ethmoid bone that separates the sinus from the anterior cranial fossa. This horizontal plate is composed of a thin medial portion named the cribriform plate and a thicker, more lateral portion named the fovea ethmoidalis, which forms the ethmoid roof. The posterior boundary of the ethmoid sinus is the anterior wall of the sphenoid sinus. The ethmoidal air cells are divided into an anterior group that drains into the ethmoidal infundibulum of the middle meatus and a posterior group that drains into the superior meatus, which is located inferior to the superior turbinate. The ethmoidal infundibulum is a three-dimensional cleft running anterosuperiorly to posteroinferiorly, and the two-dimensional opening to this cleft is the hiatus semilunaris. The bulla ethmoidalis (an anterior group of ethmoidal air cells) borders the ethmoid infundibulum posteriorly and superiorly, the lateral wall of the nose resides laterally, and the uncinate process borders anteromedially. The uncinate process is a thin semilunar piece of bone, the superior edge of which is usually free but can insert into the lamina papyracea or the fovea ethmoidalis and the posteroinferior edge of which usually lies just lateral to the maxillary sinus ostium. The ethmoid sinuses receive their blood supply from both the internal and external carotid circulations. The branches of the external carotid circulation that supply the ethmoids are the nasal branches of the sphenopalatine artery, and the branches of the internal carotid circulation are the anterior and posterior ethmoidal arteries, derived from the ophthalmic artery. Venous drainage can also be directed via the nasal veins, branches of the maxillary vein, or via the ophthalmic veins, tributaries of the cavernous sinus. The latter pathway is responsible for cavernous sinus thrombosis after ethmoid sinusitis. The sensory innervation of these sinuses is supplied by the ophthalmic and maxillary divisions of the trigeminal nerve.

Maxillary Sinuses

The size of the maxillary sinus is estimated to be 6 to 8 cm^3 at birth. The sinus then grows rapidly until three years of age and then more slowly until the seventh year. Another growth acceleration occurs then until about age 12 years. By then, pneumatization has extended laterally as far as the lateral wall of the orbit and inferiorly so that the floor of the sinus is even with the floor of the nasal cavity. Much of the growth that occurs after the twelfth year is in the inferior direction with pneumatization of the alveolar process after eruption of the secondary dentition. By adulthood, the floor of the maxillary sinus is usually 4 to 5 mm inferior to that of the nasal cavity. The maxillary sinus occupies the body of the maxilla and each sinus has a capacity of around 15 mL. Its anterior wall is the facial surface of the maxilla and the posterior wall corresponds to the infratemporal surface of the maxilla. Its roof is the inferior orbital floor and is about twice as wide as its floor, formed by the alveolar process of the maxilla. The medial wall of the sinus forms part of the lateral nasal wall and has the ostium of the sinus that is located within the infundibulum of the middle meatus, with accessory ostia occurring in 25% to 30% of individuals. Mucociliary clearance within the maxillary sinus moves secretions in the direction of the natural ostium. The major blood supply of the maxillary sinuses is via branches of the maxillary artery with a small contribution from the facial artery. Venous drainage occurs anteriorly via the anterior facial vein into the jugular vein or posteriorly via the tributaries of the maxillary vein, which also eventually drains into the jugular system. Innervation of the mucosa of the maxillary sinuses is via several branches of the maxillary nerve, which primarily carry sensory fibers. Another contribution to the innervation via the maxillary nerve are postganglionic parasympathetic secretomotor fibers originating in the facial nerve and carried to the sphenopalatine ganglion in the pterygopalatine fossa via the greater petrosal nerve and the nerve of the pterygoid canal.

Sphenoid Sinuses

At birth, the size of the sphenoid sinus is small and is little more than an evagination of the sphenoethmoid recess. By the age of seven years, the sphenoid sinuses have extended posteriorly to the level of the sella turcica. By the late teens, most of the sinuses have aerated to the dorsum sellae and some further enlargement may occur in adults. The sphenoid sinuses are frequently asymmetric because the intersinus septum is bowed or twisted. The optic nerve, internal carotid artery, nerve of the pterygoid canal, maxillary nerve, and sphenopalatine ganglion may all appear as impressions indenting the walls of the sphenoid sinuses depending on the extent of pneumatization. The sphenoid sinus drains into the sphenoethmoid recess above the superior turbinate and the ostium typically lies 10 mm above the floor of the sinus. The blood supply is via branches of the internal and external carotid arteries and the venous drainage follows that of the nasopharynx and the nasal cavity into the maxillary vein and pterygoid venous plexus. The first and second divisions of the trigeminal nerve supply the mucosa of the sphenoid sinus.

Function of the Paranasal Sinuses

Many theories exist related to the function of the paranasal sinuses. Some of these theories include imparting additional voice resonance, humidifying and warming inspired air, secreting mucus to keep the nose moist, and providing thermal insulation for the brain. While none of these theories have been supported by objective evidence, it is commonly believed that the paranasal sinuses form a collapsible framework to help protect the brain from frontal blunt trauma. Recent studies have documented significant production of nitric oxide by the nose and the paranasal sinuses and have suggested the involvement of this produced gas in regulatory and defensive effects such as contribution to nonspecific host defenses against bacterial, viral, and fungal infections and therefore helping to maintain a sterile environment within the paranasal sinuses (7). There is also evidence that nitric oxide regulates ciliary motility and that low levels of this gas are associated with impaired mucociliary function in the upper airway (8). While the function of the paranasal sinuses might not be completely understood, they are the frequent target of infections, both acute and chronic.

The middle meatus is an important anatomic area in the pathophysiology of sinus disease. It has a complex anatomy of bones and mucosal folds, often referred to as the osteomeatal unit, between which drain the frontal, anterior ethmoid, and maxillary sinuses. Anatomic abnormalities or inflammatory mucosal changes in the area of the osteomeatal complex can lead to impaired drainage from these sinuses which can, at least in part, be responsible for acute and chronic sinus disease. Endoscopic sinus surgery is targeted at restoring the functionality of this drainage system in patients with chronic sinus disease that is refractory to medical management.

NASAL MUCOSA

A thin, moderately keratinized, stratified squamous epithelium lines the vestibular region. The anterior tips of the turbinates provide a transition from

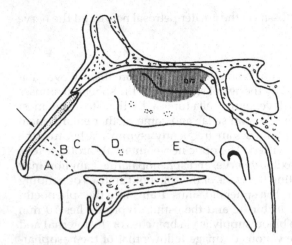

FIGURE 4 Distribution of types of epithelium along the lateral nasal wall. The hatched region represents the olfactory epithelium. The arrow represents the area of the nasal valve: A, skin; B, squamous epithelium without microvilli; C, transitional epithelium; D, pseudostratified columnar epithelium with few ciliated cells; and E, pseudostratified columnar epithelium with many ciliated cells. *Source*: From Ref. 9.

squamous to transitional and finally to pseudostratified columnar ciliated epithelium, which lines the remainder of the nasal cavity except for the roof that is lined with olfactory epithelium (Fig. 4) (4). All cells of the pseudostratified columnar ciliated epithelium contact the basement membrane, but not all reach the epithelial surface. The basement membrane separates the epithelium from the lamina propria, or submucosa.

Nasal Epithelium
Within the epithelium, three types of cells are identified: basal, goblet, and columnar, which are either ciliated or nonciliated.

Basal Cells
Basal cells lie on the basement membrane and do not reach the airway lumen. They have an electron-dense cytoplasm and bundles of tonofilaments. Among their morphologic specializations are desmosomes, which mediate adhesion between adjacent cells, and hemidesmosomes, which help anchor the cells to the basement membrane (10). These cells have long been thought to be progenitors of the columnar and goblet cells of the airway epithelium, but experiments in rat bronchial epithelium suggest that the primary progenitor cell of airway epithelium might be the nonciliated columnar cell population (11). Currently, basal cells are believed to help in the adhesion of columnar cells to the basement membrane. This is supported by the fact that columnar cells do not have hemidesmosomes and attach to the basement membrane only by cell adhesion molecules, that is, laminin.

Goblet Cells
The goblet cells arrange themselves perpendicular to the epithelial surface (12). The mucous granules give the mature cell its characteristic goblet shape, in which

only a narrow part of the tapering basal cytoplasm touches the basement membrane. The nucleus is situated basally, with the organelles and secretory granules that contain mucin toward the lumen. The luminal surface, covered by microvilli, has a small opening, or stoma, through which the granules secrete their content. The genesis of goblet cells is controversial, with some experimental studies supporting a cell of origin unrelated to epithelial cells and others supporting either the cylindrical nonciliated columnar cell population or undifferentiated basal cells as the cells of origin (12). There are no goblet cells in the squamous, transitional, or olfactory epithelia of adults, and they are irregularly distributed but present in all areas of pseudostratified columnar epithelium (12).

Columnar Cells
These cells are related to neighboring cells by tight junctions apically and, in the uppermost part, by interdigitations of the cell membrane. The cytoplasm contains numerous mitochondria in the apical part. All columnar cells, ciliated and nonciliated, are covered by 300 to 400 microvilli, uniformly distributed over the entire apical surface. These are not precursors of cilia but are short and slender fingerlike cytoplasmic expansions that increase the surface area of the epithelial cells, thus promoting exchange processes across the epithelium. The microvilli also prevent drying of the surface by retaining moisture essential for ciliary function (4). In man, ciliated epithelium lines the majority of the airway from the nose to the respiratory bronchioles, as well as the paranasal sinuses, the eustachian tube, and the parts of the middle ear.

Inflammatory Cells
Different types of inflammatory cells have been described in the nasal epithelium obtained from normal, nonallergic subjects. Using immunohistochemical staining, Winther and colleagues identified consistent anti-HLA-DR staining in the upper portion of nasal epithelium as well as occasional lymphocytes interspersed between the epithelial cells (13). There appeared to be more T than B lymphocytes and more T helper than T suppressor cells. The detection of HLA-DR antigens on the epithelium suggested that the airway epithelium may be potentially participating in antigen recognition and processing. Bradding and colleagues observed rare mast cells within the epithelial layer and no activated eosinophils (14).

Nasal Submucosa
The nasal submucosa lies beneath the basement membrane and contains a host of cellular components in addition to nasal glands, nerves, and blood vessels. In a light microscopy study of nasal biopsies of normal individuals, the predominant cell in the submucosa was the mononuclear cell, which includes lymphocytes and monocytes (15). Much less numerous were neutrophils and eosinophils (15). Mast cells were also found in appreciable numbers in the nasal submucosa as identified by immunohistochemical staining with a monoclonal antibody against mast cell tryptase (14). Winther and colleagues evaluated lymphocyte subsets in the nasal mucosa of normal subjects using immunohistochemistry (13). They found T lymphocytes to be the predominant cell type with fewer scattered B cells. The ratio of T helper cells to T suppressor cells in the lamina propria averaged 3:1 in the subepithelial area and 2:1 in the deeper vascular stroma with the overall ratio being 2.5:1, similar to the average ratio in peripheral blood. Natural killer

cells were very rare constituting less than 2% of the lymphocytes. Recent interest in inflammatory cytokines prompted Bradding and colleagues to investigate cells containing IL-4, IL-5, IL-6, and IL-8 in the nasal mucosa of patients with perennial rhinitis and normal subjects (14). The normal nasal mucosa was found to contain cells with positive IL-4 immunoreactivity, with 90% of these cells also staining positive for mast cell tryptase suggesting that they were mast cells. Immunoreactivity for IL-5 and IL-6 was present in 75% of the normal nasal biopsies, and IL-8 positive cells were found in all the normal nasal tissue samples. A median 50% of IL-5$^+$ cells and 100% of the IL-6$^+$ cells were mast cells. In contrast to the other cytokines, IL-8 was largely confined to the cytoplasm of epithelial cells.

From the above studies, it is clear that the normal nasal mucosa contains a host of inflammatory cells, the role of which is unclear. In allergic rhinitis, most of these inflammatory cells increase in number (16) and eosinophils are also recruited into the nasal mucosa (14). Furthermore, cells positive for IL-4 increase significantly in patients with allergic rhinitis compared to normal subjects (14).

Nasal Glands
There are three types of nasal glands: anterior nasal, seromucous, and intraepithelial. They are located in the submucosa and epithelium.

Anterior Nasal Glands
These serous glands have ducts (2–20 mm in length) that open into small crypts located in the nasal vestibule. The ducts are lined by one layer of cuboidal epithelium. Bojsen-Moeller found 50 to 80 crypts anteriorly on the septum and another 50 to 80 anteriorly on the lateral nasal wall (17). He suggested that these glands play an important role in keeping the nose moist by spreading their serous secretions backwards, thus moistening the entire mucosa. Tos, however, was able to find only 20 to 30 anterior nasal glands on the septum and an equal number on the lateral wall (12). He deduced that the contribution of these glands to the total production of secretions is minimal and that they represent a phylogenetic rudiment.

Seromucous Glands
The main duct of these glands is lined with simple cuboidal epithelium. It divides into two side ducts that collect secretions from several tubules lined either with serous or mucous cells. At the ends of the tubules are acini, which may similarly be serous or mucous. Submucosal serous glands predominate over mucinous glands by a ratio of about 8:1. The glands first laid down during development grow deep into the lamina propria before dividing and thus develop their mass in the deepest layers of the mucosa with relatively long ducts. The glands that develop later divide before growing down into the mucosa and thus form a more superficial mass with short ducts. Vessels, nerves, and fibers develop in between, giving rise to two glandular layers: superficial and deep. The mass of the deep glands is larger than that of the superficial ones, and the total number of these glands is approximately 90,000.

Intraepithelial Glands
These glands are located in the epithelium and consist of 20 to 50 mucous cells arranged radially around a small lumen. Many intraepithelial glands exist in

nasal polyps. Compared to seromucous glands, intraepithelial glands produce only a small amount of mucus and thus play a minor role in the physiology of nasal secretions.

MUCOSAL IMMUNITY

The nasal cavity is often the first point of contact between the airway mucosa and the external environment, and thus multiple mechanisms exist to defend the airway against the potentially harmful microbial and nonmicrobial elements found in inspired air. Mucosal immunity can generally be characterized as adaptive and innate. Broadly defined, the innate immune system includes all aspects of the host defense mechanisms that are encoded in the germ line genes of the host. These include barrier mechanisms, such as epithelial cell layers that express tight cell–cell contact, the secreted mucus layer that overlays the epithelium, and the epithelial cilia that sweep away this mucus layer. The innate response also includes soluble proteins and small bioactive molecules that are either constitutively present in biological fluids (such as the complement proteins and defensins) (18,19) or released from activated cells (cytokines, chemokines, lipid mediators of inflammation, and bioactive amines and enzymes). Activated phagocytes (including neutrophils, monocytes, and macrophages) are also part of the innate immune system.

Mucosal surfaces of the upper airway are less resistant than the skin and are thus more frequent portals for offending pathogens. The innate immune system reduces that vulnerability by the presence of various physical and biochemical factors. A good example is the enzyme lysozyme, which is distributed widely in secretions and can split the cell walls of most bacteria. If an offending organism penetrates this first line of defense, bone marrow–derived phagocytic cells attempt to engulf and destroy it. Last, the innate immune system includes cell surface receptors that bind molecular patterns expressed on the surfaces of invading microbes.

Unlike the innate mechanisms of defense, the adaptive immune system manifests exquisite specificity for its target antigens. Adaptive responses are based primarily on the antigen-specific receptors expressed on the surfaces of T and B lymphocytes. These antigen-specific receptors of the adaptive immune response are assembled by somatic rearrangement of germ line gene elements to form intact T-cell receptor and B-cell antigen receptor genes. The assembly of antigen receptors from a collection of a few hundred germ line–encoded gene elements permits the formation of millions of different antigen receptors, each with a potentially unique specificity for a different antigen.

Because the recognition molecules used by the innate system are expressed broadly on a large number of cells, this system is poised to act rapidly after an invading pathogen is encountered. The second set of responses constitutes the adaptive immune response. Because the adaptive system is composed of small numbers of cells with specificity for any individual pathogen, the responding cells must proliferate after encountering the pathogen to attain sufficient numbers to mount an effective response against the microbe. Thus, the adaptive response generally expresses itself temporally after the innate response in host defense. A key feature of the adaptive system is that it produces long-lived cells that persist in an apparently dormant state, but can re-express effector functions rapidly after repeated encounter with an antigen. This provides the adaptive response with immune memory, permitting it to contribute to a more effective host response against specific pathogens when they are encountered a second time.

Innate immune effectors are critical for effective host defense. In addition to local defenses at mucosal surfaces, such as mucus and mucociliary transport (described below), the effectors of innate immunity include Toll-like receptors (TLRs), antimicrobial peptides, phagocytic cells, natural killer cells, and complement.

Toll-Like Receptors

An important advance in our understanding of innate immunity to microbial pathogens was the identification of a human homolog of *Drosophila* Toll receptor. Mammalian TLR family members are transmembrane proteins containing repeated leucine-rich motifs in their extracellular portions. Mammalian TLR proteins contain a cytoplasmic portion that is homologous to the IL-1 receptor and can therefore trigger intracellular signaling pathways. TLRs are pattern recognition receptors that recognize pathogen-associated molecular patterns present on a variety of bacteria, viruses, and fungi. The activation of TLRs induces expression of costimulatory molecules and the release of cytokines that instruct the adaptive immune response. Finally, TLRs directly activate host defense mechanisms that directly combat the foreign invader or contribute to tissue injury (20).

TLRs were initially found to be expressed in all lymphoid tissues but are most highly expressed in peripheral blood leukocytes. Expression of TLR mRNA has been found in monocytes, B cells, T cells, and dendritic cells (21,22). The expression of TLRs on cells of the monocyte–macrophage lineage is consistent with the role of TLRs in modulating inflammatory responses via cytokine release. Some TLRs are located intracellularly, like TLR9.

Lipopolysaccharides (LPS) of gram-negative bacteria generate responses mediated via the TLR4 receptor (23,24). Microbial lipoproteins and lipopeptides have been shown to activate cells in a TLR2-dependent manner (25). Lipoproteins have been found extensively in both gram-positive and gram-negative bacteria, as well as spirochetes. Mammalian TLR9 mediates the immune response to a specific pattern in bacterial DNA, an unmethylated cytidine–phosphate–guanosine (CpG) dinucleotide with appropriate flanking regions. These CpG-DNA sequences are 20-fold more common in microbial than in mammalian DNA; thus, mammalian TLR9 is more likely to be activated by bacterial than mammalian DNA. Human TLR9 confers responsiveness to bacterial DNA via species-specific CpG motif recognition (26). Mammalian TLR5 has been shown to mediate the response to flagellin, a component of bacterial flagella (27). Mammalian TLR3 mediates the response to double-stranded RNA, a molecular pattern expressed by many viruses during infection (28). Activation of TLR3 induces IFN-α and IFN-β, the cytokines important for antiviral responses. Finally, single-stranded RNA binds to TLR7 and TLR8 (29).

TLRs and the Adaptive Immune Response

Critical proinflammatory and immunomodulatory cytokines such as IL-1, IL-6, IL-8, IL-10, IL-12, and TNF-α have been shown to be induced after activation of TLRs by microbial ligands (20). Activation of TLRs on dendritic cells triggers their maturation, leading to cell surface changes that enhance antigen presentation, thus promoting the ability of these cells to present antigen to T cells and generate TH1 responses critical for cell-mediated immunity. Therefore, activation of TLRs as part of the innate response can influence and modulate the adaptive T-cell response and modify the shaping of the TH1/TH2 balance (30).

TLR and the Host

Similar to the *Drosophila*, mammalian TLRs have been shown to play a prominent role in directly activating host defense mechanisms. For example, activation of TLR2 by microbial lipoproteins induces activation of the inducible nitric oxide synthase promoter that leads to the production of nitric oxide, a known antimicrobial agent (25). In *Drosophila*, activation of Toll leads to the NFκB-dependent induction of a variety of antimicrobial peptides (31). In a similar fashion, it has been shown that LPS induces β-defensin-2 in tracheobronchial epithelium, suggesting similar pathways in humans (32). The activation of TLRs can also be detrimental, causing tissue injury. The administration of LPS to mice can result in shock, a feature that is dependent on TLR4 (23), and microbial lipoproteins induce features of apoptosis via TLR2 (33). Thus, microbial lipoproteins have the ability to induce both TLR-dependent activation of host defense and tissue pathology. This might be one way for the immune system to activate host defenses and then downregulate the response from causing tissue injury by apoptosis.

Antimicrobial Peptides

The function of antimicrobial peptides is essential to the mammalian immune response. They participate primarily in the innate immune system and are used as a first-line immune defense. Antimicrobial peptides directly kill a broad spectrum of microbes, including gram-positive and gram-negative bacteria, fungi, and certain viruses. In addition, these peptides interact with the host itself, triggering events that complement their role as antibiotics. These secreted antimicrobials play a major role in immediate host defense against potential pathogens entering the body through the nasal mucosa. They include small cationic peptides such as the defensins and cathelicidins and larger antimicrobial proteins such as lysozyme, lactoferrin, and secretory leukocyte proteinase inhibitor. These secreted natural "antibiotics" inhibit microbial growth and allow for time to eliminate the microbial threat through mucociliary clearance or through the recruitment of phagocytic cells and the development of an adaptive immune response when necessary.

Lysozyme

Lysozyme is an enzyme directed against the peptidoglycan cell wall of bacteria and is secreted by nasal monocytes, macrophages, and epithelial cells. This antimicrobial substance is highly effective against many upper airway gram-positive bacterial species such as *Streptococci*, but requires potentiation by cofactors such as lactoferrin, antibody–complement complexes, or ascorbic acid for the destruction of gram-negative bacteria (34). Transgenic mice overexpressing lysozyme display increased resistance to lung infection with group B streptococcus or *Pseudomonas aeruginosa*, supporting the critical importance of this enzyme in the local immune system (35).

Lactoferrin
Lactoferrin is an antimicrobial product secreted by neutrophil granules and stored and released by mucosal glands, and it acts as an iron-binding protein that inhibits microbial growth by sequestering iron.

Secretory Leukocyte Proteinase Inhibitor
Secretory leukocyte proteinase inhibitor is another defense molecule found in nasal mucus that consists of two separate functional domains, the N-terminal domain, which has in vitro activity against both gram-negative and gram-positive bacteria, and the C-terminal domain, which inhibits neutrophil elastase (36).

Cathelicidins
Most cathelicidins undergo extracellular proteolytic cleavage that releases their C-terminal peptide containing the antimicrobial activity. The only known human cathelicidin hCAP-18 (human cationic antimicrobial peptide, 18 kDa) was initially identified in granules of human neutrophils (37). Its free C-terminal peptide is called LL-37 and has a broad spectrum of antimicrobial activity in vitro against *Pseudomonas aeruginosa, Salmonella typhinurium, Escherichia coli, Listeria monocytogenes*, and *Staphylococcus aureus* (38). LL-37 is a chemoattractant for mast cells (39) and human neutrophils, monocytes, and T cells (40) and induces degranulation and histamine release in mast cells (41). Thus, this antimicrobial peptide has the potential to participate in the innate immune response both by killing bacteria and by recruiting a cellular immune response and promoting tissue inflammation.

Defensins
Defensins are a broadly dispersed family of gene-encoded antimicrobials that exhibit antimicrobial activity against bacteria, fungi, and enveloped viruses (42,43). Defensins are classified into three distinct families: the α-defensins, the β-defensins, and the θ-defensins. α-**Defensins** are 29 to 35 amino acids in length. Human neutrophils express a number of distinct defensins (44). To date, six α-defensins have been identified. Of these six, four are known as α-defensins 1, 2, 3, and 4 (also referred to as human neutrophil peptides HNP 1 through 4). The other two α-defensins, known as human defensins 5 and 6 (HD-5, HD-6), are abundantly expressed in Paneth's cells of the small intestinal crypts (45,46), epithelial cells of the female urogenital tract (47), and nasal and bronchial epithelial cells (48). HNPs 1 through 4 are localized in azurophilic granules of neutrophils and contribute to the oxygen-independent killing of phagocytosed microorganisms. Furthermore, HNPs 1 through 3 can increase the expression of TNF-α and IL-1 in human monocytes that have been activated by *Staphylococcus aureus* (49). In human beings, four types of β-**defensins** have been identified thus far; they are referred to as human β-defensins 1 through 4. They have a broad spectrum of antimicrobial activity, bind to CCR6, and are chemotactic for immature dendritic cells and memory T cells (50). Human β-defensin-2 can also promote histamine release and prostaglandin D_2 production in mast cells, suggesting a role in allergic reactions (51,52). Thus, the defensins, like the cathelicidins, can contribute to the immune response by both killing bacteria and influencing the cellular innate and adaptive immune response. θ-**Defensins** have been isolated from rhesus

monkey neutrophils but no data about the presence of these molecules in different tissues are currently available.

All of the β-defensins are present in the respiratory system. These are expressed at various levels in the epithelia of the trachea and lung, as well as in the serous cells of the submucosal glands (53–55). Cathelicidins are also present in the conducting airway epithelium, pulmonary epithelium, and submucosal glands (56). Recent investigations have also documented the presence of these peptides in the nose and paranasal sinuses in health and diseases including rhinitis, chronic rhinosinusitis, and nasal polyposis (57–61).

VASCULAR AND LYMPHATIC SUPPLIES

The nose receives its blood supply from both the internal and external carotid circulations via the ophthalmic and internal maxillary arteries, respectively (Fig. 5). The ophthalmic artery gives rise to the anterior and posterior ethmoid arteries, which supply the anterosuperior portion of the septum, the lateral nasal walls, the olfactory region, and a small part of the posterosuperior region. The external carotid artery gives rise to the internal maxillary artery that ends as the sphenopalatine artery, which enters the nasal cavity through the sphenopalatine foramen behind the posterior end of the middle turbinate. The sphenopalatine artery gives origin to a number of posterior lateral and septal nasal branches. The posterolateral branches proceed to the region of the middle and inferior turbinates and to the floor of the nasal cavity. The posterior septal branches supply the corresponding area of the septum, including the nasal floor. Because it supplies the majority of blood to the nose and is often involved in severe epistaxis, the sphenopalatine artery has been called the "rhinologist's" artery. The region of the vestibule is supplied by the facial artery through lateral and septal nasal branches. The septal branches of the sphenopalatine artery form multiple anastomoses with the terminal branches of the anterior ethmoidal and facial arteries giving rise to Kiesselbach's area, located at the caudal aspect of the septum and also known as Little's area. Most cases of epistaxis occur in this region (62).

The veins accompanying the branches of the sphenopalatine artery drain into the pterygoid plexus. The ethmoidal veins join the ophthalmic plexus in the orbit. Part of the drainage to the ophthalmic plexus proceeds to the cavernous sinus via the superior ophthalmic veins and the other part to the pterygoid plexus via the inferior ophthalmic veins. Furthermore, the nasal veins form numerous anastomoses with the veins of the face, palate, and pharynx. The nasal venous system is valveless, predisposing to the spread of infections and constituting a dynamic system reflecting body position.

The subepithelial and glandular zones of the nasal mucosa are supplied by arteries derived from the periosteal or perichondrial vessels. Branches from these vessels ascend perpendicularly toward the surface, anastomosing with the cavernous plexi (venous system) before forming fenestrated capillary networks next to the respiratory epithelium and around the glandular tissue. The fenestrae always face the respiratory epithelium and are believed to be one of the sources of fluid for humidification. The capillaries of the subepithelial and periglandular network join to form venules that drain into larger superficial veins. They, in turn, join the sinuses of the cavernous plexus. The cavernous plexi, or sinusoids, consist of networks of large, tortuous,

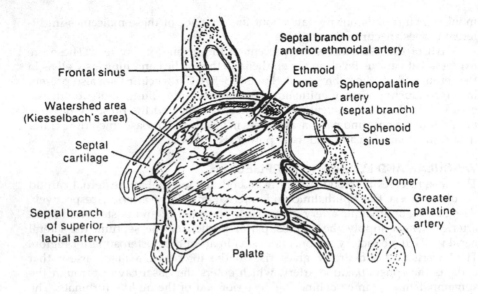

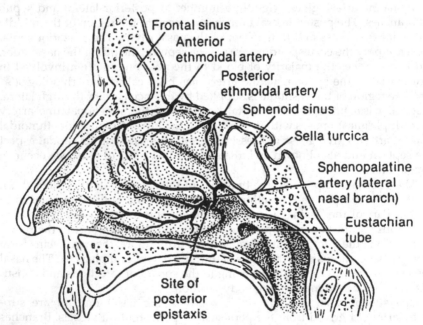

FIGURE 5 Nasal blood supply. The top panel represents the supply to the nasal septum and the bottom panel that to the lateral nasal wall. *Source*: From Ref. 6.

valveless, anastomosing veins mostly found over the inferior and middle turbinates but also in the midlevel of the septum. They consist of a superficial layer formed by the union of veins that drain the subepithelial and glandular capillaries and a deeper layer where the sinuses acquire thicker walls and assume a course parallel to the periosteum or perichondrium. They receive venous blood from the subepithelial and glandular capillaries and arterial blood from arteriovenous anastomoses. The arterial segments of the anastomoses are surrounded by a longitudinal smooth muscle layer that controls their blood flow. When the muscular layer contracts, the artery occludes; when it relaxes, the anastomosis opens, allowing the sinuses to fill rapidly with blood. Because of this function, the sinusoids are physiologically referred to as capacitance vessels. Only endothelium interposes between the longitudinal muscles and the blood stream, making them sensitive to circulating agents. The cavernous plexi change their blood volume in response to neural, mechanical, thermal, psychologic, or chemical stimulation. They expand and shrink, altering the caliber of the air passages and, consequently, the speed and volume of airflow.

Lymphatic vessels from the nasal vestibule drain toward the external nose, whereas the nasal fossa drains posteriorly. The first-order lymph nodes for posterior drainage are the lateral retropharyngeal nodes, whereas the subdigastric nodes serve that function for anterior drainage.

NEURAL SUPPLY

The nasal neural supply is overwhelmingly sensory and autonomic (sympathetic, parasympathetic, and nonadrenergic noncholinergic) (Fig. 6). The sensory nasal innervation comes via both the ophthalmic and maxillary divisions of the trigeminal nerve and supplies the septum, the lateral walls, the anterior part of

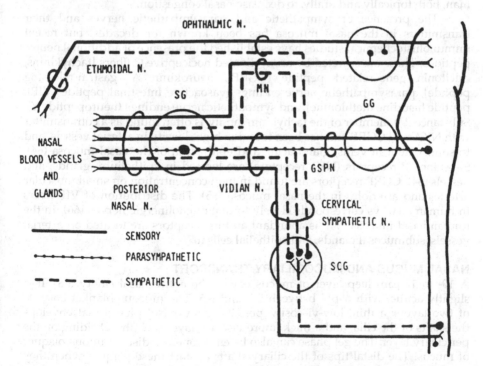

FIGURE 6 Nasal neural supply: sensory, sympathetic, and parasympathetic. *Abbreviations*: SG, sphenopalatine ganglion; MN, maxillary nerve; GG, geniculate ganglion; GSPN, greater superficial petrosal nerve; SCG, superior cervical ganglion. *Source*: From Ref. 9.

the nasal floor, and the inferior meatus. The parasympathetic nasal fibers travel from their origin in the superior salivary nucleus of the midbrain via the nervus intermedius of the facial nerve to the geniculate ganglion where they join the greater superficial petrosal nerve, which, in turn, joins the deep petrosal nerve to form the vidian nerve. This nerve travels to the sphenopalatine ganglion where the preganglionic parasympathetic fibers synapse and postganglionic fibers supply the nasal mucosa. The sympathetic input originates as preganglionic fibers in the thoracolumbar region of the spinal cord, which pass into the vagosympathetic trunk and relay in the superior cervical ganglion. The postganglionic fibers end as the deep petrosal nerve which joins the greater superficial nerve to form the vidian nerve. They traverse the sphenopalatine ganglion without synapsing and are distributed to the nasal mucosa.

Nasal glands receive direct parasympathetic nerve supply, and electrical stimulation of parasympathetic nerves in animals induces glandular secretions that are blocked by atropine. Furthermore, stimulation of the human nasal mucosa with methacholine, a cholinomimetic, produces an atropine-sensitive increase in nasal secretions (63). Parasympathetic nerves also provide innervation to the nasal vasculature and stimulation of these fibers causes vasodilatation. Sympathetic fibers supply the nasal vasculature but do not establish a close relationship with nasal glands and their exact role in the control of nasal secretions is not clear. Stimulation of these fibers in cats causes vasoconstriction and a decrease in nasal airway resistance. Adrenergic agonists are commonly used in man, both topically and orally, to decrease nasal congestion.

The presence of sympathetic and parasympathetic nerves and their transmitters in the nasal mucosa has been known for decades, but recent immunohistochemical studies have established the presence of additional neuropeptides. These are secreted by unmyelinated nociceptive C fibers [tachykinins, calcitonin gene–related peptide (CGRP), neurokinin A, gastrin-releasing peptide], parasympathetic nerve endings [vasoactive intestinal peptide (VIP), peptide histidine methionine], and sympathetic nerve endings (neuropeptide Y). Substance P, a member of the tachykinin family, is often found as a cotransmitter with NKA and CGRP and has been found in high density in arterial vessels, and to some extent in veins, gland acini, and epithelium of the nasal mucosa (64). Substance P receptors (NK1 receptors) are located in epithelium, glands, and vessels (64). CGRP receptors are found in high concentration on small muscular arteries and arterioles in the nasal mucosa (65). The distribution of VIP fibers in human airways corresponds closely to that of cholinergic nerves (66). In the human nasal mucosa, VIP is abundant and its receptors are located on arterial vessels, submucosal glands, and epithelial cells (67).

NASAL MUCUS AND MUCOCILIARY TRANSPORT

A 10- to 15-μm deep layer of mucus covers the entire nasal cavity (68). It is slightly acidic, with a pH between 5.5 and 6.5. The mucous blanket consists of two layers: a thin, low-viscosity, periciliary layer (sol phase) that envelops the shafts of the cilia, and a thick, more viscous, layer (gel phase) riding on the periciliary layer. The gel phase can also be envisioned as discontinuous plaques of mucus. The distal tips of the ciliary shafts contact these plaques when they are fully extended. Insoluble particles caught on the mucous plaques move with them as a consequence of ciliary beating. Soluble materials like droplets, formaldehyde, and CO_2 dissolve in the periciliary layer. Thus, nasal mucus effectively filters and removes nearly 100% of particles greater than 4 μm in diameter (69–71). An estimated 1 to 2 L of nasal mucus, composed of 2.5% to 3% glycoproteins, 1% to 2% salts, and 95% water, is produced per day. Mucin, one of the glycoproteins, gives mucus its unique attributes of protection and lubrication of mucosal surfaces.

The sources of nasal secretions are multiple and include anterior nasal glands, seromucous submucosal glands, epithelial secretory cells (of both mucous and serous types), tears, and transudation from blood vessels. Transudation increases in pathologic conditions as a result of the effects of inflammatory mediators that increase vascular permeability. A good example is the increased vascular permeability seen in response to allergen challenge of subjects with allergic rhinitis as measured by increasing levels of albumin in nasal lavages after provocation (72). In contrast to serum, immunoglobulins make up the bulk of the protein in mucus; other substances in nasal secretions include lactoferrin, lysozyme, antitrypsin, transferrin, lipids, histamine and other mediators, cytokines, antioxidants, ions (Cl, Na, Ca, K), cells, and bacteria. Mucus functions in mucociliary transport, and substances will not be cleared from the nose without it, despite adequate ciliary function. Furthermore, mucus provides immune and mechanical mucosal protection and its high water content plays a significant role in humidifying inspired air.

Mucociliary transport is unidirectional based on the unique characteristics of cilia. Cilia in mammals beat in a biphasic, or to-and-fro, manner. The beat consists of a rapid effective stroke during which the cilium straightens, bringing it in contact with the gel phase of the mucus, and a slow recovery phase during which the bent cilium returns in the periciliary or sol layer of the mucus, thus propelling it in one direction (Fig. 7).

Metachrony is the coordination of the beat of individual cilia, which prevents collision between cilia in different phases of motion and results in the

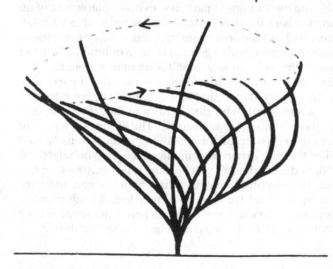

FIGURE 7 A schematic diagram of motion of a single cilium during the rapid forward beat and the slower recovery phase. *Source*: From Ref. 9.

unidirectional flow of mucus. Ciliary beating produces a current in the superficial layer of the periciliary fluid in the direction of the effective stroke. The mucous plaques move as a result of motion of the periciliary fluid layer and the movement of the extended tips of the cilia into the plaques. Thus, the depth of the periciliary fluid is the key factor in mucociliary transport. If excessive, the extended ciliary tips fail to contact mucous plaques, and the current of the periciliary fluid provides the only means of movement.

Mucociliary transport moves mucus and its contents toward the nasopharynx, with the exception of the anterior portion of the inferior turbinates, where transport is anterior. This anterior current prevents many of the particles deposited in this area from progressing further into the nasal cavity. The particles transported posteriorly toward the nasopharynx are periodically swallowed. Mucociliary transport, however, is not the only mechanism by which particles and secretions are cleared from the nose. Sniffing and nose blowing help in moving airway secretions backward and forward, respectively. Sneezing results in a burst of air, accompanied by an increase in watery nasal secretions that are then cleared by nose blowing and sniffing.

Respiratory cilia beat about 1000 times per minute, which translates to surface materials being moved at a rate of 3 to 25 mm/min. Both the beat rate and propelling speed vary. Several substances have been used to measure nasal mucociliary clearance, and the most utilized are sodium saccharin, dyes, or tagged particles. The dye and saccharin methods are similar, consisting of placing a strong dye or saccharin sodium on the nasal mucosa just behind the internal ostium and recording the time it takes to reach the pharyngeal cavity; this interval is termed nasal mucociliary transport time. With saccharin, the time is recorded when the subject reports a sweet taste, whereas with a dye, when it appears in the pharyngeal cavity. Combining the two methods reduces the disadvantages of both—namely, variable taste thresholds in different subjects when using saccharin and repeated pharyngeal inspection when using the dye—and makes them more reliable. The use of tagged particles involves placement of an anion exchange resin particle about 0.5 mm in diameter tagged with a 99Tc ion on the anterior nasal mucosa, behind the area of anterior mucociliary movement, and following its subsequent clearance with a gamma camera or multicollimated detectors. This last method permits continuous monitoring of movement.

Studies of several hundred healthy adult subjects by the tagged particle or saccharin methods have consistently shown that 80% exhibit clearance rates of 3 to 25 mm/min (average = 6 mm/min), with slower rates in the remaining 20% (9). The latter subjects have been termed "slow clearers." The findings of a greater proportion of slow clearers in one group of subjects living in an extremely cold climate raise the possibility that the differences in clearance may be related to an effect of inspired air (9). In diseased subjects, slow clearance may be due to a variety of factors, including the immotility of cilia, transient or permanent injury to the mucociliary system by physical trauma, viral infection, dehydration, or excessively viscid secretions secondary to decreased ions and water in the mucus paired with increased amounts of DNA from dying cells, as in cystic fibrosis.

NASAL AIRFLOW

The nose provides the main pathway for inhaled air to the lower airways and offers two areas of resistance to airflow (provided there are no gross deviations of the nasal septum): the nasal valve and the state of mucosal swelling of the nasal airway. The cross-sectional area of the nasal airway decreases dramatically at each nasal valve to reach 30 to 40 mm^2. This narrowed area separates the vestibules from the main airway and account for approximately half of the total resistance to respiratory airflow from ambient air to the alveoli. After bypassing this narrow area, inspired air flows in the main nasal airway, which is a broader tube bounded by the septal surface medially, and the irregular inferior and middle turbinates laterally. The variable caliber of the lumen of this portion of the airway is governed by changes in the blood content of the capillaries, capacitance vessels, and arteriovenous shunts of the lining mucosa and constitutes the second resistive segment that inspired air encounters on its way to the lungs. Changes in the blood content of these structures occur spontaneously and rhythmically, resulting in alternating volume reductions in the lumen of the two nasal cavities, a phenomenon referred to as the nasal cycle. This occurs in approximately 80% of normal individuals, and the reciprocity of changes between the two sides of the nasal cavity maintains total nasal airway resistance unchanged (73). The duration of one cycle varies between 50 minutes and 4 hours and is interrupted by vasoconstrictive medications or exercise, which leads to a marked reduction of total nasal airway resistance. Kennedy and colleagues observed the nasal passages using T2-weighted magnetic resonance imaging and demonstrated an alternating increase and decrease in signal intensity and turbinate size over time in a fashion consistent with the nasal cycle (74). The nasal cycle can be exacerbated by the increase in nasal airway resistance caused by exposure to allergic stimuli and explains why some allergic individuals complain of alternating exacerbations of their nasal obstructive symptoms.

Swift and Proctor presented a detailed description of nasal airflow and its characteristics (Fig. 8) (75). Upon inspiration, air first passes upwards into the vestibules in a vertical direction at a velocity of 2 to 3 m/sec, and then converges and changes its direction from vertical to horizontal just prior to the nasal valve, where, due to the narrowing of the airway, velocities reach their highest levels (up to 12–18 m/sec). After passing the nasal valve, the cross-sectional area increases, and velocity decreases concomitantly to about 2 to 3 m/sec. The nature of flow changes from laminar, before and at the nasal valve, to more turbulent posteriorly. As inspiratory flow increases beyond resting levels, turbulent

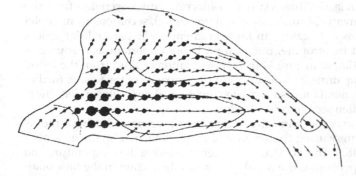

FIGURE 8 Schematic diagram of the direction and velocity of inspired air. The size of the dots is directly proportional to velocity and the arrows depict direction of airflow. *Source*: From Ref. 9.

characteristics commence at an increasingly anterior position and, with mild exercise, are found as early as the anterior ends of the turbinates. The airstream increases in velocity to 3 to 4 m/sec in the nasopharynx, where the direction again changes from horizontal to vertical as air moves down through the pharynx and larynx to reach the trachea. Turbulence of nasal airflow minimizes the presence of a boundary layer of air that would exist with laminar flow and maximizes interaction between the airstream and the nasal mucosa. This, in turn, allows the nose to perform its functions of heat and moisture exchange and of cleaning inspired air of suspended or soluble particles.

NASAL CONDITIONING OF TEMPERATURE AND HUMIDITY OF INSPIRED AIR

Inspiratory air is rapidly warmed and moistened mainly in the nasal cavities and, to a lesser extent, in the remainder of the upper airway down to the lungs (76). Inspired air is warmed from a temperature of around 20°C at the portal of entry to 31°C in the pharynx and 35°C in the trachea. This is facilitated by the turbulent characteristics of nasal airflow, which maximize the contact between inspired and expired air and the nasal mucosal surface (77). After inspiration ceases, warming of the nasal mucosa by the blood is such a relatively slow process that, at expiration, the temperature of the nasal mucosa remains lower than that of expired air. As expiratory air passes through the nose, it gives up heat to the cooler nasal mucosa. This cooling causes condensation of water vapor and, thus, a 33% return of both heat and moisture to the mucosal surface. Since recovery of heat from expiratory air occurs mainly in the region of the respiratory portal, blood flow changes that take place in the nasal mucosa affect respiratory air conditioning more markedly in this region (78).

Ingelstedt showed that the humidifying capacity of the nose is greatly impaired in healthy volunteers after a subcutaneous injection of atropine (76,79). He thus concluded that atropine-inhibitable glandular secretion is a major source of water for humidification of inspired air. In contrast, Kumlien and Drettner failed to show any effect of intranasal ipratropium bromide, another anticholinergic agent, on the degree of warming and humidification of air during passage in the nasal cavity in a small clinical trial using normal subjects and a group of patients with vasomotor rhinitis (80). In addition to glandular secretions, other sources provide water for humidification of inspired air and these include water content of ambient air, lacrimation via the nasolacrimal duct, secretion from the paranasal sinuses, salivation (during oronasal breathing), secretions from goblet cells, and passive transport against an ionic gradient in the paracellular spaces (79,81). Not inhibited by atropine, but also probably important as a source of water for humidification of inspired air, is transudation of fluid from the blood vessels of the nose. Impairment of the humidifying capacity of the nose is further accentuated when the nasal mucosa is chilled, leading, along with condensation, to the nasal drip so often seen in cold weather.

The ability to warm and humidify air has been investigated using a model system that involves measuring the amount of water delivered by the nose after inhaling cold dry air (82). This is calculated after measuring the temperature and humidity of air as it penetrates the nasal cavity and then again in the nasopharynx by using a specially designed probe. Using this model, the investigators were able to show that the ability to warm and humidify inhaled air is lower in subjects with allergic rhinitis out of season compared to normal controls. The effect of allergic inflammation on the nasal conditioning capacity of individuals with seasonal allergic rhinitis was then investigated by evaluating the ability of the nose to warm and humidify cold dry air in allergics before and after the season as well as 24 hours after allergen challenge (83). These studies showed that allergic inflammation, induced by either the allergy season or an allergen challenge, increased the ability of the nose to warm and humidify inhaled air, and the authors speculated that this was related to a change in the nasal perimeter induced by allergic inflammation. In an interesting follow-up study, the same investigators compared the ability of the following groups of subjects to warm and humidify inhaled air: patients with perennial allergic rhinitis, seasonal allergic rhinitis out of season, and normal subjects and subjects with bronchial asthma (84). They showed that subjects with perennial allergic rhinitis were comparable to normals in their ability to condition air and that subjects with asthma had a reduced ability to perform this function compared to normals. Furthermore, the total water gradient, a measure of the ability of the nose to condition air, correlated negatively with severity of asthma assessed by using two different gradings, suggesting that the ability to condition inspired air was worse in subjects with more severe asthma and suggesting that this reduced ability might contribute, at least in part, to the pathophysiology of asthma.

OLFACTION
One of the important sensory functions of the nose is olfaction. The olfactory airway is 1 to 2 mm wide and lies above the middle turbinate just inferior to the cribriform plate between the septum and the lateral wall of the nose. The olfactory mucosa has a surface area of 200 to 400 mm^2 on each side and contains numerous odor receptor cells with thin cilia that project into the covering mucus layer and increase the surface area of the epithelium (85). The olfactory mucosa also contains small, tubular, serous Bowman's glands situated immediately below the epithelium. Each receptor cell is connected to the olfactory bulb by a thin nonmyelinated nerve fiber that is slow conducting (velocity 50 m/sec) but short, making the conduction time low. The impulses from the olfactory bulb are conveyed to the piriform and entorhinal cortices, which together constitute the primary olfactory cortex.

The area where the olfactory epithelium is located is poorly ventilated as most of the inhaled air passes through the lower aspect of the nasal cavity. Therefore, nasal obstruction, as documented by elevations in nasal airway resistance, leads to an elevation in olfactory thresholds (86). This may be secondary to several reasons such as septal deviations, nasal polyposis, nasal deformities, or increased nasal congestion, one of the characteristic symptoms of allergic rhinitis. Sniffing helps the process of smell by increasing the flow rate of, and degree of turbulence of, inhaled air and, consequently, raising the proportion of air reaching the olfactory epithelium by 5% to 20%. This results in increasing the number of odorant molecules available to the olfactory receptors and proportionally enhancing odor sensation. In addition to crossing the anatomic barriers of the nose, the odorant molecules must have a dual solubility in lipids and water to be able to reach the olfactory receptors. To penetrate the mucus covering the olfactory mucosa, they solubilize to a certain extent in water. Lipid solubility, on the other hand, enhances their interaction with the receptor membrane of the olfactory epithelial cilia. Lastly, it is to be mentioned that olfactory sensitivity normally decreases with age as evidenced by a recent longitudinal study of men and women between the ages of 19 and 95 followed over a three-year period (87).

VOMERONASAL ORGAN

Many vertebrate species including many mammals have a small chemosensory structure in the nose called the vomeronasal organ (VNO), which is dedicated to detecting chemical signals that mediate sexual and territorial behaviors. A similar structure appears to exist in the human nose and is described as two small sacs about 2 mm deep that open into shallow pits on either side of the nasal septum. Jacob and colleagues performed a study to characterize the nasal opening of the nasopalatine duct (NPD), which, with the VNO, forms the vomeronasal system (88). Otolaryngologists examined the nose of normal volunteers endoscopically looking for distinct morphologic features of the NPD, including the structure's larger fossa or craterlike indentation and its smaller aperture within the fossa. The area examined for presence or absence of the duct was approximately 2 cm dorsal to the nostril opening and <0.5 cm above the junction of the nasal floor and septum. The NPD was detected in 94% of 221 nostrils and its nasal opening was consistently located 1.9 cm dorsal to the collumella, 0.2 cm above the junction of the nasal floor and septum. The authors also examined cadaver specimens and found bilateral nasopalatine fossae with apertures in every specimen and VNOs in less than half of the septal regions examined. Thus, while the authors did not establish function, their study confirms the existence of a vomeronasal system, or its anatomic remnants in humans.

In vertebrates, the pair of small sacs is lined by sensory neurons, tucked inside the vomer bone where the hard palate and nasal septum meet. In mice and rats, the VNO is connected to the brain through a neural pathway that is independent from the olfactory pathway, but it is not clear whether the human VNO is connected to the brain. There is renewed interest in researching the anatomy and function of this organ in man, and this effort is primarily funded by the perfume industry (89).

CONCLUSION

As detailed in this chapter, the nose is an intricate organ with important functions which include filtration, humidification, and temperature control of inspired air in preparation for transit to the lower airways. It is also important in providing the sense of olfaction and sensory irritation (via the trigeminal system). It has an intricate network of nerves, vessels, glands, and inflammatory cells, which all help to modulate its function. Chronic inflammation affects multiple end organs within the nasal cavity and will lead to diseases such as rhinitis (allergic and nonallergic) as well as sinusitis. The overview of the anatomy and function of the nose provided in this chapter will hopefully serve as a useful prelude to the coming chapters where different aspects of nasal toxicology are discussed.

REFERENCES

1. Drumheller GW. Topology of the lateral nasal cartilages: the anatomical relationship of the lateral nasal to the greater alar cartilage, lateral crus. Anat Rec 1973; 176:321–327.

2. Van Loosen J, Van Zanten GA, Howard CV, et al. Growth characteristics of the human nasal septum. Rhinology 1996; 34:78.
3. Gray L. Deviated nasal septum. III. Its influence on the physiology and disease of the nose and ears. J Laryngol 1967; 81:953.
4. Mygind N, Pedersen M, Nielsen M. Morphology of the upper airway epithelium. In: Proctor DF, Andersen IB, eds. The Nose. Amsterdam, The Netherlands: Elsevier Biomedical Press BV, 1982:71–97.
5. Cummings CW, Fredrickson JM, Harker LA, et al. Otolaryngology—Head and Neck Surgery, 2nd ed. St. Louis, MO: Mosby-Year Book, 1993.
6. Montgomery WW. Surgery of the Upper Respiratory System. Philadelphia, PA: Lea and Febiger, 1979.
7. Maniscalco M, Sofia M, Pelaia G. Nitric oxide in upper airways inflammatory diseases. Inflamm Res 2007; 56:58–69.
8. Lindberg S, Cervin A, Runer T. Low levels of nasal nitric oxide (NO) correlates to impaired mucociliary function in the upper airways. Acta Otolaryngol 1997; 117:728–734.
9. Proctor DF. The mucociliary system. In: Proctor DF, Andersen IB, eds. The Nose: Upper Airway Physiology and the Atmospheric Environment. Amsterdam, The Netherlands: Elsevier Biomedical Press BV, 1982:245–278.
10. Evans MJ, Plopper GG. The role of basal cells in adhesion of columnar epithelium to airway basement membrane. Am Rev Respir Dis 1988; 138:481.
11. Evans MJ, Shami S, Cabral-Anderson LJ, et al. Role of nonciliated cells in renewal of the bronchial epithelium of rats exposed to NO_2. Am J Pathol 1986; 123:126.
12. Tos M. Goblet cells and glands in the nose and paranasal sinuses. In: Proctor DF, Andersen IB, eds. The Nose. Amsterdam, The Netherlands: Elsevier Biomedical Press BV, 1982:98–144.
13. Winther B, Innes DJ, Mills SE, et al. Lymphocyte subsets in normal airway mucosa of the human nose. Arch Otolaryngol Head Neck Surg 1987; 113:59.
14. Bradding P, Feather IH, Wilson S, et al. Immunolocalization of cytokines in the nasal mucosa of normal and perennial rhinitic subjects. J Immunol 1993; 151:3853.
15. Lim MC, Taylor RM, Naclerio RM. The histology of allergic rhinitis and its comparison to cellular changes in nasal lavage. Am J Respir Crit Care Med 1995; 151:136.
16. Varney VA, Jacobson MR, Sudderick RM, et al. Immunohistology of the nasal mucosa following allergen-induced rhinitis. Am Rev Respir Dis 1992; 146:170.
17. Bojsen-Moller F. Glandulae nasales anteriores in the human nose. Ann Otol Rhinol Laryngol 1965; 74:363.
18. Frank MM, Fries LF. The role of complement in inflammation and phagocytosis. Immunol Today 1991; 12:322.
19. Yang D, Chertov O, Bykovskaia SN, et al. Beta-defensins: linking innate and adaptive immunity through dendritic and T cell CCR6. Science 1999; 286:525.
20. Modlin RL. Mammalian Toll-like receptors. Ann Allergy Asthma Immunol 2002; 88:543.
21. Akashi S, Shimazu R, Ogata H, et al. Cutting edge: cell surface expression and lipopolysaccharide signaling via the toll-like receptor 4-MD-2 complex on mouse peritoneal macrophages. J Immunol 2000; 164:3471.
22. Muzio M, Bosisio D, Polentarutti N, et al. Differential expression and regulation of toll-like receptors (TLR) in human leukocytes: selective expression of TLR3 in dendritic cells. J Immunol 2000; 164:5998.
23. Poltorak A, He X, Smirnova I, et al. Defective LPS signaling in C3H/HeJ and C57BL/10ScCr mice: mutations in Tlr4 gene. Science 1998; 282:2085.
24. Takeuchi O, Hoshino K, Kawai T, et al. Differential roles of TLR2 and TLR4 in recognition of gram-negative and gram-positive bacterial cell wall components. Immunity 1999; 11:443.
25. Brightbill HD, Libraty DH, Krutzik SR, et al. Host defense mechanisms triggered by microbial lipoproteins through toll-like receptors. Science 1999; 285:732.

26. Bauer S, Kirschning CJ, Hacker H, et al. Human TLR9 confers responsiveness to bacterial DNA via species-specific CpG motif recognition. Proc Natl Acad Sci U S A 2001; 98:9237.

27. Hayashi F, Smith K, Ozinsky A, et al. The innate immune response to bacterial flagellin is mediated by Toll-like receptor 5. Nature 2001; 410:1099.

28. Alexopoulou L, Holt AC, Medzhitov R, et al. Recognition of double stranded RNA and activation of NF-κB by Toll-like receptor 3. Nature 2001; 413:732.

29. Chinen J, Shearer WT. Advances in basic and clinical immunology in 2006. J Allergy Clin Immunol 2007; 120:263–270.

30. Abreu MT, Arditi M. Innate immunity and toll-like receptors: clinical implications of basic science research. J Pediatr 2004; 144:421–429.

31. Hoffmann JA, Kafatos F, Janeway C, et al. Phylogenetic perspectives in innate immunity. Science 1999; 284:1313.

32. Becker MN, Diamond G, Verhese MW, et al. CD14-dependent lipopolysaccharide-induced β-defensin-2 expression in human tracheobronchial epithelium. J Biol Chem 2000; 275:29731.

33. Aliprantis AO, Yang RB, Mark M, et al. Cell activation and apoptosis by bacterial lipoproteins through Toll-like receptor-2. Science 1999; 285:736.

34. Ellison RT, Giehl TJ. Killing of gram-negative bacteria by lactoferrin and lysozyme. J Clin Invest 1991; 88:1080–1091.

35. Akinbi HT, Epaud R, Bhatt H, et al. Bacterial killing is enhanced by expression of lysozyme in the lungs of transgenic mice. J Immunol 2000; 165:5760–5766.

36. Zhu J, Nathan C, Ding A. Suppression of macrophage responses to bacterial lipopolysaccharide by a non-secretory form of secretory leukocyte protease inhibitor. Biochim Biophys Acta 1999; 1451:219–223.

37. Sorensen O, Arnljots K, Cowland JB, et al. The human antibacterial cathelicidin, hCAP-18, is synthesized in myelocytes and metamyelocytes and localized to specific granules in neutrophils. Blood 1997; 90:2796–2803.

38. Ooi EH, Wormald PJ, Tan LW. Innate immunity in the paranasal sinuses: a review of nasal host defenses. Am J Rhinol 2008; 22:13–19.

39. Niyonsaba F, Iwabuchi K, Someya A, et al. A cathelicidin family of human antibacterial peptide LL-37 induces mast cell chemotaxis. Immunology 2002; 106:20.

40. De Y, Chen Q, Scmidt AP, et al. LL-37, the neutrophil granule- and epithelial cell-derived cathelicidin, utilizes formyl peptide recptor-like 1 (FPRL1) as a receptor to chemoattract human peripheral blood neutrophils, monocytes, and T cells. J Exp Med 2000; 192:1069–1074.

41. Niyonsaba F, Hirata M, Ogawa H, et al. Epithelial cell derived antibacterial peptides human beta-defensins and cathelicidin: multifunctional activities on mast cells. Curr Drug Targets Inflamm Allergy 2003; 2:224–231.

42. Ganz T, Selsted ME, Szklarek D, et al. Defensins. Natural peptide antibiotics of human neutrophils. J Clin Invest 1985; 76:1427.

43. Lehrer RI, Daher K, Ganz T, et al. Direct inactivation of viruses by MCP-1 and MCP-2, natural peptide antibiotics from rabbit leukocytes. J Virol 1985; 54:467.

44. Harwig SS, Ganz T, Lehrer RI. Neutrophil defensins: purification, characterization, and antimicrobial testing. Methods Enzymol 1994; 236:160.

45. Jones DE, Bevins CL. Defensin-6 mRNA in human Paneth cells: implications for antimicrobial peptides in host defense of the human bowel. FEBS Lett 1993; 315:187.

46. Selsted ME, Miller SI, Henschen AH, et al. Enteric defensins: antibiotic peptide components of intestinal host defense. J Cell Biol 1992; 118:929.

47. Quayle AJ, Porter EM, Nussbaum AA, et al. Gene expression, immunolocalization, and secretion of human defensin-5 in human female reproductive tract. Am J Pathol 1998; 152:1247.

48. Frye M, Bargon J, Daultbaev N, et al. Expression of human alpha-defensin 5 (HD5) mRNA in nasal and bronchial epithelial cells. J Clin Pathol 2000; 53:770–773.

49. Chaly YV, Paleolog EM, Kolesnikova TS, et al. Neutrophil alpha-defensin human neutrophil peptide modulates cytokine production in human monocytes and adhesion molecule expression in endothelial cells. Eur Cytokine Netw 2000; 11:257.
50. Yang D, Biragyn A, Kwak LW, et al. Mammalian defensins in immunity: more than just microbicidal. Trends Immunol 2002; 23:291.
51. Niyonsaba F. Evaluation of the effects of peptide antibiotics human beta-defensins-1/-2 and LL-37 on histamine release and prostaglandin D(2) production from mast cells. Eur J Immunol 2001; 31:1066.
52. Befus AD, Mowat C, Gilchrist M, et al. Neutrophil defensins induce histamine secretion from mast cells: mechanisms of action. J Immunol 1999; 163:947.
53. Harder J, Bartels J, Christophers E, et al. Isolation and characterization of human beta-defensin-3, a novel human inducible peptide antibiotic. J Biol Chem 2001; 276:5707.
54. Garcia JR, Krause A, Schulz S, et al. Human beta-defensin 4: a novel inducible peptide with a specific salt-sensitive spectrum of antimicrobial activity. FASEB J 2001; 15:1819.
55. Zhao C, Wang I, Lehrer RI. Widespread expression of beta-defensin hBD-1 in human secretory glands and epithelial cells. FEBS Lett 1996; 396:319.
56. Bals R, Wang X, Zasloff M, et al. The peptide antibiotic LL-37/hCAP-18 is expressed in epithelia of the human lung where it has broad antimicrobial activity at the airway surface. Proc Natl Acad Sci U S A 1998; 95:9541.
57. Lee SH, Lim HH, Lee HM, et al. Expression of human beta-defensin 1 mRNA in human nasal mucosa. Acta Otolaryngol 2000; 120:58.
58. Lee SH, Kim SH, Lim HH, et al. Antimicrobial defensin peptides of the human nasal mucosa. Ann Otol Rhinol Laryngol 2002; 111:135.
59. Kim ST, Cha HE, Kim DY, et al. Antimicrobial peptide LL-37 is upregulated in chronic nasal inflammatory disease. Acta Otolaryngol 2003; 123:81–85.
60. Chen PH, Fang SY. The expresion of human antimicrobial peptide LL-37 in the human nasal mucosa. Am J Rhinol 2004; 18:381–385.
61. Ooi EH, Wormald PJ, Carney AS, et al. Human cathelicidin antimicrobial peptide is upregulated in the eosinophilic mucus subgroup of chronic rhinosinusitis patients. Am J Rhinol 2007; 21:395–401.
62. Cauna N. Blood and nerve supply of the nasal lining. In: Proctor DF, Andersen IB, eds. The Nose. Amsterdam, The Netherlands: Elsevier Biomedical Press BV, 1982:45–69.
63. Baroody FM, Wagenmann M, Naclerio RM. A comparison of the secretory response of the nasal mucosa to histamine and methacholine. J Appl Physiol 1993; 74(6):2661.
64. Baraniuk JN, Lundgren JD, Mullol J, et al. Substance P and neurokinin A in human nasal mucosa. Am J Respir Cell Mol Biol 1991; 4:228.
65. Baraniuk JN, Castellino S, Merida M, et al. Calcitonin gene related peptide in human nasal mucosa. Am J Physiol 1990; 258:L81.
66. Laitinen A, Partanen M, Hervonen A, et al. VIP-like immunoreactive nerves in human respiratory tract. Light and electron microscopic study. Histochemistry 1985; 82:313.
67. Baraniuk JN, Okayama M, Lundgren JD, et al. Vasoactive intestinal peptide (VIP) in human nasal mucosa. J Clin Invest 1990; 86:825.
68. Wilson WR, Allansmith MR. Rapid, atraumatic method for obtaining nasal mucus samples. Ann Otol Rhinol Laryngol 1976; 85:391.
69. Andersen I, Lundqvist G, Proctor DF. Human nasal mucosal function under four controlled humidities. Am Rev Respir Dis 1979; 119:619.
70. Fry FA, Black A. Regional deposition and clearance of particles in the human nose. Aerosol Sci 1973; 4:113.
71. Lippmann M. Deposition and clearance of inhaled particles in the human nose. Ann Otol Rhinol Laryngol 1970; 79:519.
72. Baumgarten C, Togias AG, Naclerio RM, et al. Influx of kininogens into nasal secretions after antigen challenge of allergic individuals. J Clin Invest 1985; 76:191.
73. Hasegawa M, Kern EB. The human cycle. Mayo Clin Proc 1977; 52:28.
74. Kennedy DW, Zinreich SJ, Kumar AJ, et al. Physiologic mucosal changes within the nose and the ethmoid sinus: imaging of the nasal cycle by MRI. Laryngoscope 1988; 98:928.

75. Swift DL, Proctor DF. Access of air to the respiratory tract. In: Brain JD, Proctor DF, Reid LM, eds. Respiratory Defense Mechanisms. New York, NY: Marcel Dekker, 1977.
76. Ingelstedt S, Ivstam B. Study in the humidifying capacity of the nose. Acta Otolaryngol 1951; 39:286.
77. Aharonson EF, Menkes H, Gurtner G, et al. The effect of respiratory airflow rate on the removal of soluble vapors by the nose. J Appl Physiol 1974; 37:654.
78. Scherer PW, Hahn II, Mozell MM. The biophysics of nasal airflow. Otolaryngol Clin North Am 1989; 22:265.
79. Ingelstedt S, Ivstam B. The source of nasal secretion in normal condition. Acta Otolaryngol 1949; 37:446.
80. Kumlien J, Drettner B. The effect of ipratropium bromide (Atrovent) on the air conditioning capacity of the nose. Clin Otolaryngol 1985; 10:165.
81. Togias AG, Proud D, Lichtenstein LM, et al. The osmolality of nasal secretions increases when inflammatory mediators are released in response to inhalation of cold, dry air. Am Rev Respir Dis 1988; 137:625.
82. Rouadi P, Baroody Fm, Abbott D, et al. A technique to measure the ability of the human nose to warm and humidify air. J Appl Physiol 1999; 87:400–406.
83. Assanasen P, Baroody FM, Abbott DJ, et al. Natural and induced allergic responses increase the ability of the nose to warm and humidify air. J Allergy Clin Immunol 2000; 106:1045–1052.
84. Assanasen P, Baroody FM, Naureckas E, et al. The nasal passage of subjects with asthma has a decreased ability to warm and humidify inspired air. Am J Respir Crit Care Med 2001; 164:1640–1646.
85. Berglund B, Lindvall T. Olfaction. In: Proctor DF, Andersen IB, eds. The Nose. Amsterdam, The Netherlands: Elsevier Biomedical Press BV, 1982.
86. Rous J, Kober F. Influence of one-sided nasal respiratory occlusion of the olfactory threshold values. Arch Klin Exp Ohren Nasen Kehlkopfheilkd 1970; 196(2):374.
87. Ship JA, Pearson JD, Cruise LJ, et al. Longitudinal changes in smell identification. J Gerontol 1996; 51(2):M86.
88. Jacob S, Zelano B, Gungor A, et al. Location and gross morphology of the nasopalatine duct in human adults. Arch Otolaryngol Head Neck Surg 2000; 126:741–748.
89. Taylor R. Brave new nose: sniffing out human sexual chemistry. J NIH Res 1994; 6:47.

3 Functional Neuroanatomy of the Upper Airway in Experimental Animals

Paige M. Richards

Department of Biology, Wake Forest University, Winston-Salem, North Carolina, U.S.A.

C. J. Saunders

Neuroscience Program, University of Colorado Denver, Anschutz Medical Campus, Aurora, Colorado, U.S.A.

Wayne L. Silver

Department of Biology, Wake Forest University, Winston-Salem, North Carolina, U.S.A.

INTRODUCTION

Across species, the common function of the nasal cavities is the detection of chemical stimuli. In tetrapod vertebrates, the nasal cavities assumed another function—serving as the first part of the respiratory passages. In this capacity, they also serve to humidify, warm, and filter inspired air. This aspect of the nasal cavities is discussed in chapters 1 and 2. In fact, some animals, such as mice, rats, and the human newborn, are obligate nose breathers (i.e., can only breathe through their noses). In some mammals, this respiratory function may overshadow the detection of chemicals.

Phylogenetically, in vertebrates, the nose began solely as housing for the olfactory organ. In some fishes, the nasal cavities (or sacs) are blind-ended pits [Fig. 1(A)]. Water carrying chemical stimuli moves in and out of the sacs as the fish swims, although some fish appear to be able to actively pump water. In most teleost fish, the nasal sac develops separate incurrent and excurrent openings. Water flows across the olfactory epithelium (OE), which resides on a series of folds or lamellae in the nasal cavity [Fig. 1(B)]. Some fish may have as many as 100 lamellae per nasal cavity. In choanate fish, the excurrent opening (internal naris or choana) may open into the mouth or buccal cavity, while the external naris is open to the environment [Fig. 1(C)]. In these fishes, respiratory and olfactory currents may be coupled.

The nasal cavities of amphibians are similar to those of the choanate fishes. The OE is a flat sheet that occupies both the dorsal and ventral surfaces of the nasal cavity. One difference is that amphibians have a vomeronasal (or Jacobson's) organ (VNO). The VNO is a neuroepithelium separate and distinct from the OE [Fig. 1(D)]. The VNO resides on the ventral surface of the nasal cavity and is activated by socially relevant chemicals including those that are important in identifying home territory and sensing pheromones. The noses of some amphibians have grooves connecting the front of the mouth with each external

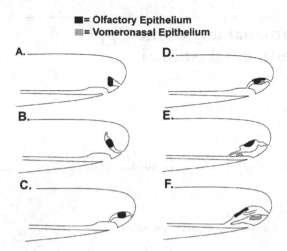

■ = Olfactory Epithelium
▨ = Vomeronasal Epithelium

A.

D.

B.

E.

C.

F.

FIGURE 1 Phylogeny of nasal cavities. (**A**) Teleost fish with one naris opening into the nasal cavity. (**B**) Teleost fish with incurrent and excurrent nares. Water flows in through the anterior nares and out the posterior nares. (**C**) Choanate fish with external nares and internal nares, which open into the mouth. (**D**) Amphibians have a vomeronasal organ that is located in the nasal cavity but separate from the olfactory organ. (**E**) Snakes have a vomeronasal organ housed in a cavity that is separate from the main nasal cavity. (**F**) Mammals have a vomeronasal organ that is housed in a capsule that is completely separated from the nasal cavity. See text for details.

naris (1). Apparently, these grooves carry aqueous material from the mouth to the VNO.

Reptilian nasal cavities are divided into an anterior vestibule, a posterior nasal chamber, and a nasopharyngeal duct. The nasopharyngeal duct connects to the internal nares opening into the oral cavity [Fig. 1(E)]. The OE lines most of the dorsal half of the nasal chamber. In some reptiles, the lateral wall of the nasal chamber projects inward to form conchae or turbinates. The VNO in reptiles, notably some snakes, may be larger than the olfactory organ. In snakes, the VNO is entirely separate from the nasal cavity [Fig. 1(E)]. In this case, the epithelium is located on the dorsal surface of a pit that opens into the oral cavity. The snake's forked tongue is used to carry chemicals to its VNO. The nasal cavities of birds are similar to those of reptiles, although they may be more elaborate with the conchae forming complicated scrolls. One of these, the olfactory tubercle, is covered in OE. Birds, like crocodilian reptiles, lack VNOs.

The most complicated nasal cavities are found in mammals and may have extensive turbinates [Fig. 1(F)]. The OE usually lies on the turbinates in the posterior of the nasal cavity. The VNO in most mammals is a blind pouch with sensory epithelium lining the medial surface of the cavity. The VNO opens to the nasal cavity directly or to the nasopharyngeal duct depending on the species. Although the classification has been called into question, mammals have long been considered microsmotic (humans, monkeys, sheep) or macrosmatic (dogs, rodents, rabbits), depending on their level of olfactory function (2). The remainder of this chapter will concentrate on the functional neuroanatomy of the two best-studied mammals—the mouse and the rat.

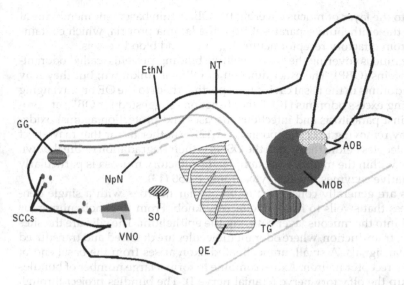

FIGURE 2 Schematic diagram of chemoreceptive organs in the rodent nasal cavity. *Abbreviations*: AOB, accessory olfactory bulb; EthN, ethmoid branch of the trigeminal nerve; GG, Grueneberg ganglion; MOB, main olfactory bulb; MOE, main olfactory epithelium; NpN, nasopalatine branch of the trigeminal nerve; NT, nervus terminalis; SCCs, solitary chemoreceptor cells; SO, septal organ; TG, trigeminal ganglion; VNO, vomeronasal organ.

OLFACTORY ORGAN

Overview

The OE houses the olfactory sensory neurons (OSNs), which are responsible for detecting odorants emanating from the external environment and the oral cavity. OSN axons form the olfactory or first cranial nerve (Cr. N. I). In rats and mice, the OE is located in the dorsal posterior region of the nasal cavity (Fig. 2). In this area, most of the multiple turbinates arising from the lateral wall of the nasal cavity contain OSNs. The most medial turbinate extends anterior into the nasal cavity and has some respiratory epithelium on the anterior region. A more thorough description of the nasal cavity in rodents can be found in chapter 1.

The OE is a pseudostratified neuroepithelium of approximately 450 mm^2 in adult rats (3) and 85 mm^2 in adult mice (4). For comparison, the OE covers areas as high as 1000 mm^2 in humans, 1000 mm^2 in rabbits, and 1500 mm^2 in some dogs (4). However, the size of the OE is not always a valid predictor of olfactory acuity, since the density of OSNs may vary from animal to animal and even between different locations within the same animal (5).

Several cell types are found in the OE. For example, there are 6 to 10 million OSNs in rodents (6). These OSNs detect odorants. Basal (progenitor cells) divide to provide a renewal of OSNs every six to eight weeks (5). Nonneuronal, glia-like, sustentacular (supporting) cells are involved in mucus secretion (7), phagocytosis of dead and dying cells (8), elimination of noxious substances, and the regulation of the extracellular ionic environment (7). They may also affect the turnover of basal cells (9). In addition, the OE contains Bowman's glands, which contribute to the layer of mucus covering the OE. A thin basement membrane at the base of the epithelium separates it from the lamina propria, which contains the axons from olfactory receptor neurons, glands, and blood vessels.

In the mucus covering the OE are soluble binding proteins, called odorant-binding proteins (OBP). The exact function of OBPs is unknown, but they may bind with odorants in the nasal cavity, transporting them to the OE or scavenging and removing excess odorants (10). Other functions suggested for OBPs are protection against parasitosis and infectious diseases and protection against oxidative stress by removing toxic compounds (11,12). Another factor that may affect odorant molecules before they reach the OSNs is their degradation or breakdown by enzymes within the mucus. For example, the olfactory mucosa is particularly rich in oxidative enzymes, such as cytochrome P-450 (13).

OSNs are generally considered to be bipolar neurons with a single dendritic process that swells to form an olfactory knob. From the knob, numerous cilia extend into the mucous layer covering the epithelium. The cilia are the sites for olfactory transduction, where odorant molecules are detected and transduced into electrical signals. A small, unmeylinated axon arises from the basal end of the olfactory receptor neuron. Axons combine to form a large number of bundles that make up the olfactory nerve (cranial nerve I). The bundles project through the bony cribriform plate to terminate in glomeruli in the olfactory bulb. In the glomeruli, a number of neuronal elements meet to process information from the OSNs and relay that information to the olfactory cortex. Several excellent reviews that discuss olfaction and other chemical sensing subsystems in the nose have recently been published (14–16).

Molecular Characterization of the OE

The OE has traditionally been thought of as housing a uniform population of OSNs expressing olfactory marker protein (OMP), a characteristic protein whose function is unknown (17). However, beginning with the discovery of olfactory receptor proteins (ORPs) by Buck and Axel (18), this idea has been changing.

Neurons in the OE Expressing the Cyclic AMP Transduction Cascade

The first signal transduction cascade ascribed to OSNs expressing ORPs involved adenylate cyclase III and cyclic AMP (cAMP) (19). The cAMP, in turn, opens a cyclic nucleotide-gated channel, leading to a depolarization of the neuron due to an influx of Na^+ and Ca^{2+} ions. The Ca^{2+} ions activate a chloride channel causing Cl^- ions to leave the cell amplifying the response to the odorant.

The ORPs that activate the G protein, G_{olf}, to begin the cAMP transduction cascade are G-protein–coupled receptors and represent the largest known gene family in mammals. About 1200 functional genes have been identified in mice. Each olfactory receptor neuron expresses only one ORP gene and hence expresses a single ORP. However, an individual olfactory receptor neuron may recognize several odorant molecules and a given ORP may be broadly tuned. Therefore, odorants are discriminated by an ensemble of responding fibers, implying the existence of a peripheral combinatorial code with central decoding (20).

All of the OSNs expressing the same ORP send their axons to two glomeruli in the main olfactory bulb (MOB). The OSNs in the medial aspect of the OE send their axons to a glomerulus in the medial portion of the olfactory bulb, while those in the lateral aspect send their axons to a glomerulus in the lateral portion of the olfactory bulb.

Early experiments suggested a topographical pattern of gene expression for the OE with four discrete expression zones organized along the dorsal–ventral and medial–lateral axes of the nasal cavity (21). More recent studies suggest that there are overlapping zones of cells expressing individual ORPs (22). Interestingly, this spatial organization of the OE appears to be correlated with the pattern of deposition of odorants during inspiratory airflow (23,24). That is, the nasal cavity could function as a gas chromatograph with specific adsorption patterns for specific odorant molecules based on their physicochemical properties (25).

In a given zone of the OE, ORPs appear to be randomly distributed (26). However, one subfamily of ORPs, "OR37s," can be found clustered in a patch of the OE on the central region of the turbinates (14). OSNs expressing "OR37s" send their axons to a single glomerulus in the ventral region of the olfactory bulb—an area that is activated in the presence of urine (27).

Neurons in the OE Expressing Guanylyl Cyclase
Not all the OSNs in the OE use the cAMP cascade to transduce chemical stimuli into electrical signals. A small subpopulation of OSNs employs the cGMP transduction cascade involving phosphodiesterase and a cGMP-sensitive cyclic nucleotide-gated channel (28). One of these neurons expressing a specific guanylyl cyclase, called the GC-D+ neuron, is found concentrated in particular regions of the OE (29) and sends its axons to a specific area of the olfactory bulb that contains the "necklace glomerular complex" (30). Stimulation of the necklace glomeruli has been associated with pheromone-like compounds from urine as well as CO_2 (30,31). Whether these stimuli work through ORPs or directly through guanylyl cyclase D is unknown.

Neurons in the OE Expressing Trace Amine–Associated Receptors
All OSNs do not express ORPs. A subpopulation of OSNs appears to use trace amine–associated receptors (TAARs) to detect chemical stimuli, such as biogenic amines (32). The distribution of these neurons throughout the OE is similar to the distribution of OSNs expressing ORPs (32). Like neurons expressing ORPs, OSNs expressing TAARs apparently use the cAMP signal transduction cascade (32). It is not yet known where these neurons project to in the olfactory bulb.

Neurons in the OE Expressing Transient Receptor Potential Channels
A subpopulation of cells in the OE expresses several transient receptor potential (TRP) channels. These include TRPC2, TRPC6, and TRPM5 channels. A more detailed account of TRP channels will be given later in this chapter. TRPC2 channels are reported in some cells in the basal layers of the OE. These cells are not thought to be mature, functioning OSNs (33). TRPC6 channels, which are activated by diacylglycerol, are found in a population of nonciliated, bipolar microvillar cells in the apical OE (34). These cells appear to use the IP_3/DAG signal transduction cascade and may account for IP_3-mediated odor transduction in the OE (35).

TRPM5 channels are expressed in a group of OSNs found in the ventrolateral OE (36). These neurons send their axons to the ventral, medial, and lateral olfactory bulb, areas that are known to process information about socially relevant odorants (36). TRPM5 channels are involved in the transduction cascade of some taste receptor cells and solitary chemoreceptor cells (SCCs) (see below) and are probably responsible for the depolarizing currents that arise upon chemical stimulation of these cells. TRPM5 channels are heat activated as well as Ca^{2+} and voltage-gated (37).

SEPTAL ORGAN

Overview
The septal organ (SO), described by Rodoflo-Masera in 1943, is an isolated patch of olfactory-like epithelium lying amidst respiratory epithelium. SOs are found on both sides of the nasal septum at the entrance to the nasopaharynx (38) (Fig. 2). SOs have only been found in rodents, rabbits, and some marsupials (39,40). Because of its location in the nasal cavity, the SO has been proposed to monitor odorants that do not reach the main OE (38), although this hypothesis has not been demonstrated experimentally. Recently, SO sensory receptor neurons (like 50% of main OSNs) were shown to respond to mechanical as well as chemical stimulation. These results suggest that SO sensory receptor neurons play a role in synchronizing the rhythmic activity in the olfactory bulb with respiration (41).

Sensory receptor neurons in the SO express OMP and send their axons through a specific portion of the cribriform plate to terminate in the posterior, ventromedial region of the MOB (42,43). The axons of some sensory neurons in the SO terminate in specific "septal" glomeruli, while others appear to innervate glomeruli which also receive input from the main OE (43).

Molecular Characterization of the SO Epithelium
SO sensory receptor neurons in mice express about 4 % to 8% of the ORPs found in the main OE (44,45). Interestingly, more than 90% of the sensory neurons in the SO express only nine ORPs. Most of the SO sensory receptor neurons express OMP as well as the same cAMP transduction cascade elements (G_{olf}, adenylate cyclase III) as receptor neurons in the OE (46). A few SO sensory receptor neurons express guanynlate cyclase and phosphodiesterase (46).

GRUENEBERG GANGLION

Overview
Although first described in 1973 by Hans Grüneberg, most of the work on the Grueneberg ganglion (GG) has been done since 2005 (47). The GG is a small cluster of 300 to 500 cells (48–52) located bilaterally, dorsolaterally at the rostral end of the nose (47,49–51,53) (Fig. 2). The GG has been identified in the mouse, rat, Syrian hamster, oriental tree shrew, pangolin, and the domestic cat, but has not been identified in the red squirrel, the guinea pig, the common shrew, the tarsier, or the raccoon (47). Evidence suggests that the ganglion appears in the mole and humans (47).

The GG comprises sensory receptor neurons located in a small pocket between the lumen of the nasal cavity and blood vessels and does not appear to have direct contact with either (47,49–51,53). In electron micrographs, the sensory receptor neurons in the GG are seen to possess cilia which project into the extracellular matrix, but which do not penetrate either the keratinized epithelium or the nasal cavity (48). Sensory receptor neurons in the GG express OMP

(49–54) and send their axons to a subset of the necklace glomeruli on the caudal side of the olfactory bulb (49,50,52,53), which is also innervated by olfactory receptor neurons expressing GC-D. Stimulation of the necklace glomeruli is associated with pheromone-like compounds from urine as well as CO_2 (30,31).

Little is known about the function of the GG, in part because it is not clear how stimuli reach the receptor neurons. However, recently, the keratinized epithelium in mice was shown to be soluble to hydrophilic compounds and the underlying sensory receptor neurons responsive to water-soluble mouse alarm pheromones (48). In vivo mouse experiments demonstrated that a functional GG was required for a normal "freezing response" to alarm pheromones (48).

Molecular Characterization of the GG

Sensory receptor neurons in the GG express the G-protein G_i (54). G_o appears to be coexpressed with G_i (54). These neurons also coexpress potassium-independent sodium/calcium ion exchangers (NCXI, NCXII, NCXIII) (55). Prenatally, a very small number of GG sensory receptor neurons express the odorant receptor mOR256-17, but they are apparently not present in adults.

Neurons in the GG Expressing V2r83

The majority of OMP-expressing cells in the GG appear to express the vomeronasal receptor V2r83, part of the V2R-C family (54 and see below) although they lack TRPC2. This subset of neurons may be responsible for temperature-mediated responses of GG sensory neurons (56,57). Temperature-mediated activation elicited an increase in the expression of cFos only in this subtype. cFos is a transcription factor used as an indirect measure of neural activity. (56,57). These GG sensory receptor neurons express a guanylyl cyclase subtype (GC-G), which is coexpressed with PDE2A, another member of the cGMP signaling cascade (57). Ligands for this particular subset of neurons could bind either V2r83 or GC-G (57). This subset of GG sensory receptor neurons may also innervate the necklace glomeruli (57).

Neurons in the GG Expressing TAARs

Approximately 15% of the OMP-positive neurons in newborns express some type of TAAR (58). After day P7, this percentage greatly decreases. With highest expression perinatally, the number of TAAR-expressing neurons declines by ~55% neonatally and ~88% postnatally (58). This subtype of neurons is clearly distinct from those expressing V2r83 (58). In the GG, TAARs 2, 4, 5, 6, and 7 are expressed; however, most TAAR-expressing neurons express either TAAR 6 or TAAR 7s (58).

VOMERONASAL ORGAN

Overview

The VNO, also referred to as the Jacobson's organ, is traditionally thought to act as an alternative olfactory system for socially relevant odorants (59) in contrast to the olfactory organ which detects odorants from the environment. However, recent work has clearly demonstrated that there is substantial overlap in the types of stimuli detected by both systems (60). A functional VNO is found in most tetrapods with the notable exception of humans, chimpanzees, and old world primates (59). While there is still some debate concerning the existence of a functional VNO in these animals, the general consensus is that humans lack working genes for VNO receptor proteins (61) and do not possess an accessory olfactory bulb (AOB) to which VNO neurons project in other animals (see Ref. 59 for an in-depth discussion).

In most rodents, the VNO consists of two symmetrical hollow tubes lying on the floor of the nasal cavity on the either side of the nasal septum (Fig. 2). The tubes open into the nasal cavity on the anterior end (62). The long axis of the VNO runs rostral to caudal and if coronal sections are examined, the lumen appears crescent shaped with the convex surface facing the nasal septum (63). The tissue lateral to the concave surface is highly vascularized and vasodilatation and constriction in this tissue result in "pumping" of the VNO lumen which is thought to aid in stimulus delivery to the neurons which innervate the sensory epithelium of the VNO (64). The convex and concave luminal surfaces of the VNO are covered in sensory and nonsensory ciliated epithelium (65).

The VNO is innervated by diverse populations of fibers some of which are not directly involved in the detection of the socially relevant odors. Fibers that contain calcitonin gene–related peptide (CGRP) and substance P, presumably nociceptive trigeminal fibers (see below), are present in both the erectile tissue and the sensory and nonsensory epithelium of the VNO (66). The pumping of the VNO is controlled by sympathetic fibers in the nasopalatine branch of the trigeminal nerve and parasympathetic fibers from the sphenopalatine ganglion (67–69). Nerve fibers innervating this erectile tissue appear to contain nitric oxide synthase and vasoactive intestinal peptide, which are found in other erectile tissues important in mammalian reproductive behavior (66).

The sensory epithelium of the VNO contains two layers of bipolar receptor neurons which extend their dendrites through a layer of supporting cells. The dendrites of these bipolar neurons do not possess cilia as do OSNs (70) but rather end in microvilli that are exposed to the environment of the VNO lumen (71). Signals from the vomeronasal sensory neurons (VSNs) are conducted by the vomeronasal nerve to the AOB. To reach its destination in the AOB, the vomeronasal nerve travels along the nasal septum and through the cribriform plate (15). Similar to the arrangement of OSNs once they reach the MOB, the axons originating in the VNO end in glomeruli in the AOB. However, the glomeruli in the AOB do not appear to be as highly organized as the glomeruli in the MOB (72,73), and the vomeronasal system may not use the combinatorial coding system found in the olfactory system (15).

Molecular Characterization of the VNO
VSNs express the TRP channel TRPC2 (33). TRPC2 is thought to be activated by DAG, which is produced when vomeronasal receptors activate the phospholipase C (PLC) signaling pathway (15). Although it has been demonstrated that TRPC2 is critical for complete VNO function (59), there are basal neurons in the VNO that still respond to chemicals when TRPC2 has been knocked out (74).

Neurons in the VNO Expressing V1R
VSNs in the apical layer of the VNO express the G-protein–coupled receptor, V1R. In the rat and mouse, there are about 115 and 190 intact genes, respectively (75). Both mice and rats also possess a number of V1R pseudogenes. A single V1R appears to be expressed in each apical VSN along with the G-protein subunit $G_{\alpha i2}$ (26). V1R-expressing VSNs respond primarily to small molecules found in conspecific urine (76) and project their axons to the anterior AOB (77).

Neurons in the VNO Expressing V2R

Members of the V2R family are expressed in each basal VSN with the G-protein subunit $G_{o\alpha}$ and innervate the posterior AOB (26). A subset of these neurons also express *H2-Mv*, nonclassical class I genes of the major histocompatibility complex (78). V2R sensory neurons not expressing *H2-Mv* reside in the upper sublayers of the basal layer of the vomeronasal epithelium and project their axons to the anterior subdomains of the posterior AOB. V2R sensory neurons that do express *H2-Mv* are found in the lower sublayers of the basal layer of the vomeronasal epithelium and project their axons to the posterior subdomains of the posterior AOB. It appears that, unlike OSNs and VSNs expressing V1R, more than one V2R may be expressed in basal VSNs (79). VSNs expressing V2Rs may respond to nonvolatile peptide and protein pheromones including major histocompatibility complex class 1 proteins (80), male-specific compounds found in lacrimal gland secretions called exocrine gland-secreting peptides (ESPs) (81), and nonvolatile major urinary proteins (82).

Neurons in the VNO Expressing ORPs

As many as 44 ORPs are expressed in some of the VSNs in the apical layer of the VNO, which project to the anterior AOB (83). However, these VSNs lack the signaling machinery found in the main olfactory system, G_{olf} and adenylate cyclase III, and express $G_{o\alpha i}$ and TRPC2, both of which are important signaling molecules in the vomeronasal system (83). These VSNs could mediate the response of the VNO to odors that are not considered to be socially relevant (83,84).

NERVUS TERMINALIS

Overview

First identified in dogfish in 1878, the nervus terminalis (NT), or 0th cranial nerve, is a diffusely organized, ganglionated nerve which is difficult to distinguish in most animals anatomically (85). For recent reviews of the NT see (86,87). While much of the research involving this nerve has been conducted in fishes, the nerve is found in species of almost every vertebrate class (88). Arising from both the septal area of the forebrain and the ventromedial surface of the telencephalon, this unmyelinated nerve runs from the nasal septum, where it is associated with the olfactory and vomeronasal nerves (38), to innervate the septal and preoptic areas of the brain (89). The nerve's intracranial course involves the formation of a loose plexus medial to the olfactory bulb (38). Along its path, the NT is associated with numerous ganglion cell clusters and is in close proximity to either intranasal or cerebral blood vessels. The largest ganglion cluster, the ganglion terminale, is found caudal to the olfactory bulb (90). While the function of the NT is largely unknown, it has been proposed that it plays a role in mediating reproductive behavior due to possible neuromodulatory mechanisms of gonadotropin-releasing hormone (GnRH) (85).

The NT is unique among cranial nerves in that it is shrouded in controversy. Both the definition of what components actually comprise the NT and the origination of the nerve cells are heavily debated. There appears to be two basic schools of thought regarding the definition of the NT (91). One definition bases classification primarily on the location and presence of central projections, which would include a wider variety of organisms, such as lampreys, as having a NT.

The other definition includes GnRH-containing cells as an indispensable criterion, excluding species such as the lamprey from being classified as having a NT.

For several decades, the GnRH cells of the NT have been thought to originate in the olfactory placode; however, recent evidence suggests that these cells may originate from the cranial neural crest (85,92). Immunocytochemistry, in situ hybridization, and a variety of other techniques have pointed to the GnRH cells associated with the NT as being of placodal origin (93–95); however, lineage tracing and genetic knockout of multiple neural crest genes simultaneously support the neural crest origin hypothesis (85,92,96,97). Results from experiments using genetically manipulated mice have given little insight into the true origin of the NT as several processes, such as placodal formation and migration from the neural crest, were disrupted (85,98,99).

Molecular Characterization of the NT
There appears to be at least two subsets of cells in the NT—GnRH- or LHRH-expressing cells (100,101) and cholinergic cells (102,103)—although other subsets of cells could exist (91). While one definition of the NT requires cells expressing GnRH, these cells can comprise as little as 10% of NT cells in some species (87,104). With a centrally placed nucleolus, these cells are rounded or fusiform in nature (100). Unlike hamsters and guinea pigs (100), rats do not contain ganglia of GnRH-expressing cells along the NT (105); however, this nerve is the principal source of GnRH in the fetal rat from days 15 to 19 of gestation (90). The accessory and MOBs may receive GnRH-containing fibers from the NT. In rodents, there is great variation in both the relative amount and distribution of GnRH (LHRH) innervation (105). Retrograde transport studies of GnRH-expressing neurons in the NT suggest that these cells are linked to the amygdala (106).

In the NT, GnRH-expressing cells are always accompanied by non-GnRH-expressing cells (102,106). These cells appear to be larger in size and may have a double nucleolus (100). Their developmental origin remains unclear (85). A subset of NT cells are known to be cholinergic due to their expression of choline acetyltransferase (102) and acetylcholinesterase (38). While several other molecules such as Spot 35-calbindin (107), vasoactive intestinal polypeptide (102), and substance P (102) have been identified as expressed by some NT cells, the expression pattern of these molecules is unknown. Additionally, several other molecules such as OMP and beta-tubulin have been reported in presumptive NT cells migrating from the OE during development (91).

TRIGEMINAL NERVE

Overview
The trigeminal, or fifth cranial, nerve (Cr. N. V) provides somatosensation for the head and face. This includes touch, temperature perception, chemesthesis, pain, and proprioception. The cell bodies for most of the trigeminal nerve fibers lie in the trigeminal (Gasserian) ganglia. These axons terminate centrally in two main nuclei: the principal (main sensory) trigeminal nucleus and the spinal (descending) trigeminal nucleus. Tactile information is transmitted largely through the principal sensory nucleus, while thermal and pain (including chemesthetic) sensations are transmitted mostly through the spinal subnucleus of the spinal trigeminal complex.

Peripherally, axons leave the trigeminal ganglia through three divisions (Fig. 2): the mandibular, maxillary, and ophthalmic. The nasal cavity is innervated by the nasopalatine nerve, a branch of the maxillary division, and the ethmoid nerve, a branch of the ophthalmic division. The nasopalatine nerve contains sensory fibers as well as parasympathetic and sympathetic postganglionic fibers. The ethmoid nerve contains sensory and some sympathetic postganglionic fibers that are distributed to the anterior nasal mucosa and the external nasal surface. Ethmoid cell bodies are clustered around the dorsal edge of the anterior trigeminal ganglion.

The nasal trigeminal nerves include small-diameter (Aδ) and medium-to large-diameter (Aα,β) myelinated afferent fibers, as well as small-diameter unmyelinated afferent fibers (i.e., C fibers). The Aα,β fibers respond to proprioceptive stimuli and touch, while the Aδ respond to mechanical, thermal, and chemesthetic stimuli. The unmyelinated fibers are activated by temperature, itch, and mechanical, thermal, and chemesthetic stimuli (108). Nociceptive fibers appear to express substance P and CGRP (109,110).

The ethmoid nerve of muskrats contains approximately 800 unmyelinated and 450 myelinated axons (111), while that of cats contains at least 1400 myelinated axons and a larger number of unmyelinated axons (112). Rats have approximately 1800 unmyelinated and 600 myelinated axons (Silver and Finger, unpublished data, 1993). The majority of unmyelinated axons are grouped in Remak bundles, a collection of axons ensheathed by a Schwann cell.

Some nociceptive ethmoid nerve cell bodies in the trigeminal ganglia send axon collaterals to the nasal epithelium, olfactory bulb, and caudal spinal trigeminal nucleus (113). The connection to the olfactory bulb is interesting. It is not clear in which direction information is traveling, to or from the bulbs, although some evidence suggests that trigeminal stimulation may inhibit olfactory bulb responses.

When the ethmoid and nasopalatine nerves reach the nasal epithelium, they branch repeatedly. Traditionally, the nerves were thought to terminate as free nerve endings a few microns below a line of tight junctions (114). This barrier should not pose a problem for lipid-soluble irritants that could diffuse across the junctional membranes to reach the nerve endings. How water-soluble trigeminal stimuli might reach the sensory nerve fibers is less clear, but it has been suggested a paracellular pathway (115) or perhaps SCCs (see below) are involved.

Molecular Characterization of the Nasal Trigeminal Nerve

Nasal trigeminal nerve fibers respond to mechanical, thermal, and nociceptive (mechanical, thermal, and chemesthetic) stimuli. A number of receptor proteins are known to be activated by chemicals found in the "inflammatory soup." These include, TRKa (nerve growth factor), B2R (bradykinin), PGE(2) (prostaglandins), 5-HT (serotonin) P2×3 (ATP), and ASIC3 (acid-sensing ion channels) (low pH) (108). In addition, nAChRs (nicotine) are located on trigeminal nerves (116). However, with the discovery of TRP channels, there has been an explosion of information about how stimuli activate trigeminal nerves (117).

In mammals, there are six TRP channels subfamilies containing 28 members (118). Each of the TRP channels contains six membrane-spanning domains as well as a cytoplasmic C- and N-terminal domain. The pore loop forming the ion channel lies between the fifth and sixth transmembrane segments. All but the TRPM channels contain a number of ankyrin repeats in their intracellular N-terminals. Ankyrin repeats are repeated motifs, which are found in a number of proteins and are thought to bind with other proteins or the cytoskeleton (119). Ankyrin repeats in TRP channels may play a role in connecting subunits together, since most functional TRP channels contain four TRP subunits. Most TRP channels are nonselective cation channels, although some may be more selective for calcium than for other cations (120). Stimulation of TRP channels causes an increase in intracellular Ca^{2+} through voltage-dependent gating of the channel, resulting in a depolarization and the generation of an action potential.

Mechanoreceptors Expressed in the Trigeminal Nerve

Low-threshold mechanoreceptors on trigeminal nerves that respond to mechanical stimuli (punctuate, stretch, or osmotic) have not yet been definitively identified, although mechanical stimulation is thought to directly activate mechanosensitive ion channels expressed on the receptive endings of these neurons. Several TRP channels, however, have been implicated as mechanoreceptors. TRPV1, TRPV2, TRPV4, TRPC1, and TrpA1 are all candidate mechanoreceptive channels, although there is no conclusive evidence for any of them (121).

In addition to TRP channels, other putative mechanosensitive channels are members of the DEG/ENaC superfamily [ASICs (122) and stomatin-like protein 3 (123)], two-pore-domain K^+ channels (TREK) (124), and purinergic receptors (P2X) (125).

Thermoreceptors Expressed in the Trigeminal Nerve

Six TRP channels (thermoTRPs: TRPV1–4, TRPM8, TRPA1) found on sensory nerve fibers are thought to play a role in thermosensation (118). Each channel responds to a different range of temperatures. TRPV1 which begins responding at ∼43°C and TRPV2 at ∼50°C detect nociceptive heat pain. Interestingly, compounds released during inflammation and found in the inflammatory soup sensitize TRPV1 channels to heat, so that even normal body temperature could activate nociceptive nerve fibers (126). TRPV3 (34–39°C) and TRPV4 (25–34°C) are considered to be warm detectors (127,128). While they are expressed at low levels in sensory nerve endings, they are heavily expressed in keratinocytes. Once activated by warmth, the keratinocytes are thought to release chemicals, which would, in turn, activate sensory nerve fibers (129).

Approximately 7% to 20% of trigeminal neurons respond to cooling temperatures (118). This is due to the presence of TRPM8 and TRPA1 channels. TRPM8 is activated by temperatures below ∼28°C (130). TRPM8 also apparently mediates the analgesic effect of moderate cooling (131). The fact that TRPM8 knockout mice are still able to detect cold stimuli, especially <15°C, suggested the presence of a second cold receptor (132). The TRPA1 channel was thought to be the cold receptor that responded to noxious cold. However, the role of TRPA1 channels as cold sensors is currently highly debated (129) and more study must be done to resolve this issue.

Chemesthetic Receptors Expressed in the Trigeminal Nerve

Chemesthesis is the detection of irritating chemicals. A number of receptors that detect compounds found in the inflammatory soup (see above) also sense environmental chemical irritants (e.g., nicotine by nAChRs; acids by ASICs) (115).

However, many of the same TRP channels that are activated by temperature are also activated by chemical irritants.

The discovery of a TRP channel that was activated by capsaicin (TRPV1) began the explosion of research into TRP channels and their responses to natural compounds (133). Besides capsaicin, TRPV1 channels are activated by numerous chemical stimuli, including the gingerol, zingerone, capsiate, eugenol, piperine, resiniferatoxin, allicin, camphor, and cyclohexanone (120,134,135). TRPV1 channels are also affected by low pH. TRPV1 channels are found primarily on substance P–containing trigeminal nerve C fibers (136). TRPV3 channels respond to camphor, carvacrol, thymol, and eugenol (137,138).

TRPM8 channels which respond to cold are also activated by menthol, eugenol, icilin, and several monterpenes (eucalyptol, geraniol, linalool, menthyl lactate, *trans* and *cis-p*-menthane-3,8-diol, l-carvone, isopulegol, and hydroxycitronellal) (134). TRPM8 channels are expressed in small-diameter primary sensory neurons found in the trigeminal ganglia (130).

TRPA1 channels respond to the largest variety of chemicals of any of the TRP channels; over 90 have been reported. TRPA1 channels appear to be expressed in a subset of neurons which also express TRPV1 (139). Among the diverse chemicals that activate TRPA1 channels are the psychoactive component in marijuana, environmental irritants, and pungent compounds such as eugenol, gingerol, methyl salicylate, allyl isothiocyanate, and cinnamaldehyde (140–142).

SOLITARY CHEMORECEPTOR CELLS

Overview

SCCs were first discovered in the skin of fish (143). These cells, which are sensory epithelial cells, respond to fish body mucus and bile and may be used to detect potential predators or competitors (144). Since their discovery, SCCs have been found in rodent oral and nasal cavities, larynx, trachea (145), and intestines (146).

In the nasal cavity, SCCs are concentrated posterior to the vestibule and the anterior ducts of the VNO and are innervated by afferent trigeminal nerve fibers containing substance P and CGRP (147,148). The SCCs have synaptic vesicles and synaptic specializations similar to the ones found in taste cells. The SCCs in the nasal cavity respond to a variety of chemical stimuli, including bitter compounds (148,149). Since the SCC microvilli project directly into the lumen of the nasal cavity, these cells might provide a mechanism by which water-soluble stimuli could stimulate trigeminal nerve ending without having to pass through the line of tight junctions (see above).

Molecular Characterization of Nasal SCCs

In the nose, SCCs have the characteristics of individual taste receptor cells which detect bitter compounds. They may contain the "bitter" taste receptor proteins T2R8 (a receptor for denatonium), T2R19, and T2R5 (a receptor for cycloheximide) (148,150). In addition, SCCs may express TRPM5 channels, the G-protein, gustducin, and PLCβ2, all found in bitter taste receptor cells. Not every SCC expresses both gustducin and TRPM5. Of approximately 5700 TRPM5-expressing SCCs in each side of the mouse anterior nasal cavity, only 15% to 20% also expressed gustducin. Some SCCs express the sweet/umami receptor protein, T1R3 (150). These receptor proteins are coexpressed with T2R8 and T2R5 mRNAs in the same SCC.

Although the SCCs contain receptor proteins for sweet/umami and bitter stimuli, the sensation elicited by stimulating these receptors would presumably be irritation, since the cells activate trigeminal nerve fibers. SCCs are thought to act as sentinels of the respiratory system by activating trigeminal protective reflexes upon inhalation of irritating substances (147). Additionally, SCCs have been suggested to detect bacterial metabolites, such as lactones and diterpenoids, which could lead to trigeminal nerve activation and trigger protective reflexes to rid of the bacteria from the respiratory system (148). Stimulation of SCCs might also increase mucociliary activity to dilute or expel bacteria (151). Nasal SCCs, therefore, could possibly serve to prevent bacterial colonization of the respiratory system.

SUMMARY/CONCLUSIONS

The nasal cavity of mammals is invested with a complex array of sensory nerves. Some neural structures (Cr. N's I and V) are shared with humans and provide a direct model in predicting human responses. Others (the VNO, NT, SO, and solitary chemoreceptor cells) are less clear analogs. What is certain, however, is that the perception of airborne environmental stimuli involves complex, interacting, and sometimes overlapping neural processes.

REFERENCES

1. Stuelpnagel JT, Reiss JO. Olfactory metamorphosis in the Coastal Giant Salamander (*Dicamptodon tenebrosus*). J Morphol 2005; 266(1):22–45.
2. Smith TD, Bhatnagar KP. Microsmatic primates: reconsidering how and when size matters. Anat Rec B New Anat 2004; 279(1):24–31.
3. Weiler E, Farbman AI. Proliferation in the rat olfactory epithelium: age-dependent changes. J Neurosci 1997; 17(10):3610–3622.
4. Adams DR. Olfactory and non-olfactory epithelium in the nasal cavity of the mouse, *Peromyscus*. Am J Anat 1972; 133(1):37–50.
5. Graziadei PPC. The olfactory mucosa in vertebrates. In: Beidler LM, ed. Chemical Senses Part 1, Vol IV. Berlin: Springer-Verlag, 1971:27–58.
6. Imai T, Sakano H. Roles of odorant receptors in projecting axons in the mouse olfactory system. Curr Opin Neurobiol 2007; 17(5):507–515.
7. Getchell ML, Getchell TV. Fine structural aspects of secretion and extrinsic innervation in the olfactory mucosa. Microsc Res Tech 1992; 23(2):111–127.
8. Suzuki Y, Takeda M, Farbman AI. Supporting cells as phagocytes in the olfactory epithelium after bulbectomy. J Comp Neurol 1996; 376(4):509–517.
9. Asson-Batres MA, Smith WB. Localization of retinaldehyde dehydrogenases and retinoid binding proteins to sustentacular cells, glia, Bowman's gland cells, and stroma: potential sites of retinoic acid synthesis in the postnatal rat olfactory organ. J Comp Neurol 2006; 496(2):149–171.
10. Pelosi P. The role of perireceptor events in vertebrate olfaction. Cell Mol Life Sci 2001; 58(4):503–509.
11. Ramoni R, Vincent F, Grolli S, et al. The insect attractant 1-octen-3-ol is the natural ligand of bovine odorant-binding protein. J Biol Chem 2001; 276(10):7150–7155.
12. Grolli S, Merli E, Conti V, et al. Odorant binding protein has the biochemical properties of a scavenger for 4-hydroxy-2-nonenal in mammalian nasal mucosa. FEBS J 2006; 273(22):5131–5142.
13. Ling G, Gu J, Genter MB, et al. Regulation of cytochrome P450 gene expression in the olfactory mucosa. Chem Biol Interact 2004; 147(3):247–258.

14. Breer H, Fleischer J, Strotmann J. The sense of smell: multiple olfactory subsystems. Cell Mol Life Sci 2006; 63(13):1465–1475.
15. Ma M. Encoding olfactory signals via multiple chemosensory systems. Crit Rev Biochem Mol Biol 2007; 42(6):463–480.
16. Munger SD, Leinders-Zufall T, Zufall F. Subsystem organization of the mammalian sense of smell. Annu Rev Physiol 2009; 71(3):3.1–3.26.
17. Monti-Graziadei GA, Margolis FL, Harding JW, et al. Immunocytochemistry of the olfactory marker protein. J Histochem Cytochem 1977; 25(12):1311–1316.
18. Buck L, Axel R. A novel multigene family may encode odorant receptors: a molecular basis for odor recognition. Cell 1991; 65(1):175–187.
19. Wong ST, Trinh K, Hacker B, et al. Disruption of the type III adenylyl cyclase gene leads to peripheral and behavioral anosmia in transgenic mice. Neuron 2000 27(3), 487–497
20. Malnic B, Hirono J, Sato T. Combinatorial receptor codes for odors. Cell 1999; 96(5):713–723.
21. Ressler KJ, Sullivan SL, Buck LB. A zonal organization of odorant receptor gene expression in the olfactory epithelium. Cell 1993; 73(3):597–609.
22. Iwema CL, Fang H, Kurtz DB. Odorant receptor expression patterns are restored in lesion-recovered rat olfactory epithelium. J Neurosci 2004; 24(2):356–369.
23. Scott JW. Sniffing and spatiotemporal coding in olfaction. Chem Senses 2006; 31(2):119–130.
24. Zhao K, Dalton P, Yang GC, et al. Numerical modeling of turbulent and laminar airflow and odorant transport during sniffing in the human and rat nose. Chem Senses 2006; 31(2):107–118.
25. Kent PF, Youngentob SL, Sheehe PR. Odorant-specific spatial patterns in mucosal activity predict perceptual differences among odorants. J Neurophysiol 1995; 74(4):1777–1781.
26. Mombaerts P. Genes and ligands for odorant, vomeronasal and taste receptors. Nat Rev Neurosci 2004; 5(4):263–278.
27. Xu F, Schaefer M, Kida I, et al. Simultaneous activation of mouse main and accessory olfactory bulbs by odors or pheromones. J Comp Neurol 2005; 489(4):491–500.
28. Gibson AD, Garbers DL. Guanylyl cyclases as a family of putative odorant receptors. Annu Rev Neurosci 2000; 23:417–439.
29. Fülle HJ, Garbers DL. Guanylyl cyclases: a family of receptor-linked enzymes. Cell Biochem Funct 1994; 12(3):157–165.
30. Juilfs DM, Fülle HJ, Zhao AZ, et al. A subset of olfactory neurons that selectively express cGMP-stimulated phosphodiesterase (PDE2) and guanylyl cyclase-D define a unique olfactory signal transduction pathway. Proc Natl Acad Sci U S A 1997; 94(7):3388–3395.
31. Hu J, Zhong C, Ding C, et al. Detection of near-atmospheric concentrations of CO_2 by an olfactory subsystem in the mouse. Science 2007; 317(5840):953–957.
32. Liberles SD, Buck LB. A second class of chemosensory receptors in the olfactory epithelium. Nature 2006; 442(7103):645–650.
33. Liman ER, Corey DP, Dulac C. TRP2: a candidate transduction channel for mammalian pheromone sensory signaling. Proc Natl Acad Sci U S A 1999; 96(10):5791–5796.
34. Elsaesser R, Montani G, Tirindelli R, et al. Phosphatidyl-inositide signalling proteins in a novel class of sensory cells in the mammalian olfactory epithelium. Eur J Neurosci 2005; 21(10):2692–2700.
35. Boekhoff I, Tareilus E, Strotmann J, et al. Rapid activation of alternative second messenger pathways in olfactory cilia from rats by different odorants. EMBO J 1990; 9(8):2453–2458.
36. Lin W, Margolskee R, Donnert G, et al. Olfactory neurons expressing transient receptor potential channel M5 (TRPM5) are involved in sensing semiochemicals. Proc Natl Acad Sci U S A 2007; 104(7):2471–2476.
37. Talavera K, Yasumatsu K, Voets T, et al. Heat activation of TRPM5 underlies thermal sensitivity of sweet taste. Nature 2005; 438(7070):1022–1025.

38. Bojsen-Moller F. Demonstration of terminalis, olfactory, trigeminal and perivascular nerves in the rat nasal septum. J Comp Neurol 1975; 159(2):245–256.

39. Kratzing JE. The olfactory apparatus of the bandicoot (*Isoodon macrourus*): fine structure and presence of a septal olfactory organ. J Anat 1978; 125(pt 3):601–613.

40. Adams DR, McFarland LZ. Septal olfactory organ in *Peromyscus*. Comp Biochem Physiol A Comp Physiol 1971; 40(4):971–974.

41. Grosmaitre X, Santarelli LC, Tan J, et al. Dual functions of mammalian olfactory sensory neurons as odor detectors and mechanical sensors. Nat Neurosci 2007; 10(3):348–354.

42. Giannetti N, Saucier D, Astic L. Organization of the septal organ projection to the main olfactory bulb in adult and newborn rats. J Comp Neurol 1992; 323(2):288–298.

43. Lèvai O, Strotmann J. Projection pattern of nerve fibers from the septal organ: DiI-tracing studies with transgenic OMP mice. Histochem Cell Biol 2003; 120(6):483–492.

44. Kaluza JF, Gussing F, Bohm S, et al. Olfactory receptors in the mouse septal organ. J Neurosci Res 2004; 76(4):442–452.

45. Tian H, Ma M. Molecular organization of the olfactory septal organ. J Neurosci 2004; 24(38):8383–8390.

46. Ma M, Grosmaitre X, Iwema CL, et al. Olfactory signal transduction in the mouse septal organ. J Neurosci 2003; 23(1):317–324.

47. Grüneberg H. A ganglion probably belonging to the N. terminalis system in the nasal mucosa of the mouse. Z Anat Entwicklungsgesch 1973; 140(1):39–52.

48. Brechbühl J, Klaey M, Broillet MC. Grueneberg ganglion cells mediate alarm pheromone detection in mice. Science 2008; 321(5892):1092–1095.

49. Fuss SH, Omura M, Mombaerts P. The Grueneberg ganglion of the mouse projects axons to glomeruli in the olfactory bulb. Eur J Neurosci 2005; 22(10):2649–2654.

50. Koos DS, Fraser SE. The Grueneberg ganglion projects to the olfactory bulb. Neuroreport 2005; 16(17):1929–1932.

51. Fleischer J, Schwarzenbacher K, Besser S, et al. Olfactory receptors and signaling elements in the Grueneberg ganglion. J Neurochem 2006; 98(2):543–554.

52. Storan MJ, Key B. Septal organ of Grueneberg is part of the olfactory system. J Comp Neurol 2006; 494:834–844.

53. Roppolo D, Ribaud V, Jungo VP, et al. Projection of the Grueneberg ganglion to the mouse olfactory bulb. Eur J Neurosci 2006; 23(11):2887–2894.

54. Fleischer J, Hass N, Schwarzenbacher K, et al. A novel population of neuronal cells expressing the olfactory marker protein (OMP) in the anterior/dorsal region of the nasal cavity. Histochem Cell Biol 2006; 125(4):337–349.

55. Pyrski M, Koo JH, Polumuri SK, et al. Sodium/calcium exchanger expression in the mouse and rat olfactory systems. J Comp Neurol 2007; 501(6):944–958.

56. Mamasuew K, Breer H, Fleischer J. Grueneberg ganglion neurons respond to cool ambient temperatures. Eur J Neurosci 2008; 28(9):1775–1785.

57. Fleischer J, Mamasuew K, Breer H. Expression of cGMP signaling elements in the Grueneberg ganglion. Histochem Cell Biol 2009; 131(1):75–88.

58. Fleischer J, Schwarzenbacher K, Breer H. Expression of trace amine–associated receptors in the Grueneberg ganglion. Chem Senses 2007; 32(6):623–631.

59. Halpern M, Martinez-Marcos A. Structure and function of the vomeronasal system: an update. Prog Neurobiol 2003; 70(3):245–318.

60. Swaney WT, Keverne EB. The evolution of pheromonal communication. Behav Brain Res 2009; 200(2):239–247.

61. Vannier B, Peyton M, Boulay G, et al. Mouse trp2, the homologue of the human trpc2 pseudogene encodes mTrp2, a store depletion-activated capacitative Ca^{2+} entry channel. Proc Natl Acad Sci U S A 1999; 96(5):2060–2064.

62. Weiler E, Farbman AI. The septal organ of the rat during postnatal development. Chem Senses 2003; 28(7):581–593.

63. Døving KB, Trotier D. Structure and function of the vomeronasal organ. J Exp Biol 1998; 201(pt 21):2913–2925.

64. Meredith M. Chronic recording of vomeronasal pump activation in awake behaving hamsters. Physiol Behav 1994; 56(2):345–354.

65. Halpern M, Shapiro LS, Jia C. Heterogeneity in the accessory olfactory system. Chem Senses 1998; 23(4):477–481.
66. Uddman R, Malm L, Cardell LO. Neurotransmitter candidates in the vomeronasal organ of the rat. Acta Otolaryngol 2007; 127(9):952–956.
67. Meredith M, O'Connell RJ. Efferent control of stimulus access to the hamster vomeronasal organ. J Physiol 1979; 286:301–316.
68. Eccles R. Autonomic innervation of the vomeronasal organ of the cat. Physiol Behav 1982; 28(6):1011–1015.
69. Matsuda H, Kusakabe T, Kawakami T, et al. Coexistence of nitric oxide synthase and neuropeptides in the mouse vomeronasal organ demonstrated by a combination of double immunofluorescence labeling and a multiple dye filter. Brain Res 1996; 712(1):35–39.
70. Adams DR, Wiekamp MD. The canine vomeronasal organ. J Anat 1984; 128 (pt 4):771–787.
71. Altner H, Muller W. Electrophysiological and electron microscopial investigation of the sensory epithelium in the vomeronasal organ in lizards (*Lacerta*). Z Vgl Physiol 1968; 60:151–155.
72. Belluscio L, Koentges G, Axel R, et al. A map of pheromone receptor activation in the mammalian brain. Cell 1999; 97(2):209–220.
73. Buck LB. The molecular architecture of odor and pheromone sensing in mammals. Cell 2000; 100(6):611–618.
74. Kelliher KR, Spehr M, Li XH, et al. Pheromonal recognition memory induced by TRPC2-independent vomeronasal sensing. Eur J Neurosci 2006; 23(12):3385–3390.
75. Zhang JJ, Huang GZ, Halpern M. Firing properties of accessory olfactory bulb mitral/tufted cells in response to urine delivered to the vomeronasal organ of gray short-tailed opossums. Chem Senses 2007; 32(4):355–360.
76. Leinders-Zufall T, Lane AP, Puche AC, et al. Ultrasensitive pheromone detection by mammalian vomeronasal neurons. Nature 2000; 405(6788):792–796.
77. Dulac C, Torello AT. Molecular detection of pheromone signals in mammals: from genes to behaviour. Nat Rev Neurosci 2003; 4(7):551–562.
78. Ishii T, Mombaerts P. Expression of nonclassical class I major histocompatibility genes defines a tripartite organization of the mouse vomeronasal system. J Neurosci 2008; 28(10):2332–2341.
79. Silvotti L, Moiani A, Gatti R, et al. Combinatorial co-expression of pheromone receptors, V2Rs. J Neurochem 2007; 103(5):1753–1763.
80. Leinders-Zufall T, Brennan P, Widmayer P, et al. MHC class I peptides as chemosensory signals in the vomeronasal organ. Science 2004; 306(5698):1033–1037.
81. Kimoto H, Haga S, Sato K, et al. Sex-specific peptides from exocrine glands stimulate mouse vomeronasal sensory neurons. Nature 2005; 437(7060):898–901.
82. Chamero P, Marton TF, Logan DW, et al. Identification of protein pheromones that promote aggressive behaviour. Nature 2007; 450(7171):899–902.
83. Levai O, Feistel T, Breer H, et al. Cells in the vomeronasal organ express odorant receptors but project to the accessory olfactory bulb. J Comp Neurol 2006; 498(4):476–490.
84. Sam M, Vora S, Malnic B, et al. Neuropharmacology: odorants may arouse instinctive behaviours. Nature 2001; 412(6843):142.
85. Whitlock KE. Development of the nervus terminalis: origin and migration. Microsc Res Tech 2004; 65(1/2):2–12.
86. Wirsig-Wiechmann CR. Function of gonadotropin-releasing hormone in olfaction. Keio J Med1 2001; 50(2):81–85.
87. Wirsig-Wiechmann CR, Wiechmann AF, Eisthen HL. What defines the nervus terminalis? Neurochemical, developmental, and anatomical criteria. Prog Brain Res 2002; 141:45–58.
88. Demski LS, Schwanzel-Fukuda M. The terminal nerve (nervus terminalis): structure, function, and evolution. Ann N Y Acad Sci 1987; 519:ix–xi
89. Larsell O. The nervus terminalis. Ann Otol Rhinol Laryngol 1950; 59(2):414–438.

90. Schwanzel-Fukuda M, Morrell JI, Pfaff DW. Ontogenesis of neurons producing luteinizing hormone-releasing hormone (LHRH) in the nervus terminalis of the rat. J Comp Neurol 1985; 238(3):348–364.

91. von Bartheld CS. The terminal nerve and its relation with extrabulbar "olfactory" projections: lessons from lampreys and lungfishes. Microsc Res Tech 2004; 65(1/2):13–24.

92. Whitlock KE, Wolf CD, Boyce ML. Gonadotropin-releasing hormone (GnRH) cells arise from cranial neural crest and adenohypophyseal regions of the neural plate in the zebrafish, *Danio rerio*. Dev Biol 2003; 257(1):140–152.

93. Schwanzel-Fukuda M, Pfaff DW. Origin of luteinizing hormone-releasing hormone neurons. Nature 1989; 338(6211):161–163.

94. Wray S, Grant P, Gainer H. Evidence that cells expressing luteinizing hormone-releasing hormone mRNA in the mouse are derived from progenitor cells in the olfactory placode. Proc Natl Acad Sci U S A 1989; 86(20):8132–8136.

95. Wray S, Nieburgs A, Elkabes S. Spatiotemporal cell expression of luteinizing hormone-releasing hormone in the prenatal mouse: evidence for an embryonic origin in the olfactory placode. Brain Res Dev Brain Res 1989; 46(2):309–318.

96. Kelsh RN, Dutton K, Medlin J, et al. Expression of zebrafish fkd6 in neural crest-derived glia. Mech Dev 2000; 93(1/2):161–164.

97. Dutton KA, Pauliny A, Lopes SS, et al. Zebrafish colourless encodes sox10 and specifies non-ectomesenchymal neural crest fates. Development 2001; 128:4113–4125.

98. Matsuo T, Osumi-Yamashita N, Noji S, et al. A mutation in the Pax-6 gene in rat smalleye is associated with impaired migration of midbrain crest cells. Nat Genet 1993; 3(4):299–304.

99. Quinn JC, West JD, Hill RE. Multiple functions for Pax6 in mouse eye and nasal development. Genes Dev 1996; 10(4):435–446.

100. Schwanzel-Fukuda M, Silverman AJ. The nervus terminalis of the guinea pig: a new luteinizing hormone-releasing hormone (LHRH) neuronal system. J Comp Neurol 1980; 191(2):213–225.

101. Garcia MS, Schwanzel-Fukuda M, Morrell JI, et al. Immunochemical studies of the location of luteinizing hormone-releasing hormone neurons in the nervus terminalis of the mouse. Ann N Y Acad Sci 1987; 519:465–468.

102. Schwanzel-Fukuda M., Morrell JI, Pfaff DW. Localization of choline acetyltransferase and vasoactive intestinal polypeptide-like immunoreactivity in the nervus terminalis of the fetal and neonatal rat. Peptides 1986; 7(5):899–906.

103. von Bartheld CS, Meyer DL. Paraventricular organ of the lungfish *Protopterus dolloi*: morphology and projections of CSF-contacting neurons. J Comp Neurol 1990; 297(3):410–434.

104. Wirsig CR, Leonard CM. Terminal nerve damage impairs mating behavior of the male hamster. Brain Res 1987; 417(2):293–303.

105. Witkin JW, Silverman AJ. Luteinizing hormone-releasing hormone (LHRH) in rat olfactory systems. J Comp Neurol 1983; 218(4):426–432.

106. Jennes L. The nervus terminalis in the mouse: light and electron microscopic immunocytochemical studies. Ann N Y Acad Sci 1987; 519:165–173.

107. Abe H, Watanabe M, Kondo H. Developmental changes in expression of calcium-binding protein (Spot 35-calbindin) in the nervus terminalis and the vomeronasal and olfactory receptor cells. Acta Otolaryngol 1992; 112(5):862–871.

108. Julius D, Basbaum AI. Molecular mechanisms of nociception. Nature 2001; 413(6852):203–210.

109. Lee Y, Kawai Y, Shiosaka S, et al. Coexistence of calcitonin gene-related peptide and substance P-like peptide in single cells of the trigeminal ganglion of the rat: immunohistochemical analysis. Brain Res 1985; 330(1):194–196.

110. Silver WL, Farley LG, Finger TE. The effects of neonatal capsaicin administration on trigeminal nerve chemoreceptors in the rat nasal cavity. Brain Res 1991; 561(2):212–216.

111. McCulloch PF, Faber KM, Panneton WM. Electrical stimulation of the anterior ethmoidal nerve produces the diving response. Brain Res 1999; 830(1):24–31.
112. Biedenbach MA, Beuerman RW, Brown AC. Graphic-digitizer analysis of axon spectra in ethmoidal and lingual branches of the trigeminal nerve. Cell Tissue Res 1975; 157(3):341–352.
113. Schaefer ML, Böttger B, Silver WL, et al. Trigeminal collaterals in the nasal epithelium and olfactory bulb: a potential route for direct modulation of olfactory information by trigeminal stimuli. J Comp Neurol 2002; 444(3):221–226 [Erratum in: J Comp Neurol 2002; 448(4):423].
114. Finger TE, St Jeor VL, Kinnamon JC, et al. Ultrastructure of substance P- and CGRP-immunoreactive nerve fibers in the nasal epithelium of rodents. J Comp Neurol 1990; 294(2):293–305.
115. Bryant BP, Silver WL. Chemesthesis: the common chemical sense. In: Finger TE, Silver WL, Restrepo D, eds. The Neurobiology of Taste and Smell. New York, NY: Wiley, 2000:73–100.
116. Alimohammadi H, Silver WL. Evidence for nicotinic acetylcholine receptors on nasal trigeminal nerve endings of the rat. Chem Senses 2000; 25(1):61–66.
117. Damann N, Voets T, Nilius B. TRPs in our senses. Curr Biol 2008; 18(18):R880–R889.
118. Talavera K, Nilius B, Voets T. Neuronal TRP channels: thermometers, pathfinders and life-savers. Trends Neurosci 2008; 31(6):287–295.
119. Hoenderop JG, Nilius B, Bindels RJ. Epithelial calcium channels: from identification to function and regulation. Pflugers Arch 2003; 446(3):304–308.
120. Venkatachalam K, Montell C. TRP channels. Annu Rev Biochem 2007; 76:387–417.
121. Raoux M, Rodat-Despoix L, Azorin N, et al. Mechanosensor channels in mammalian somatosensory neurons. Sensors 2007; 7(9):1667–1682.
122. García-Añoveros J, Samad TA, Zuvela-Jelaska L, et al. Transport and localization of the DEG/ENaC ion channel BNaC1alpha to peripheral mechanosensory terminals of dorsal root ganglia neurons. J Neurosci 2001; 21(8):2678–2686.
123. Wetzel C, Hu J, Riethmacher D, et al. A stomatin-domain protein essential for touch sensation in the mouse. Nature 2007; 445(7124):206–209.
124. Bang H, Kim Y, Kim D. TREK-2, a new member of the mechanosensitive tandem-pore K^+ channel family. J Biol Chem 2000; 275(23):17412–17419.
125. Cockayne DA, Hamilton SG, Zhu QM, et al. Urinary bladder hyporeflexia and reduced pain-related behaviour in P2x3-deficient mice. Nature 2000; 407(6807):1011–1015.
126. Tominaga M, Caterina MJ, Malmberg AB, et al. The cloned capsaicin receptor integrates multiple pain-producing stimuli. Neuron 1998; 21(3):531–543.
127. Smith GD, Gunthorpe MJ, Kelsell RE, et al. TRPV3 is a temperature-sensitive vanilloid receptor-like protein. Nature 2002; 418(6894):186–190.
128. Güler AD, Lee H, Iida T, et al. Heat-evoked activation of the ion channel, TRPV4. J Neurosci 2002; 22(15):6408–6414.
129. Caterina MJ. Transient receptor potential ion channels as participants in thermosensation and thermoregulation. Am J Physiol Regul Integr Comp Physiol 2007; 292(1):R64–R76.
130. Peier AM, Moqrich A, Hergarden AC, et al. A TRP channel that senses cold stimuli and menthol. Cell 2002; 108(5):705–715.
131. Proudfoot CJ, Garry EM, Cottrell DF. Analgesia mediated by the TRPM8 cold receptor in chronic neuropathic pain. Curr Biol 2006; 16(16):1591–1605.
132. Bautista DM, Siemens J, Glazer JM. The menthol receptor TRPM8 is the principal detector of environmental cold. Nature 2007; 448(7150):204–228.
133. Caterina MJ, Schumacher MA, Tominaga M, et al. The capsaicin receptor: a heat-activated ion channel in the pain pathway. Nature 1997; 389(6653):816–824.
134. Calixto JB, Kassuya CA, André E, et al. Contribution of natural products to the discovery of the transient receptor potential (TRP) channels family and their functions. Pharmacol Ther 2005; 106(2):179–208.

135. Silver WL, Clapp TR, Stone LM, et al. TRPV1 receptors and nasal trigeminal chemesthesis. Chem Senses 2006; 31(9):807–812.
136. Dinh QT, Groneberg DA, Peiser C, et al. Substance P expression in TRPV1 and trkA-positive dorsal root ganglion neurons innervating the mouse lung. Respir Physiol Neurobiol 2004; 144(1):15–24.
137. Moqrich A, Hwang SW, Earley TJ, et al. Impaired thermosensation in mice lacking TRPV3, a heat and camphor sensor in the skin. Science 2005; 307(5714):1468–1472.
138. Xu H, Delling M, Jun JC, et al. Oregano, thyme and clove-derived flavors and skin sensitizers activate specific TRP channels. Nat Neurosci 2006; 9(5):628–635.
139. Story GM, Peier AM, Reeve AJ, et al. ANKTM1, a TRP-like channel expressed in nociceptive neurons, is activated by cold temperatures. Cell 2003; 112(6):819–829.
140. Bandell M, Story GM, Hwang SW, et al. Noxious cold ion channel TRPA1 is activated by pungent compounds and bradykinin. Neuron 2004; 41(6):849–857.
141. Bautista DM, Jordt SE, Nikai T, et al. TRPA1 mediates the inflammatory actions of environmental irritants and proalgesic agents. Cell 2006; 124(6):1269–1282.
142. Bessac BF, Sivula M, von Hehn CA, et al. TRPA1 is a major oxidant sensor in murine airway sensory neurons. J Clin Invest 2008; 118(5):1899–1910.
143. Whitear M. Presumed sensory cells in fish epidermis. Nature 1965; 208(5011): 703–704.
144. Peters RC, Kotrschal K, Krautgartner W-D. Solitary chemoreceptor cells of Ciliata mustela (Gadidae, Teleostei) are tuned to mucoid stimuli. Chem Senses 1991; 16(1):31–42.
145. Sbarbati A, Osculati F. The taste cell-related diffuse chemosensory system. Prog Neurobiol 2005; 75(4):295–307.
146. Bezençon C, le Coutre J, Damak S. Taste-signaling proteins are coexpressed in solitary intestinal epithelial cells. Chem Senses 2007; 32(1):41–49.
147. Finger TE, Böttger B, Hansen A. Solitary chemoreceptor cells in the nasal cavity serve as sentinels of respiration. Proc Natl Acad Sci U S A 2003; 100(15):8981–8986.
148. Gulbransen BD, Clapp TR, Finger TE. Nasal solitary chemoreceptor cell responses to bitter and trigeminal stimulants in vitro. J Neurophysiol 2008; 99(6):2929–2937.
149. Lin W, Ogura T, Margolskee RF, et al. TRPM5-expressing solitary chemosensory cells respond to odorous irritants. J Neurophysiol 2008; 99(3):1451–1460.
150. Ohmoto M, Matsumoto I, Yasuoka A, et al. Genetic tracing of the gustatory and trigeminal neural pathways originating from T1R3-expressing taste receptor cells and solitary chemoreceptor cells. Mol Cell Neurosci 2008; 38(4):505–517.
151. Lindberg S, Dolata J, Mercke U. Stimulation of C fibers by ammonia vapor triggers mucociliary defense reflex. Am Rev Respir Dis 1987; 135(5):1093–1098.

4

Functional Neuroanatomy of the Human Upper Airway

Murugan Ravindran, Samantha Jean Merck, and James N. Baraniuk

Division of Rheumatology, Immunology and Allergy, Georgetown University Medical Center, Washington, D.C., U.S.A.

INTRODUCTION

Sensory nerves monitor the status of the nasal mucosal microenvironment to (*i*) initiate immediate protective mucosal responses via the axon response mechanism, (*ii*) recruit brain stem reflexes including locally active parasympathetic exocytosis and systemically active sympathetic vasoconstrictor reflexes, (*iii*) modulate the work of breathing, and (*iv*) notify cerebral mucosal sensory centers (e.g., insula, secondary somatosensory cortex) of the conditions of inhaled air. These integrated reflexes regulate epithelial, vascular, glandular, smooth muscle, and inflammatory protective defenses (1,2).

Several distinct subsets of nonmyelinated type C nociceptive neurons can now be differentiated (Fig. 1). These slowly conducting neurons convey sensations of dull, aching or parasthetic, tingling pain, itch, and probably other poorly defined sensations ("second pain") (3). These neurons may mediate sensations of major clinical interest such as nasal congestion, fullness, obstruction to airflow, postnasal drip, and midfacial headaches. Thinly myelinated Aδ neurons rapidly convey information about sharp "first pain," heat, cold, and potentially other sensations to the brain stem. Large myelinated Aβ fibers convey light touch, proprioceptive, vibratory, and other information, but in general they are not an important source of information from mucosal surfaces (3).

TRIGEMINAL AND OTHER AFFERENT NERVES

Nasal and sinus sensory nerves originate in the trigeminal ganglion (Fig. 2) (4). The ophthalmic division (cranial nerve V1) proceeds anteriorly through the cavernous sinus and superior orbital fissure to the orbit where it divides into the lacrimal, frontal, and nasociliary branches. The lacrimal branch also contains parasympathetic innervation to the lacrimal glands, and sympathetic neurons. Injury leads to lacrimation and nasal congestion. The larger frontal branch becomes the supraorbital nerve and innervates the upper eyelid, frontal sinus mucosa, and anterior scalp. This branch may contribute to cephalic pain syndromes that mimic "sinusitis."

The maxillary division (cranial nerve V2) is entirely sensory. The middle meningeal nerve branches off near the Gasserian ganglion to innervate the dura mater. Ultimately, the maxillary nerve divides into the infraorbital branch to innervate the lower lid, frontal, ethmoid and sphenoid sinus mucosae, and lateral facial skin. This distribution may explain the relationship between pain and other

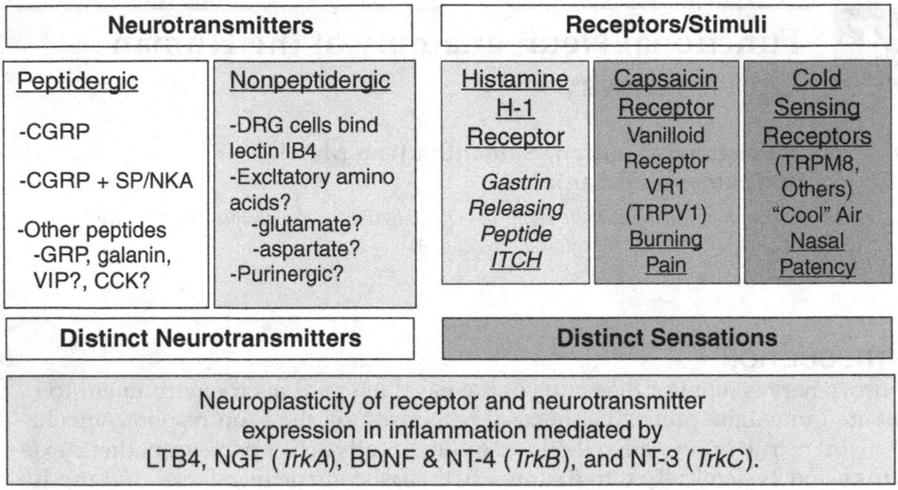

Neurotransmitters		Receptors/Stimuli		
Peptidergic	Nonpeptidergic	Histamine H-1 Receptor	Capsaicin Receptor	Cold Sensing Receptors
-CGRP	-DRG cells bind lectin IB4		Vanilloid Receptor	(TRPM8, Others)
-CGRP + SP/NKA	-Excltatory amino acids?	Gastrin Releasing Peptide	VR1 (TRPV1)	"Cool" Air Nasal
-Other peptides -GRP, galanin, VIP?, CCK?	-glutamate? -aspartate? -Purinergic?	ITCH	Burning Pain	Patency

Distinct Neurotransmitters	Distinct Sensations

Neuroplasticity of receptor and neurotransmitter
expression in inflammation mediated by
LTB4, NGF (*TrkA*), BDNF & NT-4 (*TrkB*), and NT-3 (*TrkC*).

FIGURE 1 Subsets of mucosal nociceptive neurons based on neurotransmitters, sensations conveyed, and neural plasticity in inflammation. *Source*: Courtesy of J.N. Baraniuk.

sensations in sinusitis and referred pain to specific facial regions as in Sluder's "sphenopalatine neuralgia" (5) and "midfacial pain syndrome" (6). The greater palatine branch innervates the hard and anterior soft palate, uvula, and gums. The posterior superior alveolar branch supplies gums, buccal mucosa, and molar teeth, which is of relevance since maxillary toothache is a diagnostic symptom for acute sinusitis (7).

The sensory mandibular division (cranial nerve V3) innervates the cutaneous lateral head, ear and mandibular region, mouth, lower teeth, temporomandibular joint, and anterior dura mater. The motor branches supply the muscles of mastication (Fig. 3). This nerve carries both sensory afferent and motor efferent nerves that may be active in temporomandibular disease.

Sensory nerves enter the nasal mucosa via the ethmoidal and posterior nasal nerves (4). These highly branched neurons innervate vessels, glands, and the epithelium where they extend between basal cells as fine, bare nerve terminal endings near tight junctions. Up until recently it was believed that the airway mucosa does not have specialized sensory organs as found in skin or tendons. However, the relatively recent discovery of so-called "solitary chemoreceptor cells" in the nasal mucosa of several vertebrate species has raised the possibility that humans, too, may exhibit this specialized nociceptive cell type in the upper airway (8).

The olfactory mucosa in the superior nasal cavity is innervated by cranial nerve I and the nervus terminalis, or cranial nerve XIII (9).

Central Afferent Connections

Nerve impulses are carried from epidermal and dermal free endings to layers I and II of the spinal cord substantia gelatinosa (Fig. 2). Afferent trigeminal neurons enter the pons through the sensory root, and turn caudally in the trigeminal spinal tract to terminate in the pars caudalis of the nucleus of the spinal

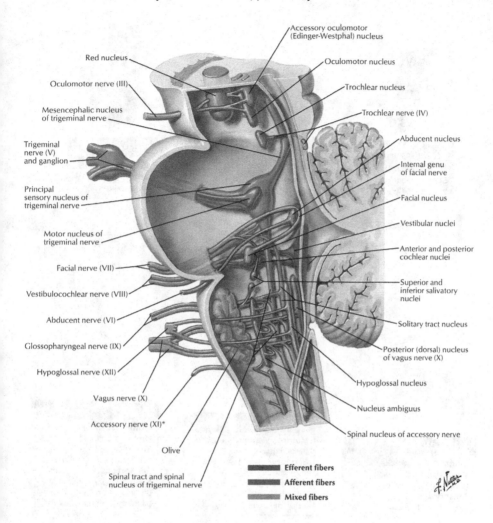

*Accessory oculomotor
(Edinger-Westphal) nucleus*

Oculomotor nucleus

Red nucleus

Oculomotor nerve (III)

Trochlear nucleus

Mesencephalic nucleus
of trigeminal nerve

Trochlear nerve (IV)

Abducent nucleus

Trigeminal
nerve (V)
and ganglion

*Internal genu
of facial nerve*

Principal
sensory nucleus of
trigeminal nerve

Facial nucleus

Vestibular nuclei

Motor nucleus of
trigeminal nerve

*Anterior and posterior
cochlear nuclei*

Facial nerve (VII)

*Superior and
inferior salivatory
nuclei*

Vestibulocochlear nerve (VIII)

Abducent nerve (VI)

Solitary tract nucleus

Glossopharyngeal nerve (IX)

*Posterior (dorsal) nucleus
of vagus nerve (X)*

Hypoglossal nerve (XII)

Hypoglossal nucleus

Vagus nerve (X)

Nucleus ambiguus

Accessory nerve (XI)*

Spinal nucleus of accessory nerve

Olive

Spinal tract and spinal
nucleus of trigeminal nerve

■■■ **Efferent fibers**
■■■ **Afferent fibers**
■■■ **Mixed fibers**

*Recent evidence suggests that the accessory nerve lacks a cranial root and has no connection to the vagus nerve.
Verification of this finding awaits further investigation

FIGURE 2 Cranial nerve nuclei in brain stem (medial dissection). *Source*: Courtesy of Elsevier
Health, Inc.

tract in the lower medulla and upper three cervical segments of the spinal cord
(4). The diverse afferent neurons release combinations of calcitonin gene–related
peptides (CGRP), somatostatin, gastrin-releasing peptide, substance P (SP) and
colocalized neurokinin A, and probably other colocalized neurotransmitters
(Table 1). Gastrin-releasing peptide may be a mediator of itch (10). Messages of
mechanicosensitive, chemosensitive, and itch neurons synapse on different sets
of dorsal horn neurons. Secondary interneurons that convey sensations of itch
and pain are located in different lateral spinothalamic pathways of the pars caudalis (11). These interneurons cross the midline to enter the trigeminothalamic

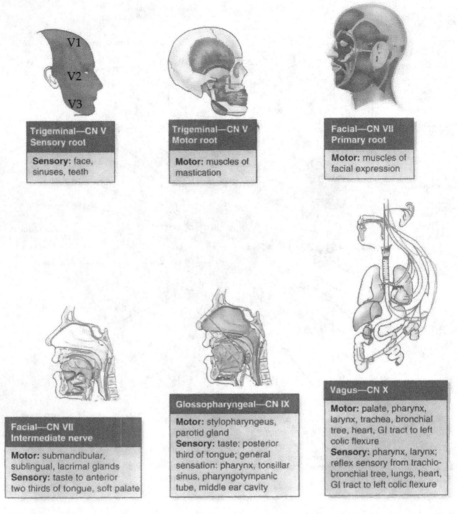

FIGURE 3 Sensory and motor functions of cranial nerves V, VII, IX, and X. *Source*: Courtesy of Wolters Kluwer.

tract and terminate in the medial part of the ventral posterior thalamic nucleus (arcuate or semilunar nucleus). Pain and itch fibers synapse in different thalamic nuclei. Intense pain can be appreciated at the thalamic level. Tertiary neural relays to the lower third of parietal cortical somesthetic areas provide greater localization of painful stimuli.

Aδ nerves (first pain) (3) rapidly transmit burning pain sensations to bilateral associative somatosensory cortex (SII) regions (12). Additional activation of the posterior parietal cortex likely orchestrates sensorimotor coordination to produce precise motor acts to avoid, reduce, or prevent the offensive agent. More slowly conducting type C neurons innervate these same regions and convey dull ache or parasthetic sensations (second pain) (3). Aδ and type C neural activation

TABLE 1 Airway Neurotransmitters and Receptors: Trigeminal Sensory Neurotransmitters and Receptors

Peptides	Receptors	Functions
Tachykinins	*Neurokinins*	
SP	NK-1	Mucus secretion, arterial dilatation, vascular permeability, leukocyte infiltration
NKA	NK-2	Bronchoconstriction
Neuropeptide K		
NKB	NK-3	Inhibitory autoreceptor
Calcitonin gene–related peptides		
CGRPα, CGRPβ	CGRP-R1, CGRP-R2 Amylin, adrenomedullin	Vasodilator
Bombesin family		
GRP	BB2-R (GRP-R)	Glandular secretagogues, cell proliferation
Neuromedin B	BB1-R (NMB-R)	
Others?		

Abbreviations: SP, substance P; NKA, neurokinin A; NKB, neurokinin B; GRP, gastrin-releasing peptide.

also stimulates the cingulate cortex (emotion and executive function) and bilateral insular cortex (visceral nociception). Therefore, first and second pain is mediated by Aδ and C-fibers that travel in a common cortical network but with different time frames. The insula serves as an integration cortex for multimodal convergence of distributed neural networks such as the somesthetic-limbic, insulo-limbic, insulo-orbito-temporal, and the prefrontal-striato-pallidal-basal forebrain (13).

The dorsal part of the trigeminal spinal tract includes a small bundle of thermosensitive and nociceptive neurons from the facial, glossopharyngeal, and vagal nerves (4). The glossopharyngeal nerve innervates the posterior third of the tongue, upper pharynx, tonsils, eustachian tube, and middle ear (Fig. 3). Glossopharyngeal afferents from the carotid sinus baroreceptors and carotid body chemoceptors regulate arterial blood pressure, and oxygen tension/ventilatory function, respectively. Vagal afferents with cell bodies in the small, superior, or nodose vagal ganglia innervate the lower pharynx, epiglottis, larynx (recurrent laryngeal nerve), trachea, main stem and subsegmental bronchi, and esophagus (Fig. 3). The vagal auricular branch innervates a portion of the external ear, auditory canal, tympanic membrane, middle ear, and eustachian tube.

Central glossopharyngeal and vagal afferents descend in the medullary solitary tract and end in the more caudal part of its nucleus (Fig. 2). From these nuclei, interneuron connections are made with the dorsal motor neurons of the vagus nerve, reticular visceral centers, and subsequently with hypothalamic nuclei. The neurotransmitters of these interneurons are unknown. However, SP receptors in the ventrolateral medulla regulate ventilatory responses (14), suggesting tachykinins as candidate neurotransmitters.

Fibers transmitting touch sensations from the posterior tongue, pharynx, and larynx may terminate in either the nucleus of the trigeminal spinal tract or the nucleus of the solitary tract (Fig. 2). These serve as the afferent limbs of the "gag reflex" (4) and cough (15). These neurons may also modify the responses from taste receptors.

EFFERENT PATHWAYS

Afferent interneurons from the nuclei of the trigeminal spinal tract and the solitary tract synapse in the nucleus ambiguous (Fig. 2) to establish the efferent limbs of the sneezing, coughing, gagging, and vomiting reflexes (Fig. 3). These have vital functions in protecting the integrity and patency of the nasal, pharyngeal, and laryngeal airways. The motor division of the accessory nerve acts as the efferent limb of the gag reflex, because activation of the nucleus ambiguus and hypoglossal nucleus leads to stimulation of the striated muscle of the soft palate, pharynx, larynx, and upper esophagus (4). Similar connections regulate parasympathetically mediated glandular secretion in the nose (superior salivatory nucleus, facial nerve, sphenopalatine ganglion), and tracheobronchial glandular and smooth muscle function (dorsal motor nuclei of the vagal nerve).

Parasympathetic preganglionic fibers exit the brain stem with the VIIth nerve. They join with postganglionic sympathetic neurons to form the Vidian nerve. After passing through the Vidian canal, the parasympathetic neurons terminate in the sphenopalatine ganglion. Postganglionic parasympathetic cell bodies may act as electrical "filters" by integrating stimulatory and inhibitory inputs before being triggered to depolarize. These cells innervate the various exocrine mucosal organs. A population of large-diameter cell bodies has axons that release acetylcholine to stimulate nasal glandular exocytosis via muscarinic M3 receptors (Table 2) (16). M1 and M2 receptors were identified to a lesser extent on glands, arteries, veins, and epithelia. M4 receptors were found around arteries. M5 receptors were identified on glands and arteries. A smaller-diameter population contains combinations of vasoactive intestinal polypeptide (VIP), PHM (Polypeptide with Histidine at the N-terminal and Methionine at the C-terminal), PACAP (pituitary adenyl cyclase activating peptide), and neuronal nitric oxide (17). VIP and PHM are derived from a single gene which explains their colocalization. They have similar vasodilatory, bronchodilatory, and secretomotor effects, but VIP is consistently more potent. VIP-immunoreactive nerves (18) increase in density in perennial allergic rhinitis (19), aspirin-sensitive rhinitis (20), chronic cigarette smokers (21), and idiopathic "vasomotor" rhinitis (22), demonstrating their plasticity of expression in airway mucosal inflammation. In general, markers of nociceptive and sympathetic neurons were not altered. These exposure-related findings suggest that other toxic irritant rhinitis syndromes induced by chemical compounds such as ozone, formaldehyde, nickel, chrome, solvents, and methylbromide may also demonstrate significant autonomic nerve dysregulation as part of the mucosal response to tissue injury.

Sympathetic, preganglionic nerve fibers originate in the intermediolateral cell column of thoracic spinal segments (4). Those destined to regulate respiratory vascular functions in the head synapse with postganglionic cells in the superior cervical ganglion, while those regulating thoracic functions synapse predominantly in the superior, middle and lower cervical ganglia, and the

TABLE 2 Airway Neurotransmitters and Receptors: Postganglionic Parasympathetic and Sympathetic Neurotransmitters and Receptors

Peptides	Receptors	Functions
Large-diameter cell bodies		
Acetylcholine	Muscarinic M1	
	M2	Inhibitory autoreceptor
	M3	Glandular secretion, bronchoconstriction
	M4	
	M5	
Small-diameter cell bodies		
VIP	VIP1-R, VIP2-R	Vasodilation, smooth muscle contraction, glandular secretagogues
VIP-like peptides: PACAP, PACAP-27[a] PHM[b] PHI[c] PHV[d] Helodermin	VIP1-R, VIP2-R, PACAP	
nNOS	Guanylyl cyclase	Vasodilator, final common mediator of exocytosis?
Postganglionic sympathetic neurotransmitters and receptors		
NE	α1A, α1B, α1D	Vasoconstriction
	α2A, α2B, α2C	Vasoconstriction, inhibitory autoreceptors
	β_1	Cardiac iono- and chrono-trope
	β_2	Bronchodilator, inhibitor autoreceptor
	β_3	Adipose regulation?
NPY	Y1, Y2, Y3, Y4, Y5	Y1—long-lasting vasoconstrictor
		Y2—inhibitory autoreceptor

[a]Pituitary adenylate cyclase activating peptide (PACAP, PACAP-27).
[b]Peptide with <u>H</u>istidine at the N-terminal and <u>M</u>ethionine at the C-terminal (PHM).
[c]Peptide with <u>H</u>istidine at the N-terminal and <u>I</u>soleucine at the C-terminal (PHI).
[d]Peptide with <u>H</u>istidine at the N-terminal and <u>V</u>aline at the C-terminal (PHV).
Abbreviations: VIP, vasoactive intestinal peptide; nNOS, neuronal nitric oxide synthase; NE, norepinephrine; NPY, neuropeptide tyrosine (Y).

upper five thoracic ganglia. Sympathetic reflexes induce vasoconstriction in respiratory mucosa and may partially attenuate bronchoconstricting parasympathetic reflexes. At least two populations of sympathetic neurons are known that contain either norepinephrine or norepinephrine plus neuropeptide Y (23). ATP may also be a sympathetic neurotransmitter.

AFFERENT Aδ NERVE FIBERS

Aδ fibers convey initial sharp pain ("first pain") (3). In the feline nose, there appear to be a set of low-threshold mechanoreceptive Aδ fibers sensitive only to light touch, and a set of chemomechanical sensitive fibers that synapse on wide dynamic range neurons in the spinal cord dorsal horn (24). The feline anterior ethmoid nerve contains 6000 C fibers, 650 Aδ, and 350 Aβ fibers (25). These Aβ fibers contain γ-aminobutyric acid and glycine and may mediate proprioceptive functions of feline nasal vibrissae (26).

A set of cold-sensitive Aδ fibers play an important afferent role in breath-to-breath regulation of respiration (27) and in the apnea and pressor responses mediated by the nucleus of the solitary tract (28). Cold temperatures below ~28°C activate the L-menthol–sensitive transient receptor potential melanostatin 8 (TRPM8) ion channel to generate a cooling sensation (29). This sensory ion channel is a member of the transient receptor potential (TRP) hexahelical transmembrane ion channel protein superfamily (30). In response to cold or menthol, the TRPM8 channel opens and calcium influx occurs to cause neuron depolarization (31). The effect may be rapid (milliseconds) but short-lived ("rapidly adapting receptor") in order to allow for breath-to-breath regulation of respiration (32). The magnitude of the "cold" stimulus may be proportional to the rate of airflow through the nostril and evaporation effects. When airflow is rapid, as occurs during normal inspiration and exhalation through patent nostrils, air is warmed and humidified by evaporation of the epithelial-lining fluid. Evaporation of the high-energy water molecules leaves the lower-energy (colder) water molecules in the fluid phase of the epithelial-lining fluid. This temperature change is sufficient to trigger this population of cold-responsive Aδ fibers. In the brain stem, they provide information that helps regulate the muscular work of inspiration and sensation of dyspnea. Application of L-menthol may reduce dyspnea by activating TRPM8. However, the pharmacological effects may be short-lived because the rapidly adapting receptor will adapt to a steady-state concentration of the drug. TRPM8 responds to phospholipid membrane fluidity (33). Cooling may be sensed by alterations in the membrane property. These effects may contribute to the differential nasal airway obstruction found in idiopathic nonallergic rhinitis following inhalation of cold dry air (34).

TYPE C NEURONS

Type C chemosensitive and mechanothermal sensitive neurons can be stimulated by inflammatory mediators such as bradykinin (B1 and B2 receptors), serotonin, ATP (P2Y2 and P2X receptors), K^+, H^+, and histamine (H1 receptor), which are released following mucosal injury or mast cell degranulation (35–37). Acidosis may be the most potent physiological stimulus. Acid-sensing ion channel 3 may play a predominant role in conjunction with numerous other acid-sensitive ion channels, receptors, and modulatory proteins. SO_2, O_3, formaldehyde, nicotine, cigarette smoke, and capsaicin can also stimulate type C neurons (38). Prostaglandins and peptidoleukotrienes modulate sensory nerve function by reducing the threshold for depolarization, so that a smaller stimulus is required to depolarize and activate these nerves. The capsaicin-sensitive TRP vanilloid 1 (TRPV1) ion channel is an important polymodal chemosensor on subsets of type C nerve fibers containing either neuropeptides or other (e.g., glutamate) neurotransmitters (39,40). In mice, TRPA1 may mediate some of bradykinin's proinflammatory effects (41). Numerous neurotrophins may regulate the expression of these diverse sensor proteins in health, injury, inflammation, and tissue repair (42).

CAPSAICIN AND THE TRPV1

TRPV1 was discovered because of its activation by capsaicin (trans-8-methyl-N-vanillyl-6-nonenamide), the hot, spicy essence of red peppers (39,43). Other agonist ligands and conditions include low-threshold heat (>43°C), pH < 5.9, ethanol, cyclohexanone, arachidonic acid metabolites that form endogenous cannabinoids such as the cannabinoid receptor 1 agonist N-arachadonoyl-dopamine (44), 12- and 15-lipoxygenase products generated by activation of bradykinin B2 and other receptors, and additional cationic lipids. A subpopulation of neurons with P2X, acid-sensing ion channel 3, and TRPV1 receptors may be very important for detection of visceral and potentially mucosal injury (45). Their sensitivities to many diverse ligands and physicomechanical stimuli may allow neural responses to both unique chemicals and more "nonspecific" mucosal irritant conditions (30).

Acute capsaicin administration activates TRPV1 pore proteins that allow an influx of Na^+ and Ca^{2+} ions to depolarize neurons (30). This depolarization leads to the axon response release of CGRP that mediates the cutaneous and nasal mucosal vasodilation and flare (46). In the skin, capsaicin induces sharply defined regions of flare, and mechanical and heat hyperalgesia. This indicates that the type C neurons responsible for CGRP release and vasodilation are also involved in the altered neuroregulation of spinal cord dorsal horn secondary and interneurons. These central effects may be mediated by SP or combinations of neurotransmitters. The flare, temperature changes, and hyperalgesia are blocked by local anesthetics. Commercial, cutaneous capsaicin "muscle" and "pain" lotions induce a mild sensation of heat that offsets perceptions of pain from injured peripheral tissue. Chronic cutaneous capsaicin treatment in humans reduces the flare and hyperalgesic responses to chemical irritants, intradermally injected histamine, anti-IgE, antigen skin tests, mechanical and heat stimuli, but does not affect local plasma exudation (wheal size), touch, pressure, vibration, cold, or other Aδ and Aβ fiber–mediated sensations. Chronic oral capsaicin is well tolerated, since some cultures consume up to 30 to 60 g/day as a culinary delight (43). Desensitization is evident since higher doses are required to maintain the sensation of piquancy.

Capsaicin applied topically to the nasal mucosa is beneficial in idiopathic, nonallergic rhinitis. Initial studies showing the ability of capsaicin pretreatment to block hypertonic saline and other provocation systems (46–52) led to its effective, topical clinical use in nonallergic, "vasomotor," idiopathic rhinitis (46,47,49–52). Beneficial effects have lasted up to six months. These findings suggested that TRPV1 was involved in idiopathic rhinitis, and that this mechanism was a predominant one in the pathophysiology of this condition. In contrast, results were inconsistent or negative in seasonal and perennial allergic rhinitis (53–59), nasal polyps (60,61), and otitis media (62). This suggests that TRPV1 mechanisms were relatively minor or did not contribute to the pathophysiology when other, overwhelmingly inflammatory upper airway illnesses were active. Difficulties with topical intranasal capsaicin treatment are that no treatment regimen has been performed twice to assess reproducibility; treatment periods have evolved from weekly dosing to multiple high doses on a single day; and that severe burning pain must be avoided during the capsaicin applications.

TRPV1 is also present on keratinocytes as well as epithelial cells of the nasal, tracheobronchial, gastrointestinal, bladder, and rectal mucosae (Table 3)

TABLE 3   Distribution of Transient Receptor Potential Vanilloid 1 Ion Channels (TRPV1; Capsaicin Receptor)

Neurons	Other relevant cell types

• DRG
  − Type C nerves (some Aδ)
  − Colocalized with:
    • *TrkA* (NGF-R)
    • *TrkB* (BNDF, NT-4)
    • Purinergic P2×3 receptor
    • Lectin IB4+
    • CGRP ± SP
    • Negative for neurofilament 200
• Spinal cord—dorsal horn
• Hypothalamus
• Hippocampus
• Substantia nigra
  − Dopaminergic cells
	• Airway epithelium
	• Keratinocytes
	• Neutrophils
	• Macrophages
	  − Downregulates *iNOS* and *COX2*

Abbreviation: DRG, dorsal root ganglia.

(63). Limited evidence suggests that pithelial cells may act as an accessory sensor location responding to capsaicin (64,65) and may release mediators such as IL-1β and MMP-1 (66) that secondarily activate TRPV1 and other sensor proteins on nociceptive neurons, mast cells, Langerhans cells, neutrophils, and macrophages. The acute toxic effects of capsaicin on epithelial TRPV1 explain acute toxic exposures in industrial chili pepper workers and victims of high doses of "pepper spray" self-defense products. Extensive activation of TRPV1 can damage keratinocytes and fibroblasts, and lead to extremely severe erythema (CGRP), edema (CGRP plus vasodilatory mediators), desquamation analogous to third-degree burns (dermal and epidermal cell death), and mucosal irritation with reflex-mediated lacrimation, rhinorrhea, cough, and potentially laryngospasm and bronchoconstriction. TRPV1 innervation is increased in breast pain syndrome and irritable bowel syndrome (67,68).

Ethanol stimulates the TRPV1 receptor (39). This may explain its burning astringent effects when applied to abraded tissues and mucosal surfaces. Ingestion may stimulate gastroesophageal vagal afferent C fibers that cause burning, and may participate in gastroesophageal reflexes and gastritis.

The functions of TRPV1, TRPA1, other sensors, and type C neurons can be modulated by G protein–coupled receptors and intracellular adapter proteins. MAS-related GPR, member X2 [Homo sapiens] (MRGPRX2; GeneID: 117194), is a human peripheral sensory neuron-specific gene involved in nociception (69). Genetic studies of widely dispersed human populations and other primates demonstrated positive Darwinian selection for four human-specific amino acid substitutions. Three were in the extracellular N-terminal region than may form a binding site for cortistatin-14 (70). The fourth substitution is in an intracellular location that could modify interactions with Gq proteins or other regulatory enzymes. Although the exact role of this receptor and its four function-altering substitutions is poorly understood, their presence suggests positive adaptive changes in pain detection or perception during human evolution.

Another genetic anomaly can occur in small-diameter dorsal root ganglion cells. These express high levels of high-affinity receptor for nerve growth factor

(*TrkA*), are sensitive to capsaicin, and express mRNA for a 157-kDa protein that contains a mitochondrial targeting signal peptide sequence and is called leucine-rich PPR-motif containing (LRPPRC [Homo sapiens]; GeneID: 10128) (71). A mutant isoform is responsible for Leigh syndrome, French-Canadian subtype, which leads to the loss of mitochondrial cytochrome c oxidase activity followed by neurodegeneration. Mutants such as these may identify novel mechanisms that lead to peripheral and visceral neuropathies, and to alterations in the generation of airway sensations that are associated with "congestion," "rhinorrhea," "sinus pain and headache," and other symptoms.

TRP THERMOMETER AND AROMATHERAPY

Temperatures in the ambient range are detected by TRPV3 (dynamic range ~31–39°C) and TRPV4 (25–35°C) (Fig. 4) (30). These two ion channels are also mechanic sensitive since they can detect cell swelling induced by hypotonic conditions and an influx of water into cells. Higher temperatures activate TRPV1, the capsaicin receptor. It is a polymodal ion channel that responds to "warm" temperatures above 43°C, pH $\leq$ 5.9, ethanol, and endovanilloid/endocannabinoid products of arachidonic acid metabolism (72). Hot temperatures $\geq$52°C may be detected by TRPV2 ion channels on Aδ nerve fibers. Cooling of the mucosa, as occurs due to evaporation during normal nasal breathing, is detected by TRPM8, L-menthol–activated ion channels (33). This cooling effect is the basis for the use of L-menthol in many proprietary cold remedies. TRPM8 responds to temperatures below 28°C, linalool, geraniol, hydroxycitronellal, and icillin, the "supercooling" topical agent found in many cosmetic and muscle strain products (73). Temperatures below 17°C may activate TRP ankyrin (TRPA1) "very cold" ion channels, although this is controversial (29,41). TRPA1 is present on

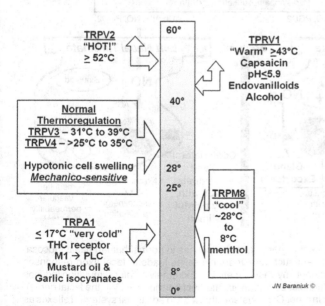

FIGURE 4 The TRP thermometer of type C and Aδ nociceptive nerves. *Source*: Courtesy of J.N. Baraniuk.

a set of TRPV1 neurons that mediate cold-induced pain. Tetrahydrocannabinol (THC), isothiocyanates found in garlic and mustard oil, acrolein (41), gingerol, cinnamaldehyde, methyl salicylate, (73) and the reactive oxidant species hypochlorite and hydrogen peroxide (74) are also agonists that stimulate TRPA1. Heterotetrameric combinations of this suite of ion channels on different subsets of Type Aδ and C neurons may provide for specific dynamic ranges of sensitivity by distinct subsets of nerves (75).

NOCICEPTIVE NERVE AXON RESPONSES

Afferent mucosal neurons are highly branched. Depolarization of one nerve ending depolarizes all the dendritic branches and causes local mucosal neuropeptide release. In the human nasal mucosa, hypertonic saline induces dose-dependent sensations of "first" and "second" pain, blockage, and rhinorrhea (76,77). Droplets of hypertonic saline have higher osmolality compared to epithelial-lining fluid. In addition, the surface chemistry of hypertonic saline droplets leads to the formation of O_3, OCl^-, and other free radicals (78). "First pain" lasts for several seconds. "Second pain" is slower developing, but lasts longer as a paresthetic or dull ache sensation. The maximum pain levels are about 3 to 4 on a 10-cm linear analog scale and are similar to sensations experienced by patients in their daily lives. SP is released in dose-dependent fashion. Exocytosis from both the serous and mucous cells of submucosal glands is increased (Fig. 5).

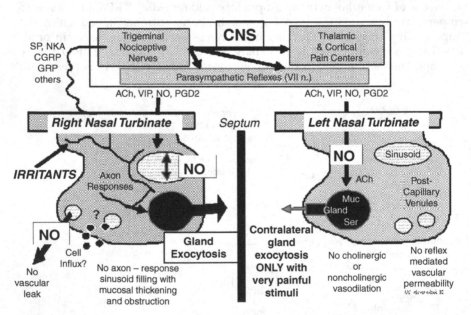

FIGURE 5 Irritant-induced nociceptive nerve axon responses in normal human nasal mucosa. Unilateral provocation with an irritant such as hypertonic saline leads to local axon responses that cause glandular secretion without any vascular effects. Extremely painful stimulation induces contralateral parasympathetic glandular secretion via efferent fibers in Cr. N. VII (the facial nerve). Nociceptive nerves activate pain in other CNS systems that lead to additional systemic reflexes as well. Vascular reflexes have been difficult to document in normal humans, but may be upregulated in severe untreated allergic rhinitis. *Source*: Courtesy of J.N. Baraniuk.

Of note, SP-preferring NK-1 receptor mRNA is present in these glands, suggesting that nociceptive nerve SP release is responsible for the glandular secretion (76). Other neurotransmitters colocalized in SP-containing neurons may also be secretagogues. In contrast, unilateral provocations leading to pain levels much higher than those encountered on a daily basis (6 to 9 out of 10) are able to recruit contralateral parasympathetic reflexes (79). Unilateral hypertonic saline provocations in acute sinusitis and active allergic rhinitis patients do not cause these high levels of pain, and induce erratic amounts of glandular exocytosis. Secretion is probably more related to the underlying disease than the provocation (77).

In stark contrast to all the data gained from rodent models (80), there is no vascular permeability change during this neurogenic response in humans (76,77). Nociceptive release of CGRP causes the flare response without any wheal in the skin and is a vasodilator in nasal mucosa, but does not stimulate mast cell degranulation in either human organ (81). Axon responses and CGRP release may not cause nasal obstruction since acoustic rhinometry showed no change in nasal patency after CGRP or hypertonic saline provocations (76,77).

CONCLUSION

Trigeminal type C sensory neurons are a heterogenous set of nonmyelinated nerves that mediate sensations of pain, itch, heat, and probably other symptoms associated with rhinitis and responses to mucosal irritants (Fig. 5) (82). Mucosal stimulation of TRPV1 and other sensor proteins leads to neuron depolarization. Understanding this complexity of epithelial cell–neuron interaction will help define the spectrum of physical and chemical stimuli that can depolarize the neuron, and possibly the sensation(s) that these nerves convey to the central nervous system. Mediators and other processes that decrease the threshold for type C neuron depolarization cause "peripheral sensitization" (see chap. 26). Type C neurons are highly branched in the nasal mucosa. The wave of depolarization releases combinations of neurotransmitters that activate glandular exocytosis. This constitutes the human airway axon response. In the dorsal horn of the spinal cord, type C neurons synapse and convey their messages to secondary ascending neurons and regulatory interneurons. Messages are usually inhibited until they reach a specific threshold. Modulation of this threshold to make sensory transmission easier leads to "central sensitization" and increased perception of the pain or other sensation being conveyed from the mucosa. Distinct pain, itch, and other pathways ascend to the brain stem, thalamus, and higher centers. Brain stem nuclei respond by regulating the work of breathing, local mucosal parasympathetic cholinergic output, and so glandular exocytosis, systemic sympathetic vasoconstriction, and other functions. Primary mucosal somatosensory regions in the insula and elsewhere lead to conscious appreciation of the sensation, and recruit affective, limbic memory, amygdala fear, and anterior cingulate and other frontal executive functions. Subsequent avoidance actions are planned and executed through motor pathways to generate sneezing, the allergic salute, and other avoidance behaviors. These events likely developed as highly integrated defense mechanisms for protecting the airway from dangers imposed by smoke, cold or dry air, and other nociceptive conditions. Allergic mechanisms co-opt many of these functions to generate itch, sneeze, sensations of nasal fullness, congestion and airflow obstruction, and to recruit parasympathetic glandular hypersecretion. Dysfunction at anywhere from the nerve ending to the cerebral circuitry may lead to the pathologically elevated perceptions and unrequired emotional and physical actions attained in idiopathic rhinitis. In this fashion, the nose represents the cross-roads between the conditions of inhaled air and our perceptions and responses to those conditions. Thus, the nasal mucosa serves as our vanguard for detection and clearance of inhaled toxic materials, and as the thin epithelial line protecting the body against airborne toxicity.

ACKNOWLEDGMENTS

Supported by U.S. Public Health Service Awards RO1-ES015382, M01-RR13297, P50 DC000214, and Department of the Army Award W81XWH-07-1-0618.

REFERENCES

1. Baraniuk JN. Neural regulation of mucosal function. Pulm Pharmacol Ther 2008; 21:442–448.
2. Kim D, Baraniuk JN. Neural aspects of allergic rhinitis. Curr Opin Otolaryngol Head Neck Surg 2007; 15:268–273.
3. Dray A, Urban L, Dickenson A. Pharmacology of chronic pain. Trends Pharmacol Sci 1994; 15:190–197.
4. Calliet R. Head and Face Pain Syndromes. Philadelphia, PA: F.A. Davis, 1992.
5. Sluder G. The role of the sphenopalatine ganglion in nasal headaches. N Y Med J 1908; 87:989–990.
6. Jones NS. The classification and diagnosis of facial pain. Hosp Med 2001; 62:598–606.
7. Low DE, Desrosiers M, McSherry J, et al. A practical guide for the diagnosis and treatment of acute sinusitis. CMAJ 1997; 156(suppl 6):S1–S14.
8. Finger TE, Böttger B, Hansen A, et al. Solitary chemoreceptor cells in the nasal cavity serve as sentinels of respiration. Proc Natl Acad Sci U S A 2003; 100(15):8981–8986.
9. Demski LS, Schwanzel-Fukuda M, eds. The terminal nerve [nervus terminalis]. Ann N Y Acad Sci 1987; 519:1–213.
10. Swain MG. Gastrin-releasing peptide and pruritus: more than just scratching the surface. J Hepatol 2008; 48:681–683.
11. Andrew D, Craig AD. Spinothalamic lamina 1 neurons selectively sensitive to histamine: a central neural pathway for itch. Nat Neurosci 2001; 4:72–77.
12. Forss N, Raij TT, Seppa M, et al. Common cortical network for first and second pain. Neuroimage 2005; 24:132–142.
13. Shelley BP, Trimble MR. The insular lobe of Reil—its anatamico-functional, behavioural and neuropsychiatric attributes in humans—a review. World J Biol Psychiatry 2004; 5:176–200.
14. Chen Z, Hedner J, Hedner T. Local effects of substance P on respiratory regulation in the rat medulla oblongata. J Appl Physiol 1990; 68:693–699.
15. Mazzone SB, McLennan L, McGovern AE, et al. Representation of capsaicin-evoked urge-to-cough in the human brain using functional magnetic resonance imaging. Am J Respir Crit Care Med 2007; 176:327–332.
16. Nakaya M, Yuasa T, Usui N. Immunohistochemical localization of subtypes of muscarinic receptors in human inferior turbinate mucosa. Ann Otol Rhinol Laryngol 2002; 111:593–597.
17. Kobzik L, Bredt DS, Lowenstein CJ, et al. Nitric oxide synthetase in human and rat lung: immunocytochemical and histochemical localization. Am J Respir Cell Mol Biol 1993; 9:371–377.
18. Baraniuk JN, Okayama M, Lundgren JD, et al. Vasoactive intestinal peptide (VIP) in human nasal mucosa. J Clin Invest 1990; 86:825–831.
19. Fischer A, Wussow A, Cryer A, et al. Neuronal plasticity in persistent perennial allergic rhinitis. J Occup Environ Med 2005; 47:20–25.
20. Groneberg DA, Heppt W, Welker P, et al. Aspirin-sensitive rhinitis-associated changes in upper airway innervation. Eur Respir J 2003; 22:986–991.

21. Groneberg DA, Heppt W, Cryer A, et al. Toxic rhinitis-induced changes of human nasal mucosa innervation. Toxicol Pathol 2003; 31:326–331.
22. Heppt W, Peiser C, Cryer A, et al. Innervation of human nasal mucosa in environmentally triggered hyperreflectoric rhinitis. J Occup Environ Med 2002; 44:924–929.
23. Lundblad L, Anggard A, Saria A, et al. Neuropeptide Y and non-adrenergic sympathetic vascular control of the cat nasal mucosa. J Auton Nerv Syst 1987; 20:189–197.
24. Lucier GE, Egizii R. Characterization of cat nasal afferents and brain stem neurones receiving ethmoidal input. Exp Neurol 1989; 103:83–89.
25. Wallois F, Gros F, Condamin M, et al. Postnatal development of the anterior ethmoidal nerve in cats: unmyelinated and myelinated nerve fiber analysis. Neurosci Lett 1993; 160:221–224.
26. Moon YS, Paik SK, Seo JH, et al. GABA- and glycine-like immunoreactivity in axonal endings presynaptic to the vibrissa afferents in the cat trigeminal interpolar nucleus. Neuroscience 2008; 152:138–145.
27. Nishino T, Tagaito Y, Sakurai Y. Nasal inhalation of l-menthol reduces respiratory discomfort associated with loaded breathing. Am J Respir Crit Care Med 1997; 156:309–313.
28. Dutschmann M, Herbert H. The medial nucleus of the solitary tract mediates the trigeminally evoked pressor response. Neuroreport 1998; 9:1053–1057.
29. McKemy DD, Neuhausser WM, Julius D. Identification of a cold receptor reveals a general role for TRP channels in thermosensation. Nature 2002; 416:52–58.
30. Yu FH, Catterall WA. The VGL-Chanome: a protein superfamily specialized for electrical signaling and ionic homeostasis. Sci STKE 2004; re15. www.stke.org/cgi/content/full/sigtrans;2004/253/re15.
31. Dhaka A, Murray AN, Mathur J, et al. TRPM8 is required for cold sensation in mice. Neuron 2007; 54:371–378.
32. Eccles R. Role of cold receptors and menthol in thirst, the drive to breathe and arousal. Appetite 2000; 34:29–35.
33. Andersson DA, Nash M, Bevan S. Modulation of the cold-activated channel TRPM8 by lysophospholipids and polyunsaturated fatty acids. J Neurosci 2007; 27:3347–3355.
34. Braat JP, Mulder PG, Fokkens WJ, et al. Intranasal cold dry air is superior to histamine challenge in determining the presence and degree of nasal hyperreactivity in nonallergic noninfectious perennial rhinitis Am J Respir Crit Care Med 1998; 157: 1748–1755.
35. Vaughan RP, Szewczyk MT Jr, Lanosa MJ, et al. Adenosine sensory transduction pathways contribute to activation of the sensory irritation response to inspirited irritant vapors. Toxicol Sci 2006; 93:411–421.
36. Gao Z, Li JD, Sinoway LI, et al. Effect of muscle interstitial pH on P2X and TRPV1 receptor-mediated pressor response. J Appl Physiol 2007; 102:2288–2293.
37. Lakshmi S, Joshi PG. Co-activation of P2Y2 receptor and TRPV channel by ATP: implications for ATP induced pain. Cell Mol Neurobiol 2005; 25:819–832.
38. Liu L, Zhu W, Zhang ZS, et al. Nicotine inhibits voltage-dependent sodium channels and sensitizes vanilloid receptors. J Neurophysiol 2004; 91:1482–1491.
39. Tominaga M, Caterina MJ, Malmberg AB, et al. The cloned capsaicin receptor integrates multiple pain-producing stimuli. Neuron 1998; 21:531–543.
40. Caterina MJ, Leffler A, Malmberg AB, et al. Impaired nociception and pain sensation in mice lacking the capsaicin receptor. Science 2000; 288:306–313.
41. Bautista DM, Jordt SE, Nikai T, et al. TRPA1 mediates the inflammatory actions of environmental irritants and proalgesic agents. Cell 2006; 124:1269–1282.
42. Spedding M, Gressens P. Neurotrophins and cytokines in neuronal plasticity. Novartis Found Symp 2008; 289:222–233.
43. Holzer P. Capsaicin: cellular targets, mechanisms of action, and selectivity for thin sensory neurons. Pharmacol Rev 1991; 43:143–201.
44. Huang S, Bisogno T, Trevisani M, et al. An endogenous capsaicin-like substance with high potency at recombinant and native vanilloid VR1 receptors. Proc Natl Acad Sci U S A 2002; 99:8400–8405.

45. Light AR, Hughen RW, Zhang J, et al. Dorsal root ganglion neurons innervating skeletal muscle respond to physiological combinations of protons, ATP, and lactate mediated by ASIC, P2X, and TRPV1. J Neurophysiol 2008; 100:1184–1201.
46. Rinder J, Stjarne P, Lundberg JM. Capsaicin de-sensitization of the human nasal mucosa reduces pain and vascular effects of lactic acid and hypertonic saline. Rhinology 1994; 32:173–178.
47. Stjarne P, Lundblad L, Lundberg JM, et al. Capsaicin and nicotine-sensitive afferent neurones and nasal secretion in healthy human volunteers and in patients with vasomotor rhinitis. Br J Pharmacol 1989; 96:693–701.
48. Bascom R, Kagey-Sobotka A, Proud D. Effect of intranasal capsaicin on symptoms and mediator release. J Pharmacol Exp Ther 1991; 259:1323–1327.
49. Rajakulasingam K, Polosa R, Lau LC, et al. Nasal effects of bradykinin and capsaicin: influence on plasma protein leakage and role of sensory neurons. J Appl Physiol 1992; 72:1418–1424.
50. Philip G, Baroody FM, Proud D, et al. The human nasal response to capsaicin. J Allergy Clin Immunol 1994; 94:1035–1045.
51. Joos GF, Germonpre PR, Pauwels RA. Neurogenic inflammation in human airways: is it important? Thorax 1995; 50:217–219.
52. Sanico AM, Atsuta S, Proud D, et al. Dose-dependent effects of capsaicin nasal challenge: in vivo evidence of human airway neurogenic inflammation. J Allergy Clin Immunol 1997; 100:632–641.
53. Greiff L, Svensson C, Andersson M, et al. Effects of topical capsaicin in seasonal allergic rhinitis. Thorax 1995; 50:225–229.
54. Roche N, Lurie A, Authier S, et al. Nasal response to capsaicin in patients with allergic rhinitis and in healthy volunteers: effect of colchicine. Am J Respir Crit Care Med 1995; 151:1151–1158.
55. Stjarne P, Rinder J, Heden-Blomquist E, et al. Capsaicin desensitization of the nasal mucosa reduces symptoms upon allergen challenge in patients with allergic rhinitis. Acta Otolaryngol 1998; 118:235–239.
56. Sanico AM, Philip G, Proud D, et al. Comparison of nasal mucosal responsiveness to neuronal stimulation in non-allergic and allergic rhinitis: effects of capsaicin nasal challenge. Clin Exp Allergy 1998; 28:92–100.
57. Kowalski ML, Dietrich-Milobedzki A, Majkowska-Wojciechowska B, et al. Nasal reactivity to capsaicin in patients with seasonal allergic rhinitis during and after the pollen season. Allergy 1999; 54:804–810.
58. Zhang F, Zhu X, Han D, et al. Lin Chuang Er Bi Yan Hou Ke Za Zhi [Clinical study of capsaicin in the treatment of allergic rhinitis] 1999; 13:499–500.
59. Gerth Van Wijk R, Terreehorst IT, Mulder PG, et al. Intranasal capsaicin is lacking therapeutic effect in perennial allergic rhinitis to house dust mite. A placebo-controlled study. Clin Exp Allergy 2000; 30:1792–1798.
60. Baudoin T, Kalogjera L, Hat J. Capsaicin significantly reduces sinonasal polyps. Acta Otolaryngol 2000; 120:307–311.
61. Filiaci F, Zambetti G, Luce M, et al. Local treatment of nasal polyposis with capsaicin: preliminary findings. Allergol Immunopathol (Madr) 1996; 24:13–18.
62. Basak S, Turkutanit S, Sarierler M, et al. Effects of capsaicin pre-treatment in experimentally-induced secretory otitis media. J Laryngol Otol 1999; 113:114–117.
63. Ko F, Diaz M, Smith P, et al. Toxic effects of capsaicin on keratinocytes and fibroblasts. J Burn Care Rehabil 1998; 19:409–413.
64. Denda M, Sokabe T, Fukumi-Tominaga T, et al. Effects of skin surface temperature on epidermal permeability barrier homeostasis. J Invest Dermatol 2007; 127(3):654–659.
65. Ständer S, Moormann C, Schumacher M, et al. Expression of vanilloid receptor subtype 1 in cutaneous sensory nerve fibers, mast cells, and epithelial cells of appendage structures. Exp Dermatol 2004; 13(3):129–139.
66. Lee YM, Li WH, Kim YK, et al. Heat-induced MMP-1 expression is mediated by TRPV1 through PKCalpha signaling in HaCaT cells. Exp Dermatol 2008; 17(10): 864–870.

67. Gopinath P, Wan E, Holdcroft A, et al. Increased capsaicin receptor TRPV1 in skin nerve fibers and related vanilloid receptors TRPV3 and TRPV4 in keratinocytes in human breast pain. BMC Womens Health 2005; 5(1):2.
68. Akbar A, Yiangou Y, Facer P, et al. Increased capsaicin receptor TRPV1-expressing sensory fibres in irritable bowel syndrome and their correlation with abdominal pain. Gut 2008; 57(7):923–929.
69. Yang S, Liu Y, Lin AA, et al. Adaptive evolution of MRGX2, a human sensory neuron specific gene involved in nociception. Gene 2005; 352:30–35.
70. Robas N, Mead E, Fidock M. MrgX2 is a high potency cortistatin receptor expressed in dorsal root ganglion. J Biol Chem 2003; 278:44400–44404.
71. Eilers H, Trilk SL, Lee SY, et al. Isolation of an mRNA binding protein homologue that is expressed in nociceptors. Eur J Neurosci 2004; 20:2283–2293.
72. van der Stelt M, Di Marzo V. Endovanilloids. Putative endogenous ligands of transient receptor potential vanilloid 1 channels. Eur J Biochem 2004; 271:1827–1834.
73. Harteneck C. Function and pharmacology of TRPM cation channels. Naunyn-Schmiedberg's Arch Pharmacol 2005; 371:307–314.
74. Bessac BF, Sivula M, Von Hehn, et al. TRPA1 is a major oxidant sensor in murine airway sensory neurons. J Clin Invest 2008; 118:1899–1910.
75. Belmone C, Viana F. Molecular and cellular limits to somatosensory specificity. Mol Pain 2008; 4:14.
76. Baraniuk JN, Ali M, Yuta A, et al. Hypertonic saline nasal provocation stimulates nociceptive nerves, substance P release, and glandular mucous exocytosis in normal humans. Am J Respir Crit Care Med 1999; 160:655–662.
77. Baraniuk JN, Petrie KN, Le U, et al. Neuropathology in rhinosinusitis. Am J Respir Crit Care Med 2005; 171:5–11.
78. Knipping EM, Lakin MJ, Foster KL, et al. Experiments and simulations of ion-enhanced interfacial chemistry on aqueous NaCl aerosols. Science 2000; 288(5464):301–306.
79. Sanico AM, Philip G, Lai GK, et al. Hyperosmolar saline induces reflex nasal secretions, evincing neural hyperresponsiveness in allergic rhinitis. J Appl Physiol 1999; 86:1202–1210.
80. McDonald DM. Neurogenic inflammation in the rat trachea. I. Changes in venules, leukocytes and epithelial cells. J Neurocytol 1998; 17:605–628.
81. Rinder J. Sensory neuropeptides and nitric oxide in nasal vascular regulation. Acta Physiol Scand Suppl 1996; 632:1–45.
82. Schmelz M, Schmidt R, Weidner C, et al. Chemical response pattern of different classes of C-nociceptors to pruritogens and algogens. J Neurophysiol 2003; 89: 2441–2448.

5

Nasal Enzymology and Its Relevance to Nasal Toxicity and Disease Pathogenesis

John B. Morris

Department of Pharmaceutical Sciences, School of Pharmacy,
University of Connecticut, Storrs, Connecticut, U.S.A.

Dennis J. Shusterman

Division of Occupational and Environmental Medicine, University of California,
San Francisco, California, U.S.A.

INTRODUCTION

Biotransformation of xenobiotics can profoundly influence their toxicity. Enzymes of virtually all toxicant biotransformation pathways have been detected in the nose including phase I hydrolysis (esterase, peptidase, epoxide hydrolase), reduction (carbonyl reduction), and oxidation [alcohol dehydrogenase, aldehyde dehydrogenase, flavin-monooxygenase, and cytochrome P450 monooxygenase (CYP)] pathways. Biotransformation enzymes of phase II pathways (glucuronide conjugation, glutathione conjugation, etc.) are present as well. Substrates for phase II biotransformation, in particular glutathione, are present in nasal tissues in concentrations similar to that in the liver in rodents as well as in man (1). Since phase I and phase II enzymes often catalyze sequential biotransformations, it is the balance of their activities which is critical in influencing the toxic response.

The high expression levels of multiple biotransformation enzymes led Jeffrey et al. (2) to state that the nose is a "veritable second liver" located at a portal of entry. With this perspective, it becomes clear that the toxicological concepts derived for hepatic metabolism and toxicity can also be applied to the nose. There are many detailed reviews of nasal xenobiotic metabolizing enzymes. The reader is particularly referred to those authored by XinXin Ding and/or Alan Dahl (3–6). The aim of the current review is not to repeat this information, but to provide an integrated approach to nasal metabolism and its importance in nasal toxicity and disease pathogenesis. Included in this chapter are CYP (styrene and naphthalene as examples), carboxylesterase (vinyl acetate and ethyl acrylate), and aldehyde dehydrogenase (acetaldehyde). Since sensory nerves and neuropeptides are thought to be important in human disease, this chapter also includes discussions of the metabolic pathways relevant to these processes (neutral endopeptidase) and enzymes important in experimental models for sensory nerve stimulation (carbonic anhydrase).

From a risk assessment perspective, extensive metabolism of toxicants provides many challenges with respect to extrapolation of animal toxicity data. For example, the exposure levels used in toxicity testing are often sufficiently high to result in saturation and/or capacity limitation of metabolic pathways. Thus, simple extrapolation of such data to lower, nonsaturating concentrations can be highly problematic. Considerable species differences often exist in expression of xenobiotic biotransformation enzymes. This is particularly true for CYPs for which different forms are expressed in different species (3,4,7). Several types of epithelium line the nose: respiratory, transitional, and olfactory (chap. 1). Thus, the distribution of biotransformation enzymes among these mucosa as well as the distribution in superficial (epithelial) versus deep (submucosal) sites is critical. For this reason, measurement of metabolic activities in whole mucosal homogenates often provides inadequate information for integrated quantitative assessment. In metabolic studies the cuboidal and respiratory epithelium are rarely separated. In this chapter the term "respiratory" mucosa is used generically to describe the ventral and anterior nonolfactory (nonsquamous mucosa) of the nasal passages and includes both respiratory and cuboidal mucosa. Metabolizing enzyme expression in the mucosae of the nose can differ markedly, resulting in highly specific patterns of injury. Thus, sampling of a single mucosal type, by biopsy for example, may provide tissue for assessment of activities in that mucosa, but typically will not provide information on metabolic potential in other areas of the nose. Thus, critical review of the literature requires careful consideration of the precise location of tissue sampling. By careful selection of example nasal toxicants, this review attempts to provide concrete examples of each of the issues.

SPECIFIC EXAMPLES

CYP

Prediction of hazard to humans represents the primary goal of animal toxicity testing. For metabolically activated compounds, such prediction requires knowledge of metabolic activation rates in animals and man. As a family, CYPs are likely the most important enzymes relative to xenobiotic biotransformation and toxicity. Only limited data exist on CYP expression in human nasal tissue, especially olfactory mucosa, which is difficult to obtain (8). CYP mRNAs are expressed in human respiratory mucosa (7) and CYP-related activities in human nasal respiratory mucosa have been reported as well (9). It appears that CYP levels in respiratory mucosa are relatively low whereas the expression of phase II and/or nonoxidative enzymes (glutathione-S-transferase, epoxide hydrolase, etc.) are similar to the liver (9). This suggests that the human respiratory mucosa may be relatively resistant to most metabolically activated compounds. There are virtually no data available on CYP activities in olfactory mucosa (8), although CYP including CYP2A6 and CYP2A13 can be detected by immunochemical means (10,11). Two compounds for which data are available for respiratory and olfactory mucosa in the human and/or primate are styrene and naphthalene. This is the exception rather than the rule. These compounds provide good case studies of the role of CYP in nasal toxicity and extrapolation of animal data to man for metabolically activated compounds.

Styrene

Styrene is a pungent liquid that is used extensively in the polymer industry. Metabolic activation by CYP450 has long been thought to be critical in the systemic toxicity of this compound (12). Recent inhalation toxicity studies have revealed that styrene is toxic to the nose as well. Specifically, acute, subchronic, and chronic inhalation exposure to styrene results in olfactory mucosal degeneration in both the rat and mouse, without apparent injury to the respiratory mucosa (13–15). The mouse is considerably more sensitive than the rat as reflected by the presence of nasal lesions in mice exposed for 13 weeks to 50 ppm of styrene compared to the absence of lesions in rats exposed to 200 ppm (13). In both species, injury was maximal in the dorsal medial meatus and the dorsal ethmoturbinates, along the major airstream pathway passing over the olfactory regions of the nose (chap. 6). This pattern of injury is common and suggests that delivery via the air phase (as opposed to blood) is critical relative to the ensuing pattern of injury. Morris (16) has shown that styrene is scrubbed from the airstream with moderate efficiency in the rat and mouse and, importantly, that nasal extraction of inspired styrene is strongly diminished by pretreatment with the CYP inhibitor metyrapone. These data provide strong evidence that inspired styrene is indeed extensively metabolized within the nose via local CYP. Styrene uptake efficiency demonstrated nonlinear kinetics with decreased efficiency being observed at high concentration. This was abolished by inhibitor pretreatment, indicating that saturation and/or capacity limitation of nasal metabolism was responsible.

Much evidence suggests that metabolic activation of styrene is critical to its nasal toxicity. Inhalation studies with radiolabeled styrene indicate that label is retained in the nasal olfactory (but not respiratory) mucosa of rats and mice for at least 40 hours after single 6-hour inhalation exposure, leading the authors to conclude this was reflective of tissue binding of styrene metabolites (17). Interestingly, the levels of radioactivity at this time were higher in the nose than in any other tissue. The strongest evidence for the role of metabolic activation of styrene is provided by Green et al. (18), who showed that pretreatment with the CYP suicide inhibitor 5-phenyl-1-pentyne prevented the development of nasal lesions in styrene-exposed mice. Metabolic transformation of styrene is complex and includes CYP oxidation, epoxide hydrolysis, and glutathione conjugation pathways.

Styrene is metabolized by CYP via side-chain oxidation to the 7,8 oxide (either the R or S enantiomer) and also by ring oxidation to styrene-3,4-oxide, which is converted to 4-vinylphenol (19). The 4-vinylphenol is a minor metabolite but is a more potent cytotoxicant and may, therefore, be important toxicologically. The styrene-7,8-oxide is the major active metabolite and has received greater attention. The primary CYPs involved in styrene oxidation are CYP2E1 and CYP2F2 (in mice) or CYP2F4 (in rats), with CYP2E1 preferentially producing the S enantiomer and CYP2F the R enantiomer. Styrene oxides are further metabolized by epoxide hydrolases and by glutathione-S-transferases; thus the balance of these pathways would be expected to be a critical determinant of toxicity (18).

In vitro studies with nasal microsomal fractions from the rat and mouse revealed that CYP activity, as evidenced by formation of styrene-7,8-oxide, was present in both respiratory and olfactory tissue of both species with the activity being roughly twice as high in the olfactory than in the respiratory tissue (18). The R:S ratio was ~3, suggesting that CYP2F predominates over CYP2E1. (When expressed per milligram microsomal protein, activity in the olfactory mucosal microsomes exceeded that in hepatic microsomes.) The specific activity for styrene-7,8-oxide formation in the rat and mouse were roughly similar suggesting that the formation of the active metabolite may not account for the species difference in sensitivity to styrene. There were, however, significant species differences in activity of detoxifying enzymes. In olfactory epithelium of the rat, epoxide hydrolase activity was approximately 15–20-fold higher than that in the mouse and glutathione-*S*-transferase activity was approximately five-fold higher than that in the mouse. Thus, the lower potency of styrene in the rat is likely reflective, not of major differences in metabolic activation, but of increased ability to detoxify the active metabolite via epoxide hydrolase and/or glutathione transferase.

The study of Green et al. (18) is important because it also includes metabolic studies with human nasal tissues. Collection of nasal tissues is difficult, and preservation of enzyme activities in tissues obtained postmortem is highly prob-lematic. The strength of this study is that tissues (of normal morphological appearance) were obtained during surgery and immediately frozen in liquid nitrogen. The authors noted that both respiratory and olfactory tissues were present in each specimen, but in variable amounts. This amount of detail is critical in interpreting metabolic data. Given the known differences in enzyme expression levels, use of samples collected from unidentified sites within the nose leads to data that are very difficult to interpret. Styrene-7,8-oxide formation was not detected in these human tissues; the limits of detection were approxi-mately 100-fold lower than the activities observed in rodent samples. Epoxide hydrolase and glutathione-*S*-transferase activity (using styrene-7,8-oxide as a substrate) were present in the human tissues at activity levels roughly similar to those in the mouse. That the activity of activation pathways is quite low in the human compared to rodent, while the detoxification capacity appears to be simi-lar, suggests that the styrene-induced olfactory degeneration in rodents may not be predictive of the human. In this regard, it should be noted that epidemiologi-cal studies have failed to detect consistent styrene-related olfactory dysfunction in workers occupationally exposed to styrene (20). Humans do develop a persis-tent olfactory dysfunction following H_2S exposures, indicating that this response can occur in man, but it does not appear to occur as a result of styrene exposure (chap. 21).

Styrene provides an interesting example relative to many aspects of nasal metabolism. First, the nose is a highly metabolically capable organ possessing both activating (CYP450) and detoxifying (epoxide hydrolase, glutathione-*S*-transferase) pathways. Second, the balance of these pathways can be critical in determining the toxic outcome. Third, it is possible to obtain human nasal tissues for metabolic studies, but great care must be taken in the collection and storage of samples. Fourth, significant species differences exist in metabolic potential, with the human, at least for this compound, possessing low CYP activity. Fifth, although not the focus of this chapter, the experience with styrene suggests that rodent toxicity studies may not be directly predictive of risks to the human with respect to olfactory function. In the case of styrene, this may well relate to species-specific differences in biotransformation.

Naphthalene
A variety of compounds are cytotoxic to the nose after systemic administration (2). Naphthalene is one such compound. Exposure to naphthalene can occur either at home (via moth ball usage) or occupationally, as it is present in jet fuels. Naphthalene produces nasal damage whether administered via the inhalation or

parenteral routes (21). It is also a rodent respiratory tract carcinogen in both the rat and mouse (22). Our knowledge of naphthalene toxicity in the lower airways (23), coupled with recent detailed studies on the nose, indicate that metabolic activation in situ in the nose is a critical determinant of its toxicity. Specifically, it has been shown that olfactory cytotoxicity of naphthalene is blocked by inhibition of CYP activation pathways (24). The necessary first step in naphthalene metabolism is the formation of an unstable 1,2-epoxide. This is followed by further epoxidation and/or glutathione conjugation (23). The initial step is likely catalyzed by CYP2F family. In the mouse, CYP2F2 is a high-affinity pathway ($K_m = 3$ μm). Inhaled naphthalene is extracted from the air with moderate efficiency, with extraction being dramatically reduced in animals pretreated with 5-phenyl-1-pentyne, a suicide inhibitor of CYP (25). As for styrene, these data provide strong evidence that naphthalene is extensively metabolized within nasal mucosa via CYP pathways. In the concentration range of 1 to 30 ppm, uptake kinetics demonstrated saturation, a phenomenon abolished by inhibitor pretreatment, indicating it was due to saturation and/or capacity limitation of the major CYP metabolic pathway.

Rates of naphthalene metabolism to diol and glutathione adducts (to assess CYP activation potential) have been measured in vitro in nasal mucosa of several species. In the mouse, rat, and hamster, activity in the olfactory mucosa exceeded that in respiratory mucosa by two- to ten-fold (26). Activity was highest in the mouse olfactory mucosa and was approximately 2- and 20-fold lower in olfactory mucosa of the rat and hamster, respectively. Immunoblot techniques for CYP2F suggest that expression in the monkey olfactory epithelium is 10- and 20-fold less than the rat and mouse, respectively (27). Naphthalene metabolism (via measurement of diol and glutathione adducts) is not detectable in olfactory mucosa of the monkey (Buckpitt, personal communication, 2008).

After intraperitoneal (ip) administration, naphthalene causes diffuse injury throughout the nasal olfactory mucosa of the rat; in contrast, olfactory injury following inhalation exposure is confined to the dorsal medial meatus (21). To determine if this pattern may reflect differences in local activation within the nasal mucosa, Lee et al. (21) measured naphthalene metabolism rates in olfactory mucosa dissected from the dorsal medial (septal) meatus and the ethmoturbinates from more distal regions. They determined that the CYP expression (267 and 278 pmol P450/mg protein, respectively) and specific activity of naphthalene activation (16.8 and 18.6 nmol/min-mg microsomal protein, respectively) were identical for both regions of the olfactory mucosa. The uniform distribution of CYP expression and activity is consistent with the uniform olfactory injury induced by parenterally administered naphthalene. The predominant dorsal medial injury following inhalation exposure correlated to patterns of airflow, providing evidence that airflow delivery is an important factor for metabolically activated vapors as well as for direct acting vapors (chap. 6).

Summary
Both the examples discussed above are aromatic volatile organic compounds that are metabolized by CYP2F in the rodent. These examples highlight the importance of regional activation and detoxification, the potential for nasal toxicity following systemic administration of metabolically activated compounds, and the importance of airflow delivery of vapor even for those compounds that are

metabolically activated. Based on nasal uptake efficiency data, both naphthalene and styrene exhibit metabolic saturation and/or capacity limitation behavior at the exposure concentrations associated with nasal injury in inhalation toxicity studies. Due to nonlinear metabolic relationships, extrapolation of such data to lower, nonsaturating exposure concentrations is quite difficult. In addition, both examples highlight the potentially large species differences in expression and/or activity of CYP. A variety of CYP forms are expressed in the rodent and/or human nose (7,8), including members of the CYP1A, 2A, 2E, and 2G families. While the general patterns is that of higher expression in olfactory than respiratory mucosa, this is not always the case. For example, CYP4B is expressed in higher levels in respiratory than in olfactory mucosa (28). It is critical to recognize that the regional toxicity of activated compounds within the nose will be reflective of the regional distribution of the precise CYP forms that metabolize the compound, the regional level of phase II enzymes as well as the regional ventilation patterns (for inspired materials). Significant metabolic activation of styrene or naphthalene was not observed in human or monkey olfactory mucosa. Whether this represents a generalized trend, or is specific to CYP2F substrates, is not known. Finally, it should be noted that in many tissues, CYPs are inducible; this is also true for the nose (8). This represents an important consideration in interpreting chronic toxicity data, as CYP expression levels may well change as exposures progress.

Carboxylesterase

Vinyl Acetate and Ethyl Acrylate

Nasal olfactory injury has been shown to result from inhalation exposure to at least eight volatile aliphatic esters (29). For all esters, the injury is confined to the dorsal medial meatus, suggesting that airflow delivery is an important aspect of the toxic response. This pattern of injury does not preclude the potential for a critical role of metabolic activation. Indeed, hydrolysis of esters to their respective acid is thought to be essential for the toxicity of esters, perhaps the best studied esters in this regard are ethyl acrylate and vinyl acetate (30,31). Evidence for the critical role of acid metabolites includes the identical pattern of injury induced by the acid metabolites of esters (acrylic, acetic acid), the fact that exposure of nasal epithelial cells in vitro to esters results in intracellular acidification (32), and the effects of carboxylesterase inhibitors relative to protection of ester-induced injury (33). Inspired esters are extensively metabolized within nasal tissues as revealed by the large diminution in nasal extraction by carboxylesterase inhibitors of several esters including ethyl acrylate (34,35), ethyl acetate (36), and vinyl acetate (37). Moreover, acetaldehyde, the carboxylesterase metabolite of vinyl acetate, is observed in air exiting the nose of humans exposed to vinyl acetate, indicating that physiologically relevant levels of carboxylesterase are present in the human nose (38).

Carboxylesterases are hydrolytic enzymes that split esters into their respective alcohols and acids. They demonstrate a broad substrate specificity. Carboxylesterase represents a high capacity, low-affinity enzymatic pathway in nasal tissues. Activity levels in excess of 5 μmol/min (on a whole nose basis) have been reported for the rat nose (36,39). At a concentration of 100 ppm and a minute ventilation of 250 mL/min, a rat inhales 1 μmol/min; thus, it can be appreciated that the potential exists for significant first pass clearance of esters. Indeed, Morris (36) estimated that as much as 60% of deposited ethyl acetate is hydrolyzed in situ in the nose of the rat or hamster. Unlike styrene or naphthalene, nasal uptake of most esters does not exhibit nonlinear (saturation) kinetics (35,36), likely because carboxylesterase is a high capacity–low affinity metabolic pathway in the nose.

Nasal carboxylesterase activity has been observed in nasal tissues of all laboratory species examined to date. This includes the mouse, rat, hamster, rabbits, dogs, and monkeys (30,34–36,40). Activities in the nasal mucosa approximated those in the liver. Activities are roughly similar in the respiratory versus olfactory mucosa, although in some species (e.g., the mouse) activity tends to be higher in the olfactory mucosa. Using an in vitro gas uptake technique, Bogdanffy et al. (39) obtained estimates of carboxylesterase activity using vinyl acetate as a substrate in human nasal mucosa. Nasal samples were obtained postmortem (within two hours of death) and were kept on ice until assay of activity. Tissues sampling included sites of the nasal cavity lined by respiratory and olfactory epithelium, allowing for assessment of activity in both sites. Activity was present in human samples and approximated that observed in the rat. Estimated activities in human respiratory and olfactory epithelium were roughly similar to each other. This is unlike CYP, for which the limited database suggests that the human activity is considerably lower than that of the rodent.

It is important to recognize that the nasal mucosa is complex, consisting of multiple cell types distributed throughout several mucosal types. Therefore, data obtained on whole tissue homogenates is insufficient to provide a comprehensive understanding of biotransformation as it relates to nasal toxicity. The distribution of carboxylesterase throughout nasal mucosa has been assessed by immunologic techniques in several species. Bogdanffy et al. (40) were the first to report on cellular distribution in the mouse and rat, there results were confirmed and extended to the dog and human by Lewis et al. (41). In respiratory mucosa, carboxylesterase is expressed in the epithelium, but not in the subepithelium. In the olfactory mucosa, it is highly expressed in the subepithelial Bowman's Gland and moderately expressed in the apical portion of the sustentacular cells. The same pattern is seen in all species. Because the epithelium is closest to the airspace, activity in that site is more likely to be of quantitative importance for inspired vapors than activity deep within the mucosa.

An integrated understanding of the toxicological significance of biotransformation of volatile toxicants within the nose requires integration of a variety of data. Vinyl acetate provides a good example. One important factor is the air flow delivery patterns (chap. 6). The nose is not uniformly ventilated, with only ~15% of the air passing over the olfactory mucosa, with the remainder passing over the extensive portions of the respiratory mucosa. Overall carboxylesterase activities are roughly similar between respiratory and olfactory mucosa (as measured in vitro). Thus, 15% of the airstream passes over the olfactory mucosa (with $^1/_2$ the activity), while 85% passes over extensive portions of the respiratory mucosa (with $^1/_2$ the activity), suggesting that ester vapor depositing in the olfactory mucosa is much more extensively metabolized than that which deposits in the respiratory mucosa. Expression levels versus depth of tissue represent another critical factor relative to biotransformation of inspired vapors. A diffusion limitation exists relative to penetration of vapor into deep tissues of the nose (chap. 6). Thus, expression of carboxylesterase in superficial tissues (e.g., epithelium) would be anticipated to be of greater impact than expression in deep tissues (e.g., Bowman's Gland). Physiologically based nasal dosimetry models incorporate regional airflow patterns and, regional carboxylesterase expression patterns included depth within the tissue. Such models have successfully described uptake and metabolism of ethyl acetate (42), ethyl acrylate (30), and vinyl acetate (31) in the rodent nose. Importantly, this approach has recently been extended to describe nasal uptake and metabolism of vinyl acetate in the human nose (38), indicating that the model structure applies to the human nose as well. This result is important in that it provides strong evidence that insights gained from animal experimentation relative to regional airflow, regional expression, and tissue depth also apply to the humans.

In summary, carboxylesterase provides an example of a high capacity biotransformation enzyme that is expressed in nasal tissues of multiple species including man. Experience with this biotransformation enzyme highlights how immunohistochemical techniques and biochemical activity measurements are both needed to allow integrated understanding of biotransformation in inspired vapors. Finally, although not the subject of this chapter, PBPK models have successfully been used to provide insights on ester vapor dosimetry in both laboratory animals and man.

Acetaldehyde Dehydrogenase

Acetaldehyde

Nasal injury results from inhalation exposure to a variety of aldehydes including formaldehyde and acetaldehyde (chap. 18). Formaldehyde and acetaldehyde, but not isobutyl aldehyde, are nasal carcinogens in the rodent. Repeated inhalation exposure to acetaldehyde produces injury in both the respiratory and olfactory epithelium. Exposure concentrations of 1000 ppm are required for respiratory mucosal injury whereas olfactory injury results from repeated exposure to concentrations of 400 ppm (43) indicating that the olfactory epithelium is the more sensitive site.

Because of its importance in ethanol toxicity, acetaldehyde metabolism has been extensively studied. Aldehyde dehydrogenase (AlDH) catalyzes the conversion of acetaldehyde to acetic acid, a biotransformation that results in the formation of two protons. Multiple forms of AlDH exist, some cytosolic, some mitochondrial (44). Alcohol (e.g., acetaldehyde) intolerance is associated with a genetic polymorphism resulting in diminished AlDH (specifically AlDH2) capacity. Acetaldehyde is directly reactive and may adduct macromolecules (45). Therefore, it might be anticipated that AlDH represents a detoxification pathway. Indeed, knockout mice lacking AlDH2 (a mitochondrial isoform) are more sensitive to acetaldehyde-induced nasal respiratory mucosal injury (46). Interestingly, sensitivity of the olfactory mucosa was not significantly modulated in these knockout mouse. It is known that the olfactory epithelium is quite sensitive to acids (see above). It has been suggested, based on dosimetric considerations, that acid formation via AlDH is critical to the olfactory toxicity of acetaldehyde (47). In addition, pretreatment with an aldehyde dehydrogenase inhibitor diminished nasal sensory nerve–mediated responses to acetaldehyde (48). Thus, it appears that both the parent (acetaldehyde) and metabolite (acetic acid) may contribute to the nasal response with their roles differing in respiratory versus olfactory mucosa.

Inhalation dosimetry studies (49–51) indicate that inspired acetaldehyde deposits and is extensively metabolized in nasal tissues of a variety of laboratory animal species. Uptake efficiency demonstrated nonlinear kinetics with diminished uptake efficiencies being observed at high (100, 1000 ppm) compared to low (1, 10 ppm) exposure concentrations. Uptake efficiency was significantly lower in animals pretreated with an AlDH inhibitor, and the inhibitor abolished the nonlinear uptake behavior, providing strong evidence that enzyme saturation and/or capacity limitation was the cause (49).

Perhaps the first to report on acetaldehyde dehydrogenase activities in the nasal mucosa were Casanova-Schmitz et al. (52). Both low affinity–high capacity ($K_m \sim 20$ mM) and high affinity–low capacity ($K_m < 0.2$ mM) forms were detected. Total activities of $\sim$0.2 µmol/min/mg protein were observed. Activity was approximately fourfold higher in respiratory than olfactory tissues. This pattern differs from that observed for most CYPs or carboxylesterase (see above). Whole nasal tissue metabolism rates have been measured in the mouse, hamster, and rat and averaged 0.04, 0.16, 0.56 µmol/min/whole tissue, respectively (52). The total activity was roughly equal across species when normalized to minute ventilation rates. Both high affinity–low capacity and low affinity–high capacity pathways were detected in these species. The guinea pig lacked the low affinity–high capacity form and demonstrated much lower metabolic capacity. In an integrated examination of the dosimetry and metabolism of acetaldehyde in the rat, Morris (47) suggested that delivered dosage rates of acetaldehyde will exceed to total metabolic capacity in the rat at exposure concentrations of 300–600 ppm. This correlates with the nonlinear kinetics of nasal uptake of acetaldehyde (see above) and suggests that capacity limitation occurs at these high exposure concentrations. This is an important observation as it indicates that the metabolic disposition of acetaldehyde at the exposure concentrations necessary to produce nasal tumors ($\geq$750 ppm) differs significantly from that at lower, more environmentally relevant levels (<1 ppm) (50,52).

The regional and cellular distribution of AlDH has been characterized by Bogdanffy et al. (53) using immunohistochemical methods. It is widely expressed in respiratory epithelial cells. In the olfactory mucosa, it is expressed only in basal cells and to a limited degree in Bowman's glands. Thus, its pattern of expression in olfactory tissue differs from that of carboxylesterase. By including information on expression levels throughout nasal respiratory and olfactory mucosa with biochemical characterization of the high affinity–low capacity and low affinity–high capacity pathway, it was possible to describe the inhalation dosimetry patterns for acetaldehyde that were observed in the rat (54). As for carboxylesterase, this result highlights that our conceptual understanding of the role of metabolism relative to nasal dosimetry of inspired vapors is adequate, and, importantly, that detailed information on distribution of metabolic capacity is essential for such an understanding.

Information is also available on human nasal expression of AlDH. Because acetaldehyde is a metabolite of vinyl acetate, aldehyde dehydrogenase activities were assessed via in vitro gas uptake methodologies by Bogdanffy et al. (39). Activity was predominantly present in the respiratory epithelium, with 97% of the total nasal cavity activity estimated to be present in that site. (Olfactory mucosa accounts for only ~5% of the nasal mucosa in the human compared to ~50% in the rodent.) Activity, expressed per epithelial cell volume, was similar in the human and rat. These activities were used to predict inhalation dosimetry and metabolism patterns for acetaldehyde in the human nose (54). Direct data are not available to validate the approach in the human (it was validated for the rat), however, the vinyl acetate model (which included AlDH for downstream metabolism of acetaldehyde) has some success in predicting acetaldehyde levels in air exiting the human nose during vinyl acetate exposure (38).

In summary, AlDH is expressed in nasal tissues of multiple species including the human. It is expressed at higher levels in the respiratory than olfactory mucosa. It is widely expressed in the epithelium of the respiratory mucosa but in olfactory epithelium, it is expressed only in basal cells. Thus, its distribution differs from that of most CYPs and carboxylesterase. Dosimetry studies indicate that AlDH extensively metabolizes acetaldehyde that deposits in nasal tissues during inhalation exposure. Capacity limitation occurs at concentrations known to induce toxicity and carcinogenicity (>400 ppm), but does not occur at typical environmental levels (<1 ppm). This highlights the caution needed for extrapolation of high concentration toxicity studies to predict effects at lower environmentally relevant concentrations. Interestingly, AlDH appears to be a detoxification pathway in respiratory mucosa, but may serve as an activation pathway (via formation of acetic acid) in olfactory mucosa. This highlights that respiratory and olfactory mucosa need to be considered separately relative to metabolism and mechanism of toxicity.

ENZYMES RELEVANT TO SENSORY PROCESSES
In addition to carrying out xenobiotic metabolism relevant to toxicokinetics, nasal mucosal enzymes are also involved in sensory signaling (i.e., involving endogenous neuropeptides), in the activation of at least one sensory irritant (carbon dioxide), and in modulating the effects of oxidant air pollutants on the local immune response (e.g., diesel exhaust particles). Each of these endpoints is considered briefly in the following paragraphs.

Proteases
The respiratory tract is heavily invested with peptide-containing nociceptive afferent nerves, chiefly C-fibers. In response to noxious stimuli, these fibers can release substance P (SP) and calcitonin gene-related peptide (CGRP), both secretagogues and vasodilators. Release of neuropeptides from afferent nerve fibers is referred to as the axon reflex (chap. 4). Neuropeptides can also be released from autonomic efferent fibers: vasoactive intestinal peptide (VIP), a vasodilator from the parasympathetic nervous system, and neuropeptide Y (NPY), a vasoconstrictor from the sympathetic (55). Since the response repertoire of the nose is limited (irritation, obstruction, secretion, itching, and sneezing), factors influencing the peptidergic system can have significant physiologic consequences. In addition to neuropeptides, bradykinin—another endogenous peptide—can be liberated as part of the inflammatory cascade and/or as a direct result of tissue damage. Xenobiotics can, in fact, influence these peptidergic systems at several levels, including alterations in the production, release, and degradation of signaling peptides.

Several investigators have described neuroplasticity in afferent nerves, including induction of neuropeptide synthesis in both C-fibers (increased production) and A-δ fibers (de novo production, or "phenotypic switching"). Wilfong and Dey (56), for example, documented increases in nerve growth factor (NGF), followed by increases in SP content, in nasal lavage fluid from rats exposed to toluene diisocyanate (TDI). Increases in SP, but not NGF, were blocked by pre-administration of a tyrosine kinase (i.e., NGF receptor) inhibitor, indicating that the modulatory action of TDI on SP expression was likely NGF-mediated. Increases in peripheral nerve—and trigeminal ganglion—SP content were also documented using immunohistochemical staining. Both augmented SP production in C-fibers and de novo production of neuropeptides in A-δ fibers have been described in allergic inflammation (57). These phenomena, collectively referred to as "neuromodulation," are reviewed in detail in chapter 26.

Mucosal irritants can also trigger the release of neuropeptides, although the evidence points to a high threshold for such release. As reviewed in chapter 4, selected noxious stimuli, such as hypertonic saline or allergen (in sensitized individuals), can release SP from the human nasal mucosa (58,59). However, other noxious stimuli, including dry mannitol powder and chlorine gas, do not augment SP content in human nasal lavage fluid (60,61). Thus, the role of the axon reflex/neuropeptide release in explaining irritant-induced nasal reflexes is unclear at this time.

By contrast, mucosal peptidases are clearly recognized to degrade endogenous peptides such as bradykinin and neuropeptides (and therefore antagonize their actions). Identified peptidases in human nasal respiratory epithelium include neutral endopeptidase (NEP), angiotensin-converting enzyme (ACE), carboxypeptidase, and aminopeptidase (62–66). Should a toxic insult interfere with protease activity, signaling peptides may persist and/or accumulate abnormally, leading to pathological states in the affected tissues. In an extensive series of experiments conducted by Nadel et al., it was established that NEP is the dominant mucosal enzyme involved in SP and bradykinin degradation in the rat nasal mucosa (67,68). It was further established that the degree of antigen-induced plasma extravasation in the nasal mucosa of sensitized guinea pigs was augmented by pharmacologic inhibition of NEP (69). Finally, mucosal insults that resulted in epithelial cell damage—including viral infection and toxic insult (from cigarette smoke or TDI)—decreased NEP activity and potentiated neuropeptide action (70,71).

In sum, respiratory toxicants can induce the production of neuropeptides, in some cases trigger their release, and finally inhibit their degradation by interfering with peptidase (particularly NEP) function. Thus, peptidases—and the effects of inhaled toxicants thereon—are important in understanding the function of airway nerves and the reflexes for which they are responsible.

Carbonic Anhydrase

Another enzyme of relevance to sensory processes is carbonic anhydrase (CA). CA catalyzes the hydration in CO_2 in mucous membrane water, thereby generating carbonic acid. Carbonic acid, in turn, dissociates to bicarbonate and H^+. Nasal glandular CA may be important in buffering nasal mucus and in providing a defense against acid air pollutants. In addition, H^+ derived from intermittent high-level exogenous CO_2 can stimulate acid-sensitive ion channels (ASIC) and the vanilloid receptor, TRPV1, resulting in transient tingling, stinging, or burning.

CA has been documented in the olfactory epithelium of a variety of vertebrate species, including amphibians, reptiles, and mammals (72). The human nasal respiratory epithelium expresses transcriptional message for 10 of 11 CA isoenzymes, and based on limited immunocytochemical data, specific CA isoenzymes appear to be differentially expressed by cell type (73). In behavioral assays in mice, animals are able to detect atmospheric CO_2 at less than twice normal background levels (i.e., 660 ppm vs. 380 ppm). CO_2 discrimination tasks by these animals reflect the presence of CA activity in a subset of olfactory receptor neurons and are inhibited by chemical lesioning of the olfactory mucosa, as well as being essentially absent in CA-II knockout mice (74). CO_2 detection by rats is similarly impaired after topical administration of the CA-inhibiting drug, methazolamide (75).

The toxicologic significance of CA activity in the human nasal mucosa derives, at least in part, from the use of CO_2 as a "pure" trigeminal stimulant (i.e., devoid of odor) in psychophysics experiments (76–78). Phasic mucosal pH changes in response to CO_2 stimuli (presumably CA-mediated) have been documented in real-time in the human nose, and predict the intensity of transient CO_2-induced subjective nasal irritation (79). As a model system, the use of CO_2 has enabled researchers to explore inter-individual variability in trigeminal sensitivity, concentration–time stimulus effects, and odor–irritant interactions in psychophysical, peripheral electrophysiologic, and functional CNS imaging studies (80–83).

Glutathione Transferases

Both diesel exhaust particles (DEP) and sidestream tobacco smoke (STS) can act as adjuvant or "priming" agents, predisposing the immune system toward immediate sensitivity reactions (i.e., allergic rhinitis) in the upper airway (84). Phase-II detoxification of DEP or STS via the glutathione transferase (GST) pathway acts to modulate this effect (85–87). As a consequence, both inherited factors (GST variant enzymes) and acceleration of phase II processes (e.g., administration of supplemental sulfhydral compounds) have been explored as modulators of DEP and STS adjuvant/priming effects, as reviewed in detail in chapter 27. As noted above (see styrene), differences in phase II metabolism are often critical in influencing the nasal response to metabolized toxicants.

SUMMARY AND CONCLUSIONS

In summary, the nose is a highly metabolically active organ, expressing significant levels of a variety of biotransformation enzymes. These include CYP, glutathione transferase, carboxylesterase, aldehyde dehydrogenase, neutral endopeptidase, and carbonic anhydrase. These enzymes are important in toxicant biotransformation, both for environmental toxicants (e.g., CYP, etc.) and experimental agents (e.g., carbonic anhydrase), as well as for degradation of important physiological mediators within the nasal mucosa (NEP).

The cellular distribution within the nasal mucosa is highly enzyme specific with some enzymes being expressed in higher proportions in olfactory and others in respiratory mucosa. The regional expression pattern profoundly affects regional injury patterns and is, therefore, an important consideration relative to nasal toxicity. As in any tissue, the nose has a finite metabolic capacity. The exposure concentrations necessary to produce marked nasal injury in rodents are often high enough to result in saturation and/or capacity limitation of in situ biotransformation pathways. Thus, direct (linear) extrapolation of animal toxicity data from high concentrations used in toxicity studies to low, more environmentally relevant, exposure concentrations may be inappropriate, resulting in either over- or under-estimation of risk, depending upon whether activation or detoxication enzymes are operating in a saturated mode. Most enzymes studied to date appear to be expressed in both rodent and human nasal mucosa, but often at widely different levels and/or with differing isoform patterns (e.g., CYP). This represents another complexity that must be considered relative to predicting human risk from animal data. Current state-of-the-art approaches for high-to-low concentration and animal-to-man extrapolation often rely on physiologically based pharmacokinetic modeling. Such models have been successfully developed for nasal toxicants and may provide a useful paradigm for incorporating anatomic and enzymological data into risk assessment analyses. Irrespective of the precise approaches used to evaluate nasal biotransformation and toxicity, it is now clear that the nose is highly complex and metabolically active. Thus, the nasal passages should not be considered as inert deadspace, but rather as a highly complex tissue or, as expressed by Jeffrey et al. (2) as a "veritable second liver."

REFERENCES

1. Potter DW, Finch L, Udinsky JR. Glutathione content and turnover in rat nasal epithelia. Toxicol Appl Pharmacol 1995; 135(2):185–191.
2. Jeffrey AM, Iatropoulos MJ, Williams GM. Nasal cytotoxic and carcinogenic activities of systemically distributed organic chemicals. Toxicol Pathol 2006; 34:827–852.
3. Ding X, Dahl AR, Olfactory mucosa: composition, enzymatic localization and metabolism. In: RL Doty, ed. Handbook of Olfaction and Gustation, 2nd ed. New York, NY: Marcel Dekker, 2003:51–73.
4. Ding X, Kaminsky LS. Human extrahepatic cytochromes P450: function in xenobiotic metabolism and tissue-selective chemical toxicity in respiratory and gastrointestinal tracts. Annu Rev Pharmacol Toxicol 2003; 43:149–173.
5. Thornton-Manning JR, Dahl AR. Metabolic capacity of nasal tissue, interspecies comparisons of xenobiotic-metabolizing enzymes. Mutat Res 1997; 380:43–59.
6. Dahl AR, Hadley WM. Nasal cavity enzymes involved in xenobiotic metabolism: effects on the toxicity of inhalants. CRC Crit Rev Toxicol 1991; 21(5):345–372.
7. Zhang X, Zhang Q-Y, Liu D, et al. Expression of cytochrome P450 and other biotransformation genes in fetal and adult human nasal mucosa. Drug Metab Dispos 2005; 33:1423–1428.
8. Ling, G, Gu J, Genter MB, et al. Regulation of cytochrome P450 gene expression in the olfactory mucosa. Chem Biol Interact 2004; 147:247–258.
9. Gervasi PG, Longo V, Naldi F, et al. Xenobiotic-metabolizing enzymes in human respiratory nasal mucosa. Biochem Pharmacol 1991; 41:177–184.
10. Chen, Y, Liu Y-Q, Su T, et al. Immunoblot analysis and immunohistochemical characterization of CYP2A expression in human olfactory mucosa. Biochem Pharmacol 2003; 66:1245–1251.
11. Getchell ML, Chen Y, Ding X, et al. Immunohistochemical localization of a cytochrome P450 isozyme in human nasal mucosa: age-related trends. Otol Rhinol Laryngol 1993; 102:368–374.

12. Bond J. Review of the toxicology of styrene. CRC Crit Rev Toxicol 1989; 19:227–249.
13. Cruzan G, Cushman JR, Andrews LS, et al. Subchronic inhalation studies of styrene in CD rats and CD-1 mice. Fundam Appl Toxicol 1997; 35:152–165.
14. Cruzan G, Cushman JR, Andrews LS, et al. Chronic toxicity/oncogenicity study of styrene in CD rats by inhalation exposure for 104 weeks. Toxicol Sci 1998; 46: 266–281.
15. Cruzan G, Cushman JR, Andrews LS, et al. Chronic toxicity/oncogenicity study of styrene in CD-1 mice by inhalation exposure for 104 weeks. J Appl Toxicol 2001; 21:185–198.
16. Morris JB. Uptake of styrene in the upper respiratory tract of the CD mouse and Sprague-Dawley rat. Toxicol Sci 2000; 54:222–228.
17. Boogaard PJ, deKloe KP, Sumner SCJ, et al. Disposition of [Ring-U-^{14}C]styrene in rats and mice exposed by recirculating nose-only inhalation. Toxicol Sci 2000; 58:161–172.
18. Green T, Lee R, Toghill A, et al. The toxicity of styrene to the nasal epithelium of mice and rats: studies on the mode of action and relevance to humans. Chem Biol Interact 2001; 137:185–202.
19. Cruzan G, Carlson GP, Johnson KA, et al. Styrene respiratory tract toxicity and mouse lung tumors are mediated by CYP2F-generated metabolites. Regul Toxicol Pharmacol 2002; 35:308–319.
20. Dalton P, Lees PSH, Gould M, et al. Evaluation of long-term occupational exposure to styrene vapor on olfactory function. Chem Senses 2007; 32:739–747.
21. Lee MG, Phimister A, Morin D, et al. In situ naphthalene bioactivation and nasal airflow cause region-specific injury patterns in the nasal mucosa of rats exposed to naphthalene by inhalation. J Pharmacol Exp Ther 2005; 314(1):103–110.
22. North DW, Abdo KM, Benson JM, et al. A review of whole animal bioassays of the carcinogenic potential of naphthalene. Regul Toxicol Pharmacol 2008; 51:S6–S14.
23. Buckpitt AR, Boland B, Isbell M, et al. Naphthalene-induced respiratory tract toxicity: metabolic mechanisms of toxicity. Drug Metab Rev 2002; 34(4):791–820.
24. Genter MB, Marlowe J, Kerzee JK, et al. Napththalene toxicity in mice and aryl hydrocarbon receptor-mediated CYPs. Biochem Biophys Res Commun 2006; 348: 120–123.
25. Morris JB, Buckpitt AR. Upper respiratory tract uptake of naphthalene. Toxicol Sci 2009; 111:383–391.
26. Buckpitt AR, Buonarati M, Avey LB, et al. Relationship of cytochrome P450 activity to Clara cell cytotoxicity. II. Comparison of stereoselectivity of naphthalene epoxidation in lung and nasal mucosa of mouse, hamster, rat and rhesus monkey. J Pharmacol Exp Ther 1992; 261(1):364–372.
27. Baldwin RM, Jewell WT, Fanucchi MB. Comparison of pulmonary/nasal CYP2F expression levels in rodents and rhesus macaque. J Pharmacol Exp Ther 2004; 309: 127–136.
28. Genter, MB, Yost GS, Rettie AE. Localization of CYP4B1 in the rat nasal cavity and analysis of CYPs as secreted proteins. J Biochem Mol Toxicol 2006; 20(3):139–141.
29. Hardisty JF, Garman RH, Harkema JR, et al. Histopathology of nasal olfactory mucosa from selected inhalation toxicity studies conducted with volatile chemicals. Toxicol Pathol 1999; 27(6):618–627.
30. Frederick CB, Lomax, LG, Black KA, et al. Use of a hybrid computational fluid dynamics and physiologically based inhalation model for interspecies dosimetry comparisons of ester vapors. Toxicol Appl Pharmacol 2002; 183:23–40.
31. Plowchalk DR, Andersen ME, Bogdanffy MS. Physiologically based modeling of vinyl acetate uptake, metabolism, and intracellular pH changes in the rat nasal cavity. Toxicol Appl Pharmacol 1997; 142(2):386–400.
32. Lantz RC, Orozco J, Bogdanffy MS. Vinyl acetate decreases intracellular pH in rat nasal epithelial cells. Toxicol Sci 2003; 75(2):423–431.
33. Kuykendall JR, Taylor ML, Bogdanffy MS. Cytotoxicity and DNA-protein crosslink formation in rat nasal tissues exposed to vinyl acetate are carboxylesterase-mediated. Toxicol Appl Pharmacol 1993; 123(2):283–292.

34. Stott WT, McKenna MJ. The comparative absorption and excretion of chemical vapors by the upper, lower, and intact respiratory tract of rats. Fundam Appl Toxicol 1984; 4(4):594–602.
35. Morris JB, Frederick CB. Upper respiratory tract uptake of acrylate ester and acid vapors. Inhal Toxicol 1995; 7:557–574.
36. Morris JB. First-pass metabolism of inspired ethyl acetate in the upper respiratory tracts of the F344 rat and syrian hamster. Toxicol Appl Pharmacol 1990; 102: 331–345.
37. Bogdanffy MS, Manning LA, Sarangapani R. High-affinity nasal extraction of vinyl acetate vapor is carboxylesterase dependent. Inhal Toxicol 1999; 11(10):927–941
38. Hinderliter PM, Thrall KD, Corley RA, et al. Validation of human physiologically based pharmacokinetic model for vinyl acetate against human nasal dosimetry data. Toxicol Sci 2005; 85(1):460–467.
39. Bogdanffy MS, Sarangapani R, Kimbell JS, et al. Analysis of vinyl acetate metabolism in rat and human nasal tissues by an in vitro gas uptake technique. Toxicol Sci. 1998; 46(2):235–246.
40. Bogdanffy MS, Randall HW, Morgan KT. Biochemical quantitation and histochemical localization of carboxylesterase in the nasal passages of the Fischer-344 rat and B6C3F1 mouse. Toxicol Appl Pharmacol 1987; 88:183–194.
41. Lewis JL, Nikula KJ, Novak R, et al. Comparative localization of carboxylesterase in F344 rat, beagle dog, and human nasal tissue. Anat Rec 1994; 239(1):55–64.
42. Morris JB, Hassett DN, Blanchard KT. A physiologically based pharmacokinetic model for nasal uptake and metabolism of non-reactive vapors. Toxicol Appl Pharmacol 1993; 123:120–129.
43. Appelman LM, Woutersen RA, Feron VJ. Inhalation toxicity of acetaldehyde in rats I. Acute and subacute effects. Toxicol 1982; 23:293–297
44. Deitrich RA, Petersen D, Vasiliou V. Removal of acetaldehyde from the body. Novartis Found Symp 2007; 285:23–40; discussion 40–51, 198–199.
45. Lam CW, Casanova M, Heck HD-A. Decreased extractability of DNA from proteins in the rat nasal mucosa after acetaldehyde exposures. Fundam Appl Toxicol 1986; 6:541–550.
46. Oyama T, Isse T, Ogawa M, et al. Susceptibility to inhalation toxicity of acetaldehyde in Aldh2 knockout mice. Front Biosci 2007; 12:1927–1934.
47. Morris JB. Dosimetry, toxicity and carcinogenicity of inspired acetaldehyde in the rat. Mutat Res 1997; 380:113–124.
48. Stanek J. Symanowicz PT, Olsen JE, et al. Sensory nerve mediated nasal vasodilatory response to inspired acetaldehyde and acetic acid vapors. Inhal Toxicol 2001; 13: 807–822.
49. Stanek JJ, Morris JB. The effect of inhibition of aldehyde dehydrogenase on nasal uptake of inspired acetaldehyde. Toxicol Sci 1999; 49:225–231.
50. Morris JB, Blanchard KT. Upper respiratory tract deposition of inspired acetaldehyde. Toxicol Appl Pharmacol 1992; 114:140–146.
51. Morris JB. Uptake of acetaldehyde and aldehyde dehydrogenase levels in the upper respiratory tracts of the mouse, rat, hamster and guinea pig. Fundam Appl Toxicol 1997; 35:91–100.
52. Casanova-Schmitz M, David RM, Heck HD. Oxidation of formaldehyde and acetaldehyde by NAD^+-dependent dehydrogenases in rat nasal mucosal homogenates. Biochem Pharmacol 1984; 33:1137–1142.
53. Bogdanffy MS, Randall HW, Morgan KT. Histochemical localization of aldehyde dehydrogenase in the respiratory tract of the Fischer-344 rat. Toxicol Appl Pharmacol 1986; 82(3):560–567.
54. Teeguarden JG, Bogdanffy MS, Covington TR, et al. A PBPK model for evaluating the impact of aldehyde dehydrogenase polymorphisms on comparative rat and human nasal tissue acetaldehyde dosimetry. Inhal Toxicol 2008; 20(4):375–390.
55. Baraniuk JN, Kaliner M. Neuropeptides and nasal secretion. Am J Physiol 1991; 261 (4 Pt 1):223–235.

56. Wilfong ER, Dey RD. Nerve growth factor and substance P regulation in nasal sensory neurons after toluene diisocyanate exposure. Am J Respir Cell Mol Biol 2004; 30(6):793–800.
57. Carr MJ, Undem BJ. Inflammation-induced plasticity of the afferent innervation of the airways. Environ Health Perspect 2001; 109(suppl 4):567–571.
58. Mosimann BL, White MV, Hohman RJ, et al. Substance P, calcitonin gene-related peptide, and vasoactive intestinal peptide increase in nasal secretions after allergen challenge in atopic patients. J Allergy Clin Immunol 1993; 92(1 Pt 1): 95–104.
59. Baraniuk JN, Ali M, Yuta A, et al. Hypertonic saline nasal provocation stimulates nociceptive nerves, substance P release, and glandular mucous exocytosis in normal humans. Am J Respir Crit Care Med 1999; 160(2):655–662.
60. Koskela H, Di Sciascio MB, Anderson SD, et al. Nasal hyperosmolar challenge with a dry powder of mannitol in patients with allergic rhinitis. Evidence for epithelial cell involvement. Clin Exp Allergy 2000; 30(11):1627–1636.
61. Shusterman D, Balmes J, Murphy MA, et al. Chlorine inhalation produces nasal airflow limitation in allergic rhinitic subjects without evidence of neuropeptide release. Neuropeptides 2004; 38(6):351–358.
62. Baraniuk JN, Ohkubo K, Kwon OJ, et al. Identification of neutral endopeptidase mRNA in human nasal mucosa. J Appl Physiol 1993; 74(1):272–279.
63. Ohkubo K, Baraniuk JN, Hohman RJ, et al. Aminopeptidase activity in human nasal mucosa. J Allergy Clin Immunol 1998; 102(5):741–750.
64. Ohkubo K, Baraniuk JN, Hohman RJ, et al. Human nasal mucosal neutral endopeptidase (NEP): location, quantitation, and secretion. Am J Respir Cell Mol Biol 1993; 9(5):557–567.
65. Ohkubo K, Baraniuk JN, Merida M, et al. Human nasal mucosal carboxypeptidase: activity, location, and release. J Allergy Clin Immunol 1995; 96(6 Pt 1):924–931.
66. Ohkubo K, Lee CH, Baraniuk JN, et al. Angiotensin-converting enzyme in the human nasal mucosa. Am J Respir Cell Mol Biol 1994; 11(2):173–180.
67. Bertrand C, Geppetti P, Baker J, et al. Role of peptidases and NK1 receptors in vascular extravasation induced by bradykinin in rat nasal mucosa. J Appl Physiol 1993; 74(5):2456–2461.
68. Petersson G, Bacci E, McDonald DM, et al. Neurogenic plasma extravasation in the rat nasal mucosa is potentiated by peptidase inhibitors. J Pharmacol Exp Ther 1993; 264(1):509–514.
69. Ricciardolo FL, Nadel JA, Bertrand C, et al. Tachykinins and kinins in antigen-evoked plasma extravasation in guinea-pig nasal mucosa. Eur J Pharmacol 1994; 261(1–2):127–132.
70. Sheppard D, Thompson JE, Scypinski L, et al. Toluene diisocyanate increases airway responsiveness to substance P and decreases airway neutral endopeptidase. J Clin Invest 1988; 81(4):1111–1115.
71. Nadel JA. Neutral endopeptidase modulates neurogenic inflammation. Eur Respir J 1991; 4(6):745–754.
72. Coates EL. Olfactory CO_2 chemoreceptors. Respir Physiol 2001; 129(1–2):219–229.
73. Tarun AS, Bryant B, Zhai W, et al. Gene expression for carbonic anhydrase isoenzymes in human nasal mucosa. Chem Senses 2003; 28(7):621–629.
74. Hu J, Zhong C, Ding C, et al. Detection of near-atmospheric concentrations of CO2 by an olfactory subsystem in the mouse. Science 2007; 317(5840):953–957.
75. Ferris KE, Clark RD, Coates EL. Topical inhibition of nasal carbonic anhydrase affects the CO_2 detection threshold in rats. Chem Senses 2007; 32(3):263–271.
76. Cometto-Muniz JE, Cain WS. Perception of nasal pungency in smokers and nonsmokers. Physiol Behav 1982; 29(4):727–731.
77. Shusterman DJ, Balmes JR. A comparison of two methods for determining nasal irritant sensitivity. Am J Rhinol 1997; 11(5):371–378.
78. Hummel T, Kraetsch HG, Pauli E, et al. Responses to nasal irritation obtained from the human nasal mucosa. Rhinology 1998; 36(4):168–172.

79. Shusterman D, Avila PC. Real-time monitoring of nasal mucosal pH during carbon dioxide stimulation: implications for stimulus dynamics. Chem Senses 2003; 28(7):595–601.

80. Livermore A, Hummel T, Kobal G. Chemosensory event-related potentials in the investigation of interactions between the olfactory and the somatosensory (trigeminal) systems. Electroencephalogr Clin Neurophysiol 1992; 83(3):201–210.

81. Shusterman D, Murphy MA, Balmes J. Differences in nasal irritant sensitivity by age, gender, and allergic rhinitis status. Int Arch Occup Environ Health 2003; 76(8):577–583.

82. Wise PM, Radil T, Wysocki CJ. Temporal integration in nasal lateralization and nasal detection of carbon dioxide. Chem Senses 2004; 29(2):137–142.

83. Wise PM, Wysocki CJ, Radil T. Time-intensity ratings of nasal irritation from carbon dioxide. Chem Senses 2003; 28(9):751–760.

84. Saxon A, Diaz-Sanchez D. Air pollution and allergy: you are what you breathe. Nat Immunol 2005; 6(3):223–226.

85. Gilliland FD, Li YF, Saxon A, et al. Effect of glutathione-S-transferase M1 and P1 genotypes on xenobiotic enhancement of allergic responses: randomised, placebo-controlled crossover study. Lancet 2004; 363(9403):119–125.

86. Gilliland FD, Li YF, Gong H Jr., et al. Glutathione s-transferases M1 and P1 prevent aggravation of allergic responses by secondhand smoke. Am J Respir Crit Care Med 2006; 174(12):1335–1341.

87. Wan J, Diaz-Sanchez D. Phase II enzymes induction blocks the enhanced IgE production in B cells by diesel exhaust particles. J Immunol 2006; 177(5):3477–3483.

6 Upper Airway Dosimetry of Gases, Vapors, and Particulate Matter in Rodents

John B. Morris

Department of Pharmaceutical Sciences, School of Pharmacy, University of Connecticut, Storrs, Connecticut, U.S.A.

Bahman Asgharian

Applied Research Associates, Raleigh, North Carolina, U.S.A.

Julia S. Kimbell

Otolaryngology/Head and Neck Surgery, University of North Carolina, Research Triangle Park, North Carolina, U.S.A.

INTRODUCTION

The nasal cavity is an extremely common site of injury following inhalation exposure in rodents. The significance of nasal lesions in rodents relative to the potential risk to humans is often poorly understood. Experience has indicated that the relationships between airborne concentration and delivered dose to nasal tissues are often critical to the interpretation of animal toxicity data. This is not surprising given that dose–response is one of the most fundamental principles of toxicology. From this perspective, an understanding of the delivered dose relationships of an inhaled material is fundamental to assessment of its hazard. It is these relationships that are the focus of the current chapter.

Nasal dosimetry is the term used to describe the relationships between airborne concentration and the dose of inhaled material delivered to nasal target sites. Nasal dosimetric relationships are complex and species-specific. For this reason, useful interpretation and extrapolation of laboratory animal inhalation toxicity data can be difficult. For inhaled materials, the site selectivity of damage is due to regional toxicokinetic (delivered dose) and/or toxicodynamic (tissue sensitivity) properties of the toxicant and airways. While the focus of this chapter is on nasal dosimetry, it is essential to recognize that nasal responses in the rodent may be sentinels for potential lower airway injury in humans simply because of toxicokinetic (delivered dose) issues. Thus, nasal rodent toxicity data are best interpreted in a much broader sense than merely being reflective of risk to nasal tissues alone in man.

The inhalation toxicity of hydrofluoric acid vapor provides a useful example of the importance of this phenomenon. Rodents are obligate nose breathers, whereas humans are capable of mouth breathing. Indeed, it has been estimated that as much as 40% of the population respires, at least in part, through their mouth (1). Acute inhalation exposure of rats to hydrofluoric acid vapor results in nasal injury with no damage in the lower airways (2,3). It has been estimated that greater than 99.7% of inspired hydrofluoric acid vapor is scrubbed from the airstream in the upper respiratory tract of the rat with less than 0.3% penetrating to the lower airways, providing a toxicokinetic (delivered dose) basis for its regional toxicity (4). If rats are forced to breathe through their mouth via insertion of an oropharyngeal cannula, acute hydrofluoric acid vapor exposure results in tracheobronchial rather than nasal injury. This elegantly simple experiment demonstrates the importance of inhalation dosimetry in determining the site of injury (3). In humans, lower airway injury results from hydrofluoric acid exposure (e.g., 5) indicating, at least for this irritant, that nasal injury in the rodent was indeed predictive of lower airway risk in man. The columnar mucociliated epithelium of the nasal passages is structurally similar to that of the large lower airways (chap. 2) suggesting the sensitivity of these tissues might be similar. Little information is available in this regard; however, recent studies suggest the nasal and tracheal epithelium are equally responsive to the irritant 2,3-butanedione (6). While the relationship of nasal versus tracheobronchial tissue sensitivity is an area in need of much research, it is nonetheless clear that nasal injury in rodents may be predictive of lower airway risk in mouth-breathing humans.

The goal of this chapter is to highlight the basic principles and the current state-of-the-art relative to the nasal dosimetry of inspired gases, vapor, and particles. Since the physical factors that control dosimetry of gases and vapors differ from those controlling particle dosimetry, they are discussed separately. Both experimental and computational approaches to define dosimetric relationships are discussed. Perhaps not surprisingly, both approaches lead to similar predictions. The strengths, weaknesses, and potential usefulness of these approaches are highlighted along with the integration of experimental and computational data.

DOSIMETRY OF GASES AND VAPORS

Basic Principles

The ability of the upper respiratory tract to scrub gases and vapors from the airstream has long been appreciated (7). The absorption of gas or vapor molecules into airway mucosa is a dynamic reversible process in which gas molecules may be absorbed (typically during inspiration) and desorbed (typically during expiration). It is the balance between these processes that determines the net transfer of gas molecules to airway tissues. Various terms have been used to describe this phenomenon in living animals including: nasal uptake, nasal extraction, nasal deposition, and nasal scrubbing. In the most simplistic sense, the volume and/or flow rate of air and the mass of nasal tissues will be important factors relative to nasal uptake. The nose of the rodent is not uniformly ventilated, with differing regions of the nose receiving widely differing fractions of the inspired air (8,9). Thus, it can be appreciated that the deposition patterns and tissue doses will differ regionally throughout the nose.

There have been a number of publications in recent years describing the basic principles of gas and vapor uptake in the respiratory tract (e.g., 10–12). Chemical engineering principles in particular provide useful insights (10,13). The uptake process consists of three steps: convection and diffusion of airborne gas or vapor molecules to the air:tissue interface, transfer across that interface, and diffusion of tissue-borne gas or vapor away from the interface. From the engineering perspective, these steps can be quantified on the basis of air:phase, tissue:phase, and overall mass transfer coefficients (13,14). These coefficients relate to the speed and efficiency of each of these processes. The use of these coefficients is discussed below.

A fundamental physicochemical principle governing gas or vapor uptake is Henry's Law, which states that the concentration (or partial pressure) of gas or vapor in a liquid phase at equilibrium is directly proportional to the concentration (or partial pressure) in the gas phase above that liquid. Conceptually, this means that airway tissues have the potential to become saturated with gas or vapor molecule. The proportionality constant can be expressed in many ways, but the most common approach is via a tissue:air partition coefficient, which is reflective of the ratio of the concentration at equilibrium of gas or vapor in tissue divided by that in air.

The importance of tissue solubility in influencing regional uptake of vapors has long been realized and was perhaps first discussed from a theoretical and pragmatic perspective nearly a century ago by Haggard (15). *It is absolutely essential, however, to realize that this concept holds only for parent gas or vapor molecules.* If the gas or vapor is converted to another chemical species, either by direct chemical reaction or metabolism, then direct application of Henry's Law is inappropriate. Some vapors are so highly reactive that they are instantly removed from tissue. In such circumstances, tissue concentrations are negligible and the tissue acts as an infinite sink. Tissue factors are often unimportant in such cases and vapor or gas uptake is "air-driven" and is critically dependent only on air-phase phenomenon. Formaldehyde offers an example of such a case (16). The US EPA terms these gases "category 1" gases (13). For all other gases or vapors, the concentration of vapor in tissue (e.g., "backpressure") will influence and/or control the uptake process. As discussed below, differing theoretical modeling approaches are typically used for these two types of gases or vapors. Nasal vapor uptake efficiency can range from essentially 0% for low solubility (low partition coefficient) vapors or gases to essentially 100% for highly soluble, highly reactive vapors or gases.

For gases or vapors for which tissue backpressure influences uptake, it is important to understand the tissue factors that are involved. In essence, tissue clearance pathways serve to reduce the concentration of vapor in tissue and will enhance the uptake process as additional vapor or gas molecules will be transferred to tissue to maintain Henry's Law equilibrium (provided there is sufficient time). Experimentally, a steady state occurs in which the uptake rate equals the tissue clearance rate. Removal of vapor via the bloodstream is one such clearance pathway. In fact, the use of vapor uptake rates to measure airway perfusion rates is not new (103). Biotransformation enzymes are highly expressed in nasal tissue (chap. 5), metabolic clearance represents another important clearance pathway. Direct reaction is the third common pathway for removal of parent vapor molecules from airway tissues.

In summary, gas or vapor uptake is a dynamic process in nasal tissues that is dependent upon the gas or vapor solubility as measured by the tissue:air partition coefficient, the rate at which vapor directly reacts with tissue substrates and the rate at which vapor is metabolized by biotransformation pathways and the airway perfusion rate. As exposures progress these clearance pathways can change. For example, substrates for reaction (e.g., glutathione for electrophilic vapors) can become depleted altering uptake efficiencies (17). Physiological factors (e.g., nasal vasodilation) can occur (98,100). Uptake should not be viewed as static but as a process that can change quickly as biochemical and/or physiological conditions in the nose change.

Experimental Dosimetry Data

Since nasal tissues of laboratory animals are easily accessible, measurements of nasal vapor uptake efficiencies date back nearly 50 years (18). Typically, an endotracheal tube is inserted in an anterior direction and vapor-laden air is drawn through the nose either under unidirectional or cyclic conditions. Measurement of vapor concentration in air entering the nose versus that in the endotracheal tube itself provides an estimate of uptake efficiency. An advantage of this approach is that definitive measures of uptake efficiency under defined airflow conditions are possible. A disadvantage is that it relies upon nonphysiological flow conditions in anesthetized animals.

A rich database exists on nasal uptake efficiency of gases and vapors, many of these data have been generated in the laboratory of one of the authors (JBM). Data are available on such diverse agents as the inorganic gases ozone (19), sulfur dioxide (20), nitrogen dioxide, inorganic and organic acids (4,17,21), the solvents ethanol, acetone and styrene (22,23,99), esters (24), and the reactive volatile organics acrolein (17,25,96), propylene oxide (26,27) and formaldehyde (28,29). Data are available on numerous species including the dog (20), rabbit (30), guinea pig, rat, hamster, and mouse (91,92,94,97). Experimental data obtained using physiological inspiratory flow rates in rats indicate low (10% or less) uptake efficiencies for low partition coefficient (blood:air partition coefficient <50), nonreactive, slowly metabolized vapors, to extremely high (>90%) for water-soluble highly reactive vapors (e.g., sulfur dioxide, chlorine, formaldehyde, ionizable acids). Intermediate uptake efficiencies are observed for vapors of intermediate partition coefficient (50–2000) that are slowly reactive and/or slowly metabolized.

As noted above, metabolism within nasal tissues serves to lower tissue concentrations and, therefore, enhances vapor uptake. A biological approach to confirm this phenomenon is to measure nasal uptake efficiency in control and metabolically inhibited animals. Stott and McKenna (31) were perhaps the first to document this phenomena by showing that the carboxylesterase inhibitor TOCP diminished nasal uptake efficiency of inspired ethyl acetate. This phenomenon has been studied extensively by Morris, who has showed via use of metabolic inhibitors that several biotransformation enzymes including carboxylesterase (93), alcohol dehydrogenase (95), aldehyde dehydrogenase (91), and cytochrome P450 mixed functions oxidases (23,95) can serve to enhance nasal scrubbing efficiency. It should be recognized that the toxicological importance of metabolism relates not only to its effect on nasal dosimetry but also on whether the biotransformation reflects an activation or detoxification pathway (chap. 5).

Other experimental dosimetry work has focused on the measurement of specific markers of dose called dosimeters in tissues. Direct measurement of local tissue dose is often difficult. However, dosimeters based on potential modes of action underlying tissues responses can be very useful both for understanding the biological basis for the response and for extrapolating dose–response relationships from animals to people. One example of such a dosimeter is found in studies on DNA–protein crosslinks (DPX), which form in tissue exposed to formaldehyde gas (32). Correlation of high DPX levels with formaldehyde-induced lesion locations in rats and primates (33,34) led to the use of DPX to extrapolate animal data to people in subsequent human health risk assessment studies of formaldehyde gas (35,36).

Dosimetry Modeling

Three approaches have been used to model nasal gas and vapor uptake in experimental animals: the computational fluid dynamic (CFD) approach developed by Kimbell (9), the physiologically based pharmacokinetic (PBPK) approach developed by Morris (37), and the hybrid CFD–PBPK approach, an insightful contribution first proposed by Frederick (38,39). The hybrid model represented a significant advance and provided a PBPK-based modeling approach superior to that first proposed by Morris et al. (37). The three modeling approaches are all theoretically and experimentally valid. They have differing strengths and weaknesses; selection of the most appropriate approach is dependent on the physical chemical properties of the gas or vapor of interest and how much anatomical detail is needed in modeling outcomes.

The development of the CFD modeling approach was based on observations of the site specificity of formaldehyde-induced nasal tumors in rats. These tumors invariably appeared in selected regions of the nose, leading Morgan et al. (40) to hypothesize that high delivery of formaldehyde to these regions (e.g., hotspots) occurred. Since formaldehyde is highly soluble and reactive, nasal tissue likely represents an infinite sink; therefore, this phenomena must be reflective of factors occurring in the air phase. To examine this possibility, streamlines were first assessed in nasal casts. These studies suggested a concordance between major streams of inspired air and lesion locations in both rats and primates (41) but did not provide local dose estimates.

The level of anatomical detail available in nasal lesion data drove the concurrent development of three-dimensional (3D) anatomically accurate dose prediction models for inhaled gases in the nasal passages of rats, rhesus monkeys, and a number of humans (8,9,42–46). These models use a CFD approach to predict inhaled gas flux from nasal airspaces to airway walls. The CFD models have been used to study mechanisms of toxicity for ozone (47,48), formaldehyde (9,36,41,49–52), hydrogen sulfide (53–55), and acrolein (56) and to support risk assessment activities for formaldehyde (28,29,35,57–60), acidic vapors (38), methyl methacrylate (61), acrylic acid (62), and esters (39).

The CFD studies revealed that the nasal passages are not uniformly ventilated. For example, only ~12% of the inhaled airstream in the rat passes over a dorsal medial pathway to the olfactory-lined ethmoturbinates. [Concurrently a similar estimate of olfactory mucosal airflow was independently obtained by a PBPK approach (37).] This knowledge, coupled with chemical engineering mass transfer principles, resulted in the first estimation of intranasal regional deposition patterns of an inspired vapor, formaldehyde (9). As shown in Figure 1, flux (mass of formaldehyde entering tissue per surface area) is not uniformly distributed throughout the nose but is localized in hotspots. The location of these hotspots associated closely with the location of formaldehyde-induced tumors (51), thus supporting the original hypothesis regarding the importance of localized delivery of formaldehyde within the nose.

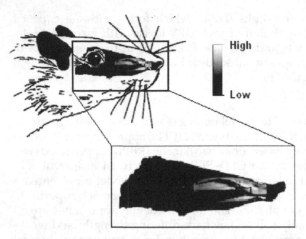

FIGURE 1 Lateral view of rat head showing nasal passages. Inset shows computational model of the nasal passages shaded by flux (rate of gas uptake from air to tissue surface at the tissue surface).

The success of the CFD approach has led to its use to model other vapors including hydrogen sulfide and acrolein. For both agents, a close association between lesion location and flux was observed (54,56). A strong advantage of this approach is that it provides a means of estimating delivered dosage rates at specific regions within the nose as well as providing a means of examining the effect of anatomic changes on delivered dosage rates. This approach has been extended to the human nose to extrapolate animal toxicity data to the humans, not on the basis of inspired concentration, but on the basis of regional flux to precise locations within the nose (54). Another significant contribution of the CFD modeling effort is that it provides estimates of gas-phase mass transfer coefficients. These coefficients are essential components of the current CFD–PBPK models. A disadvantage of the CFD approach is that it is not easy to carry out automated optimizations of fitted model parameters for tissue uptake; previous studies have used a combined CFD–PBPK approach to conduct optimizations (55).

PBPK models provide an approach in which to explicitly incorporate physiologically and biologically constrained data into a dosimetry model. The first PBPK nasal dosimetry model was developed by Morris et al. (37). In this approach, air is allowed to pass over tissue, which is modeled as stacks of 10-µm thick mucus, epithelial and submucosal compartments. Blood perfuses the submucosal compartments and vapor transfer between compartments occurs on the basis of tissue diffusivity. The overall depth of the tissue layers is based on anatomical measurements. Tissue metabolism rates, based on in vitro measurements, can be assigned independently to respiratory or olfactory mucosa and, within each mucosal type, separately to epithelial or submucosal compartments. In this first modeling effort, uptake data from a variety of vapors were fit to provide estimates of nasal perfusion rate and air flow patterns within the nose. The best fit estimated that 8% of the inspired air penetrates over the olfactory epithelium, a value similar to that obtained independently in Ref. 10.

A significant disadvantage of the PBPK model approach is that it did not explicitly incorporate air phase diffusion/mass transfer phenomena. Frederick et al. (38) provided a mechanism to incorporate these phenomena into the model structure with the hybrid CFD–PBPK model. In the hybrid model, vapor is allowed to enter (or desorb from) the mucus lining layer in accordance with an overall mass transfer coefficient. This is calculated on the basis of the air-phase mass transfer coefficient that is provided by the CFD models and a tissue phase mass transfer coefficient estimated from the tissue:air partition coefficient and the molecular diffusivity. With this approach, truly first principle models were developed for ethyl acrylate and acrylic acid (38,39), meaning that every parameter in the model was pre-assigned based on independent biological or physical measurement. Model output was then compared to experimental data and an extraordinarily close correspondence was seen. This approach has been successfully extended to a variety of vapors (6,63–65) and to describe uptake in the human (66).

Thus, the combined CFD–PBPK modeling efforts reported in the literature have taken one of two forms. In one form, enough information was available to calibrate the CFD model for the specific gas. Gas-specific flux estimates made by the CFD models were then handed off to PBPK models that estimated tissue dose (36,50,67). In the other form, the CFD models were used to calculate air-phase mass transfer coefficients, which depend only on air-phase diffusivity and airflow patterns. The coefficients are easily scalable to other air-phase diffusivities and can incorporate localized airflow effects when calculated regionally. Regional air-phase mass transfer coefficients derived from the rat and human nasal CFD models have been incorporated into PBPK models to estimate tissue dose when modeling goals do not require the complexity of the nasal CFD model (38,39,61,62).

The accuracy of inspiratory nasal airflow patterns predicted by rat, monkey, and human CFD models has been confirmed by comparison with experimental measurements (9,44,45). Confirmation of inhaled gas transport and uptake simulations using rat and monkey nasal CFD models has been demonstrated for formaldehyde (36,67) using formaldehyde-induced DNA–protein crosslinks measured in exposed F344 rats and rhesus monkeys (32–34).

The advantages of the CFD–PBPK modeling approaches are that it allows explicit incorporation of tissue-specific parameters into its structure. Since biotransformation enzymes are nonuniformly distributed between respiratory and olfactory mucosa (chap. 5), this is particularly important for vapors which are metabolized within the nose. A disadvantage of this approach is that it does not provide a means of estimating delivered dosage patterns within small regions of the nose (e.g., hotspots). Additionally, this modeling approach is ill suited to describe nasal dosimetry of highly soluble reactive gases and vapors because air-phase, not tissue-phase, factors control nasal uptake. Overall, it can be appreciated that the CFD and CFD–PBPK approaches are complimentary efforts, with the former being suited for localized dosimetry of water-soluble reactive vapors and the CFD–PBPK being suited for less soluble–less reactive vapors. Both approaches have been validated with experimental data, the needs of the study dictating the selection of modeling approach.

Integration of Experimental and Computational Approaches

Over the last 50 years, a rich database on nasal vapor and gas uptake has been developed. These data indicate that nasal vapor uptake can vary from essentially 0% to 100% depending on the properties of the gas or vapor. Critical vapor properties are solubility (tissue:air partition coefficient), and reactivity (either direct or enzymatically mediated). These data also provided critical quantitative insights into the factors which control the uptake process and ultimately to the development of mathematical models to describe uptake. The original modeling efforts were fit to uptake data to allow parameterization (8,37). These models have now become sufficiently advanced that they are developed a priori, with uptake measurements being made subsequently to validate the model-based predictions (e.g., 6,39). These models are extraordinarily useful as they facilitate extrapolation of data. For example, models that are validated with data obtained under nonphysiological flow conditions can be used to estimate uptake patterns during normal breathing and/or exercise. Such models can also be used to extrapolate nasal dosimetric relationships in animals exposed to high concentrations, to those in humans exposed to low concentrations. Delivered dose relationships are critical in determining the ultimate toxic response to inspired gases and vapors, thus dosimetric modeling is an essential aspect of evaluation and extrapolation of animal inhalation toxicity data.

DOSIMETRY OF PARTICLES

Basic Principles

Transfer of particles from the airstream to the tissue surface (termed deposition) is an irreversible process. Once deposited, insoluble particles can be removed by mechanical means including sneezing and mucociliary transport. Particles may also enter cells via dissolution and diffusion or by active (e.g., pinocytotic) pathways. These later processes are considered to constitute clearance mechanisms and are not considered herein. The fundamental processes determining transfer of particulates from a moving airstream to a tissue surface are well understood and include sedimentation, inertial impaction, and diffusion. Sedimentation is not thought to be important in the nose. Inertial impaction occurs due to the mass of the particle in motion, causing it to leave the airstream as the airstream bends. Inertial impaction increases with ρd^2 (particle density times particle diameter squared), and is important for fine and coarse particles (>0.5 μm in diameter). Diffusion (Brownian movement) results from the energy imparted by collisions of air molecules with particles and is related to $1/d$ (the inverse of the particle diameter) and is important for small particles (<0.1 μm). The diffusive process is particularly important for nanoparticles. Models for total respiratory tract particle dosimetry (e.g., ICRP) include explicit discussions of nasal particle deposition. Reviews are available summarizing data on regional particle deposition in animals and man (101,102).

Experimental Data

In Vivo Studies
Data from a limited number of experiments are available on measurements of aerosol deposition in the respiratory tracts of nonhuman primates and rats. A

recent study by Cheng et al. (68) reported measurements of head and lung deposition in anesthetized cynomolgus monkeys. The animals were exposed to 2.3- and 5.1-μm particles and head deposition fractions were found to be 39% and 58% for the two particle sizes. In rodents, inhalation exposures are conducted in nose-only chambers for which there is a good control over exposure concentration and delivery airflow rate. In addition, animal breathing parameters can be monitored accurately when animals are placed in a body plethysmograph. Particles can be sampled at one port of the nose-only tower for size analysis. Following the exposure, animals are sacrificed and dissected according to physiologically defined regions (head, tracheobronchial, pulmonary, etc.) and particle deposition is measured in each region.

Regional deposition of ultrafine particles have been reported by Chen et al. (69) for cigarette smoke particles, Wolff et al. (70) for aggregate 67Ga2O3 particles in Fischer-344 rats, and Dohlback and Eirefelt (71) and Dohlback et al. (72) for particle deposition in male Sprague-Dawley rats. Raabe et al. (73,74) measured deposition of fine and coarse particles in the head and lungs of Long-Evans rats. However, minute ventilations of the test animals were not measured and instead allometric equations were used for lung parameters. Therefore, uncertainty existed with the calculated deposition fractions. Asgharian et al. (75) repeated the study while the ventilation parameters were monitored throughout the exposure. Particle sizes ranged between 0.9 and 4.2 μm. Deposition for particles greater than 3 μm tended to decline, which may have been due to limitations with the inhalability of these particles.

Since the focus of inhalation exposure studies has primarily been on the measurements of lung deposition, limited data on head deposition have emerged from the above studies. Experiments to measure the deposition of particles in the nasal passages of rodents and nonhuman primates are scarce due to the difficulty and cost associated with conducting such experiments and also because it was not realized until very recently that the nasal airways serve as a potential route of transport of materials to the brain (76,77). Gerde et al. (78) devised a technique in which a catheter was inserted in the trachea distal to the nasal airways of anesthetized male Fischer 344 rats to allow measurements of ultrafine particle concentrations at the proximal and distal ends of the nasal region and calculate deposition fraction during steady inspiration and expiration. Quantification of particles in the airstream instead of on the nasal tissues eliminated uncertainty due to fast mucociliary clearance of deposited particles on the nasal surfaces. Measured deposition for particles between 5 nm and 100 nm at steady breathing rates of 200 mL/min to 600 mL/min was found to be similar to values measured in casts (79).

Kelly et al. (80) employed the same technique to measure deposition of fine and coarse aerosols in the nasal airways of female Long-Evans rats during steady and pulsatile breathing. Partial obstruction of the flow through the nasal airways due to the positioning of the catheter with respect to the head and possible leaks imposed a challenge to reliable and reproducible runs and may have contributed to variability in measurements. Deposition measurements were higher for pulsatile breathing than for steady breathing due to enhanced particle inertia at the peak flow. It was also observed that more particles were deposited during exhalation than during inhalation.

Studies in Nasal Molds

Difficulty with in vivo measurements has led investigators to build hollow molds of the nasal airways in which to conduct particle deposition experiments. Hollow molds have been created from nasal specimens, serial sections, and computer reconstructions using stereolithography techniques. The physical nasal molds can be used to measure particle deposition in situ, which can be used to confirm computer model (in silico) predictions. In addition, multiple runs may be made by varying the flow rate and particle size in the same setup and hence can collect more information in a short time when compared with in vivo measurements.

Kelly et al. (81) used a nasal mold to measure deposition fractions of 1–10 μm particles in the nasal passages of a 12-kg, male rhesus monkey for steady inspiration flow rates of 2–7 LPM. Particle deposition was described by a single inertial parameter that varied between 0% and 100% and showed a sigmoidal shape. Cheng et al. (79) measured pressure drop and deposition of ultrafine aerosol particles in a mold of the nasal passages of an adult male Fischer 344 rat. Measured nonlinear flow-pressure drop relationship indicated the presence of nonlaminar regimes in the flow. Measured depositions of 43 nm to 210 nm particles in the nasal mold were in agreement with in vivo measurements from inhalation studies (69,70,73).

Measurements of fine and coarse particle deposition, for which deposition in the nasal passages is by inertial impaction, are reported by Kelly et al. (82). Particles with sizes between 0.48 μm and 4.18 μm were passed through the nasal mold of a male 344 Fischer rat at flow rates of 100 to 900 mL/min for inspiratory and expiratory breathings. Collected data were in agreement with in vivo measurements and similar trends of deposition with flow type and direction were observed in in vivo measurements (80). Comparison of in vivo and in situ measurements suggested that good estimates of total nasal deposition in rats can be obtained by using nasal molds as surrogates for live animals (83).

Dosimetry Modeling

Empirical Studies

Losses of particles in the nasal airways are significant for two reasons. First, particles may deposit and accumulate locally and elicit toxicological response in the nasal airways or deposit and transport to the brain via the olfactory nerves. Second, to study toxicity of inhaled particles in the lung, one must determine the amount that passes through the nose and become available for deposition in the lung. Therefore, it is necessary to develop predictive models of particle deposition in the head based on the physics of airflow and particle transport. Consequently, models of particle deposition efficiency are sought as a function of parameters that control particle movement and deposition.

There are two main mechanisms that influence particle deposition in the nasal passages. Small particles are deposited by Brownian diffusion while fine and coarse particles are removed from the inhaled air by inertial impaction. A different mathematical model is required for each mechanism. Cheng et al. (79) postulated that due to nonlaminar characteristics of the flow, turbulent diffusion was responsible for losses of ultrafine particles in the nasal passages and proposed the following relationship:

$$\eta = 1 - e^{\alpha D^\beta Q^\gamma} \tag{1}$$

where η is the deposition efficiency, D is the diffusion coefficient, and Q is the airflow rate through the nose, and α, β, and γ are fitted constants. Equation (1) exhibits increasing deposition efficiency with increasing diffusion parameter ($D^\beta Q^\gamma$) but decreasing efficiency with increasing particle size. Coefficients α, β, and γ were obtained by fitting equation (1) to deposition measurements made in a nasal mold (79). Additionally, Asgharian et al. (84) found by examining the governing equations of transport for the air and particles that deposition efficiency depends on D and Q, and proposed

$$\eta = \alpha D^\beta Q^\gamma \tag{2}$$

Similarly, coefficients α, β, and γ were found by fitting equation (2) to the measurements reported by Wong et al. (85).

Deposition efficiency by inertial impaction can be correlated with impaction parameter $d_\alpha^2 Q$, where d_α denotes particle aerodynamic diameter. Extending the mathematical model suggested by Rodulf et al. (86), Zhang and Yu (87) employed the following relationship to predict nasal deposition efficiency:

$$\eta = \left[\frac{(d_\alpha^2)^\alpha}{\beta + (d_\alpha^2)^\alpha} \right]^\gamma \tag{3}$$

Coefficients α, β, and γ were obtained by fitting equation (3) to the measurements by Raabe et al. (73,74) from a nose-only inhalation exposure conducted in Fischer 344 rats. The same model was employed by Kelly et al. (83) to predict nasal deposition efficiency in Long-Evans rats. Measured deposition fraction in live animals (80) and nasal mold (82) was used to find parameters α, β, and γ. Different models were obtained for inspiratory and expiratory steady and pulsatile breathings. Table 1 lists fitted parameters for equations 1–3 in different studies.

TABLE 1 Calculated Values of α, β, and γ Used in Equations 1–3

Reference	Experiment	Breathing	α	β	γ
(79)	Nasal mold	Inspiration, steady flow	−16.8	0.517	−0.234
		Expiration, steady flow	−14.0	0.517	−0.234
(84)	Nasal mold	Inspiration, steady flow	7.351	−0.2438	0.402
(87)	Nose-only inhalation	Normal	2.553	10^5	0.627
(83)	Nose-only inhalation	Inspiration, steady flow	2.42	10^5	0.27
		Expiration, steady flow	2.59	10^5	0.279
		Inspiration, pulsatile flow	2.88	10^5	0.304
		Expiration, pulsatile flow	3.26	10^5	0.262
(83)	Nasal mold	Inspiration, steady flow	3.00	10^5	0.907
		Expiration, steady flow	2.98	10^5	0.659
		Inspiration, pulsatile flow	3.78	10^5	0.539
		Expiration, pulsatile flow	3.88	10^5	0.447

The regional deposition of smaller particles, between 1 and 100 nm in diameter, has been studied in the rat nose computationally by Garcia et al. (88). In this study, estimates were developed of nanoparticle deposition in the nasal and, more specifically, olfactory regions of the rat. The rat CFD model of Kimbell et al. (8) was employed to simulate inhaled airflow and to calculate nasal deposition efficiency. Simulations predicted that olfactory deposition is maximum at 6–9% of inhaled material for 3–4 nm particles. The spatial distribution of deposited particles was predicted to change significantly with particle size, with 3-nm particles depositing mostly in the anterior nose, while 30-nm particles were more uniformly distributed throughout the nasal passages.

CONCLUSIONS

Dose–response is among the most fundamental principles of toxicology. This chapter has highlighted the current state of the art relative to nasal dosimetry of gases, vapors, and particles. Such evaluations are best performed in the context of risk to the entire respiratory tract, not merely risk to nasal tissues alone as inhaled materials can produce damage in the nasal passages, tracheobronchial tree, and/or the alveoli. Indeed, nasal damage in rodents may be the sentinel for lower airway risk in man. Regardless of the site of injury, delivered dosage rates are critically important in determining the toxic response to inhaled materials and, therefore, represent an essential aspect in inhalation risk assessment. With respect to nasal dosimetry, the physical/chemical factors influencing inhalation dosimetry are well understood and mathematic inhalation dosimetry models are sufficiently advanced to allow prediction of nasal dosimetry based on the physical/chemical properties of the gas, vapor, or particle of interest.

REFERENCES

1. DeSesso JM. The relevance to humans of animal models for inhalation studies of cancer in the nose and upper airways. Qual Assur 1993; 2:213–231.
2. Rosenholtz MJ, Carson TR, Weeks MH, et al. A toxicopathologic study in animals after brief single exposure to hydrogen fluoride. Am Ind Hyg Assoc J 1963; 24:253–261.
3. Stavert DM, Archuleta DC, Behr MJ, et al. Relative acute toxicities of hydrogen fluoride, hydrogen chloride and hydrogen bromide in nose- and pseudo-mouth-breathing rats. Fundam Appl Toxicol 1991; 16:636–655.
4. Morris JB, Smith FA. Regional deposition and absorption of inhaled hydrogen fluoride in the rat. Toxicol Appl Pharmacol 1982; 62:81–89.
5. Bennion JR, Franzblau A. Chemical pneumonitis following household exposure to hydrofluoric acid. Am J Ind Med 1997; 31:474–478.
6. Morris JB, Hubbs AF. Inhalation dosimetry of diacetyl and butyric acid, two components of butter flavoring vapors. Toxicol Sci 2009; 108:173–183.
7. Jordan EO, Carlson AJ. Ozone: its bactericidal, physiologic and deodorizing action. J Am Med Assoc 1913; 61:1008–1012.
8. Kimbell JS, Godo MN, Gross EA, et al. Computer simulation of inspiratory airflow in all regions of the F344 rat nasal passages. Toxicol Appl Pharmacol 1997a; 145:388–398.
9. Kimbell JS, Gross EA, Joyner DR, et al. Application of computational fluid dynamics to regional dosimetry of inhaled chemicals in the upper respiratory tract of the rat. Toxicol Appl Pharmacol 1993a; 121:253–263.
10. Dahl AR. Contemporary issues in toxicology: Dose concepts for inhaled vapors and gases. Toxicol Appl Pharmacol 1990; 103:185–197.

11. Miller FJ, Kimbell JS. Regional dosimetry of inhaled reactive gases. In: Henderson R, McClellan RO, eds. Concepts in Inhalation Toxicology, 2nd ed. Washington, DC: Taylor and Francis, 1995:257–287.
12. Ultman JS. Dosimetry modeling: Approaches and issues. Inhal Toxicol 1994; 6(suppl):59–71.
13. U.S. Environmental Protection Agency. Methods for Derivation of Inhaled Reference Concentrations and Application of Inhalation Dosimetry. Office of Research and Development, Washington, DC. EPA/600/8–90–066F, October, 1994.
14. Cussler EL. Diffusion: Mass Transfer in Fluid Systems, 2nd ed. New York, NY: Cambridge University Press, 1999:580.
15. Haggard HW. The absorption, distribution and elimination of ethyl ether. V. The importance of the volume of breathing during the induction and termination of anesthesia. J Biol Chem 1924; 59:795–802.
16. Morgan KT, Monticello TM. Airflow, gas deposition, and lesion distribution in the nasal passages. Environ Health Perspect 1990; 85:209–218.
17. Morris JB, Frederick CB. Upper respiratory tract uptake of acrylate ester and acid vapors. Inhal Toxicol 1995; 7:557–574.
18. Dalhamn T, Sjoholm J. Studies on SO2, NO2 and NH3: effect on ciliary activity in rabbit trachea of single in vitro exposure and resorption in rabbit nasal cavity. Acta Physiol Scand 1963; 58:287–291.
19. Miller FJ, McNeal CA, Kirtz JM, et al. Nasopharyngeal removal of ozone in rabbits and guinea pigs. Toxicology 1979; 14:273–281.
20. Frank NR, Yoder RE, Brain JD, et al. SO2 (35S labeled) absorption by the nose and mouth under conditions of varying concentration and flow. Arch Environ Health 1969; 18: 315–322.
21. Morris JB, Symanowicz PT, Olsen JE, et al. Immediate sensory-nerve mediated respiratory responses to irritants in healthy and allergic airway diseased mice. J Appl Physiol 2003; 94:1563–1571.
22. Morris JB, Clay RJ, Cavanagh DG. Species differences in upper respiratory tract deposition of acetone and ethanol vapors. Fundam Appl Toxicol 1986; 7:671–680.
23. Morris JB. Uptake of styrene vapor in the upper respiratory tracts of the CD mouse and Sprague-Dawley rat. Toxicol Sci 1999; 54:222–228.
24. Bogdanffy MS, Manning LA, Sarangapani R. High-affinity nasal extraction of vinyl acetate vapor is carboxylesterase dependent. Inhal Toxicol 1999; 11:927–941.
25. Struve MF, Wong VA, Marshall MW, et al. Nasal uptake of inhaled acrolein in rats. Inhal Toxicol 2008; 20(3):217–225.
26. Morris JB, Banton M, Pottenger LH. Uptake of inspired propylene oxide in the upper respiratory tract of the F344 rat. Toxicol Sci 2005; 81:216–224.
27. Morris JB, Pottenger LH. Nasal NPSH depletion and propylene oxide uptake in the upper respiratory tract of the mouse. Toxicol Sci 2006; 92:228–234.
28. Kimbell JS, Subramaniam RP, Gross EA, et al. Dosimetry modeling of inhaled formaldehyde: Comparisons of local flux predictions in the rat, monkey, and human nasal passages. Toxicol Sci 2001a; 64:100–110.
29. Kimbell JS, Overton JH, Subramaniam RP, et al. Dosimetry modeling of inhaled formaldehyde: Binning nasal flux predictions for quantitative risk assessment. Toxicol Sci 2001b; 64:111–121.
30. Miller FJ, McNeal CA, Kirtz JM et al. Nasopharyngeal removal of ozone in rabbits and guinea pigs. Toxicol 1979; 14:273–281.
31. Stott MJ, McKenna MJ. The comparative absorption and excretion of chemical vapors by the upper, lower and intact respiratory tract of rats. Fundam Appl Toxicol 1984; 4:594–602.
32. Casanova M, Deyo DF, Heck Hd'A. Covalent binding of inhaled formaldehyde to DNA in the nasal mucosa of Fischer 344 rats: Analysis of formaldehyde and DNA by high-performance liquid chromatography and provisional pharmacokinetic interpretation. Fundam Appl Toxicol 1989; 12:397–417.

33. Casanova M, Morgan KT, Steinhagen WH, et al. Covalent binding of inhaled formaldehyde to DNA in the respiratory tract of rhesus monkeys: Pharmacokinetics, rat-to-monkey interspecies scaling, and extrapolation to man. Fundam Appl Toxicol 1991; 17:409–428.
34. Casanova M, Morgan KT, Gross EA, et al. DNA-protein cross-links and cell replication at specific sites in the nose of F344 rats exposed subchronically to formaldehyde. Fundam Appl Toxicol 1994; 23:525–536.
35. Conolly RB, Kimbell JS, Janszen DB, et al. Human respiratory tract cancer risks of inhaled formaldehyde: Dose-response predictions derived from biologically-motivated computational modeling of a combined rodent and human dataset. Toxicol Sci 2004; 82(1):279–296.
36. Conolly RB, Lilly PD, Kimbell JS. Simulation modeling of the tissue disposition of formaldehyde to predict nasal DNA-protein cross-links in F344 rats, rhesus monkeys, and humans. Environ Health Perspect 2000; 108(suppl 5):919–924.
37. Morris JB, Hassett DN, Blanchard KT. A physiologically based pharmacokinetic model for nasal uptake and metabolism of non-reactive vapors. Toxicol Appl Pharmacol 1993; 123:120–129.
38. Frederick CB, Bush ML, Lomax LG, et al. Application of a hybrid computational fluid dynamics and physiologically-based inhalation model for interspecies dosimetry extrapolation of acidic vapors in the upper airways. Toxicol Appl Pharmacol 1998; 152:211–231.
39. Frederick CB, Lomax LG, Black KA, et al. Use of a hybrid computational fluid dynamics and physiologically-based inhalation model for interspecies dosimetry comparisons of ester vapors. Toxicol Appl Pharmacol 2002; 183:23–40.
40. Morgan KT, Jiang XZ, Starr TB, et al. More precise localization of nasal tumors associated with chronic exposure of F-344 rats to formaldehyde gas. Toxicol Appl Pharmacol 1986; 82:264–271.
41. Morgan KT, Kimbell JS, Monticello TM, et al. Studies of inspiratory airflow patterns in the nasal passages of the F-344 rat and rhesus monkey using nasal molds: Relevance to formaldehyde toxicity. Toxicol Appl Pharmacol 1991; 110:223–240.
42. Garcia GJM, Bailie N, Martins DA, et al. Atrophic rhinitis: a CFD study of air conditioning in the nasal cavity. J Appl Physiol 2007; 103:1082–1092.
43. Garcia GJ M, Schroeter JD, Segal RA, et al. Dosimetry of nasal uptake of soluble and reactive gases: A first study of inter-human variability. Inhal Toxicol 2009a; 21:607–618.
44. Kepler GM, Richardson RB, Morgan KT, et al. Computer simulation of inspiratory nasal airflow and inhaled gas uptake in a rhesus monkey. Toxicol Appl Pharmacol 1998; 150:1–11.
45. Subramaniam RP, Richardson RB, Morgan KT, et al. Computational fluid dynamics simulations of inspiratory airflow in the human nose and nasopharynx. Inhal Toxicol 1998; 10:91–120.
46. Segal RA, Kepler GM, Kimbell JS. Effects of differences in nasal anatomy on airflow distribution: A comparison of four individuals at rest. Ann Biomed Engr 2008; 36(11):1870–1882.
47. Hotchkiss JA, Herrera LK, Harkema JR, et al. Regional differences in ozone-induced nasal epithelial cell proliferation in F344 rats: Comparison with computational mass flux predictions of ozone dosimetry." Inhal Toxicol 1994; 6(suppl):390–392.
48. Kimbell JS, Morgan KT, Hatch GE, et al. Regional nasal uptake and toxicity of ozone: Computational mass flux predictions, 18O-ozone analysis, and cell replication. Toxicologist 1993b; 13:259.
49. Cohen Hubal EA, Kimbell JS, Fedkiw PS. Incorporation of nasal-lining mass-transfer resistance into a CFD model for prediction of ozone dosimetry in the upper respiratory tract. Inhal Toxicol 1996; 8:831–857.
50. Georgieva AV, Kimbell JS, Schlosser PM. A distributed-parameter model for formaldehyde uptake and disposition in the rat nasal lining. Inhal Toxicol 2003; 15:1435–1463.

51. Kimbell JS, Gross EA, Richardson RB, et al. Correlation of regional formaldehyde flux predictions with the distribution of formaldehyde-induced squamous metaplasia in F344 rat nasal passages. Mut Res 1997b; 380:143–154.
52. Monticello TM, Swenberg JA, Gross EA, et al. Correlation of regional and nonlinear formaldehyde-induced nasal cancer with proliferating populations of cells. Cancer Res 1996; 56:1012–1022.
53. Moulin FJ, Brenneman KA, Kimbell JS, et al. Predicted regional flux of hydrogen sulfide correlates with distribution of nasal olfactory lesions in rats. Toxicol Sci 2002; 66:7–15.
54. Schroeter JD, Kimbell JS, Andersen ME, et al. Use of a pharmacokinetic-driven computational fluid dynamics model to predict nasal extraction of hydrogen sulfide in rats and humans. Toxicol Sci 2006a; 94(2):359–367.
55. Schroeter JD, Kimbell JS, Bonner AM, et al. Incorporation of tissue reaction kinetics in a computational fluid dynamics model for nasal extraction of inhaled hydrogen sulfide in rats. Toxicol Sci 2006b; 90(1):198–207.
56. Schroeter JD, Kimbell JS, Gross EA, et al. Application of physiological computational fluid dynamics models to predict interspecies nasal dosimetry of inhaled acrolein. Inhal Toxicol 2008; 20(3):227–243.
57. Conolly RB, Kimbell JS, Janszen DB, et al. Dose-response for formaldehyde-induced cytotoxicity in the human respiratory tract. Regul Toxicol Pharmacol 2002; 35: 32–43.
58. Conolly RB, Kimbell JS, Janszen DB, et al. Biologically motivated computational modeling of formaldehyde carcinogenicity in the F344 rat. Toxicol Sci 2003; 75:432–447.
59. Overton JH, Kimbell JS, Miller FJ. Dosimetry modeling of inhaled formaldehyde: The human respiratory tract. Toxicol Sci 2001; 64:122–134.
60. Schlosser PM, Lilly PD, Conolly RB, et al. Benchmark dose risk assessment for formaldehyde using airflow modeling and a single-compartment DNA-protein cross-link dosimetry model to estimate human equivalent doses. Risk Anal 2003; 23:473–487.
61. Andersen ME, Sarangapani R, Frederick CB, et al. Dosimetric adjustment factors for methyl methacrylate derived from a steady-state analysis of a physiologically based clearance-extraction model. Inhal Toxicol 1999; 11:899–926.
62. Frederick CB, Gentry PR, Bush ML, et al. A hybrid computational fluid dynamics and physiologically based pharmacokinetic model for comparison of predicted tissue concentrations of acrylic acid and other vapors in the rat and human nasal cavities following inhalation exposure. Inhal Toxicol 2001; 13:359–376.
63. Csanady GA, Filser JG. A physiological toxicokinetic model for inhaled propylene oxide in rat and human with special emphasis on the nose. Toxicol Sci 2007; 95: 37–62.
64. Saragapani R, Teeguarden JG, Cruzan G, et al. Physiologically based pharmacokinetic modeling of styrene and styrene oxide respiratory-tract dosimetry in rodents and humans. Inhal Toxicol 2002; 14:789–834.
65. Teeguarden JG, Bogdanffy MS, Covington TR, et al. A PBPK model for evaluating the impact of aldehyde dehydrogenase polymorphisms on comparative rat and human nasal tissue acetaldehyde dosimetry. Inhal Toxicol 2007; 20:375–390.
66. Hinderliter PM, Thrall KD, Corley RA, et al. Validation of human physiologically based pharmacokinetic model for vinyl acetate against human nasal dosimetry data. Toxicol Sci 2005; 85:460–467.
67. Cohen Hubal, EA, Schlosser PM, Conolly RB, et al. Comparison of inhaled formaldehyde dosimetry predictions with DNA-protein cross-link measurements in the rat nasal passages. Toxicol Appl Pharmacol 1997; 143:47–55.
68. Cheng YS, Irshad H, Kuehl P, et al. Lung Deposition of Droplet Aerosols in Monkeys. Inhal Toxicol 2008; 20:1029–1036.
69. Chen BT, Weber RE, Yeh HC, et al. Deposition of cigarette smoke particles in the rat. Fundam Appl Toxicol 1989; 13:429–438

70. Wolff RK, Kanapilly GM, Gray RH, et al. Deposition and retention of inhaled aggregate ^{67}Ga$_2$O$_3$ particles in Beagle dogs, Fischer-344 rats, and CD-1 mice Am Ind Hyg Assoc J 1984; 45(6):377–381

71. Dahlbäck M, Eirefelt S. Total deposition of fluorescent monodisperse particles in rats. Ann Occup Hyg 1994; 38(suppl 1):127–134.

72. Dohlbacket al. 1989.

73. Raabe OG, Al-Bayati MA, Teague SV, et al. Regional deposition of inhaled monodisperse coarse and fine aerosol particles in small laboratory animals. Ann Occup Hyg 1988; 32(suppl):53–63.

74. Raabe OG, HS Yeh, GJ Newton, et al. Deposition of Inhaled monodisperse aerosols in small rodents in Inhaled Particles IV (eds. WH Walton and B McGovern) NY: Pergamon Press, 1977:3–20.

75. Asgharian B, Kelly JT, Rewksbury EW. Respiratory deposition and inhalability of monodisperse aerosols in Long-Evans rats. Toxicol Sci 2003; 71:104–111.

76. Brenneman KA, Wong BA, Buccellato MA, et al. Direct Olfactory Transport of Inhaled Manganese (^{54}MnCl$_2$) to the Rat Brain: Toxicokinetic Investigations in a Unilateral Nasal Occlusion Model. Toxicol Appl Pharmacol 2000; 169:238–248.

77. Oberdoerster G, Sharp Z, Atudorei V, et al. Translocation of inhaled ultrafine particles to the brain. Inhal Toxicol 2004; 16:437–445.

78. Gerde P, Cheng YS, Medinsky MA. In vivo deposition of ultrafine aerosols in the nasal airways of the rat. Fundam Appl Toxicol 1991; 16:330–336.

79. Cheng YS, GK Hansen, YF Su, et al. Deposition of ultrafine aerosols in rat nasal molds. Toxicol Appl Pharmacol 1990; 106:222–233,376

80. Kelly JT, Bobbitt CM, Asgharian B. In vivo measurement of fine and coarse aerosol deposition in the nasal airways of female Long-Evans rats. Toxicol Sci 2001a; 64:253–258.

81. Kelly JT, Asgharian B, Wong BA. Inertial particle deposition in a monkey nasal mold compared with that in human nasal replicas. Inhal Toxicol 2005; 17:823–830.

82. Kelly JT, Kimbell JS, Asgharian B. Deposition of fine and coarse aerosols in a rat nasal mold. Inhal Toxicol 2001b; 13:577–588.

83. Kelly JT, Asgharian B. Nasal molds as predictors of fine and coarse particle deposition in rat nasal airways. Inhal Toxicol 2003; 15:859–877.

84. Asgharian B, Price OT, Wong BA, et al. Model of nanoparticle transport and deposition in the nasal and lung airways of humans and rats. Toxicologist, 2008; 62:204.

85. Wong B, Tewksbury AW, Asgharian B. Nanoparticle deposition efficiency in rat and human nasal replicas. Toxicologist. 2008; 62:310.

86. Rudolf G, Kobrich R, Stahlhofen W, et al. In proceedings, 7th International Symposium on Inhaled particles, September 16–20, 1991, Edinburgh, Scotland.

87. Zhang L and Yu CP. Empirical equations for nasal deposition of inhaled particles in small laboratory animals and humans. Aerosol Sci Technol 1993; 19:51–56.

88. Garcia GJM, Kimbell JS. Deposition of inhaled nanoparticles in the rat nasal passages: Dose to the olfactory region. Inhal Toxicol, 2009; 21:1165–1175.

89. Dahl AR. Snipes MB, Gerde P. Sites for uptake of inhaled vapors in beagle dogs. Toxicol Appl Pharmacol 1991; 109:263–275.

90. Dahlbäck M, Eirefelt S, Karlbergm I-B, et al. Total deposition of evans blue in aerosol exposed rats and guinea pigs. J Aerosol Sci 1989; 20(8):1325–1327.

91. Morris JB, Clay RJ, Cavanagh DG. Species differences in upper respiratory tract deposition of acetone and ethanol vapors. Fundam Appl Toxicol 1986; 7:671–680.

92. Morris JB, Cavanagh DG. Metabolism and deposition of propanol and acetone vapors in the upper respiratory tract of the hamster. Fundam Appl Toxicol 1987; 9:34–40.

93. Morris JB. First-pass metabolism of inspired ethyl acetate in the upper respiratory tracts of the F344 rat and syrian hamster. Toxicol Appl Pharmacol 1990; 102:331–345.

94. Morris JB. Deposition of acetone vapor in the upper respiratory tract of the B6C3F1 mouse. Toxicol Lett 1991; 56:187–196.

95. Morris JB. Upper respiratory tract metabolism of inspired alcohol dehydrogenase and mixed function oxidase substrate vapors under defined airflow conditions. Inhal Toxicol 1993; 5:203–221.
96. Morris JB. Uptake of acrolein in the upper respiratory tract of the F344 rat. Inhal Toxicol 1996; 8:387–403.
97. Morris JB. Uptake of acetaldehyde and aldehyde dehydrogenase levels in the upper respiratory tracts of the mouse, rat, hamster and guinea pig. Fundam Appl Toxicol 1997; 35:91–100.
98. Morris JB, Stanek J, Gianutsos G. Sensory nerve-mediated immediate nasal responses to inspired acrolein. J Appl Physiol 1999; 87:1877–1886.
99. Morris JB. Uptake of styrene vapor in the upper respiratory tracts of the CD mouse and Sprague-Dawley rat. Toxicol Sci 1999; 54:222–228.
100. Morris JB, Wilkie WS, Shusterman DJ. Acute respiratory responses of the mouse to chlorine. Toxicol Sci 2005; 83:380–387.
101. Schlesinger RB. Comparative deposition of inhaled aerosols in experimental animals and humans: a review. J Toxicol Environ Health 1985; 15:197–214.
102. Schreider JP. Nasal airway anatomy and ihalation deposition in experimental animals and people. In: Reznik G, Stinson SF, eds. Nasal Tumors in Animals and Man, Vol. III. Boca Raton FL: CRC Press, 1983:1–26.
103. Wanner A, Mendes ES, Atkins ND. A simplified noninvasive method to measure airway blood flow in humans. J Appl Physiol 2006; 100:1674–1678.

Vapor Dosimetry in the Nose and Upper Airways of Humans

Karla D. Thrall

Pacific Northwest National Laboratory, Richland, Washington, U.S.A.

INTRODUCTION

A number of methodologies have been reported for measuring vapor uptake efficiencies in the upper respiratory tract of experimental animals (1). Hybrid computational fluid dynamic (CFD) and physiologically based pharmacokinetic (PBPK) models, as described by Frederick et al. (2), that incorporate information on the anatomy of both rats and humans have been used to improve interspecies dosimetric corrections for human health risk assessments. However, validation of these models requires sufficient experimental data, and robust data defining the role of the upper respiratory tract in modulating the absorption of gases and vapors in human volunteers are lacking.

A survey of the available literature shows a limited number of experimental studies to evaluate the dosimetry of vapors in the nose and upper airways of humans. The scarcity of literature data undoubtedly reflects the complication of conducting controlled studies in human volunteers, and with the exception of a few limited studies, little experimental data is available. This chapter highlights studies specific for nasal dosimetry data from humans and briefly reviews modeling approaches for predictive extrapolations from animal data.

HUMAN UPPER RESPIRATORY TRACT VAPOR STUDIES

Speizer and Frank (3) conducted early studies in which healthy human volunteers were exposed to approximately 25 ppm sulfur dioxide (SO_2) by inhalation wearing a close-fitting plastic mask over the mouth and nose. The mask had one port for breathing and another for passage of sampling tubes into the interior. Samples via tygon tubing were taken in the following four locations: (*i*) in the respiratory tubing just before the mask, representing concentration inhaled; (*ii*) within the mask and at 2 cm in front of the nose, representing the inspiratory sample; (*iii*) within the nose, approximately 1 to 2 cm beyond the flaring of the nostrils, representing the nose inspiratory and nose expiratory sample; and (*iv*) as far back into the oropharynx as the volunteer could tolerate, representing the pharynx, both inspiratory and expiratory. Volunteers participated in a control period, an exposure period, and a recovery period.

Measurement of samples indicated that virtually all of the SO_2 in inspired air was removed by the nose during the 30-minute exposures, with approximately 15% desorbed from the nasal mucosa during expiration. Therefore, assuming the minute ventilation remained constant during exposure and recovery, the net uptake of SO_2 by the body was about 85% of the exposure concentration. Similar studies conducted in dogs (4) confirm the effectiveness of the nose and portions of the pharynx in absorbing SO_2 and support the suggestion that some desorption of the gas from the nasal mucosa occurs early in the period of recovery following the exposure phase.

A study by Guyatt et al. (5) exposed 12 human volunteers (8 male and 4 female) to a mixture of 0.3% carbon monoxide (CO) and 9.6% argon by mouth breathing and a single volunteer by nasal breathing. For these studies, the volunteer breathed to and from a bag-in-box system fitted with a spirometer through an electronically controlled solenoid tap system. Carbon monoxide concentration was measured continually using a rapid infrared analyzer.

As part of these evaluations, a pilot nasal study was carried out in which the volunteer inserted into his nostril the two ends of a plastic "Y" piece rounded at the tips with adhesive tape to provide good fit. The other end of the "Y" was connected to a 60-mL plastic bag filled with the CO/argon mixture and sealed off with a spring clip. When the clip was released, the volunteer inhaled through his nose to empty the bag, then after a breath holding interval of 0, 10, or 20 seconds, exhaled back into the bag. The analysis of the bag indicates that no measurable uptake of CO occurs in the upper airways.

A study by Utell et al. (6) exposed 12 volunteers (8 male and 4 female) to octamethylcyclotetrasiloxane (D_4) at 10 ppm by mouth breathing and 8 volunteers (6 males and 2 females) by nasal exposure. Exposures were divided into three rest periods of 10, 20, and 10 minutes, respectively, and two exercise periods, each of 10-minute duration. Exercise was designed to approximately triple resting minute ventilation using an appropriate workload on a bicycle ergometer. Exhaled air, blood, and urine samples were collected before and after exposure. The siloxane vapor was delivered by an infusion pump to a J-tube generator, where it was vaporized and passed through a 120-L bag reservoir to the subject's mouthpiece. Three-way valves were used to switch the subject between air and vapor, and could be used for collection of expired air into a Tedlar sampling bag. Nasal and mouthpiece exposure systems differed only by the substitution of a nosepiece for the mouthpiece. For nasal breathing, volunteers were exposed for 16 minutes via the nasal device, followed by 10 minutes off the device, and then for 16 minutes by mouthpiece to compare the deposition of D_4 during the two different breathing routes of exposure.

For nasal breathing, the average total intake and estimated uptake was 11.6 ± 1.9 mg and 1.1 ± 0.4 mg, respectively. In comparison, mouthpiece breathing was slightly higher, averaging 14.9 ± 3.3 mg for total intake and 2.0 ± 0.9 mg for estimated uptake. However, after correcting for exposure system loss and differences in minute ventilation, the mean deposition fraction was found to be 0.12 ± 0.01 mg for both exposure modes.

More recently, Thrall et al. (7) developed a methodology for evaluating nasal wash-in, wash-out phenomena of compounds in human volunteers. In a study evaluating the nasal deposition and clearance of acetone, volunteers were administered a nasal topical anesthetic and decongestant aerosol spray, and a small-diameter flexible probe was inserted approximately 9 cm into one nostril such that the tip of the probe was positioned in the nasopharyngeal region. The placement of the probe was visually verified using a fiber-optic flexible nasopharyngoscope. The exterior portion of the probe was threaded through a port on a face mask containing two one-way non-rebreathing valves. The port was sealed with Teflon tape to prevent leakage. The distal end of the probe was

TABLE 1   Experimental Protocol

Breathing rate (breaths/min)	Study 1 (nose inhale/nose exhale)	Study 2 (nose inhale/mouth exhale)	Study 3 (mouth inhale/nose exhale)
Normal (12–15)	2 min	2 min	2 min
Deep (6–8)	2 min	2 min	2 min
Normal (12–15)	2 min	2 min	2 min
Rapid (20)	1 min	1 min	1 min
Normal (12–15)	2 min (room air)	2 min (room air)	2 min (room air)

Source: Adapted from Ref. 7.

attached to a Teflon transfer line that fed directly into an ion-trap mass spectrometer for continual analysis. A separate valve from the face mask was connected to a second ion-trap mass spectrometer for continual analysis of exhaled breath. Breathing frequency and depth were monitored using a breathing monitor belt with an inflatable rubber bladder connected to an absolute pressure sensor. The sensor monitored the change in air pressure within the inflatable bladder as the lung expanded and contracted. Heart rate was assessed using a wireless heart-rate monitor. An acoustic rhinometry system was used to measure the cross-sectional area and volume of the nasal cavity. Comparison of acoustic rhinometer measurements of the nasal cavity before and after application of the topical anesthetic and decongestant indicated that nasal volume was not significantly altered by use of the spray.

The study protocol simultaneously involved monitoring the volunteer by both the nasopharyngeal probe and the exhaled breath for background levels of $[^{13}C]$-acetone levels for 5 minutes prior to initiation of inhalation exposure. The exposure atmosphere was prepared in a 500-L foil-lined bag at a target concentration of approximately 1 ppm. During the exposure phase, volunteers were instructed to vary their breathing pattern, as shown in Table 1.

In the methodology reported by Thrall et al. (7), $[^{13}C]$-acetone concentrations in the nasopharyngeal region and the exhaled breath were separately measured and graphically compared to provide insight into nasal absorption versus desorption during inhalation and exhalation. A representative time-sequence overlay of the nasopharyngeal $[^{13}C]$-acetone concentration (points, thin dashed line) with data collected using the breathing rate monitor (bold line) illustrates that acetone concentrations corresponded with the inhalation/exhalation cycle (Fig. 1). As shown in Figure 1, the lowest $[^{13}C]$-acetone concentrations regularly occur in the trough between the exhalation and inhalation cycle. A visual comparison of breath-to-breath peaks indicated that roughly half of the inhaled acetone was absorbed in the nasal cavity before reaching the point of the nasopharyngeal probe and nearly all of that absorbed was released back into the breath stream during the exhalation phase. Overall, the results indicate that between 40% and 75% of the inhaled acetone is absorbed, depending on breathing maneuver (nose inhale/nose exhale; nose inhale/mouth exhale; mouth inhale/nose exhale). This fractional deposition corresponds well with prior reports that the respiratory uptake of acetone in humans ranges from 45% to 57% (8–10).

Using the same procedural approach as described in Thrall et al. (7), Hinderliter et al. (11) conducted human studies with $[^{13}C]$-vinyl acetate. Male

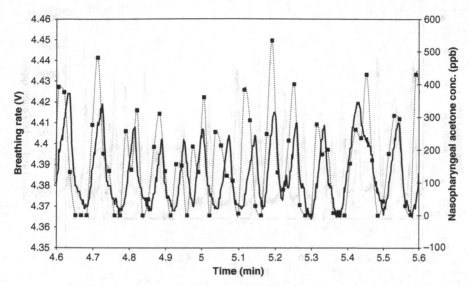

FIGURE 1 Time course of breathing cycle (*bold solid line*) and concentration of [^{13}C]-acetone measured in the nasopharyngeal region (*points, thin dashed line*) for a study volunteer during the nose inhale/nose exhale breathing cycle. The volunteer was a 57-year-old Caucasian male (173 cm height, 83.9 kg weight) exposed to [^{13}C]-acetone at a concentration of 862 ppb.

and female volunteers were exposed to vinyl acetate at 1, 5, and 10 ppm concentrations, and nasal concentrations of [^{13}C]-vinyl acetate and the metabolite, [^{13}C]-acetaldehyde, were measured before, during, and after the exposure. Exhaled breath was sampled for [^{13}C]-vinyl acetate. The experimental protocol consisted of triplicate cycles of exposure to vinyl acetate under a resting breathing rate for three minutes followed by three minutes of exposure to vinyl acetate under light exercise on a stationary bicycle at 50 W.

A representative time-sequence overlay of the nasopharyngeal [^{13}C]-vinyl acetate concentration (points, thin dashed line) with data collected using the breathing rate monitor (bold line) illustrates that vinyl acetate concentrations corresponded with the inhalation/exhalation cycle (Fig. 2).

These experiments presented the first opportunity to validate a human nasal dosimetry PBPK model described by Bogdanffy et al. (12). The Bogdanffy et al. model separates nasal tissue into four compartments (three respiratory and one olfactory), each containing compartments representing the mucus, epithelial cells, basal cells, and submucosa. Model simulations assumed that inhaled vinyl acetate was extracted into the nasal tissue, diffused through the various layers, and was metabolized. The model further assumes that the bidirectional airflow in the human nose only contains vinyl acetate from inhalation and not from release from the lower respiratory tract; little to no contribution is expected from the exhaled breath due to expected absorption in the lower respiratory tract. All model parameters used to simulate the experiments described in Hinderliter et al. (11) were unchanged from those reported by Bogdanffy et al. (12), including nasal dimensions.

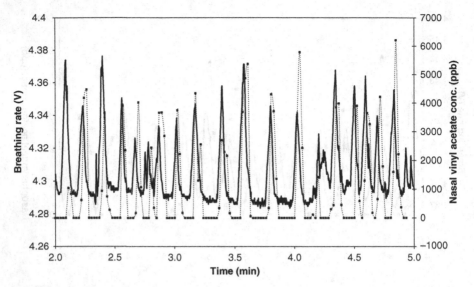

FIGURE 2 Time course of breathing cycle (*bold solid line*) and concentration of [^{13}C]-vinyl acetate measured in the nasopharyngeal region (*points, thin dashed line*) for a study volunteer during the nose inhale/nose exhale breathing cycle. The volunteer was a 30-year-old Caucasian male (183 cm height, 85 kg weight) exposed to [^{13}C]-vinyl acetate at a concentration of 9353 ppb.

Based on the resulting model simulation comparisons to experimental data, the Bogdanffy et al. (12) model structure appears to be an appropriate representation of the upper respiratory tract for vinyl acetate. Simulations demonstrated that the model reasonably predicted the experimental observations with regard to nasal disposition of vinyl acetate and wash-out of acetaldehyde in a concentration range evaluated.

Overall, nasal dosimetry models have become increasingly quantitative as insights into tissue deposition and clearance and computational fluid dynamics have become available. The available scientific literature contains models that describe nasal airflow, mucosal diffusion, clearance–extraction, and metabolism within different epithelial layers. Examples include dosimetry modeling of inhaled formaldehyde in human nasal passages (13), nasal tissue dosimetry of methyl methacrylate and vinyl acetate (14), prediction of nasal extraction of hydrogen sulfide (15), propylene oxide dosimetry in respiratory nasal passages (16), and a nasal dosimetry model for ethyl acrylate (17), just to name a few. Collectively, these types of models can help identify and quantify interspecies differences in airflow and uptake patterns and provide a basis for extrapolating nasal experimental data in rodents to estimate acceptable toxicity levels in humans.

Modeling efforts have contributed toward refining and extending the conceptual guidance for nasal modeling initially proposed in the U.S. EPA reference concentration documentation (18). However, a major obstacle in model validation has been the lack of data obtained from controlled in vivo human studies. Perhaps the experimental approaches described here, particularly methodologies that characterize nasal concentrations on a breath-by-breath basis will enable further studies to be conducted in order to provide insight into the deposition and clearance of vapors in humans for low-level, nontoxic materials. These human validation studies, coupled with parallel studies using laboratory animals will significantly increase the confidence in species extrapolations using hybrid CFD-PBPK modeling approaches.

REFERENCES

1. Morris JB. Overview of upper respiratory tract vapor uptake studies. Inhal Toxicol 2001; 13:335–345.
2. Frederick CB, Gentry PR, Bush ML, et al. A hybrid computational fluid dynamics and physiologically based pharmacokinetic model for comparison of predicted tissue concentrations of acrylic acid and other vapors in the rat and human nasal cavities following inhalation exposure. Inhal Toxicol 2001; 13:359–376.
3. Speizer FE, Frank NR. The uptake and release of SO_2 by the human nose. Arch Environ Health 1966; 12:725–728.
4. Hodgman CD, ed. Handbook of Chemistry and Physics, 36th ed. Cleveland, OH: Chemical Rubber Company, 1954–1955:1609.
5. Guyatt AR, Holmes MA, Cumming G. Can carbon monoxide be absorbed from the upper respiratory tract in man? Eur J Respir Dis 1981; 62:383–390.
6. Utell MJ, Gelein R, Yu CP, et al. Quantitative exposure of humans to an octamethylcyclotetrasiloxane (D_4) vapor. Toxicol Sci 1998; 44:206–213.
7. Thrall KD, Schwartz RE, Weitz KK, et al. A real-time method to evaluate the nasal deposition and clearance of acetone in the human volunteer. Inhal Toxicol 2003; 15:523–538.
8. Landahl HD, Herrmann RG. Retention of vapors and gases in the human nose and lung. Arch Ind Hyg Occup Med 1950; 1:36–45.
9. Schrikker ACM, de Vries WR, Zwart A, et al. Uptake of highly soluble gases in the epithelium of the conducting airways. Pflugers Arch 1985; 405:389–394.
10. Wigaeus E, Holm S, Åstrand I. Exposure to acetone: Uptake and elimination in man. Scand J Work Environ Health 1981; 7:84–94.
11. Hinderliter PM, Thrall KD, Corley RA, et al. Validation of human physiologically based pharmacokinetic model for vinyl acetate against human nasal dosimetry data. Toxicol Sci 2005; 85:460–467.
12. Bogdanffy MS, Sarangapani R, Plowchalk DR, et al. A biologically based risk assessment for vinyl acetate-induced cancer and noncancer inhalation toxicity. Toxicol Sci 1999, 51:19–35.
13. Kimbell JS, Subramaniam RP, Gross EA, et al. Dosimetry modeling of inhaled formaldehyde: Comparisons of local flux predictions in the rat, monkey, and human nasal passages. Toxicol Sci 2001; 64:100–110.
14. Andersen ME, Green T, Frederick CB, et al. Physiologically based pharmacokinetic (PBPK) models for nasal tissue dosimetry of organic esters: Assessing the state-of-knowledge and risk assessment applications with methyl methacrylate and vinyl acetate. Reg Toxicol Pharmacol 2002; 36:234–245.
15. Schroeter JD, Kimbell JS, Andersen ME, et al. Use of a pharmacokinetic-driven computational fluid dynamics model to predict nasal extraction of hydrogen sulfide in rats and humans. Toxicol Sci 2006; 94:359–367.
16. Csanády GA, Filser JG. A physiological toxicokinetic model for inhaled propylene oxide in rat and human with special emphasis on the nose. Toxicol Sci 2007; 95:37–62.
17. Sweeney LM, Andersen ME, Gargas ML. Ethyl acrylate risk assessment with a hybrid computational fluid dynamics and physiologically based nasal dosimetry model. Toxicol Sci 2004; 79:394–403.
18. U.S. EPA (U.S. Environmental Protection Agency). Methods for derivation of inhalation reference concentrations and applications of inhalation dosimetry. EPA/600/8-90-066F, 1994, Washington, DC: U.S. EPA.

8 Particle Dosimetry in the Nose and Upper Airways of Humans

Owen R. Moss

POK Research, Apex, North Carolina, U.S.A.

INTRODUCTION

During inhalation exposures, the dose delivered to airway tissues is determined by the aerosol regional deposition. Particle deposition patterns in the human upper respiratory tract (including the nose) are well understood, both from a theoretical and an experimental perspective (1–8). These deposition patterns are also correlated with aerosol deposition patterns in the rodent upper respiratory tract (9–14). Such human-to-rodent correlations are made based on the key assumption that rodent inhalation exposures are a useful surrogate to human exposures. This chapter addresses this key assumption by asking the following question: Just how useful are rodent data in gaining insight into human response from particle dose to the nose and upper airways?

For those of us involved in inhalation toxicology and pharmaceutical research, our reply is "of course, rodent data is useful as a translational tool between theoretical considerations, risk assessment, and clinical applications." However, this reply is not sufficient: In order to answer the question, we must find parallel measurements on rodent-response and human-response to the same inhaled particles. We do so in this chapter by highlighting an example where upper-respiratory-tract-particle-deposition experiments and computational models were relevant, not only in mice, but also provided basic insights into human dose–response. The response is upper-airway hyperresponsiveness and the inhaled particles are airborne particles of a smooth muscle constrictor, methacholine.

Hyperresponsiveness is one of five components of allergic asthma: airway collagen and smooth muscle remodeling; serum immunoglobin E (IgE) concentration increase; pulmonary eosinophilia; excessive airway-mucus production; and airway hyperresponsiveness (15–18). Upper respiratory tract hyperresponsiveness is associated with a rapid increase in nasal and upper-airway resistance; the exposed subject experiences an increase in the effort needed to breathe. In nasal airways, where the initial response is associated with excessive airway-mucus production and tissue swelling, resistance in these airways can significantly increase in several minutes to a half-hour (19): Following the end of exposure, recovery can take much longer. However, in the upper airways of the lung where smooth muscle is associated with the airway wall, during inhalation of a smooth muscle constrictor, airway resistance can significantly increase in less than a minute (20): Following the end of exposure, recovery can occur within the same time frame of under a minute. It is this difference in recovery times for nasal tissue and for upper-airway tissue that allows comparison of parallel exposures in mice and in humans.

For inhaled particles of methacholine, a smooth muscle constrictor, the pulmonary dose–response in nose-breathing mice can be compared to the pulmonary dose–response in nose or mouth breathing humans.

The protocol, called the methacholine challenge, is used as one measure of hyperresponsiveness in humans. The protocol (21) consists of a sequence of exposures or challenges. In the first challenge, the base-line challenge, a methacholine-free liquid solution is generated as an aerosol. The subject inhales the mist for two minutes, followed by a quantifiable maneuver that allows an estimation to be made of how difficult it is to breathe: This maneuver is normally the maximum amount of air that can be exhaled in one second (the forced expiratory volume in one second, FEV_1).

For each subsequent challenge in the sequence, the concentration of methacholine in the solution is increased from a starting minimum concentration, X. For example, if $X = 0.025$ mg/mL, subsequent concentrations may increase as follows: $1X$, $10X$, $100X$, $400X$, and $1000X$. This sequence is followed until the subject has either inhaled the aerosol produced from the highest concentration level or until the subject's FEV_1 has decreased by a predetermined percentage, ΔP_0 (normally 10% or 20%). The data from these challenges is plotted as $FEV_{1,i}$ verses the methacholine concentration, C_i. From these data points is extracted the provocative concentration, C_P, that would produce a $\Delta P_0\%$ decrease in FEV_1.

In this procedure, there are two cases where C_P cannot be estimated: These are when the subject is either very responsive or nonresponsive. For the very responsive subject, upon inhalation of the lowest methacholine challenge, the FEV_1 changes by more than $\Delta P_0\%$ and the procedure is stopped. For the nonresponsive subject, upon inhalation of the highest methacholine challenge, there is no significant change in FEV_1.

When the methacholine challenge is given to rodents, FEV_1 cannot directly be measured, instead the degree of difficulty in breathing (airway resistance) is measured indirectly as a lag time, Δt, between the start of a breath (in the thorax) and the movement of air (in the nose). As an indicator of change in airway resistance, this lag time method is sufficiently reproducible to allow the ranking of mouse varieties according to hyperresponsiveness, and this rank ordering is assumed to be indicative of similar degrees of hyperresponsiveness in humans.

However, for mice and humans, these comparisons depend on the underlying assumption that the provocative concentration, C_P, is an accurate indicator of the effective dose that produces the $\Delta P_0\%$ change in difficulty to breathe. Unfortunately, this assumption is not supported by recent analysis of the dosimetry—this is the analysis reviewed in this chapter.

REANALYSIS OF THE METHACHOLINE CHALLENGE

Until 2002, the inhalation dosimetry of the methacholine challenge was of little interest. There was no apparent need to question the use of the provocative concentration as a rough indication of hyperresponsiveness. However, 2002 marked the beginning of the following sequence of events:

1. Within a technique paper that involved the use of the methacholine challenge, the measurements reconfirmed the relative hyperresponsiveness of four varieties of mice (20).
2. Independent of the first paper, a second paper verified that two of these mouse varieties differed in the size of their upper airways: The most hyperresponsive mouse had the smaller airways (22).
3. Authors from these two publications collaborated to determine the extent that the difference in hyperresponsiveness could be explained by the difference in the amount of methacholine deposited on airway epithelium (9): The dosimetry explained the difference and implied that the respiratory isthmus (airway generations 5 and 6) might be the most sensitive region of the lung.
4. A pilot study was conducted to test, in human subjects, whether there was a correlation between hyperresponsiveness (as classified by the methacholine challenge) and respiratory isthmus volume: Preliminary results indicated that the most sensitive subjects had the smallest respiratory-isthmus volumes (23).

This sequence of events represents an example of the application of currently available particle dosimetry tools to the estimation of tissue dose in the rodent and human nose and upper airways: When response is correlated with tissue dose and the physiology of airway constriction, measurements from the methacholine challenge provide insight into the size of subject airways and their relative sensitivity to any challenge that causes a change in the volume of the respiratory isthmus. In addition, the revised dose–response relations provide insight into how to distinguish subjects who have a biochemical-based sensitivity from subjects who do not.

Data-set Part 1: Measuring Mouse Hyperresponsiveness

This sequence of events began with four varieties of female mice (A/J, BALB/c, CD-1, and B6C3F1). In a paper (20) in which they describe the use of a mouse–methacholine challenge protocol to assess two plethysmographic methods, Delorme and Moss also ranked these mice according to hyperresponsiveness. Their ranking agreed with previously published differences in pulmonary function; in the order shown above, the A/J was most sensitive and the B6C3F1 least sensitive. In particular, the Balb/c mouse was 12 times more sensitive than the B6C3F1 mouse.

For each methacholine challenge, a mouse was placed in a plethysmograph chamber. A nebulizer was used to generate an aerosol of methacholine, which was delivered to the breathing zone of the mouse. The provocative concentration, C_P, was calculated as described above, and the change in airway resistance (the dependent variable) was plotted against the concentration of methacholine in the liquid being nebulized (the independent variable). No further dosimetry was done.

Data-set Part 2: Measuring the Dimensions of Mouse Airways

However, in a separate, independent, sequence of experiments, Oldham et al. (10,22) reported morphometric measurements on the lungs of male Balb/c and B6C3F1 mice. In these experiments, replicate lung cases were made of silicone rubber (Silastic E, Dow-Corning, Midland, MI). For the first six generations of the tracheobronchial region (where the trachea equals generation 1), morphometric measurements were made of the most complete lung casts from each mouse variety ($n \geq 3$). For each airway, four measurements were made: length, diameter, branch angle, and gravity angle (the inclination of the airway to the gravity force vector). The measurements demonstrated that, for the B6C3F1 mouse, the first six airway generations have average diameters approximately 1.6 times that of the same airway generations in the Balb/c mouse. The B6C3F1 mouse has big airways: The Balb/c mouse has small airways.

Revisiting the Dosimetry

The first two data sets showed that the most hyperresponsive mouse also had the smallest airways. Because of this, two of the authors (9) collaborated to rework the dosimetry. Their goal was to determine how much of the $12x$ difference in hyperresponsiveness could be explained by the $1.6x$ difference in airway diameters. In other words, they addressed the question "Were the most sensitive mice merely the most efficient in depositing inhaled material?"

To do this they had to redo the dose–response curves from the methacholine challenge. In these original curves, the change in airway resistance was plotted as a function of the concentration of methacholine in the liquid being nebulized. Although the analysis was common practice at the time, this concentration was not an accurate representation of the cellular-dose most directly responsible for the change in airway resistance. In the methacholine challenge, at least nine key factors were ignored; these factors were as follows:

1. the aerosol size distribution
2. the exposure-atmosphere concentration
3. the minute ventilation
4. the inhaled mass
5. the respiratory system morphometry
6. the airway-specific deposition of methacholine
7. the relation between methacholine deposition and smooth muscle constriction
8. the physiology of smooth muscle constriction and airway pressure drop, and
9. the contribution of change in individual airway pressure drop to the change in total lung pressure drop.

Each of these factors had to be considered and incorporated into estimating the effective cellular dose associated with the provocative concentration (C_P).

For each concentration C_i, the aerosol size distribution was measured with a cascade impactor or similar instrument. And as expected, the size distributions shifted as the concentrations increased. There were two reasons for the shift: First, in each nebulized droplet, the mass of methacholine increased; and second, at the highest methacholine concentration the droplet size decreased due to a decrease in the surface tension of the liquid.

This change, in droplet content and size, impacted the mass output of the nebulizer—which, in turn, impacted the exposure concentration (the mass concentration of the aerosol in the breathing zone of the subject).

The inhaled mass of methacholine (M_i) was calculated from the product of the exposure concentration, the duration of the challenge, and the minute ventilation of the subject. This inhaled mass formed the starting input for using inhalation dosimetry models to estimate the effective cellular dose.

At the time (around 2005), there were three inhalation-dosimetry models where the software was available for use by most investigators:

- the National Council on Radiation Protection and Measurement (NCRP) deposition model (7)
- the International Council on Radiation Protection and Measurement (ICRP) deposition model (8), and
- the Multiple Path Particle Deposition (MPPD) model (11–14,24).

The models produce similar results. In addition, each model reviews the existing literature and supporting respiratory system particle deposition data. The NCRP and ICRP each present these reviews in the form of a report (7,8). The decisions and interpretations discussed in the reports are also implemented in the software programs. And each report also provides information for users on how to obtain copies of, or have access to, this software.

The MPPD model takes a different approach: The most recent version of the software is available to download for free (11). Within this software package, MPPD is just one of several deposition models: By the use of a graphical user interface, what is made available to the user is a selection of historical and current deposition programs. For each deposition model, the supporting discussion and citations are contained in an expanded "Help" function. Users not only can quickly get up to speed on the history of particle deposition but they also can run the programs and compare, for themselves, the results of different ways of thinking about lung structure and where particles land.

Although any of the three software programs would have worked to address the question of whether the most sensitive mice were merely the most efficient in depositing inhaled material, the NCRP particle deposition model was used in the reanalysis of the methacholine challenges (9). The main reason was that the NCRP software provided the easiest method for inserting the new lung morphometry data.

When the NCRP deposition model was used to calculate the dosimetry, the following was observed: For the least sensitive mouse (the B6C3F1), in order to produce the same change in resistance to flow, as seen in the most sensitive mouse (the Balb/c), the mass of methacholine in each airway had to be greater by a factor of three. By considering dosimetry alone, it appeared that, instead of being 12 times more hyperresponsive, the Balb/c mouse was only 3 times more hyperresponsive than the B6C3F1.

However, dosimetry alone does not account for the relation between change in airway resistance and smooth muscle constriction. The smooth muscle, surrounding the circumference of larger airways, needs to constrict more in order to produce the same percent change in pressure drop seen when smaller airways constrict. The more the smooth muscle needs to constrict, the greater is the required dose of methacholine. In particular, when airway diameters differ by a factor of 1.6, the methacholine dose needs to be greater by a factor of 3 (9).

When estimations of site-specific dose were matched with the physiology of airway constriction and airway pressure drop, the factor of 3- difference in hyperresponsiveness disappeared. Other than airway size and efficiency in deposition of inhaled material, there appeared to be no differences between the two mouse varieties.

This physiologically based dosimetry also produced a secondary observation. The analysis indicated the region of the lung where slight changes in airway diameter produce the greatest change in airway resistance. For both varieties of mice, this region encompassed airway generations 5 and 6. The region acted like a "respiratory isthmus." Furthermore, preliminary calculations showed that the respiratory isthmus appeared to be the same for humans (9).

Verification in Humans
The mouse methacholine challenge dosimetry indicated that for normal healthy subjects, the responsiveness or nonresponsiveness to the methacholine challenge would be correlated with airway dimensions and thus with efficiency in depositing inhaled particles. This possibility has been addressed in a pilot study (23) in which 14 human subjects received a high-resolution computed tomograph (CT) scan just prior to working through the methacholine challenge. (The study was conducted under the approval of an Institutional Review Board.)

For the first six airway-generations of each subject, the CT scans allowed measurements to be made of airway length, diameter, branch angle, and gravity angle. As for the mice, these measurements were incorporated into the physiologically based, methacholine dosimetry. Preliminary analysis indicated that the six nonresponsive subjects had airways that were larger than the eight responsive subjects. The observations made in mice appeared to carry forward to humans (23).

SUMMARY
In this chapter, we highlight an example, for mice and for humans, where equivalent particle exposures were analyzed by computational models. We show that these particle deposition models are relevant for translating mouse and human dose–response relations.

The highlighted example involves the exposure of rodents or humans to airborne particles of a smooth muscle constrictor, methacholine. The analysis provides insight into the relation between the size of subject airways and the relative sensitivity of the upper respiratory system. The calculation of particle deposition also provides a plausible explanation for most between-subject differences in respiratory system sensitivity: The sensitivity may primarily be due to differences in the volume of the respiratory isthmus.

REFERENCES
1. Shanley KT, Zamankhan P, Ahmadi G, et al. Numerical simulations investigating the regional and overall deposition efficiency of the human nasal cavity. Inhal Toxicol 2008; 20(12):1093–1100.
2. Tian ZF, Inthavong K, Tu JY. Deposition of inhaled wood dust in the nasal cavity. Inhal Toxicol 2007; 19(14):1155–1165.
3. Kimbell JS, Segal RA, Asgharian B, et al. Characterization of deposition from nasal spray devices using a computational fluid dynamics model of the human nasal passages. J Aerosol Med 2007; 20(1):59–74.
4. Zwartz GJ, Guilmette RA. Effect of flow rate on particle deposition in a replica of a human nasal airway. Inhal Toxicol 2001; 13(2):109–127.

5. Ménache MG, Miller FJ, Raabe OG. Particle inhalability curves for humans and small laboratory animals. Ann Occup Hyg 1995; 39(3):317–328.
6. Schlesinger RB. Comparative deposition of inhaled aerosols in experimental animals and humans: a review. J Toxicol Environ Health 1985; 15(2):197–214.
7. National Council on Radiation Protection and Measurements [NCRP]. Deposition, Retention and Dosimetry of Inhaled Radioactive Substances. Bethesda, MD: NCRP, 1997. NCRP Report No. 125.
8. ICRP [International Council on Radiation Protection and Measurement]. Human Respiratory Tract Model for Radiological Protection. Bethesda, MD: ICRP, 1994. ICRP Publication No. 66, Annals of the ICRP 24(1–3).
9. Moss OR, Oldham MJ. Dosimetry counts: Molecular hypersensitivity may not drive pulmonary hyperresponsiveness. J Aerosol Med 2006; 19(4):555–564.
10. Oldham MJ, Phalen RF, Schum GM, et al. Predicted nasal and tracheobronchial particle deposition efficiencies for the mouse. Ann Occup Hyg 1994; 38:135–141.
11. Asgharian B, Price O. Multiple Path Particle Deposition Model, (MPPD Version 2.0) Software (with graphical user interface) is available for free download from Bahman Asgharian, Ph.D. at Applied Research Associates, 2009. http://www.ara.com/products/mppd.htm. Accessed December 1, 2009.
12. National Institute for Public Health and the Environment (RIVM). Multiple Path Particle Dosimetry Model (MPPD v 1.0): A model for Human and Rat Airway Particle Dosimetry. The Netherlands: Bilthoven, 2002. RIVA Report 650010030.
13. Asgharian B, Hofmann W, Bergmann R. Particle deposition in a multiple-path model of the human lung, Aerosol Sci Technol 2001; 34:332–339.
14. Anjilvel S, Asgharian B. A multiple-path model of particle deposition in the rat lung. Fundam Appl Toxicol 1995; 28:41–50.
15. Taube C. Dakhama A, Gelfand EW. Insights into the pathogenesis of asthma utilizing murine models. Int Arch Allergy Immunol 2004; 135:173–186.
16. Ramos-Bardon D, Ludwig MS, Martin JG. Airway remodeling: lessons from animal models. Clin Rev Allergy Immunol 2004; 27:3–21.
17. Wohlsen A, Uhlig S, Martin C. Immediate allergic response in small airways. Am J Respir Crit Care Med 2001; 163:1462–1469.
18. Piechuta H, Smith ME, Share NN, et al. The respiratory response of sensitized rats to challenge with antigen aerosols. Immunology 1979; 38:385–392.
19. Moss OR, Tewksbury EW. Three minute inhalation of methacholine by B6C3F1 or Balb/c female mice produces no change in pressure drop across the isolated upper respiratory tract. The Toxicologist 2006; 90(S-1):1697.
20. DeLorme MP, Moss OR. Pulmonary function assessment by whole-body plethysmography in restrained versus unrestrained mice. J Pharmacol Toxicol Methods 2002; 47:1–10.
21. American Thoracic Society. Guidelines for Methacholine and Exercise Challenge Testing-1999. Am J Respir Crit Care Med 2000; 161:309–329.
22. Oldham MJ, Phalen RF. Dosimetry implications of upper tracheobronchial airway anatomy in two mouse varieties. Anat Rec 2002; 268:59–65.
23. Moss, Clinkenbeard, Doolittle. Study report released, November 2009, for submittal to Environmental Health Perspectives.
24. Asgharian B, Price OT, Hofmann W. Prediction of particle deposition in the human lung using realistic models of lung ventilation. J Aerosol Sci 2006; 37:1209–1221.

9 Exposure and Recording Systems in Human Studies

Thomas Hummel

Smell & Taste Clinic, Department of Otorhinolaryngology, University of Dresden Medical School, Dresden, Germany

Dennis J. Shusterman

Division of Occupational and Environmental Medicine, University of California, San Francisco, California, U.S.A.

INTRODUCTION

Inhalation toxicology studies of experimental animals, in general, utilize a relatively well-established set of exposure and recording methodologies. Studies may be acute, subacute, or chronic, and may involve whole-body or nose-only exposures (with the occasional study involving cannulated, anesthetized animals). While subacute and chronic exposures generally involve sacrificing animals to document pathological and/or biochemical changes, acute exposure studies may involve either pathologic or functional alterations (e.g., electrophysiologic responses; changes in respiratory behavior). Whereas the emphasis of this chapter will be on the range of techniques employed in human, as opposed to experimental animal, studies, readers interested in the latter are encouraged to consult one of several excellent reviews (1–3).

In human studies, a number of novel exposure strategies, endpoints, and types of instrumentation have been utilized. Because of the ability of humans to respond rapidly to psychophysical tasks, acute exposures ("stimuli") may last as little as a fraction of a second. Although chamber studies may last several hours, they infrequently involve multiday exposures. Thus, with the exception of observational studies carried out in occupational settings ("natural experiments"), essential all human experimental studies would be classified as "acute" in nature. Study endpoints may include psychophysical, behavioral (including changes in respiratory behavior), physiologic, electrophysiologic, biochemical, anatomical (involving biopsies or exfoliative cytology), or molecular biologic in nature. Given the ability of human subjects to follow complex instructions, the range of potential human physiologic study techniques is quite wide. The goal of this chapter is to introduce a spectrum of human study techniques that may be relatively unfamiliar to the inhalation toxicology community.

EXPOSURE APPARATUS

As noted above, controlled human exposure studies can range in duration from stimuli each lasting a second or less (psychophysical and electrophysiologic endpoints) to exposures lasting hours (physiologic and biochemical endpoints). Exposures may be presented as nasal-only or, alternatively, utilize exposure hoods or whole-body chambers (allowing the simultaneous documentation of upper airway and eye irritation). Nasal-only exposures may be occlusive (i.e., utilizing facial masks) or open (involving exposure ports, canulae, squeeze bottles, solid adsorbents, or microencapsulated odorants). In the case of nonocclusive exposures, although the stimulus concentration may be precisely documented, the degree of dilution during inhalation (or "sniffing") may be variable. For that reason, some experimenters deliver stimuli directly into the nose with canulae during breath-holding. This chapter reviews stimulus/exposure apparatus in increasing order of complexity.

Microencapsulated Odorants

The best-known example of use of microencapsulated odorants is the University of Pennsylvania Smell Identification Test (UPSIT). Depending upon the version of the test, it may utilize one or several booklets with "scratch-n-sniff" panels associated with a multiple-choice odor identification task. The UPSIT has been well-standardized vis-à-vis such variables as age and gender, and in general, correlates well with the results of quantitative odor detection tests (4). As a consequence, this test is often used in field studies for which stability and portability of stimuli is essential.

Solid Adsorbants

Odorants may also be presented in solid absorbants, for example, the "Sniffin' Sticks" (5) or the "Smell diskettes" (6). Basis of the "Sniffin' Sticks" test system is the use of commercially available felt-tip pens that are meant to be used repetitively. The pens have a length of approximately 14 cm, and the inner diameter of the cylindric pens is 1.3 cm. Instead of liquid dye, the tampon is filled with liquid odorants. For odor presentation, the cap is removed by the experimenter and the pen's tip is placed in front of the nostrils, with the subject being instructed to "sniff." These pens are easy to handle, do not require much space, have a long shelf life, and provide odors with the same constancy as pens normally would provide ink. This system is used not only for suprathreshold tasks (odor identification, odor discrimination, odor memory), but also for assessment of odor thresholds (7).

Squeeze Bottles

Squeeze bottles are frequently used to present odors. The so-called CCCRC (Connecticut Chemosensory Clinical Research Center) test (8) is based on such devices. Presentation of odorants for threshold testing is performed by means of squeezable plastic bottles that are partially filled with various concentrations of odorant dissolved in an odorless liquid carrier such as water or mineral oil. At equilibrium (and at a specified ambient temperature), the vapor headspace in these bottles is predictable, and can be made to follow a predetermined concentration series (see "psychophysical testing," below). Although easy to use, these systems are plagued with the air-puff (i.e., mechanical stimulation) that accompanies each and every odor presentation (for discussion see below; see also Ref. 9). Another critical aspect is the dependency of odor concentration on the frequency of use of the bottles. That is, when a bottle is used only every 30 minutes, the headspace concentration over the odorant/carrier mix will closely approximate equilibrium before the next stimulus is released. When a bottle is used every 30 seconds, on the other hand, the headspace concentration may vary dramatically. In addition, the solvent carrier also plays a role in the release of the odorant (10,11). These phenomena related to headspace concentration are further accentuated when the headspace is relatively small (the typical ratio of headspace to liquid being between 4:1 and 6:1). In rigorous experimental studies, headspace vapor concentration is often documented on a regular basis during the course of the experiment using an instrumental technique such as gas chromatography.

Dynamic Olfactometry

In some experiments it is necessary to apply odorous stimuli with a rectangular time profile (i.e., with rapid onset, and with controlled duration and intensity). In addition, chemical stimulation should not activate sensory systems other than chemosensory (e.g., mechanoreceptors). Based on the principles of air-dilution olfactometry (12) such systems have been developed (13,14). Often odors are either applied intranasally by means of a canula (inner diameter of 4 mm) or in front of the nostrils using a mask (15). Canulas are typically inserted for approximately 1 cm into the nostril such that its opening lies beyond the nasal valve. Odor pulses are embedded in a constantly flowing air stream (typically 2–8 L/min—which relates to a physiological airflow, e.g., during sniffing).

Embedding of the chemosensory stimuli is achieved by directing two airstreams toward the outlet of the olfactometer. Both airstreams have the same flow-rate, the same temperature, and the same humidity. One of the two airstreams contains odor in a certain concentration (Odor = O plus dilution = D), the other air stream is odorless (control = C). Immediately in front of the stimulator's outlet, a vacuum is attached such that during the interstimulus interval the odor-containing air-stream is sucked away and only odorless air enters the subject's nose. For stimulation, the vacuum is attached to the other side such that the odorless air is sucked away and the odor-containing air is led into the subject's nose. This allows switching between odorless and odor-containing air within less than 20 ms. In this system, different odor concentrations are generated by means of air dilution meaning that a pre-established, fully odor-saturated air-stream (O) is mixed with an odorless air-stream (D). While the sum of the two airstreams is kept constant (equal to the control air stream C) different O:D ratios produce different stimulus concentrations.

The constant air-flow directed into the subject's nose requires humidification ($\geq 80\%$ relative humidity) and thermo-stabilization (36°C) because dry, cool air produces nasal congestion, mucus discharge, and pain (16), which interferes with olfaction in many ways (17). This is less important when odors are presented either at very low air-flows or into a mask in front of the nostrils, because in these situations dry and cold air will be humidified and warmed up by the nose. In case that the olfactory stimulator is an open, valveless device (13), cross-contamination of odors needs to be prevented. A separate system of pressure and vacuum is applied such that a small current of odorless air prevents molecules from some odorized tubing to be drawn to other tubing of the open system (Fig. 1).

Most commercially available olfactometers use mass-flow controllers (MFCs), which allow the olfactometer to be effectively controlled by a computer. In addition, other than systems without such MFCs, the flow will be provided

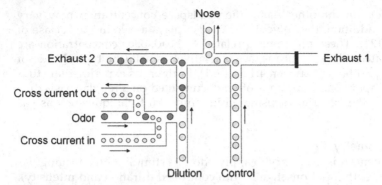

FIGURE 1 Schematic drawing of odor switching mechanism according to Kobal (1981). Two air-streams air directed toward the outlet of the olfactometer (odorless air-streams: Control; Dilution; odorized air-stream: Odor). One of the two air-streams contains odor in a certain concentration (Odor plus Dilution). In front of the stimulator's outlet, a vacuum (Exhaust 2) is attached such that during the interstimulus interval the odor-containing air-stream is vacuumed away and only odorless air enters the subject's nose. For stimulation, the vacuum is switched to Exhaust 1 while Exhaust 2 is blocked. Cross-contamination of odors is prevented with attachment of a pressure/vacuum system (Cross current in/out) to the odor channel. Arrows indicate direction of air-flow. Please see text for further information.

independent of airway resistance. This is important when presenting stimuli in a lateralized fashion to the left and right nostrils which may have very different resistances, for example, dependant on the nasal cycle (18).

When chemosensory stimuli are not embedded in a constant air-stream, but when they are puffed into the nose, a mixed activation of both, the trigeminal/olfactory system results in addition to mechanosensory stimulation. Such combined activation of the mechanosensory and chemosensory systems leads to numerous interactions at various levels of neuronal processing (19,20). This nonlinear interaction between two sensory systems is difficult to be corrected by mathematical procedures (21).

Chambers/Whole Body Exposures

Exposure chambers allow for the most realistic exposures, along with control of potentially confounding variables, such as temperature, humidity, and ambient air pollution. However, this level of control comes at a considerable cost in terms of space, monetary expenditure, and energy utilization. Basic design parameters include chamber volume, air exchange rate, degree of air recirculation, desired air flow pattern, desired control of particulate matter and volatile air pollutants, target temperature and humidity, as well as range of source air conditions. These parameters determine the thermodynamic demands to be placed on the system, which typically vary over the course of the year due to changes in source air temperature and humidity. While heating, cooling, and humidification of air are straightforward, dehumidification requires making a choice between two strategies: either precooling and then reheating source air, or use of solid sorbants to remove water vapor. Injection and mixing of test agents are additional engineering tasks to be addressed. Recent advances in HVAC (heating, ventilation, and air conditioning) technology allow for proportional control of active elements, "tuning" of the system, and minimization of the natural tendency to overshoot or oscillate.

PSYCHOPHYSICAL TESTING

Psychophysics can be defined as "the scientific study of the relation between stimulus and sensation" (22). Psychophysical procedures are divided between threshold measures (detection of low-level or barely perceptible stimuli), and suprathreshold measures. The latter include rating of the intensity of clearly perceptible stimuli (also known as "psychophysical scaling") as well as identification of, or discrimination between, qualitatively different stimuli. Much of psychophysics derives from the Weber–Fechner Law, which in effect states that either: (*i*) perception equal intervals of intensity or (*ii*) perception of minimal change (the so-called JND or "just noticeable difference") require equal changes in actual stimulus intensity on a *ratio* scale (23). Thus, stimuli utilized for psychophysical experiments typically follow logarithmic concentration series. Similarly, presentations of dose–response relationships for suprathreshold series (so-called "psychophysical functions") normally utilize log-linear scales.

Odor Threshold Testing

Odor threshold testing establishes the lowest concentration that an individual (or group of individuals) can reliably detect an odorant. Because of differences in so-called "decision criteria" (essentially, degree of caution in making choices), methods generally involve making forced choices to avoid this particular source of bias between subjects (22). Depending upon the number of alternate stimuli (generally, 2 or 3), chance responses may yield anywhere from 33% to 50% correct answers. Thus, any single correct discrimination between an odorant stimulus and a control (or "blank") does not, by itself, define a perceptual threshold. Many techniques require a criterion number of correct answers (e.g., 5 in a row, yielding a probability of chance response of <0.05 in a binary series) to establish the threshold for a given trial. Rigorous technique requires that the presentation interval between stimuli be fixed, including inter-stimulus interval ("ISI") and inter-trial interval ("ITI"). Further, repeat trials at a given stimulus level are the rule rather than the exception. Most experimenters utilize an ascending concentration series of stimuli in order to minimize odor fatigue (in the case of olfactory testing) or desensitization (in the case of trigeminal testing) as a biasing factor.

Combining these factors, the simplest method for odor threshold testing is referred to as the "ascending series, method of limits." In this technique, stimuli are alternated with (vehicle) blanks, and presented in an ascending concentration series, beginning below the known (or suspected) odor threshold. An incorrect discrimination response (i.e., short of the criterion number of correct responses) results in progression to the next higher stimulus concentration. A threshold is defined when a subject provides a criterion number of correct answers (22). Multiple trials are typically conducted to define a mean detection threshold for each individual. A frequently used variant utilizes the "staircase" technique. Like the method of limits, the staircase technique begins with an ascending series. However, once the criterion number of correct responses is achieved, this is followed by one or more trials at a lower concentration (i.e., following a "descending series"). A threshold is defined when a criterion number of "reversals" (i.e., between ascending and descending concentration series) has been reached (22). A final variant involves recording the proportion of correct responses as a function of stimulus concentration and plotting a so-called "psychometric function." A point on that probability curve, correcting for chance responses, is then selected as representing a criterion level of detection for purposes of either comparing individuals or comparing group responses to different odorants/irritants (22,24).

Suprathreshold Measures

Symptom rating scales are frequently utilized on a longitudinal basis over the course of controlled human exposure experiments. Scales vary from categorical (e.g., Leikert) to continuous (visual analog scales or VAS), with the latter involving varying degrees of linguistic "tethering." Although all such scales involve ordered data, statistical assumptions regarding the ability to manipulate these data are frequently abused. To remedy this situation, psychophysicists devised a variety of VAS known as the "labeled magnitude scale" (LMS). This scale has been demonstrated to have "ratio properties" [i.e., an equal numeric ratio on the scale corresponds to an equal ratio of subjective intensities (25)].

Suprathreshold scaling of such sensations as eye, nose, or throat irritation, odor, taste, or astringency suffer from inter-subject variability, but can be transformed to a common scale across subjects, yielding similar dose–response characteristics. In a psychophysical experiment utilizing the *method of magnitude estimation*, for example, subjects are first presented a standard stimulus, and then all subsequent stimuli are rated relative to the reference value ("modulus"). When the rating of the standard stimulus is numerically equalized across subjects (generating, in effect, perceived intensity ratios), the psychophysical slope (or dose–response) tends to be very similar, regardless of the individual rater (23). Of note, the LMS (referenced above) yields similar results to the method of magnitude estimation when applied under controlled conditions (25).

Odor Discrimination Testing

For assessment of odor discrimination frequently *3-alternative forced choice procedures* (3-AFC) are used (26,27). Subjects are presented with three odorants and the task is to identify the one sample that has a different smell. To prevent visual detection of the target, subjects are often blindfolded. For odor discrimination, criteria for the selection of odorants include the following: (*i*) odors in a triplet have to be similar with regard to intensity; (*ii*) odors should also be similar in their hedonic tone; (*iii*) correct discrimination of individual odorants should be above 75% in healthy subjects. Presentation of triplets should be separated by at least 30 seconds to prevent olfactory fatigue. The interval between presentation of individual odor probes could be approximately three seconds. More recent data also indicate that odor discrimination has a strong memory component to it (28,29), rendering it an olfactory function that seems to be influenced by cognitive factors to a much higher degree than, for example, odor thresholds (30,31).

Trigeminal Threshold Testing

Psychophysical approaches to trigeminal activation include assessment of thresholds (32,33), the subjects' ratings, or the assessment of the subjects' ability to localize chemosensory stimuli (34–38), which seems to be largely, but probably not entirely, determined by trigeminal activation (39). Threshold measures are thought to be closely related to trigeminal function at the receptor level (40). For the intranasal trigeminal system, the specific challenge is that most odors produce olfactory-mediated sensations first while thresholds for trigeminally mediated sensations are typically much higher. This means, that trigeminally mediated sensations are biased by the olfactory mediated sensations.

In a comparative study (41), trigeminal threshold measures were obtained in two ways: In a first experiment threshold concentrations were identified at which a "trigeminal" sensation appeared, based on the subjects' verbal assessment of trigeminal function. In a second experiment concentrations were determined at which the stimulated nostril could be determined after lateralized stimulus presentation (34,36,38,42). In both experiments, thresholds were determined using an initially ascending single staircase method (43). Thresholds determined in the "verbal" experiment were found to be approximately five dilution steps higher than those obtained in the "lateralization" experiment meaning that subjects appeared to be more sensitive in the verbal task. Results from the two experiments were not significantly correlated with each other. Thus, although both methods are reliable (regarding test–retest correlation), they measure different aspects of the trigeminal system. The differences may, among others, relate to a stronger bias of verbal ratings by the odor accompanying the chemosensory stimuli. They may also reflect the idea that reliable lateralization of chemosensory stimuli is only possible at clearly suprathreshold levels.

To avoid the confounding of trigeminal thresholds by odorous sensations, apart from the assessment of lateralization several approaches have been proposed. For example, trigeminal thresholds may be determined in subjects with a lack of olfactory function (44) [a problem here is that trigeminal sensitivity appears to be altered as a consequence of the loss of the sense of smell (45)]. Other approaches include use of the odorless irritant, carbon dioxide (46–49) or the assessment of trigeminal sensitivity of the conjunctiva (50,51), which seems to be comparable to intranasal mucosal sensitivity. Yet further approaches relate to electrophysiological measures of trigeminal function (52) using the negative mucosal potential (see below).

Thus, several techniques are available which seem to reflect various aspects of trigeminal sensitivity. Preference of one method over the other depends on the question being asked (compare Ref. 53).

PHYSIOLOGIC INSTRUMENTATION

Researchers concerned with the lower airway have a well-validated set of physiologic study tools. Many of these are sufficiently standardized to allow their use in clinical settings. For example, a clinical pulmonary function evaluation may involve screening spirometry (lung mechanics) or full pulmonary function testing including lung mechanics, lung volumes, and carbon monoxide diffusion capacity.

The situation for the upper airway, on the other hand, is considerably less settled. Nasal patency (e.g., nasal airway resistance), for example, cannot be predicted by variables commonly used to normalize pulmonary function tests (i.e., age, gender, height, or ethnicity). Given this high degree of intrinsic variability, many nasal physiologic tests are robustly interpretable only if utilized in a longitudinal or exposure-control study design. Thus, many "nasal function tests" have yet to make the transition from bench to bedside, and will most likely be encountered in experimental study designs. Similarly, most electrophysiological or functional imaging tools applied to olfaction and nasal trigeminal function are considered to be research—rather than clinical—tools.

Electrophysiology

Electro-Olfactogram

First electrophysiological investigations of the olfactory epithelium were performed (54), ex-vivo, in the olfactory epithelium of the dog. Ottoson (55,56) extensively studied the frog's olfactory epithelium to characterize the electro-olfactogram (EOG) which is thought to reflect the summated generator potentials of olfactory receptor neurons (ORNs). Getchell et al. extended these findings through their work in the salamander (57). Among many significant findings, they also showed that it appears highly unlikely that the EOG can be recorded from the skin overlying the olfactory epithelium (58).

First recordings of the human EOG were published in 1969 (59). Later work by Kobal was based on the development of an olfactometer, which presents stimuli without eliciting mechanically induced sensations (13) (see above). To elicit a visible EOG, odorants have to reach the epithelium with a high concentration gradient. In addition, only the use of an adequate olfactometer and "pure" olfactory stimulants would guarantee exclusive activation of ORNs.

The typical EOG recording electrode (56) is insulated, for example, through a Teflon tube. A so-called "agar-bridge" (1% Ringer-Agar) ascertains that the odorant has no direct contact with the electrode. A Ag/AgCl electrode is typically used for recording; the reference is placed at the bridge of the nose, contralateral to the recording site (13). To minimize artifacts produced by muscular contractions or eye movements, it is helpful to seat the subjects as comfortable as possible. Typically, a frame similar to lensless glasses is used to keep the EOG electrode in place; repeated nasal endoscopy helps to improve the quality of recordings (60). To avoid artifacts due to respiratory movements of the electrode, subjects should breathe in a specific manner in order to avoid movement of respiratory air in the nasal cavity, the so-called "velopharyngeal closure" which avoids air-flow due to the air stream in the nose during breathing (13,61).

The Negative Mucosal Potential

It is also possible to record potentials from the respiratory epithelium (62). This so-called negative mucosa potential (NMP) is believed to represent peripheral trigeminal activation (44,63,64). It is similar to EOGs (see above) in terms of both shape and response properties (13,62). The NMP is thought to reflect functional aspects of trigeminal chemosensors (i.e., nociceptors) (64). For instance, when stimuli are applied repetitively at short interstimulus intervals, NMP amplitudes always decrease; the NMP is greater the longer the ISI (65). The decrease of amplitudes is accompanied by the simultaneous prolongation of the latency of its onset. This behavior of the NMP is in line with observations made on the adaptive properties of both C-fibers and Adelta-fibers (63).

The NMP can be recorded from the human nasal respiratory epithelium after stimulation with chemicals, which produce stinging or burning; in contrast, these responses are not detectable when the stimuli produce sensations that are predominantly, if not exclusively, mediated by the olfactory nerve (13). The NMP is a function of both stimulus duration and stimulus concentration (47,66); it is not related to autonomic changes such as changes in blood flow (66). Importantly, it has been shown that the NMP's amplitude and its latencies correspond to stimulus intensity (13,47,66).

Cortical Event-Related Potentials to Trigeminal Stimuli (tERP)

Trigeminal event-related potentials (tERP) are direct correlates of cerebral neuronal activation. They have a high temporal resolution in the range of microseconds, allow the investigation of the sequential processing of sensory information, and can be obtained largely independently of the subject's response bias, for example, they can also be obtained in children. tERP peaks can be divided in "exogenous" components (P1, N1) that are largely determined by characteristics of the stimulus like intensity or quality, and "endogenous" components that reflect more the subjective meaning of the stimulus (67). However, in many situations "exogenous" and "endogenous" phenomena cannot be clearly separated. It can be assumed that all components are subject to modulation by cognitive processes (68); in other words, control of experimental conditions and contextual situations are highly important in terms of the outcome of tERP measurements.

ERP need to be extracted from the background EEG activity. Improvement of the signal-to-noise ratio typically is performed with averaging of individual responses to olfactory stimuli such that random activity would cancel itself out while all nonrandom activation would still be left. To increase the signal-to-noise (S/N) ratio, stimuli should be presented with a steep onset in a well-controlled, monotonous environment such that stimulus onset synchronizes the activity of as many cortical neurons as possible (compare functional MRI). The averaging process also requires stimulus repetition which, in turn, brings on issues like changes of the subjects' vigilance during long experiments and adaptation/habituation phenomena.

The frequency spectrum of tERPs ranges between 1 and 8 Hz; both filtering and sampling frequency must be set accordingly. Eight records are regarded to be the absolute minimum for averaging a response (13) (compare Ref. 69), best S/N ratio is reached when 60–80 responses are averaged (70). Considering an interstimulus interval of 30 seconds, the duration of one session would be at least 30 minutes. This long time required for investigation introduces other sources of artifacts, for example, changing levels of vigilance during recording. tERP is recorded all over the scalp; amplitudes exhibit a centro-parietal maximum (71–73).

During recording stable environmental conditions are important. This includes the visual and acoustical shielding of subjects (drapes, white noise). Further, a defined task has proven most helpful. Specifically, many labs use a tracking task where subjects are requested to keep a small square controlled by a joystick inside a larger one which randomly moves on a screen at eye-level, at a distance of approximately 1.5–2 m from the subject's eyes. Whenever adequate subjects are trained in a specific breathing technique, the velopharyngeal closure (see above). This technique helps to prevent respiratory flow through the nose by lifting the soft palate which is under voluntary control.

Functional Imaging of Trigeminally Induced Brain Activation

fMRI

Functional magnetic resonance imaging (fMRI) allows one to visualize cerebral changes that are related to neuronal activity. The most frequently used technique is the blood oxygen level dependent (BOLD) contrast (74). A typical functional neuroimaging experiment consists of acquiring a series of brain images while a subject is performing a task or receives some stimulation (75,76). Multiple slices covering the whole brain are collected repeatedly over several minutes while the paradigm switches between rest ("OFF") and activation ("ON"). Because changes of the BOLD effect are in the range of 1–10% above noise level, averaging is needed to increase the S/N ratio of the signal. Overall, fMRI provides unique information about degree and location of neuronal activation. However, its temporal resolution is relatively low, especially when compared to EEG- or MEG-related measures.

As a research tool, fMRI has become more and more important over the last years. With regard to trigeminal intranasal activation, it has helped to reveal the close relationship between the olfactory and the trigeminal systems. Specifically, chemical stimulation of the intranasal trigeminal system not only activates areas related to the processing of pain (e.g., SI, SII, anterior cingulate cortex, thalamus) (77–82), but also the piriform cortex which is thought to be "primary" olfactory cortex (83,84).

Positron Emission Tomography

Positron Emission Tomography (PET) allows one to obtain quantitative maps of numerous physiological parameters including blood flow, protein synthesis, or glucose metabolic rate. Molecules are labeled with short-lived, positron-emitting nuclei before intravenous injection. When the nucleus undergoes nuclear transition, a positron is emitted. The positron combines with a free electron which ultimately produces a pair of antiparallel photons of 511 keV energy each. These photons are detected by the positron tomograph (85). Hemodynamic tracers are used to map cerebral blood flow [e.g., ^{15}O (86)], metabolic tracers are used to map the metabolism of glucose [e.g., ^{18}F-deoxyglucose (87)].

Major disadvantages of PET are its costs, its low temporal resolution, and the subjects' exposure to radioactivity (88). Advantages include its potential to provide absolute values of physiological parameters, its suitability to investigate in vivo neurotransmission (89), and its relative freedom from the susceptibility artifacts that are seen in fMRI.

Magnetoencephalography

Based on the measurement of magnetic fields on the surface of the scalp, magneto-encephalography (MEG) can be used to localize generators of these magnetic fields in the brain (90). Thus, in contrast to PET and fMRI, MEG allows one to directly assess the neuronal activity involved in the processing of sensory information. In addition, MEG has the same temporal resolution as EEG so that it is possible to track the development of neuronal processes over time. The link between the magnetically defined current dipoles provided by MEG and anatomical data provided by MRI allows one to identify brain areas where the activation is generated. A number of studies have been performed to investigate neuronal activation induced by painful stimulation in the nasal cavity (91–95). For example, work by Hari et al. strongly suggested a greater role of the right hemisphere in pain processing (92).

Nasal Physiologic Measurements

Considered here are measures of nasal patency and nasal mucociliary clearance. For a discussion of nasal mucosal blood flow, the reader is referred to a separate review of this topic (96).

Measurement of Nasal Patency

Acute changes in nasal patency generally reflect vascular processes, including dilation of venous capacitance vessels in the nasal mucosa and/or extravasation of plasma from fenestrated capillaries. This phenomenon is of interest because it is frequently observed, not only in allergy, but also as a reflex response to irritant provocation (97–100). Physiologic consequences of nasal obstruction may include loss of upper airway filtering and air conditioning (due to mouth breathing) and obstruction of the ostia of the Eustachean tubes and paranasal sinuses (potentially contributing to the pathogenesis of otitis media or sinusitis, respectively).

Nasal Inspiratory Peak Flow

Nasal inspiratory peak flow (PIFn) is, technically speaking, the least demanding measure of nasal patency. Measurement of PIFn, in general, involves the use of mechanical devices (101). The flow meter is attached to a facemask that covers the nose. The subject exhales maximally, places the mask over the nose, and then inspires with maximal effort to total lung capacity. The reading is noted, the pointing device reset, and the procedure repeated. Typically, a total of three measurements are made, with the highest flow rate being recorded. PIFn measurement has as its principal advantage its portability, as well as the simplicity (and economy) of the apparatus involved. The measure is, however, effort-dependent, so that inadvertent or deliberate changes in inspiratory force could produce artifactual results. Finally, the considerable negative airway pressures involved in the PIFn maneuver can partially collapse the external airway, potentially confounding those changes in nasal patency due to mucosal swelling which are the actual phenomena of interest.

In terms of measurement stability, Cho et al. (102) compared the reproducibility of individual PIFn values obtained with two variants of the procedure: maximal inspiration from full exhalation (residual volume or RV method) and from the end of a normal exhalation (functional residual capacity or FRC method). The investigators obtained a coefficient of variation (CV) of 10.1% for the RV method and 12.1% for the FRC method. In terms of validity, PIFn has been compared with a more technically demanding technique (rhinomanometry), and the results correlate well (103).

Rhinomanometry

Rhinomanometry involves the simultaneous measurement of trans-nasal pressure (P) and flow (V), yielding a direct index of nasal airway resistance (NAR):

$$NAR = P/V$$

Rhinomanometry is divided into active and passive, the former involving active air movement by the subject being tested (usually during quiet tidal breathing) and the latter air movement induced by an externally imposed pressure. For all practical purposes, commercially available rhinomanometers all involve *active* rhinomanometry.

The procedure is further divided into anterior and posterior rhinomanometry. Anterior rhinomanometry involves the separate measurement of NAR in each of the two hemi-nasal cavities, whereas posterior rhinomanometry yields a global measure of upper airway resistance. Biophysically, the two nasal cavities act like parallel resistors, such that the combined resistance is less than either of the individual resistances:

$$NAR_T = (NAR_R \times NAR_L)/(NAR_R + NAR_L)$$

Where NAR_T is combined NAR, and the subscripts R and L refer the NAR on the right and left, respectively. When NAR is equal on both sides, NAR_T is one-half of the unilateral value. (Another way of conceptualizing this is that the unilateral conductances—defined as the reciprocal of the resistances—are additive).

Anterior and posterior rhinomanometry have distinct technical and patient cooperation demands. Anterior rhinomanometry involves alternate occlusion of each nostril with a pressure tap, and measurement of flow through the opposite nostril. This metric is dependent upon two conditions being satisfied: (*i*) the occluded hemi-nasal cavity (i.e., the one in which there is no flow) is patent (i.e., it communicates with the nasopharynx); and (*ii*) the two hemi-nasal cavities do not communicate except at the level of the nasopharynx (i.e., that there is no septal perforation). As neither of these assumptions involves the cooperation of the subject, an advantage of anterior rhinomanometry is that it can be performed on essentially any individual. Another advantage, at least from the standpoint of the otolaryngologist, is that unilateral NAR measurements can be correlated with anatomical abnormalities (e.g., the presence of a deviated septum, polyp, or other occluding mass).

Both environmental scientists and allergists, on the other hand, are generally more interested in a global measure of airway occlusion/mucosal swelling than in unilateral measures. For studies of the effect of the environment on the nose (with the exception of unilateral nasal challenges), measurement of total NAR using posterior rhinomanometry is both speedier and more direct (104). Nevertheless, there are limitations to the procedure that render its use problematic in some cases (see below).

Posterior rhinomanometry (Fig. 2) involves the use of a mask to measure total nasal airflow, and a flexible tube, held firmly between the lips and terminating in the space between the palate and tongue, as a pressure tap (105). In order for this technique to generate meaningful data, it is essential that the oral cavity communicate with the oropharynx/nasopharynx (i.e., the soft palate needs to be elevated away from the tongue). This is apparently problematic for the approximately 10–15% of subjects who cannot produce useful pressure-flow tracings with this technique despite extensive coaching. Thus, although posterior rhinomanometry produces data of greater direct utility to environmental scientists than does anterior rhinomanometry, it suffers from limitations due to problems with subject cooperation.

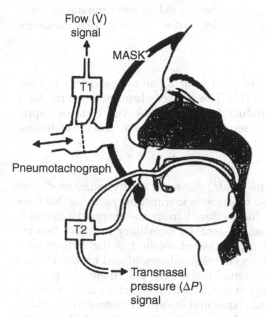

Flow (V̇) signal

MASK

T1

Pneumotachograph

T2

Transnasal pressure (ΔP) signal

FIGURE 2 Schematic showing posterior rhinomanometry. *Source*: From Ref. 105.

In terms of measurement characteristics, both anterior and posterior rhinomanometry produce results with coefficients of variation in the 12–16% range (104).

Acoustic rhinometry
Acoustic rhinometry (AR) is a technique in which sharply demarcated pulses of sound ("clicks") are directed along a tube ("sound tube") into each nostril individually. The apparatus then records reflected sound energy with a microphone in such a way that a map is generated of cross-sectional area of each heminasal cavity as a function of distance from the termination of the nosepiece of the sound tube. Integrating area over distance then yields nasal volume (106). Although there is a general expectation that nasal volume and NAR will relate in an inverse manner, the actual observed relationship between these two variables is somewhat more complex.

Although patient cooperation requirements for AR are minimal (consisting solely of a brief period of breath-holding), the technique is nevertheless fraught with potential technical pitfalls. Chief among these is a tendency for considerable within-subject variability that is apparently related to: (*i*) the adequacy of the seal achieved between the nosepiece and nares, and (*ii*) variability of geometric orientation of the sound tube relative to the nasal cavity. Guidelines for quality control in AR have been put forth by members of the Standardization Committee on Acoustic Rhinometry (107).

Similar to anterior rhinomanometry, acoustic rhinometry provides unilateral data, and hence can be used by surgeons to document anatomical abnormalities (e.g., deviated septum; polyposis), as well as to assess the results of surgical intervention (108). AR has also been used to discriminate between allergic rhinitic and nonrhinitic subjects by virtue of their pharmacologic response to nasal decongestants (109).

Rhinostereometry
Rhinostereometry is a relatively new method utilizing an optical device to visualize swelling of the nasal turbinates (110). The subject's head must be precisely immobilized in order to yield reproducible results, however. Data on reproducibility, both within and between observers, are still sparse, and the technique has enjoyed only limited application to date.

Nasal Mucociliary Clearance
Posterior to the tip of the inferior turbinate, the nasal cavity is invested with ciliated respiratory epithelium whose function is to transport particles that have been removed (by the process of "impaction") from the inspired airstream. Various techniques have been devised to assess mucociliary function; two are mentioned here. Perhaps the most sophisticated method is the radiographic visualization of the transit of radiotagged microspheres placed on the inferior turbinate (111). A much simpler technique involves the placement of a grain of saccharine just behind the anterior tip of the inferior turbinate, with notation of the time between instillation of the substance and the appearance of a sweet taste in the mouth of the subject (the so-called "saccharine transit time"). While inhibition of mucociliary clearance is a hallmark of primary ciliary dyskinesia (an inherited condition), attempts to document systematic alterations in mucociliary function after exposure to air pollutants have been of variable success (112,113).

Direct and Indirect Measures of Nasal Inflammation
Inflammatory conditions of the human upper airway (e.g., rhinitis) are discussed in Chapters 26 and 27 of this volume. Chemical irritation of the nasal mucosa can occur against a backdrop of either a diseased or healthy upper respiratory tract. As detailed elsewhere, the types of inflammatory cells and mediators activated by chemical irritants may differ from—and yet interact with—those activated by allergic processes.

Nasal Lavage
Nasal lavage ("NL") has been the most widely used technique for noninvasive documentation of nasal inflammation. It is based upon the premise that the flux of cells and mediators across the intact nasal mucosa—as well as the content of secretions from subepithelial glands and superficial goblet cells—will reflect the inflammatory state of mucosa. Supporting this thesis are a large number of studies documenting the fact that predicted cell types (e.g., eosinophils), mediators (e.g., histamine, tryptase, platelet-activating factor), and markers (e.g., albumin) are increased in NL fluid after nasal allergen challenge in a condition with known pathophysiology—allergic rhinitis (114). Extrapolating the validity of this technique, researchers have applied NL to investigating the nasal inflammatory effects of chemical pollutants and other nonallergic stimuli (115–117).

A number of operational variables must be addressed before applying NL in laboratory or field studies. The least problematic of these is the choice of lavage

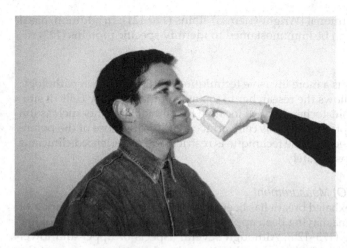

FIGURE 3 Nasal lavage, using the spray method.

fluid. The fluid used in most studies is "physiologic" saline (or 0.9% NaCl solution), as this formulation poses minimal osmotic stress on the mucosa. Alternatively, phosphate-buffered saline ("PBS") poses neither an osmotic nor a pH challenge. The technique of instillation of the lavage fluid, however, can be quite variable. Two main techniques are employed, "bolus" or "spray" (Fig. 3), with the former having several variants.

The relative strengths of the bolus and spray methods were evaluated systematically as part of a series of experiments conducted at the US EPA's Health Effects Research Laboratory [the "CompNose" study (118)]. The two methods had comparable yields (% volume recovery), total cell counts, and neutrophil counts, but the spray method yield fluid with significantly higher concentrations of albumin (in fact, over 25-fold higher levels). Since albumin is a marker of intravascular fluid leak, these data are consistent with microtrauma/mechanically induced capillary transudation. A separate study documented microepistaxis (minor bleeding) with the spray technique, and advocated a cutoff for contaminated specimens based upon the free hemoglobin concentration in lavage fluid (119).

Nasal Cytology

Superficial nasal cytologic specimens can be obtained using a variety of collection techniques, including swabs, curettes, and brushes (120). Nasal swabs (e.g., using a cotton- or calcium alginate-tipped applicator) have the advantage of producing the least discomfort, but on the other hand produce the lowest cell yield. A commercially available disposable plastic curette, the RhinoProbe® (Arlington Scientific, Springvale, UT) is well tolerated and yields quantities of cells sufficient for cytologic smears and RNA extraction. Some investigators have used modified brushes designed for bronchoscopy to perform superficial nasal sampling, but this technique appears to offer no advantage over the use of curettes and may produce slightly more discomfort. Semiquantitative scoring of inflammatory cell populations has been performed on nasal cytologic specimens processed with conventional [Wright-Giemsa] stains (120,121). In addition, nasal cytologic specimens can be immunostained to identify specific proteins (122).

Nasal Biopsy

Nasal mucosal biopsy is a more invasive technique for studying nasal pathology. On the one hand, it allows the researcher to visualize inflammatory cells in situ, and utilizing immunohistochemistry, to localize functional elements such as ion channels/ligand receptors, as well as cell surface markers. Because of the potential, but low, risk of bleeding, the technique is restricted to experienced clinicians who can perform biopsies safely.

Nasal Nitric Oxide (NO) Measurement

Nitric oxide (NO) in exhaled breath has been established as an indirect marker of inflammation in asthma, having been applied in both clinical and epidemiologic studies of the disease (123–125). Although several aspects of upper and lower airway inflammation may be linked in rhinitis and asthma, the mucosae of the upper vs. lower airways have distinct NO production dynamics. Specifically, NO production rates in the upper airway are typically much higher than those in the lower airways, necessitating collection techniques that separate the two sources (126–128). Furthermore, the paranasal sinuses (particularly, maxillary) appear to act as a reservoir for the upper airway, but forfeit that function when the sinus loses communication with the nasal cavity (as in polyposis or osteomeatal complex disease). Thus, although upper airway NO has, with some variation, been found to be increased in pure rhinitis, it is more consistently decreased in sinusitis, as well as being diminished with cystic fibrosis and primary ciliary dyskinesia (126).

In terms of measurement characteristics, one population study of normal adults found nasal NO to be roughly normally distributed, with no age or gender effect, nor any evidence of diurnal variation. Coefficients of variation (COV) for repeated measures were in the 7–10% range, with between-day COV being approximately 12% (129).

A variant of nasal NO has been used as an index of paranasal sinus osteal patency and hence the likelihood of the development of sinusitis. Maniscalco et al. observed that humming produces a dramatic increase ("spike") in nasal NO levels, but only when the osteomeatal complex appears patent by nasal endoscopy (130). This may account for reports of decreased nasal NO in sinusitis. Various attempts have been made to standardize the technique for this sampling variant (130–132).

SUMMARY/CONCLUSIONS

The human upper airway is a challenging target for experimental study. Unlike the case for the lower airway, study tools are only partially standardized, and both inter- and intraindividual variability must be controlled for in the study design. The range of target phenomena includes sensory processes (olfaction; trigeminal chemoreception), biophysical measurements of airway dynamics, and direct and indirect measures of mucosal inflammation. Instrumentation for upper airway studies spans the range from PET and MRI scanners to hand-held peak flow meters, along with several orders of magnitude difference in complexity, cost, and portability. Upper airway study techniques are currently in a state of flux, with much room for innovation.

REFERENCES

1. McFarland HN. Design and operational characteristics of inhalation exposure equipment – A review. Fundam Appl Toxicol 1983; 3:603–613.
2. Costa DL, Tepper JS. Approaches to lung function assessment in small mammals. In: Gardner DE, Crapo JD, Massaro EJ, eds. Toxicology of the Lung. New York, NY: Raven Press, Ltd., 1988:147–174.
3. Roggli VL, Brody AR. Imaging techniques for application to lung toxicology. In: Gardner DE, Crapo JD, Massaro EJ, eds. Toxicology of the Lung. New York, NY: Raven Press, Ltd, 1988:117–146.
4. Doty RL, Frye RE, Agrawal U. Internal consistency reliability of the fractionated and whole University of Pennsylvania Smell Identification Test. Percept Psychophys 1989; 45:381–384.
5. Kobal G, Hummel T, Sekinger B, et al. "Sniffin' Sticks": Screening of olfactory performance. Rhinology 1996; 34:222–226.
6. Simmen D, Briner HR, Hess K. Screeningtest des Geruchssinnes mit Riechdisketten. Laryngorhinootologie 1999; 78:125–130.
7. Hummel T, Kobal G, Gudziol H, et al. Normative data for the "Sniffin' Sticks" including tests of odor identification, odor discrimination, and olfactory thresholds: an upgrade based on a group of more than 3,000 subjects. Eur Arch Otorhinolaryngol 2007; 264:237–243.
8. Cain WS, Gent JF, Goodspeed RB, et al. Evaluation of olfactory dysfunction in the Connecticut Chemosensory Clinical Research Center (CCCRC). Laryngoscope 1988; 98:83–88.
9. Elsberg CS, Brewer ED, Levy I. The sense of smell. IV. Concerning conditions which may temporarily alter normal olfactory acuity. Bull Neurol Inst NY 1935; 4:31–34.
10. Pierce JD Jr., Doty RL, Amoore JE. Analysis of position of trial sequence and type of diluent on the detection threshold for phenyl ethyl alcohol using a single staircase method. Percept Mot Skills 1996; 82:451–458.
11. Tsukatani T, Miwa T, Furukawa M, et al. Detection thresholds for phenyl ethyl alcohol using serial dilutions in different solvents. Chem Senses 2003; 28:25–32.
12. Prah JD, Sears SB, Walker JC. Modern approaches to air dilution olfactometry. In: Doty RL, ed. Handbook of olfaction and gustation. New York, NY: Marcel Dekker, 1995:227–255.
13. Kobal G. Elektrophysiologische Untersuchungen des menschlichen Geruchssinns. Stuttgart, Germany: Thieme Verlag, 1981.
14. Kobal G, Hummel C. Cerebral chemosensory evoked potentials elicited by chemical stimulation of the human olfactory and respiratory nasal mucosa. Electroencephlogr Clin Neurophysiol 1988; 71:241–250.
15. Sobel N, Prabhakaran V, Desmond JE, et al. A method for functional magnetic resonance imaging of olfaction. J Neurosci Methods 1997; 78:115–123.
16. Lötsch J, Ahne G, Kunder J, et al. Factors affecting pain intensity in a pain model based upon tonic intranasal stimulation in humans. Inflamm Res 1998; 47:446–450.
17. Hummel T, Livermore A. Intranasal chemosensory function of the trigeminal nerve and aspects of its relation to olfaction. Int Arch Occup Enviorn Health 2002; 75:305–313.
18. Kayser R. Die exacte Messung der Luftdurchgängigkeit der Nase. Arch f Laryng 1895; 3:110–115.
19. Stone H, Rebert CS. Observations on trigeminal olfactory interactions. Brain Res 1970; 21:138–142.
20. Hummel T, Kobal G. Olfactory event-related potentials. In: Simon SA, Nicolelis MAL, eds. Methods and frontiers in chemosensory research. Boca Raton, FL: CRC Press, 2001:429–464.

21. Swandulla D. Einige Aspekte der klinischen Anwendung olfaktorisch evozierter Potentiale (OEP). Physiology. Erlangen, Germany: Erlangen-Nürnberg, 1986.
22. Gescheider GA. Psychophysics: Method, Theory, and Application. NewYork, NY: L. Erlbaum Associates, 1985.
23. Stevens SS. Psychophysics. Introduction to its Perceptual, Neural, and Social Prospects. New Brunswick, NJ: Transaction Publishers, 1986.
24. Cometto-Muñiz JE, Abraham MH. Human olfactory detection of homologous n-alcohols measured via concentration-response functions. Pharmacol Biochem Behav 2008; 89:279–291.
25. Green BG, Dalton P, Cowart B, et al. Evaluating the 'Labeled Magnitude Scale' for measuring sensations of taste and smell. Chem Senses 1996; 21:323–334.
26. Laska M. Olfactory discrimination ability for aliphatic c6 alcohols as a function of presence, position, and configuration of a double bond. Chem Senses 2005; 30: 755–760.
27. Linschoten MR, Harvey LO Jr., Eller PM, et al. Fast and accurate measurement of taste and smell thresholds using a maximum-likelihood adaptive staircase procedure. Percept Psychophys 2001; 63:1330–1347.
28. Stevenson RJ, Mahmut M, Sundqvist N. Age-related changes in odor discrimination. Dev Psychol 2007; 43:253–260.
29. Wilson DA, Stevenson RJ. Olfactory perceptual learning: the critical role of memory in odor discrimination. Neurosci Biobehav Rev 2003; 27:307–328.
30. Lötsch J, Reichmann H, Hummel T. Different odor tests contribute differently to the diagnostics of olfactory loss. Chem Senses 2008; 33:17–21.
31. Dalton P. Psychophysical methods in the study of olfaction and respiratory tract irritation. AIHAJ 2001; 62:705–710.
32. Cometto-Muniz JE, Cain WS, Hudnell HK. Agonistic effects of airborne chemicals in mixtures: odor, nasal pungency, and eye irritation. Percept Psychophys 1997; 59: 665–674.
33. Lötsch J, Nordin S, Hummel T, et al. Variation of nociceptive and olfactory thresholds in healthy young adults: circadian aspects. Chem Senses 1997; 22:210.
34. Berg J, Hummel T, Huang G, et al. Trigeminal impact of odorants assessed with lateralized stimulation. Chem Senses 1998; 23:587.
35. Hummel T, Futschik T, Frasnelli J, et al. Effects of olfactory function, age, and gender on trigeminally mediated sensations: a study based on the lateralization of chemosensory stimuli. Toxicol Lett 2003; 140–141:273–280.
36. Kobal G, Van Toller S, Hummel T. Is there directional smelling? Experientia 1989; 45:130–132.
37. Shusterman D, Murphy M-A, Balmes J. Differences in nasal irritant sensitivity by age, gender, and allergic rhinitis status. Int Arch Occup Environ Health 2003; 76: 577–583.
38. Wysocki CJ, Cowart BJ, Radil T. Nasal trigeminal chemosensitivity across the adult life span. Percept Psychophys 2003; 65:115–122.
39. Porter J, Craven B, Khan RM, et al. Mechanisms of scent-tracking in humans. Nat Neurosci 2007; 10:27–29.
40. Hornung DE, Kurtz DB, Bradshaw CB, et al. The olfactory loss that accompanies an HIV infection. Physiol Behav 1998; 15:549–556.
41. Frasnelli J, Hummel T. Intranasal trigeminal thresholds in healthy subjects. Environ Toxicol Pharmacol 2005; 19:575–580.
42. von Skramlik E. Über die Lokalisation der Empfindungen bei den niederen Sinnen. Z Sinnesphysiol 1925; 56:69–140.
43. Ehrenstein WH, Ehrenstein A. Psychophysical methods. In: Windhorst U, Johansson H, eds. Modern techniques in neuroscience research. Berlin: Springer, 1999: 1211–1241.
44. Cain WS, Lee NS, Wise PM, et al. Chemesthesis from volatile organic compounds: Psychophysical and neural responses. Physiol Behav 2006; 88:317–324.
45. Frasnelli J, Hummel T. Interactions between the chemical senses: Trigeminal function in patients with olfactory loss. Int J Psychophysiol 2007; 65:177–181.

46. Cometto-Muniz JE, Cain WS. Perception of nasal pungency in smokers and non-smokers. Physiol Behav 1982; 29:727–731.
47. Hummel T, Kraetsch H-G, Pauli E, et al. Responses to nasal irritation obtained from the human nasal mucosa. Rhinology 1998; 36:168–172.
48. Shusterman DJ, Balmes JR. A comparison of two methods for determining nasal irritant sensitivity. Am J Rhinol 1997; 11:371–378.
49. Wise PM, Wysocki CJ, Radil T. Time-intensity ratings of nasal irritation from carbon dioxide. Chem Senses 2003; 28:751–760.
50. de Wijk R, Cain WS, Pilla-Caminha G. Human psychophysical and neurophysiological measurements on ethanol. Chem Senses 1998; 23:586.
51. Vesaluoma M, Muller L, Gallar J, et al. Effects of oleoresin capsicum pepper spray on human corneal morphology and sensitivity. Invest Ophthalmol Vis Sci 2000; 41:2138–2147.
52. Frasnelli J, Hummel T. Age related decline of intranasal trigeminal sensitivity: Is it a peripheral event? Brain Res 2003; 987:201–206.
53. Doty RL, Cometto-Muñiz JE, Jalowayski AA, et al. Assessment of upper respiratory tract and ocular irritative effects of volatile chemicals in humans. Crit Rev Toxicol 2004; 34:85–142.
54. Hosoya Y, Yoshida H. Über die bioelektrischen Erscheinungen an der Riech-schleimhaut. J Med Sci III Biophys 1937; 5:22.
55. Ottoson D. Sustained potentials evoked by olfactory stimulation. Acta Physiol Scand 1954; 32:384–386.
56. Ottoson D. Analysis of the electrical activity of the olfactory epithelium. Acta Physiol Scand 1956; 35:1–83.
57. Getchell TV, Getchell ML. Peripheral mechanisms of olfaction: biochemistry and neurophysiology. In Finger TE, Silver WL, eds. Neurobiology of taste and smell. Malabar: Krieger Publishing Company, 1991:91–123.
58. Getchell TV. Analysis of intracellular recordings from salamander olfactory epithelium. Brain Res 1977; 123:275–286.
59. Osterhammel P, Terkildsen K, Zilsdorff K. Electro-olfactograms in man. J Laryngol 1969; 83:731–733.
60. Knecht M, Hummel T. Recording of the human electro-olfactogram. Physiol Behav 2004; 83:13–19.
61. Nagel WA. Einige Bemerkungen über nasales Schmecken. Ztschr f Psychol 1904; 25:268.
62. Kobal G. Pain-related electrical potentials of the human nasal mucosa elicited by chemical stimulation. Pain 1985; 22:151–163.
63. Hummel T. Assessment of intranasal trigeminal function. Int J Psychophysiol 2000; 36:147–155.
64. Thürauf N, Friedel I, Hummel C, et al. The mucosal potential elicited by noxious chemical stimuli: is it a peripheral nociceptive event. Neurosci Lett 1991; 128:297–300.
65. Hummel T, Schiessl C, Wendler J, et al. Peripheral electrophysiological responses decrease in response to repetitive painful stimulation of the human nasal mucosa. Neurosci Lett 1996; 212:37–40.
66. Thürauf N, Hummel T, Kettenmann B, et al. Nociceptive and reflexive responses recorded from the human nasal mucosa. Brain Res 1993; 629:293–299.
67. Chen ACN, Chapman CR, Harkins SW. Brain evoked potentials are functional correlates of induced pain in man. Pain 1979; 6:365–374.
68. Pause BM, Krauel K. Chemosensory event-related potentials (CSERP) as a key to the psychology of odors. Int J Psychophysiol 2000; 36:105–122.
69. Pause BM, Sojka B, Krauel K, et al. The nature of the late positive complex within the olfactory event-related potential. Psychophysiology 1996; 33:168–172.
70. Boesveldt S, Haehner A, Berendse HW, et al. Signal-to-noise ratio of chemosensory event-related potentials. Clin Neurophysiol 2007; 118:690–695.
71. Lorig TS, Matia DC, Pezka JJ, et al. The effects of active and passive stimulation on chemosensory event-related potentials. Int J Psychophysiol 1996; 23:199–205.

72. Murphy C, Wetter S, Morgan CD, et al. Age effects on central nervous system activity reflected in the olfactory event-related potential. Evidence for decline in middle age. Ann N Y Acad Sci 1998; 855:598–607.

73. Pause B, Sojka B, Ferstl R. The latency but not the amplitude of the olfactory event-related potential (OERP) varies with the odor concentration. Chem Senses 1996; 21:485.

74. Ogawa S, Lee TM, Kay AR, et al. Brain magnetic resonance imaging with contrast dependent on blood oxygenation. Proc Natl Acad Sci U S A 1990; 87:9868–9872.

75. Bandettini PA, Jesmanowicz A, Wong EC, et al. Processing strategies for time-course data sets in functional MRI of the human brain. Magn Reson Med 1993; 30: 161–173.

76. Friston KJ, Holmes AP, Poline JB, et al. Analysis of fMRI time-series revisited. Neuroimage 1995; 2:45–53.

77. Albrecht J, Kopietz R, Frasnelli J, et al. The neuronal correlates of intranasal trigeminal function – An ALE meta-analysis of human functional brain imaging data [published online ahead of print November 11, 2009]. Brain Res Rev. doi:10.1016/j.brainresrev.2009.11.001.

78. Albrecht J, Kopietz R, Linn J, et al. Activation of olfactory and trigeminal cortical areas following stimulation of the nasal mucosa with low concentrations of S(-)-nicotine vapor-An fMRI study on chemosensory perception. Hum Brain Mapp 2008; 30:699–710.

79. Boyle JA, Heinke M, Gerber J, et al. Cerebral activation to intranasal chemosensory trigeminal stimulation. Chem Senses 2007; 32:343–353.

80. Iannilli E, Gerber J, Frasnelli J, et al. Intranasal trigeminal function in subjects with and without an intact sense of smell. Brain Res 2007; 1139:235–244.

81. Oertel BG, Preibisch C, Wallenhorst T, et al. Differential opioid action on sensory and affective cerebral pain processing. Clin Pharmacol Ther 2008; 83:577–588.

82. Savic I, Gulyas B, Berglund H. Odorant differentiated pattern of cerebral activation: comparison of acetone and vanillin. Hum Brain Mapp 2002; 17:17–27.

83. Gottfried JA. Smell: central nervous processing. Adv Otorhinolaryngol 2006; 63:44–69.

84. Sobel N, Prabhakaran V, Desmond JE, et al. Sniffing and smelling: separate subsystems in the human olfactory cortex. Nature 1998; 392:282–286.

85. Bendriem JC, Frackowiak RSJ, Herholz K, et al. Use of PET methods for measurement of cerebral energy metabolism and hemodynamics in cerebrovascular disease. J Cereb Blood Flow Metab 1989; 9:723–742.

86. Hummel T, Oehme L, van den Hoff J, et al. PET-based investigation of cerebral activation following intranasal trigeminal stimulation. Hum Brain Mapp 2009; 30:1100–1104.

87. Walitt B, Roebuck-Spencer T, Esposito G, et al. The effects of multidisciplinary therapy on positron emission tomography of the brain in fibromyalgia: a pilot study. Rheumatol Int 2007; 27:1019–1024.

88. Crivello F, Mazoyer B. Positron emission tomography of the human brain. In: Windhorst U, Johansson H, eds. Modern techniques in neuroscience research. Berlin: Springer, 1999.

89. Frost JJ, Wagner HN Jr. Neuroreceptors, neurotransmitters and enzymes. New York, NY: Raven Press, 1990.

90. Cohen D. Magnetoencephalography: detection of the brain's electrical activity with a superconducting magnetometer. Science 1972; 175:664–666.

91. Hari R, Kaukoranta E, Reinikainen K, et al. Neuromagnetic localization of cortical activity evoked by painful dental stimulation in man. Neurosci Lett 1982; 42:77–82.

92. Hari R, Portin K, Kettenmann B, et al. Right-hemisphere preponderance of responses to painful CO_2 stimulation of the human nasal mucosa. Pain 1997; 72:145–151.

93. Huttunen J, Kobal G, Kaukoronta E, et al. Cortical responses to painful CO_2-stimulation of nasal mucosa: a magnetencephalographic study in man. Electroencephlogr Clin Neurophysiol 1986; 64:347–349.

94. Kettenmann B, Hummel C, Stefan H, et al. Magnetoencephalographical recordings: separation of cortical responses to different chemical stimulation in man. Funct Neurosci (EEG Suppl) 1996; 46:287–290.
95. Mäkelä JP, Kirveskari E, Seppä M, et al. Three-dimensional integration of brain anatomy and function to facilitate intraoperative navigation around the sensorimotor strip. Hum Brain Mapp 2001; 12:180–192.
96. Druce HM. Nasal blood flow. Ann Allergy 1993; 71:288–291.
97. McLean JA, Mathews KP, Solomon WR, et al. Effect of ammonia on nasal resistance in atopic and nonatopic subjects. Ann Otol Rhinol Laryngol 1979; 88(2 Pt 1):228–234.
98. Bascom R, Kulle T, Kagey-Sobotka A, et al. Upper respiratory tract environmental tobacco smoke sensitivity. Am Rev Respir Dis 1991; 143:1304–1311.
99. Shusterman D, Murphy M, Balmes J. Seasonal allergic rhinitic and non-rhinitic subjects react differentially to provocation with chlorine gas. J Allergy Clin Immunol 1998; 101:732–740.
100. Shusterman D, Tarun A, Murphy M-A Morris J. Seasonal allergic rhinitic and normal subjects respond differentially to nasal provocation with acetic acid vapor. Inhal Toxicol 2005; 17:147–152.
101. Youlten LJF. The peak nasal inspiratory flow meter: A new instrument for the assessment of the response to immunotherapy in seasonal allergic rhinitis. Allergol Immunopahol 1980; 8:344–347.
102. Cho SI, Hauser R, Christiani DC. Reproducibility of nasal peak inspiratory flow among healthy adults: assessment of epidemiologic utility. Chest 1997; 112:1547–1553.
103. Holmstrom M, Scadding G.K, Lund VJ, et al. Assessment of nasal obstruction. A comparison between rhinomanometry and nasal inspiratory peak flow. Rhinology 1990; 28:191–196.
104. Shelton DM, Eiser NM. Evaluation of active anterior and posterior rhinomanometry in normal subjects. Clin Otolaryngol 1992; 17:178–182.
105. Solomon WR, McLean JA, Cookingham C, et al. Measurement of nasal airway resistance. J Allergy Clin Immunol 1965; 36:62–69.
106. Hilberg O, Jackson AC, Swift DL, et al. Acoustic rhinometry: evaluation of nasal cavity geometry by acoustic reflection. J Appl Physiol 1989; 66:295–303.
107. Hilberg O, Pedersen OF. Acoustic rhinometry: recommendations for technical specifications and standard operating procedures. Rhinol Suppl 2000; 16:3–17.
108. Shemen, L Hamburg R. Preoperative and postoperative nasal septal surgery assessment with acoustic rhinometry. Otolaryngol Head Neck Surg 1997; 117:338–342.
109. Corey JP, Kemker BJ, Nelson R, et al. Evaluation of the nasal cavity by acoustic rhinometry in normal and allergic subjects. Otolaryngol Head Neck Surg 1997; 117:22–28.
110. Ellegard E. Practical aspects on rhinostereometry. Rhinology 2002; 40:115–117.
111. Tafaghodi M, Abolghasem Sajadi Tabassi S, Jaafari MR, et al. Evaluation of the clearance characteristics of various microspheres in the human nose by gamma-scintigraphy. Int J Pharm 2004; 280:125–135.
112. Andersen I, Lundqvist GR, Proctor DF, et al. Human response to controlled levels of inert dust. Am Rev Respir Dis 1979; 119:619–627.
113. Bascom R, Kesavanathan J, Fitzgerald TK, et al. Sidestream tobacco smoke exposure acutely alters human nasal mucociliary clearance. Environ Health Perspect 1995; 103:1026–1030.
114. Wang DY, Yeoh KH. The significance and technical aspects of quantitative measurements of inflammatory mediators in allergic rhinitis. Asian Pac J Allergy Immunol 1999; 17:219–228.
115. Graham DE, Koren HS. Biomarkers of inflammation in ozone-exposed humans. Comparison of the nasal and bronchoalveolar lavage. Am Rev Respir Dis 1990; 142:152–156.
116. Koren HS, Devlin RB. Human upper respiratory tract responses to inhaled pollutants with emphasis on nasal lavage. Ann N Y Acad Sci 1992; 641:215–224.

117. Peden DB. The use of nasal lavage for objective measurement of irritant-induced nasal inflammation. Regul Toxicol Pharmacol 1996; 24:S76–S78.
118. Harder S, Peden DB, Koren H, et al. Comparison of nasal lavage techniques for analysis of nasal inflammation. Am J Respir Crit Care Med 1995; 151(4, Part 2):A (Abstract).
119. Park YJ, Repka-Ramirez MS, Naranch K, et al. Nasal lavage concentrations of free hemoglobin as a marker of microepistaxis during nasal provocation testing. Allergy 2002; 57:329–335.
120. Meltzer EO, Jalowayski AA. Nasal cytology in clinical practice. Am J Rhinol 1988; 2:47–54.
121. Meltzer EO. Quality of life in adults and children with allergic rhinitis. J Allergy Clin Immunol 2001; 108:S45–S53.
122. Tarun AS, Bryant B, Zhai W, et al. Gene expression for carbonic anhydrase isoenzymes in human. Chem Senses 2003; 28:621–629.
123. Bates CA, Silkoff PE. Exhaled nitric oxide in asthma: from bench to bedside. J Allergy Clin Immunol 2003; 111:256–262.
124. Covar RA, Szefler SJ, Martin RJ, et al. Relations between exhaled nitric oxide and measures of disease activity among children with mild-to-moderate asthma. J Pediatr 2003; 142:469–475.
125. Olin AC, Alving K, Toren K. Exhaled nitric oxide: relation to sensitization and respiratory symptoms. Clin Exp Allergy 2004; 34:221–226.
126. Jorissen M, Lefevere L, Willems T. Nasal nitric oxide. Allergy 2001; 56:1026–1033.
127. American Thoracic Society. Recommendations for standardized procedures for the on-line and off-line measurement of exhaled lower respiratory nitric oxide and nasal nitric oxide in adults and children-1999. Am J Respir Crit Care Med 1999; 160: 2104–2117.
128. American Thoracic Society, European Respiratory Society. ATS/ERS recommendations for standardized procedures for the online and offline measurement of exhaled lower respiratory nitric oxide and nasal nitric oxide, 2005. Am Respir Crit Care Med 2005; 171:912–930.
129. Bartley J, Fergusson W, Moody A, et al. Normal adult values, diurnal variation, and repeatability of nasal nitric oxide measurement. Am J Rhinol 1999; 13:401–405.
130. Maniscalco M, Sofia M, Weitzberger E, et al. Humming-induced release of nasal nitric oxide for assessment of sinus obstruction in allergic rhinitis: Pilot study. Eur J Clin Invest 2004; 34:555–560.
131. Shusterman D, Jansen K, Weaver E, et al. Documentation of the nasal nitric oxide response to humming: Methods evaluation. Eur J Clin Invest 2007; 37:746–752.
132. de Winter-de Groot KM, van der Ent CK. Measurement of nasal nitric oxide: evaluation of six different sampling methods. Eur J Clin Invest 2009; 39:72–77.

10 Biomarkers of Nasal Toxicity in Experimental Animals

Mary Beth Genter

Department of Environmental Health, University of Cincinnati, Cincinnati, Ohio, U.S.A.

INTRODUCTION

"Biomarker" is a general term for a specific measurement of an interaction between a biological system and an environmental agent (1). In human epidemiological studies, the ideal biomarker is highly specific for the agent of concern and, further, can be measured by noninvasive means, such as measuring chemical metabolites in urine or hair. In terms of biomarkers of nasal toxicity in experimental animals, this chapter discusses several in vivo measurements that can properly be considered biomarkers. Also discussed are endpoints that do not fit the classical definition of biomarkers because they are assessed on tissues after sacrifice of the animals, but are nonetheless indicative of exposure to various compounds. In vivo biomarkers will include imaging and behavioral endpoints, whereas postmortem "biomarkers" of nasal epithelial responses to environmental agents will include induction or inhibition of various proteins; changes in the structure of the nasal turbinates; alterations in gene expression and protein expression; and DNA and protein adducts.

BIOMARKERS ASSESSED IN VIVO IN EXPERIMENTAL ANIMALS

Imaging

The use of imaging techniques as biomarkers of toxicant-induced damage in the nasal epithelia is a promising and evolving topic of study. An exciting finding in the field of manganese (Mn) neurotoxicity was the observation that magnetic resonance imaging (MRI), specifically high signal intensities on T1-weighted MRI, can be used as a biomarker of Mn exposure and brain levels of Mn in humans (2). Obvious advantages of such an approach are that a living subject can be noninvasively imaged repeatedly, and the effects of interventions, such as chelation therapy, can be assessed. More recently, Mn-enhanced MRI has been used in rodents in vivo to study axonal transport rates in the brains of aging rats as well as in a transgenic mouse model of Alzheimer's disease (AD) (3,4). Serial, noninvasive monitoring of the migration of neural progenitor cells (NPCs) from the subventricular zone to the olfactory bulb was accomplished in vivo by labeling of NPCs with fluorescent, micron-sized iron particles and visualizing by MRI (5).

Efforts are also being directed at developing imaging techniques that can be used to noninvasively assess olfactory epithelial degeneration in rodents, with the similar advantage that lesion development and resolution can be monitored in the same animals over time. Both small animal micro-computed tomography

(CT) and MRI have been explored for their utility in imaging of rodent nasal cavities for lesion development. A map of the "normal" mouse paranasal sinuses has been developed using CT, with the goal of standardizing descriptions of phenotypic changes in sinonasal architecture, particularly in new transgenic and gene knockout mice (6). Techniques for imaging the respiratory tract by MRI were developed to improve the speed and accuracy of geometric data collection for mesh reconstruction. MRI resolution is comparable to that obtained by manual measurements but with much greater speed and accuracy (7).

Historically, histologic and morphometric techniques have been used to determine what fraction of a particular species' nasal epithelia was covered by the respective epithelial types (typically respiratory vs. olfactory). Imaging techniques can now be used for this purpose. A recent study detailing MRI imaging and surface area calculations of the nasal airways of a dog [female Labrador retriever mixed-breed canine, weighing approximately 29.5 kg (66 lb)] revealed that the surface area of the left nasal cavity was approximately 411 cm^2, with the surface area of the maxilloturbinates and ethmoidal airways calculated to be approximately 120 and 210 cm^2, respectively (8).

Using 3-methylindole (3-MI) as a model compound, MRI showed that the cross-sectional area of the sinus airspaces increased by 1.7-fold in mice injected with 200 mg/kg 3-MI and 2.6-fold in mice injected with 300 mg/kg at 3 days after injection. Alterations in the nasal turbinates lined by olfactory mucosa were identified 1, 3, and 6 days posttreatment. Lesion distribution was confirmed by postmortem histological examination of the nasal tissues (9). Three-dimensional mapping of ozone-induced injury in the nasal airways of monkeys using MRI has also been conducted, and confirmed using morphometric techniques (10). Thus, these and other imaging techniques should prove to be very useful, and greatly reduce the numbers of animals needed per experiment, in future toxicological evaluations of respiratory tract toxicity.

Behavioral Biomarkers

The Physiology of the "Sniff"

Sniffing has been described as a rhythmic inhalation and exhalation of air through the nose and is proposed to play a critical role in shaping odorant representation by the nervous system (11). Rats trained on an odor-localization task can localize odors accurately in one or two sniffs. Bilateral sampling was essential for accurate odor localization, with internasal intensity and timing differences as directional cues (12). Although mice have become a prominent model for studying olfaction, sniffing behavior in mice has not been extensively characterized. A recent study found that respiration frequency in quiescent mice ranged from 3 to 5 Hz, which is higher than that reported for rats. During exploration, sniff frequency increased to ~12 Hz and exhibited rapid changes in frequency, amplitude, and waveform. Sniffing behavior was also shown to differ between tasks; for example, mice performing a digging task showed little increase in sniff frequency prior to digging, whereas mice performing a nose poke task showed robust increases. Mice also showed large increases in sniff frequency in anticipation of reward delivery (13). Olfactory discrimination accuracy and time were evaluated using a go/no-go paradigm, and mice discriminated simple odors very rapidly (in <200 ms), but required 70 to 100 ms longer to discriminate highly similar binary mixtures (14). Exploratory sniffing behavior similarly revealed that rats discriminate and initiate a response to a novel odorant in as little as 140 ms (11).

Tests of Olfactory Function

Methods for testing olfactory function in rodents can range from simple measures, such as variants on a buried food pellet test, to more sophisticated tests that evaluate odor discrimination and novel odorant discrimination. Using the buried food pellet test, the degree of olfactory impairment in rats induced by 3,3-iminodipropionitrile, dichlobenil, and methimazole was positively correlated with the extent of olfactory epithelial damage (15). 3-MI-treated rats demonstrated a treatment-related deficit in acquiring olfactory learning tasks (not related to cognitive function), as assessed by an olfactory discrimination task (16). Olfactory deficits resulting from exposure to 3-MI or zinc gluconate have been characterized using rodent "olfactometers" (17,18). An olfactory discrimination procedure based on an acquisition/reversal paradigm for mice has been described (19). The olfactory tubing maze, based on olfactory-reward associations, has been described as an apparatus for studying learning and memory in mice (20). The contribution of subclasses of cholinergic receptors in olfactory discrimination was studied in mice following olfactory bulb infusion of scopolamine (muscarinic antagonist), mecamylamine HCl (nicotinic antagonist), or the acetylcholinesterase inhibitor neostigmine in an olfactory cross-habituation task or a rewarded, forced-choice odor-discrimination task (21).

Olfactory Behavioral Effects in Genetically Modified Mice

When olfactory function is tested in genetically modified mice, subtle to severe olfactory deficits are often found. Olfactory deficits were identified in mice lacking a functional vasopressin (*Avp1a*) receptor, as assessed by habituation/dishabituation testing, as well as in the ability to discriminate female mouse urine from male mouse urine in an operant testing paradigm (22). Mice null for the expression of olfactory marker protein (OMP), which is abundantly expressed in mature olfactory neurons, displayed altered odorant quality perception based on the results of a 5-odorant identification confusion matrix task (23). Gene trap mutant mice *Stam(gt1Gaj)* were evaluated to elucidate the in vivo role of Stam2 (signal transducing adaptor molecule 2), which is highly expressed in brain regions related to olfaction. While the mutant mice displayed normal olfactory morphology, mutants needed more time, and failed more frequently, to find buried chocolate (24).

Several mouse models that mimic human diseases show profound olfactory deficits. In humans, hypomorphic mutations of the cilia-centrosomal protein CEP290/NPHP6 are associated with early onset retinal dystrophy, as well as severely abnormal olfactory function. Electroolfactogram recordings from mice with hypomorphic mutations in CEP290 revealed anosmia, despite normal appearing cilia and normal trafficking of CEP290 to olfactory dendritic knobs (25). Similarly, in a mouse model of cystic fibrosis, "CF mice" (*Ctfr-/-*, lacking the cystic fibrosis transmembrane conductance receptor), the olfactory epithelium appears normal at birth, but by 30 days of age, electroolfactograms revealed a 45% decrease in odor-evoked response, which progressed with age. By six months of age, olfactory mucosal histology was markedly abnormal, with altered appearance of sustentacular cells and markedly decreased neuronal numbers (26). Accumulation of alpha-synuclein in central and peripheral system neurons is a hallmark of sporadic Parkinson's disease (PD), and transgenic mice that overexpress human wild-type alpha-synuclein exhibited deficits in three behavioral testing paradigms, consistent with early olfactory deficits in humans who develop PD (27). In contrast, mice heterozygous for the insulin receptor kinase gene performed at the same level as wild-type mice in a nonlearning based task to test for gross anosmia, despite in vitro evidence that mitral cells cultured from heterozygous and homozygous knockout mice display decreased peak amplitude compared to wild-type mice (28).

Several genetic modifications have actually resulted in improved olfactory function. Mean food finding time (buried food) was approximately 10 times faster in leptin (ob/ob) and leptin receptor (db/db) mutant mice than in wild-type mice, and i.p. administration of leptin normalized performance of both mutant mouse strains (29). Mice lacking the voltage-gated potassium channel Kv1.3 have a 1000–10,000-fold lower threshold for odorant detection and an increased ability to discriminate between odorants. Deletion of Kv1.3 was associated with increased expression of scaffolding proteins that normally regulate the channel (30).

Analysis of Olfactory Biopsies

Although the process of taking biopsies of tissue from live rodents for disease diagnosis or evaluation of toxicant-induced damage has not become routine, clinicians and olfactory researchers have taken advantage of the fact that olfactory sensory neurons project into the nasal cavity, and are therefore accessible for biopsy in larger species, including humans. This procedure typically involves administration of local anesthesia to the patient, followed by use of an endoscope to locate the area to be biopsied and removal of a small (approximately 3 mm^3 in volume) piece of tissue, typically from the nasal septum, using cutting punch forceps (31–33). While olfactory epithelium is generally found on the upper nasal septum and superior turbinate in humans, in reality, olfactory epithelium is irregular and patchy in its distribution, with invasion of respiratory epithelium into areas normally occupied by olfactory epithelium, generally increasing as a function of age (34). Authors of a study that detailed biopsy results from 12 individuals commented that multiple (4–6) samples had to be taken from each patient in order to find one that contained olfactory epithelium (31). Follow-up studies of patients who have undergone olfactory biopsies indicate no subsequent compromise of their olfactory function (32,33,35).

Interest in the use of olfactory biopsies for diagnosis of neurodegenerative diseases expanded rapidly with the publication of results showing unique pathological changes in morphology, distribution, and immunoreactivity of neuronal structures in olfactory biopsies of AD patients (36). Although some of the changes reported in olfactory biopsies from AD patients were subsequently also found in olfactory biopsies from unaffected, elderly individuals, other investigators rapidly expanded this observation to include characterization of histological and immunohistochemical changes in olfactory biopsies of patients with other neurodegenerative diseases (37–39).

Evaluation of olfactory epithelial biopsies from patients with Rett syndrome and Creutzfeldt-Jakob (C-J) disease have also provided important clues about the etiologies and impacts of these vastly divergent conditions. Nasal biopsies from patients with Rett syndrome, a neurodevelopmental disorder caused by mutations in the gene encoding "methyl CpG binding protein 2" (MeCP2), contained far fewer mature olfactory receptor neurons (ORNs) and significantly greater numbers of immature neuron-specific tubulin-positive ORNs, suggesting that maturation of ORNs is impeded prior to the time of synapse formation (40,41). Evaluation of autopsy specimens from patients who died with C-J disease revealed significant involvement of the olfactory cortex and the olfactory tract, and a subsequent study showed prion protein immunoreactivity in olfactory biopsy tissue 45 days after onset of C-J disease (42,43).

Further insights into olfactory signal transduction and various disease states have been obtained by analysis of cultures established from olfactory biopsy tissue. ORNs isolated from human olfactory biopsies responded to odorant stimulus with an increase in intracellular calcium concentration, an observation that has led to significant advances in the molecular mechanisms of olfactory signal transduction (44). Evaluation of calcium signaling in ORNs cultured from patients with bipolar disorder revealed exactly the opposite, that is, levels of intracellular calcium decreased upon odorant stimulus, leading to the conclusion that altered calcium signaling in ORNs may be a trait associated with bipolar disorder (33,45). Olfactory cultures derived from patients with schizophrenia contained significantly more mitoses than from control patients, and microarray analysis of biopsied olfactory epithelium revealed alterations of cell cycle and phosphatidylinositol signaling in schizophrenia and bipolar I disorder, respectively (46). In addition, neurons cultured from olfactory biopsies of AD patients displayed a number of markers of lipid peroxidation and oxidative stress, showing that manifestations of oxidative imbalance observed at autopsy were also detectable in neurons cultured from olfactory epithelial biopsies (47).

POSTMORTEM "BIOMARKERS" OF EXPOSURE TO ENVIRONMENTAL AGENTS

Protein Changes in Nasal Epithelia

Xenobiotic Metabolizing Enzymes
It has long been recognized that both nasal respiratory and olfactory epithelia have considerable metabolic capabilities, including many enzymes that contribute to the detoxification and/or bioactivation of inhaled xenobiotics. Exposure to various environmental agents can induce or inhibit many of these metabolic enzymes, but, in general, the magnitude of enzyme induction in nasal epithelia is less than that occurring in the liver with the same inducing agent (48,49). Xenobiotic-induced changes in metabolic enzymes have been characterized by histochemistry, immunohistochemistry, western blot analysis, and protein purification, to name a few. For example, cytochrome P450 2E1 (CYP2E1) has been studied extensively in nasal epithelia, as it is recognized as an enzyme that bioactivates many inhaled and systemically administered compounds to tissue-reactive metabolites. CYP2E1 immunoreactivity in olfactory mucosa is dramatically decreased by inhalation exposure to chloroform in rats and mice (50), probably due to its activity as a suicide substrate for the enzyme. CYP2E1 can be

induced in olfactory mucosa by calorie restriction, or by treatment with pyrazole or acetone (48,49,51).

2,3,7,8-Tetrachlorodibenzo-*p*-dioxin (TCDD) is a potent inducer of genes under regulation of the aryl hydrocarbon receptor (AhR) in the liver, and induction of some of the AhR-regulated enzymes (CYP1A1, CYP1A2, microsomal epoxide hydrolase) has been reported in rat olfactory mucosa. CYPs 2B and 2C were also induced by the same treatment in olfactory mucosa. Induction of nasal epithelial enzymes by TCDD resulted in an increased rate of metabolism of lidocaine to monoethyl glycine xylidide (MEGX) by olfactory mucosal microsomes (52).

Inhalation exposure to pyridine was shown to induce the expression of carboxylesterase (CE) in F344 rats. While CE is not thought to be involved in metabolism of pyridine itself, CEs are important in metabolism of acrylate and acetate esters. Therefore, CE induction may also serve as a more general biomarker for nasal exposure to a broad range of solvents (53). CE immunoreactivity was studied in F344 rats, beagle dogs, and human nasal tissue, and the distribution of CE was very similar in all three species. Interestingly, immunostaining was dramatically reduced in human hyperplastic nasal lesions and virtually eliminated in areas of squamous metaplasia (54). Further, CE activity in human nasal homogenates was threefold lower than in rats, and 12-fold lower than in hamster, which led to the conclusion that humans may in fact be less susceptible to respiratory tract toxicity of methyl methacrylate than rodents (55).

CYP2A5 is expressed in many extrahepatic tissues in mice, with expression in olfactory mucosa at the highest level of any tissue examined (56). CYP2A5 is induced by pyrazole in the liver, kidney, and olfactory mucosa. Interestingly, CYP2J was induced by the same pyrazole treatment, indicating that lack of CYP2A5 induction in olfactory mucosa was not due to insufficient bioavailability of pyrazole at this site (51,57). CYP2J isoforms are also inducible in the intestine, and have been implicated as catalysts of arachidonic acid metabolism and vitamin D hydroxylation. CYP2Js have also been implicated in xenobiotic metabolism, as activity toward the antihistamine astemizole has been demonstrated (59–60).

Metal Transporters

The olfactory mucosa contains transport proteins that facilitate uptake into or exclusion of certain molecules from the tissue (61–63). Metal transporters, including divalent metal transporter (DMT1) and ZIP14, are likely responsible for the uptake of metals that are both essential (e.g., zinc) and toxic (e.g., manganese) into the olfactory mucosa, with the potential for further translocation into the brain (61,63,64). In several tissues, most notably the intestine, anemia is associated with the upregulation of DMT1, apparently as a physiological response in order to increase iron uptake (65). Anemia also caused upregulation of DMT1 in olfactory mucosal sustentacular cells in rats; DMT1 upregulation in the olfactory epithelium, in turn, was associated with increased uptake of manganese from the nasal cavity into the brain (61).

Protein Biomarkers of Oxidative Stress

Oxidative stress, whether chemically induced or disease-related, can be assessed at the protein, gene expression, and small molecule levels. At the protein level, induction of metallothioneins (MTs), superoxide dismutases (SODs), and heme oxygenases (HOs) are often measured to characterize oxidative stress responses (47,66). MT is highly inducible by a number of factors, including metals and other agents that induce oxidative stress (67,68). MT is a cysteine-rich protein that "detoxifies" metals by binding them to reduce their bioavailability. MT also functions to reduce oxidative stress by sequestration of intracellularly generated hydroxyl and superoxide radicals (66). Following intranasal administration of cadmium (Cd) to naïve rats, both reduced glutathione (GSH)–Cd and MT–Cd conjugates were recovered. Rats pretreatment with Cd had higher levels of MT, resulting in conjugation of a subsequent dose of ^{109}Cd exclusively as the MT conjugate (69). The nasal epithelia of dogs living in an environment with high ambient air pollution exhibited strong MT immunoreactivity (using an antibody recognizing MT-1 and MT-2) in olfactory neurons and axon bundles, as well as in sustentacular cells (70).

SODs provide enzymatic defense against cytotoxic reactive oxygen species and oxidative stress, which have been implicated in AD (71). Immunohistochemical analysis revealed that the combined activities of the Cu, Zn, and Mn forms of SOD were higher in control rat olfactory epithelium than nasal respiratory epithelium (72). SOD activity was markedly induced in sinus mucosa of rabbits treated with a killed suspension of *Staphylococcus aureus* to induce maxillary sinusitis (73). In addition, a pronounced increase in SOD immunoreactivity was noted in the olfactory epithelium of AD subjects (71). While SOD immunoreactivity and/or activity studies have not been widely used in studies of rodent nasal oxidative stress, these observations suggest that this could be a useful marker for such studies.

HO is an antioxidant enzyme that catabolizes heme to produce carbon monoxide (CO) and biliverdin, and upregulation of HO gene and protein expression has been validated as a marker of oxidative stress (74). Three isoforms of HO are known (HO-1, -2, and -3), and while measurement of HO protein levels in rodent studies has only infrequently been described, HO-1 levels were markedly increased in the nasal tissues of humans with allergic rhinitis inflammation (75).

Low-Molecular-Weight Molecules as Indicators of Oxidative Stress

Low molecular weight markers that can serve as indicators of oxidative stress responses include reduced glutathione (GSH), ascorbate, hydroxynonenal (as an indicator of lipid peroxidation), N(epsilon)-(carboxymethyl)lysine (indicative of glycoxidative and lipid peroxidation), and 3-nitrotyrosine (47,76,77). With the exception of GSH and ascorbate, the markers listed here have not been used appreciably in animal studies of nasal epithelial oxidative stress. Alachlor-treated rats (126 mg/kg/day in the diet) displayed an initial depletion, and then recovery to super-basal levels of both GSH and ascorbate in the olfactory mucosa, suggesting that oxidative stress contributes to the olfactory carcinogenicity of this herbicide (78). Acrolein is a highly reactive aldehyde that is a major component of cigarette smoke and is associated with GSH depletion. While naïve rats displayed decreased GSH levels following an initial exposure to acrolein, rats exposed repeatedly to acrolein had elevated levels of GSH in nasal respiratory epithelium, perhaps acting as an adaptive response to prevent tissue damage from subsequent exposure (79).

While nitrotyrosine analysis has not been reported in rodent nasal epithelia, immunohistochemical evaluation of olfactory epithelium from AD patients showed markedly higher levels of 3-nitrotyrosine than non-AD subjects (80). In addition, nitrotyrosine immunoreactivity was noted in the nasal respiratory epithelium of patients diagnosed with vasomotor rhinitis, but not in control patients (81). Nasal polyps also displayed an increase in nitrotyrosine (82). Both hydroxynonenal and N(epsilon)-(carboxymethyl)lysine, in addition to HO-1, were detected immunohistochemically at higher levels in cells cultured from the olfactory mucosa of AD subjects than from control subjects (47). These observations in human clinical samples strongly suggest that these markers could be used to expand analyses of oxidative stress endpoints in rodent studies.

Changes in the Structure of the Nasal Turbinates

Changes in nasal turbinate bone structure have been reported in response to several inhaled toxicants. In general, the toxicants that induce proliferative changes in turbinate bones (chloroform, 3-methylindole, 3-methylfuran, hexamethylphosphoramide) are bioactivated by cytochrome P450 enzymes found in the Bowman's glands of the olfactory mucosa (50,83). However, new bone formation is not a universal biomarker for olfactory toxicants that are bioactivated by enzymes in Bowman's glands, as this response has not been reported in rodents treated with other inhaled nasal toxicants such as naphthalene. The changes in turbinate structure described to date have been observed in histological sections, but this endpoint should eventually be assessable in animals in vivo using imaging techniques such as those previously described.

Alterations in Gene and Protein Expression

Several research teams have used high-throughput methods such as gene chips and microarrays to characterize gene expression patterns in normal rodent nasal epithelia (84–87). These techniques have also been used to characterize toxicant-induced effects on nasal epithelia. Administration of toluene diisocyanate to mice induces an allergic rhinitis that is phenotypically similar to that observed in exposed humans, as confirmed by evaluation of the cytokine gene expression profile induced in mice (88). Genomic analysis also confirmed that the altered gene expression profile induced by cigarette smoke exposure in the lung is identical to genomic alterations observed in nasal and buccal epithelia (89), suggesting that changes in gene expression in these more peripherally located tissues can serve as a biomarker for effects occurring in the lung.

Changes in the gene expression during the course of the development of olfactory tumors in response to exposure to the herbicide alachlor (administered through the diet at 126 mg/kg/day) revealed a dramatic acute upregulation of HO-3, which is an isoform of HO found predominantly in the brain (90,91). More significantly, genes related to olfactory structure and function (e.g., CYP2A3, CYP2F1, OMP, insulin-like growth factor receptor, and rhodanese) were downregulated following acute exposures, which is consistent with the observation that olfactory mucosal de-differentiation occurs prior to the development of alachlor-induced olfactory mucosal tumors (91,92). Acute alachlor exposure also upregulated a battery of extracellular matrix-related genes, including matrix metalloproteinase (MMP)-2, MMP-9, carboxypeptidase Z, and tissue inhibitor of metalloproteinase-1 (91). This observation led to the hypothesis that inhibition of MMP-2 and/or MMP-9 might inhibit alachlor-induced olfactory mucosal tumor formation. Indeed, administration of the dual MMP2/9 inhibitor Ro 28–2653 significantly reduced the tumor burden in alachlor-induced rats, compared to rats receiving alachlor alone (93).

Aging is also associated with altered gene expression in the olfactory system. The Harlequin (Hq) mouse has been recently described as a model for accelerated neurodegeneration, as it carries an X-linked recessive mutation in the "apoptosis-inducing factor, mitochondrion-associated-1" (*Aifm1*) gene, which encodes the AIF protein. The AIF protein has two distinct cellular functions mediated by two molecular domains: (*i*) maintenance of proper mitochondrial function and low levels of oxidative stress, and (*ii*) induction of apoptosis. The Hq mouse has an intronic proviral insertion in the oxidoreductase domain, which causes an 80% decrease in AIF protein and a concomitant increase in markers of oxidative stress. Hq mice exhibit progressive retinal degeneration beginning at about 3 months of age and go on to develop cerebellar degeneration at ~4 months of age, with ataxia observed beginning at 5 months of age, and progressing with age (94). More recent microarray studies characterized altered regulation of key cell cycle and oxidative stress-related genes, as well as increased expression of biomarkers of oxidative stress such as 8-hydroxydeoxyguanosine in the olfactory mucosa (95).

The aging olfactory system has also been the subject of proteomic analysis, which is a technique that provides high-throughput analysis of proteins with aging or following toxicant treatment. Both levels of expression and posttranslational modifications can be evaluated using this technique. While proteomic analysis of the olfactory system has not been widely used to date, proteins identified as differentially expressed in the olfactory epithelium and olfactory bulb of old versus young mice may prove to be useful biomarkers of aging in the future (96).

DNA and Protein Adducts

DNA adducts are widely regarded as contributors to mutagenesis and carcinogenesis. For example, propylene oxide (PO) is carcinogenic to the nasal respiratory epithelium of F344 rats. Direct alkylation of DNA by PO leads to the formation of a major DNA adduct, N:7-(2-hydroxypropyl)guanine (7-HPG). While 7-HPG adducts could be found in a large number of nontarget tissues for PO carcinogenicity, the 7-HPG adduct levels were considerably higher in the target nasal respiratory epithelium than in nontarget tissues, including the adjacent olfactory epithelium. Interestingly, in rats killed 7 hours or 3 days after the last of 20 exposures (500 ppm, 6 hr/day) to PO, the adducts in respiratory epithelium were approximately twice as high as in olfactory epithelium (97), suggesting that there is a clear threshold for PO DNA adduct levels that contribute to a carcinogenic response. Table 1 presents results of other studies that have examined DNA adducts in nasal epithelia following carcinogen exposure, and in general, DNA adduct formation in nasal epithelia correlates with carcinogenicity. The olfactory carcinogen alachlor is an interesting exception. 2,6-Dimethylamine (also called 2,6-xylidine), a known rodent olfactory carcinogen (98), and 2,6-diethylaniline, an alachlor metabolite (99), generate both olfactory and liver DNA adducts when injected intraperitoneally into rats. In contrast, alachlor, when administered at

TABLE 1 Assessment of DNA Adducts in Nasal Epithelia of Laboratory Rodents Treated with the Respective Compounds

Chemical	Nasal carcinogen (y/n)	Adducts correlate with tumor distribution?	Reference
2,6-Dimethylaniline	y	y	(101)
2-Methylaniline	y	y	(101)
Dimethyl sulfate	y	y	(102)
Propylene oxide	y	y	(97)
Propylene	n	No (adducts were detected; not a nasal carcinogen)	(103)
Formaldehyde	y	y	(104)
Nickel subsulfide	y	No (adducts were not detected)	(105)
4,4′-Methylenedianiline	n	No (adducts were detected; not a nasal carcinogen)	(106)
Tobacco-specific nitrosamines	y	y	(107)[a]
Wood dust extracts	y	No (adducts were not detected)	(108)
Alachlor	y	No (adducts were not detected)	(100)
Beta-propiolactone	y	y	(109)
Methylmethane sulfonate	y	y	(109)
Dimethylcarbamyl chloride	y	y	(109)

[a]Representative publication; many confirmatory papers have been published on this topic.

the same daily dose, did not result in detectable olfactory adducts, but did yield DNA adducts in liver, a nontarget tissue for alachlor carcinogenesis (100).

In several instances, it is likely that target tissue metabolism results in generation of protein-reactive metabolites that contribute to a compound's cytotoxicity. Phenacetin, an olfactory carcinogen, was shown to bind irreversibly in a dose-related manner to the Bowman's glands, suggesting that the carcinogenic metabolite is produced in the target tissue (110). Acetaminophen (APAP) is metabolized to N-acetyl-p-benzoquinoneimine, which binds with a high degree of selectivity to cysteinyl thiol groups, including GSH and cysteine residues on proteins. Immunohistochemical localization of APAP and CYP2E1 were identical to the sites of lesion formation, implicating CYP2E1 as the enzyme that bioactivates APAP in situ (111). APAP appears to bind preferentially to two proteins, one with a weight of ~44 kDa and the other ~58 kDa (112). The 44 kDa protein has been identified as a subunit of glutamine synthase (113), whereas the 58 kDa protein has been mapped to mouse chromosome 3 and has been putatively identified as LPSB (liver protein, selenium binding). *Lpsb* mRNA is present in all tissues in which APAP binding and toxicity have been reported, whereas *Lpsb2*, which is in close proximity on chromosome 3, is only expressed in liver (114).

Naphthalene (NP) similarly depletes intracellular GSH and modifies cysteine residues of cellular proteins, which likely contributes to its cytotoxicity. In this case, the protein adduct is hypothesized to result from P450-mediated formation of 1-naphthol, with subsequent conversion to quinone metabolites (115–117). Antibodies raised toward 1,2-naphthoquinone protein adducts have been used to identify proteins that are preferentially adducted by NP metabolites. An as-yet-unidentified 22 kDa protein has been shown to be a major target (118).

SUMMARY

The use of biomarkers for nasal toxicity is rapidly evolving. Noninvasive functional tests of olfactory function can detect changes in expression of single proteins as well as more profound changes induced by olfactory toxicants. Imaging techniques show great promise, as they can be used to assess the distribution of epithelial cell types and changes in nasal architecture noninvasively and longitudinally during chronic exposure studies. Analysis of tissue samples may provide biomarkers of ongoing disease pathogenesis. Changes in biotransformation enzyme activity and metal transporter expression have promise in this regard. Protein and low molecular weight biomarkers of oxidative stress, alterations in gene and protein expression, and DNA or protein adducts have all been shown to provide key insights into pathogenesis of toxicant-induced nasal lesions. Thus, while classically defined biomarkers for study of the effects of toxicant-induced nasal damage are not currently in wide use, behavioral and imaging methods to study toxicant- or gene-targeted effects are evolving and show great promise.

REFERENCES

1. Pastenbach DJ, Madl AK. The practice of exposure assessment. In: Hayes AW, ed. Principles and Methods of Toxicology, 5th edn. Boca Raton, FL: CRC Press, 2008:522.
2. Kim Y. High signal intensities on T1-weighted MRI as a biomarker of exposure to manganese. Ind Health 2004; 42:111–115.
3. Cross DJ, Flexman JA, Anzai Y, et al. Age-related decrease in axonal transport measured by MR imaging in vivo. Neuroimage 2008; 39(3):915–926.
4. Minoshima S, Cross D. In vivo imaging of axonal transport using MRI: Aging and Alzheimer's disease. Eur J Nucl Med Mol Imaging 2008; 35(suppl 1):S89–S92.
5. Shapiro EM, Gonzalez-Perez O, Manuel García-Verdugo J, et al. Magnetic resonance imaging of the migration of neuronal precursors generated in the adult rodent brain. Neuroimage 2006; 32:1150–1157.
6. Jacob A, Chole RA. Survey anatomy of the paranasal sinuses in the normal mouse. Laryngoscope 2006; 116:558–563.
7. Timchalk C, Trease HE, Trease LL, et al. Potential technology for studying dosimetry and response to airborne chemical and biological pollutants. Toxicol Ind Health 2001; 17:270–276.
8. Craven BA, Neuberger T, Paterson EG, et al. Reconstruction and morphometric analysis of the nasal airway of the dog (Canis familiaris) and implications regarding olfactory airflow. Anat Rec (Hoboken) 2007; 290:1325–1340.
9. Wiethoff AJ, Harkema JR, Koretsky AP, et al. Identification of mucosal injury in the murine nasal airways by magnetic resonance imaging: Site-specific lesions induced by 3-methylindole. Toxicol Appl Pharmacol 2001; 175:68–75.
10. Carey SA, Minard KR, Trease LL, et al. Three-dimensional mapping of ozone-induced injury in the nasal airways of monkeys using magnetic resonance imaging and morphometric techniques. Toxicol Pathol 2007; 35:27–40 [Erratum in: Toxicol Pathol 2007; 35:620].
11. Wesson DW, Donahou TN, Johnson MO, et al. Sniffing behavior of mice during performance in odor-guided tasks. Chem Senses 2008; 33(7):581–596.

12. Rajan R, Clement JP, Bhalla US. Rats smell in stereo. Science 2006; 311(5761):666–670.
13. Wesson DW, Carey RM, Verhagen JV, et al. Rapid encoding and perception of novel odors in the rat. PLoS Biol 2008; 6(4):e82.
14. Abraham NM, Spors H, Carleton A, et al. Maintaining accuracy at the expense of speed: Stimulus similarity defines odor discrimination time in mice. Neuron 2004; 44:865–876.
15. Genter MB, Owens DM, Carlone HB, et al. Characterization of olfactory deficits in the rat following administration of 2,6-dichlorobenzonitrile (dichlobenil), 3,3'-iminodipropionitrile, or methimazole. Fundam Appl Toxicol 1996; 29:71–77.
16. Peele DB, Allison SD, Bolon B, et al. Functional deficits produced by 3-methylindole-induced olfactory mucosa damage revealed by a simple olfactory learning task. Toxicol Appl Pharmacol 1991; 107:191–202.
17. Owens JG, James RA, Moss OR, et al. Design and evaluation of an olfactometer for the assessment of 3-methylindole-induced hyposmia. Fundam Appl Toxicol 1996; 33:60–70.
18. Slotnick B, Sanguino A, Husband S, et al. Olfaction and olfactory epithelium in mice treated with zinc gluconate. Laryngoscope 2007; 117:743–749.
19. Mihalick SM, Langlois JC, Krienke JD, et al. An olfactory discrimination procedure for mice. J Exp Anal Behav 2000; 73(3):305–318.
20. Roman FS, Marchetti E, Bouquerel A, et al. The olfactory tubing maze: A new apparatus for studying learning and memory processes in mice. J Neurosci Methods 2002; 117(2):173–181.
21. Mandairon N, Ferretti CJ, Stack CM, et al. Cholinergic modulation in the olfactory bulb influences spontaneous olfactory discrimination in adult rats. Eur J Neurosci 2006; 24(11):3234–3244.
22. Wersinger SR, Caldwell HK, Martinez L, et al. Vasopressin 1a receptor knockout mice have a subtle olfactory deficit but normal aggression. Genes Brain Behav 2007; 6(6):540–551.
23. Youngentob SL, Margolis FL, Youngentob LM. OMP gene deletion results in an alteration in odorant quality perception. Behav Neurosci 2001; 115(3):626–631.
24. Furić Cunko V, Mitrecić D, Mavrić S, et al. Expression pattern and functional analysis of mouse Stam2 in the olfactory system. Coll Antropol 2008; 32(suppl 1):59–63.
25. McEwen DP, Koenekoop RK, Khanna H, et al. Hypomorphic CEP290/NPHP6 mutations result in anosmia caused by the selective loss of G proteins in cilia of olfactory sensory neurons. Proc Natl Acad Sci U S A 2007; 104(40):15917–15922.
26. Grubb BR, Rogers TD, Kulaga HM, et al. Olfactory epithelia exhibit progressive functional and morphological defects in CF mice. Am J Physiol Cell Physiol 2007; 293(2):C574–C583.
27. Fleming SM, Tetreault NA, Mulligan CK, et al. Olfactory deficits in mice overexpressing human wildtype alpha-synuclein. Eur J Neurosci 2008; 28(2):247–256.
28. Das P, Parsons AD, Scarborough J, et al. Electrophysiological and behavioral phenotype of insulin receptor defective mice. Physiol Behav 2005; 86(3):287–296.
29. Getchell TV, Kwong K, Saunders CP, et al. Leptin regulates olfactory-mediated behavior in ob/ob mice. Physiol Behav 2006; 87(5):848–856.
30. Fadool DA, Tucker K, Perkins R, et al. Kv1.3 channel gene-targeted deletion produces "Super-Smeller Mice" with altered glomeruli, interacting scaffolding proteins, and biophysics. Neuron 2004; 41(3):389–404.
31. Lovell MA, Jafek BW, Moran DT, et al. Biopsy of human olfactory mucosa. An instrument and a technique. Arch Otolaryngol 1982; 108(4):247–279.
32. Lanza DC, Deems DA, Doty RL, et al. The effect of human olfactory biopsy on olfaction: a preliminary report. Laryngoscope 1994; 104(7):837–840.
33. Hahn CG, Han LY, Rawson NE, et al. In vivo and in vitro neurogenesis in human olfactory epithelium. J Comp Neurol 2005; 483(2):154–163.
34. Paik SI, Lehman MN, Seiden AM, et al. Human olfactory biopsy. The influence of age and receptor distribution. Arch Otolaryngol 1992; 118(7):731–738.

35. Winstead W, Marshall CT, Lu CL, et al. Endoscopic biopsy of human olfactory epithelium as a source of progenitor cells. Am J Rhinol 2005; 19(1):83–90.
36. Talamo BR, Rudel R, Kosik KS, et al. Pathological changes in olfactory neurons in patients with Alzheimer's disease. Nature 1989; 337(6209):736–739.
37. Trojanowski JQ, Newman PD, Hill WD, et al. Human olfactory epithelium in normal aging, Alzheimer's disease, and other neurodegenerative disorders. J Comp Neurol 1991; 310(3):365–376.
38. Tabaton M, Cammarata S, Mancardi GL, et al. Abnormal tau-reactive filaments in olfactory mucosa in biopsy specimens of patients with probable Alzheimer's disease. Neurology 1991; 41(3):391–394.
39. Perry G, Castellani RJ, Smith MA, et al. Oxidative damage in the olfactory system in Alzheimer's disease. Acta Neuropathol 2003; 106(6):552–556.
40. Johnston MV, Blue ME, Naidu S. Rett syndrome and neuronal development. J Child Neurol 2005; 20(9):759–763.
41. Ronnett GV, Leopold D, Cai X, et al. Olfactory biopsies demonstrate a defect in neuronal development in Rett's syndrome. Ann Neurol 2003; 54(2):206–218.
42. Zanusso G, Ferrari S, Cardone F, et al. Detection of pathologic prion protein in the olfactory epithelium in sporadic Creutzfeldt-Jakob disease. New Engl J Med 2003; 348(8):711–719.
43. Tabaton M, Monaco S, Cordone MP, et al. Prion deposition in olfactory biopsy of sporadic Creutzfeldt-Jakob disease. Ann Neurol 2004; 55(2):294–296.
44. Restrepo D, Okada Y, Teeter JH, et al. Human olfactory neurons respond to odor stimuli with an increase in cytoplasmic Ca2+. Biophys J 1993; 64(6):1961–1966.
45. Hahn CG, Gomez G, Restrepo D, et al. Aberrant intracellular calcium signaling in olfactory neurons from patients with bipolar disorder. Am J Psychiatry 2005; 162:616–618.
46. McCurdy RD, Féron F, Perry C, et al. Cell cycle alterations in biopsied olfactory neuroepithelium in schizophrenia and bipolar I disorder using cell culture and gene expression analyses. Schizophr Res 2006; 82(2–3):163–173.
47. Ghanbari HA, Ghanbari K, Harris PL, et al. Oxidative damage in cultured human olfactory neurons from Alzheimer's disease patients. Aging Cell 2004; 3(1): 41–44.
48. Longo V, Ingelman-Sundberg M, Amato G, et al. Effect of starvation and chlormethiazole on cytochrome P450s of rat nasal mucosa. Biochem Pharmacol 2000; 59:1425–1432.
49. Genter MB, Deamer NJ, Cao Y, et al. Effects of P450 inhibition and induction on the olfactory toxicity of beta,beta'-iminodipropionitrile (IDPN) in the rat. J Biochem Toxicol 1994; 9:31–39.
50. Méry S, Larson JL, Butterworth BE, et al. Nasal toxicity of chloroform in male F-344 rats and female B6C3F1 mice following a 1-week inhalation exposure. Toxicol Appl Pharmacol 1994; 125:214–227.
51. Su T, He W, Gu J,et al. Differential xenobiotic induction of CYP2A5 in mouse liver, kidney, lung, and olfactory mucosa. Drug Metab Dispos 1998; 26(8):822–824.
52. Genter MB, Apparaju S, Desai PB. Induction of olfactory mucosal and liver metabolism of lidocaine by 2,3,7,8-tetrachlorodibenzo-p-dioxin (TCDD) J Biochem Mol Toxicol 2002; 16:128–133.
53. Nikula KJ, Novak RF, Chang IY, et al. Induction of nasal carboxylesterase in F344 rats following inhalation exposure to pyridine. Drug Metab Dispos 1995; 23: 529–535.
54. Lewis JL, Nikula KJ, Novak R, et al. Comparative localization of carboxylesterase in F344 rat, beagle dog, and human nasal tissue. Anat Rec 1994; 239:55–64.
55. Mainwaring G, Foster JR, Lund V, et al. Methyl methacrylate toxicity in rat nasal epithelium: studies of the mechanism of action and comparisons between species. Toxicology 2001; 158(3):109–118.
56. Su T, Sheng JJ, Lipinskas TW, et al. Expression of CYP2A genes in rodent and human nasal mucosa. Drug Metab Dispos 1996; 24(8):884–890.

57. Xie Q, Zhang QY, Zhang Y, et al. Induction of mouse CYP2J by pyrazole in the eye, kidney, liver, lung, olfactory mucosa, and small intestine, but not in the heart. Drug Metab Dispos 2000; 28:1311–1316.
58. Scarborough PE, Ma J, Qu W, et al. P450 subfamily 2J and their role in the bioactivation of arachadonic acid in extrahepatic tissues. Drug Metab Rev 1999; 31:205–234.
59. Matsumoto S, Hirama T, Matsubara T, et al. Involvement of CYP2J on the intestinal first-pass metabolism of antihistamine drug, astemizole. Drug Metab Dispos 2002; 30:1240–1245.
60. Aiba I, Yamasaki T, Shinki T, et al. Characterization of rat and human CYP2J enzymes as Vitamin D 25-hydroxylases. Steroids 2006; 71:849–856.
61. Thompson K, Molina RM, Donaghey T, et al. Olfactory uptake of manganese requires DMT1 and is enhanced by anemia. FASEB J 2007; 21:223–230.
62. Kandimalla KK, Donovan MD. Localization and differential activity of P-glycoprotein in the bovine olfactory and nasal respiratory mucosae. Pharm Res 2005; 22(7):1121–1128.
63. Genter MB, Kendig EL, Knutson MD. Uptake of materials from the nasal cavity into the blood and brain: Are we finally beginning to understand these processes at the molecular level? Ann NY Acad Sci. 2009; 1170:623–628.
64. Gianutsos G, Morrow GR, Morris JB. Accumulation of manganese in rat brain following intranasal administration. Fundam Appl Toxicol 1997; 37:102–105.
65. Tchernitchko D, Bourgeois M, Martin ME, et al. Expression of the two mRNA isoforms of the iron transporter Nramp2/DMTI in mice and function of the iron responsive element. Biochem J 2002; 363(Pt 3):449–455.
66. Thornalley PJ, Vasák M. Possible role for metallothionein in protection against radiation-induced oxidative stress. Kinetics and mechanism of its reaction with superoxide and hydroxyl radicals. Biochem Biophys Acta 1985; 827(1):36–44.
67. Bauman JW, Liu J, Liu YP, et al. Increase in metallothionein produced by chemicals that induce oxidative stress. Toxicol Appl Pharmacol 1991; 110:347–354.
68. Bremner I, Davies NT. The induction of metallothionein in rat liver by zinc injection and restriction of food intake. Biochem J 1975; 149:733–738.
69. Tallkvist J, Persson E, Henriksson J, et al. Cadmium-metallothionein interactions in the olfactory pathways of rats and pikes. Toxicol Sci 2002; 67:108–113.
70. Calderón-Garcidueñas L, Maronpot RR, Torres-Jardon R, et al. DNA damage in nasal and brain tissues of canines exposed to air pollutants is associated with evidence of chronic brain inflammation and neurodegeneration. Toxicol Pathol 2003; 31(5):524–538.
71. Kulkarni-Narla A, Getchell TV, Schmitt FA, et al. ML Manganese and copper-zinc superoxide dismutases in the human olfactory mucosa: Increased immunoreactivity in Alzheimer's disease. Exp Neurol 1996; 140:115–125.
72. Reed CJ, Robinson DA, Lock EA. Antioxidant status of the rat nasal cavity. Free Radic Biol Med 2003; 34:607–615.
73. Uslu C, Taysi S, Bakan N. Lipid peroxidation and antioxidant enzyme activities in experimental maxillary sinusitis. Ann Clin Lab Sci 2003; 33(1):18–22.
74. Chen J, Tu Y, Moon C, et al. Heme oxygenase-1 and heme oxygenase-2 have distinct roles in the proliferation and survival of olfactory receptor neurons mediated by cGMP and bilirubin, respectively. J Neurochem 2003; 85:1247–1261.
75. Elhini A, Abdelwahab S, Ikeda K. Heme oxygenase (HO)-1 is upregulated in the nasal mucosa with allergic rhinitis. Laryngoscope 2006; 116:446–450.
76. Williamson JM, Boettcher B, Meister A. Intracellular cysteine delivery system that protects against toxicity by promoting glutathione synthesis. Proc Natl Acad Sci U S A 1982; 79:6246–6249.
77. Jain A, Mårtensson J, Mehta T, et al. Ascorbic acid prevents oxidative stress in glutathione-deficient mice: Effects on lung type 2 cell lamellar bodies, lung surfactant, and skeletal muscle. Proc Natl Acad Sci U S A 1992; 89:5093–5097.
78. Burman DM, Shertzer HG, Senft AP, et al. Antioxidant perturbations in the olfactory mucosa of alachlor-treated rats. Biochem Pharmacol 2003; 66:1707–1715.

79. Struve MF, Wong VA, Marshall MW, et al. Nasal uptake of inhaled acrolein in rats. Inhal Toxicol 2008; 20:217–225.
80. Getchell ML, Shah DS, Buch SK, et al. 3-Nitrotyrosine immunoreactivity in olfactory receptor neurons of patients with Alzheimer's disease: Implications for impaired odor sensitivity. Neurobiol Aging 2003; 24:663–673.
81. Giannessi F, Ursino F, Fattori B, et al. Immunohistochemical localization of 3-nitrotyrosine in the nasal respiratory mucosa of patients with vasomotor rhinitis. Acta Otolaryngol 2005; 125:65–71.
82. Cannady SB, Batra PS, Leahy R, et al. SC Signal transduction and oxidative processes in sinonasal polyposis. J Allergy Clin Immunol 2007; 120:1346–1353.
83. Keller DA, Marshall CE, Lee KP. Subchronic nasal toxicity of hexamethylphosphoramide administered to rats orally for 90 days. Fundam Appl Toxicol 1997; 40: 15–29.
84. Genter MB, Van Veldhoven PP, Jegga AG, et al. Microarray-based discovery of highly expressed olfactory mucosal genes: Potential roles in the various functions of the olfactory system. Physiol Genomics 2003; 16:67–81.
85. Roberts ES, Soucy NV, Bonner AM, et al. Basal gene expression in male and female Sprague-Dawley rat nasal respiratory and olfactory epithelium. Inhal Toxicol 2007; 19:941–949.
86. Sammeta N, Yu TT, Bose SC, et al. Mouse olfactory sensory neurons express 10,000 genes. J Comp Neurol 2007; 502:1138–1156.
87. Liu N, Crasto CJ, Ma M. Integrated olfactory receptor and microarray gene expression databases. BMC Bioinformatics 2007; 8:231.
88. Johnson VJ, Yucesoy B, Reynolds JS, et al. Inhalation of toluene diisocyanate vapor induces allergic rhinitis in mice. J Immunol 2007; 179:1864–1871.
89. Sridhar S, Schembri F, Zeskind J, et al. Smoking-induced gene expression changes in the bronchial airway are reflected in nasal and buccal epithelium. BMC Genomics 2008; 9:259.
90. Uchida Y, Takio K, Titani K, et al. The growth inhibitory factor that is deficient in the Alzheimer's disease brain is a 68 amino acid metallothionein-like protein. Neuron 1991; 7(2):337–347.
91. Genter MB, Burman DM, Vijauakumar S, et al. Genomic analysis of alachlor-induced oncogenesis in rat olfactory mucosa. Physiol Genomics 2002; 12:35–45.
92. Genter MB, Burman DM, Dingeldein MW, et al. Evolution of alachlor-induced nasal neoplasms in the Long-Evans rat. Toxicol Pathol 2000; 28(6):770–781.
93. Genter MB, Warner BM, Krell H-W, et al. Reduction of alachlor-induced olfactory mucosal neoplasms by the matrix metalloproteinase inhibitor Ro 28–2653. Toxicol Pathol 2005; 33:593–599.
94. Klein JA, Longo-Guess CM, Rossmann MP, et al. The harlequin mouse mutation downregulates apoptosis-inducing factor. Nature 2002; 419(6905):367–374.
95. Vaishnav RA, Getchell ML, Huang L, et al. Cellular and molecular characterization of oxidative stress in olfactory epithelium of Harlequin mutant mouse. J Neurosci Res 2008; 86:165–182.
96. Poon HF, Vaishnav RA, Butterfield DA, et al. Proteomic identification of differentially expressed proteins in the aging murine olfactory system and transcriptional analysis of the associated genes. J Neurochem 2005; 94:380–392.
97. Ríos-Blanco MN, Faller TH, Nakamura J, et al. Quantitation of DNA and hemoglobin adducts and apurinic/apyrimidinic sites in tissues of F344 rats exposed to propylene oxide by inhalation. Carcinogenesis 2000; 21:2011–2018.
98. Haseman JK, Hailey JR. An update of the National Toxicology Program database on nasal carcinogens. Mutat Res 1997; 380(1–2):3–11.
99. Galati R, Federico A, Cortese G, et al. Determination of serum levels of 2,6 diethylaniline in laboratory animal treated with alachlor. Anticancer Res 1998; 18(2A): 979–982.
100. Duan J-D, Genter MB, Jeffrey AM, et al. Formation of DNA adducts in F344 rat nasal tissue by 2,6-dimethylaniline and diethylaniline, but not alachlor. *The Toxicologist,*

Abstract #1901, 43rd Annual Meeting of the Society of Toxicology; March 21–25, 2004; Baltimore, MD.

101. Duan JD, Jeffrey AM, Williams GM. Assessment of the medicines lidocaine, prilocaine, and their metabolites, 2,6-dimethylaniline and 2-methylaniline, for DNA adduct formation in rat tissues. Drug Metab Dispos 2008; 36:1470–1475.

102. Sarangapani R, Teeguarden JG, Gentry PR, et al. Interspecies dose extrapolation for inhaled dimethyl sulfate: A PBPK model-based analysis using nasal cavity N7-methylguanine adducts. Inhal Toxicol 2004; 16:593–605.

103. Pottenger LH, Malley LA, Bogdanffy MS, et al. Evaluation of effects from repeated inhalation exposure of F344 rats to high concentrations of propylene. Toxicol Sci 2007; 97(2):336–347.

104. Casanova M, Deyo DF, Heck HD. Covalent binding of inhaled formaldehyde to DNA in the nasal mucosa of Fischer 344 rats: analysis of formaldehyde and DNA by high-performance liquid chromatography and provisional pharmacokinetic interpretation. Fundam Appl Toxicol 1989; 12:397–417.

105. Mayer C, Klein RG, Wesch H, et al. Nickel subsulfide is genotoxic in vitro but shows no mutagenic potential in respiratory tract tissues of BigBlue rats and Muta Mouse mice in vivo after inhalation. Mutat Res 1998; 420:85–98.

106. Vock EH, Hoymann HG, Heinrich U, et al. ^{32}P-postlabeling of a DNA adduct derived from 4,4'-methylenedianiline, in the olfactory epithelium of rats exposed by inhalation to 4,4'-methylenediphenyl diisocyanate. Carcinogenesis 1996; 17:1069–1073.

107. Belinsky SA, White CM, Boucheron JA, et al. Accumulation and persistence of DNA adducts in respiratory tissue of rats following multiple administrations of the tobacco specific carcinogen 4-(N-methyl-N-nitrosamino)-1-(3-pyridyl)-1-butanone. Cancer Res 1986; 46:1280–1284.

108. Nelson E, Zhou Z, Carmichael PL, et al. Genotoxic effects of subacute treatments with wood dust extracts on the nasal epithelium of rats: Assessment by the micronucleus and 32P-postlabelling. Arch Toxicol 1993; 67:586–589.

109. Snyder CA, Garte SJ, Sellakumar AR, et al. between the levels of binding to DNA and the carcinogenic potencies in rat nasal mucosa for three alkylating agents. Cancer Lett 1986; 33:175–181.

110. Brittebo EB. Metabolic activation of phenacetin in rat nasal mucosa: Dose-dependent binding to the glands of Bowman. Cancer Res 1987; 47:1449–1456.

111. Hart SG, Cartun RW, Wyand DS, et al. Immunohistochemical localization of acetaminophen in target tissues of the CD-1 mouse: Correspondence of covalent binding with toxicity. Fundam Appl Toxicol 1995; 24:260–274.

112. Hoffmann KJ, Streeter AJ, Axworthy DB, et al. Identification of the major covalent adduct formed in vitro and in vivo between acetaminophen and mouse liver proteins. Mol Pharmacol 1985; 27:566–573.

113. Bulera SJ, Birge RB, Cohen SD, et al. Identification of the mouse liver 44-kDa acetaminophen-binding protein as a subunit of glutamine synthetase. Toxicol Appl Pharmacol 1995; 134:313–320.

114. Navarro CL, Cohen SD, Khairallah EA. Genes encoding the acetaminophen and selenium binding proteins map to mouse chromosome 3. Mamm Genome 1996; 7(12):919–920.

115. Warren DL, Brown DL Jr., Buckpitt AR. Evidence for cytochrome P-450 mediated metabolism in the bronchiolar damage by naphthalene. Chem Biol Interact 1982; 40:287–303.

116. Tingle MD, Pirmohamed M, Templeton E, et al. An investigation of the formation of cytotoxic, genotoxic, protein reactive and stable metabolites from naphthalene by human liver microsomes. Biochem Pharmacol 1993; 46:1529–1538.

117. Wilson AS, Davis CD, Williams DP, et al. Characterization of the toxic metabolite(s) of naphthalene. Toxicology 1996; 114:233–242.

118. Zheng J, Hammoock BD. Development of polyclonal antibodies for detection of protein modification by 1,2-naphthoquinone. Chem Res Toxicol 1996; 9:904–909.

11 Biomarkers of Nasal Toxicity in Humans

T. I. Fortoul, V. Rodríguez-Lara, N. López-Valdez,
C. I. Falcón-Rodríguez, L. F. Montaño, and M. C. Ávila-Casado

*Department of Cellular and Tissular Biology, School of Medicine,
National University of Mexico, Mexico City, Mexico*

WHAT IS A BIOMARKER?

Biomarkers are defined by the National Academy of Sciences as xenobiotically induced alterations in cellular or biochemical components or processes, structures, or functions that are measurable in a biological system or sample. Kendall et al. (1) suggested adding to this list xenobiotically induced alterations in behavior. Therefore, biomarkers can be broadly categorized as markers of exposure, effects, or susceptibility (1).

- *Biomarkers of exposure*: The presence of a xenobiotic substance or its metabolite(s) or the product of an interaction between a xenobiotic agent and some target molecule, or cell that is measured within a compartment of an organism can be classified as a biomarker of exposure (1). In general, biomarkers of exposure are used to predict the dose received by an individual, which can then be related to changes resulting in a disease state (1).
- *Biomarkers of effect*: Measurable biochemical, physiologic, behavioral, or other alterations within an organism that, depending on their magnitude, can be recognized as an established or potential health impairment or disease (1). Ideally, a biomarker result must be able to stand alone. As such, it does not need chemical analysis or additional biological tests for confirmation (1). Single-cell gel electrophoresis (SCGE), or comet assay, has been proposed as an early biomarker of effect. SCGE is a very sensitive test for genotoxic damage and this technique has been widely used in human biomonitoring studies, especially in nasal epithelium (2).
- *Biomarkers of susceptibility*: End points that are indicative of an altered physiologic or biochemical state that may predispose the individual to impacts of chemical, physical, or infectious agents (2).

To identify alterations in the nasal epithelium and the application of different biomarkers of early effect, we will give an introduction to the structure of the nose and finally we will include some of our scientific experience.

THE HUMAN NOSE

The nasal cavity is the anatomical structure in which the inhaled air is moistened, warmed, and filtered; in addition, it is the place at which the olfaction takes place. Collectively with the lower airways, the nasal cavity maintains humidity and temperature at optimal conditions for the main function of the distal structures of the airway: the alveoli, the place where exchange of carbon dioxide and oxygen takes place (3). Every day about 10,000 to 20,000 L of air enters through the nostrils and the air is filtered in order to eliminate gases, particles, microorganism, and allergens which are suspended in the air; a healthy nose could retain and eliminate the majority of the contaminants suspended in the inhaled air and allows that a clean, warm, moist, and almost sterile air arrives at the alveolar surface to interact with the epithelium (4).

In the upper respiratory tract, the deposition of the inhaled particles depends on a variety of characteristics from the nasal cavity structure, as well as the size, speed, and turbulence of the airstream. It is important to mention that in the anterior section of the nose, about 45% of inspired particles are deposited; this is an area devoid of ciliated cells, and as a consequence, the clearance is achieved by wiping or blowing the nose (5,6). Particles deposited on the ciliated epithelium are rapidly removed by the mucociliary clearance mechanisms, while particles deposited in the nonciliated region are removed relatively slowly (7). The largest depositions of inhalable particles—those with a diameter larger than 10 µm—as well as the smallest ones—less than 1.0 µm—are deposited on the ciliated epithelium, transported by the mucociliary movement for its elimination by swallowing, nose blowing, or sneezing.

Substances that are dissolved may be subsequently translocated into the bloodstream following movement within intercellular pathways between epithelial cell tight junctions or by active or passive transcellular transport mechanisms (5,8).

The main area of physical contact with xenobiotics is the turbinates that are designed to induce turbulence in the inspired air, which in turn favors the contact of the exogenous agents with the respiratory epithelium. These structures (consisting of a bony and cartilaginous framework covered by thick vascular tissue) increase the inner surface area of the nose, a situation that is important for the functions of filtering, humidification, and warming of the inspired air. The turbinated, main chamber of the human nose is only about 5 to 8 cm long, and its surface area is approximately 150 to 200 cm^2, about four times that of the trachea (4) [Fig. 1(A)].

NASAL EPITHELIUM

As referenced by Proctor and Andersen (4) and Hakerma et al. (8), the nose acts as an "air conditioner," and for this function, it will need different types of epithelia with a wide variety of cells (9):

- *Basal cells*: These cells replace the population and contribute to the pseudostratified appearance of the epithelium in the diverse sections of the respiratory tree. Also, it is thought that basal cells play a role in the attachment of the superficial cells to the airway basement membrane (10) [Fig. 1(C)].
- *Ciliated cells*: This cell population acts to filter and move mucus and materials adhered to it. Ciliated cells exhibit a well-synchronized beating of the cilia, and propel the mucus at different speeds and directions depending on its intranasal location (8) [Fig. 1(C)].
- *Goblet cells*: To moist with a watery and sticky substance, goblet cells produce glycoproteins—called mucus—with physical and chemical properties well suited for its role as an airway defense mechanism [Fig. 1(C)].

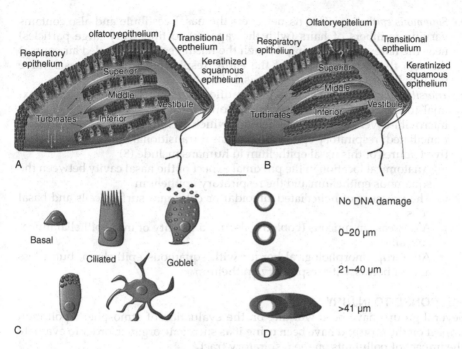

FIGURE 1 (**A**) Normal human nasal structure with the differences in epithelia distribution. (**B**) Sustained environmental insult to the nasal epithelium will change the functionality of the cells, modifying the normal respiratory epithelium and observing a squamous or mucous metaplasia. (**C**) Schematic representation of respiratory cell's epithelium. (**D**) Single cell electrophoresis (SCGE) or comet assay identifies DNA damage (fragmentation) by the length of the "tail."

- *Serous cells*: These are the primary secretory cells in the remainder of the respiratory epithelium. These cells show abundant smooth endoplasmic reticulum, which support its secretory function and metabolic capacities for several xenobiotic agents (11). In particular, carboxylesterase, aldehyde dehydrogenase, cytochrome P-450, epoxide hydrolase, and glutathione S-transferases have been localized by histochemical techniques. The distribution of these enzymes appears to be cell type specific, and the presence of the enzyme may predispose particular cell types to enhanced susceptibility or resistance to chemical-induced injury (12) [Fig. 1(C)].
- *Dendritic cells*: These cells primarily were described as Langerhans cells identified by the presence of CD1a and langerin (CD207) immunohistochemically and are identified by electron microscopy as well based on the presence of Birbeck granules. Recently, three phenotypically and functionally different pulmonary dendritic cell subsets, such as those circulating in the human blood, have been described: type 1 or myeloid CD1c, type 2 or myeloid CD141, and plasmocytoid Cds. Myeloid and plasmacytoid dendritic cells perform different functions in both innate and adaptive immunity. At the same time, they have functional plasticity to induce appropriate T-cell responses depending on the type of stimuli (13) [Fig. 1(C)].

- *Squamous epithelium*: This tissue covers the nasal vestibule and also contains varying numbers of hairs (with the capacity to filter very large particles) near the nares. After passing through the nasal vestibule, inhaled air courses through the narrowest part of the entire respiratory tract, the nasal valve (ostium internum), into the main nasal chamber.
- *Transitional epithelium*: Distal to the stratified squamous epithelium and proximal to the ciliated respiratory epithelium is a narrow zone of nonciliated, microvilli-covered surface epithelium, which has been referred to as nasal, nonciliated, respiratory epithelium, or nasal transitional epithelium. Distinctive features of this nasal epithelium in humans include (8)
  - Anatomical location in the proximal aspect of the nasal cavity between the squamous epithelium and the respiratory epithelium
  - The presence of nonciliated cuboidal or columnar surface cells and basal cells
  - A decrease in mucous (goblet) cells and a paucity of intraepithelial mucosubstances
  - An abrupt morphological border with squamous epithelium, but a less abrupt border with respiratory epithelium

RESPONSE TO INJURY

Several groups have been working on the evaluation of atmospheric pollution impact on the nose and have been using it as surrogate organ in order to evaluate the impact of pollutants on the respiratory tract.

The fact that the nasal cavity is easily approachable and that some studies have reported that the nasal epithelium behaves as the bronchial epithelium cells (14) gives us the opportunity to extrapolate the nasal data about the impact of xenobiotics into the whole respiratory tract.

A vast scope of techniques is attainable to researchers to assess and quantify nasal mucosal damage in human subjects. There are several methods for obtaining nasal cells and different techniques for sampling different compartments of the nasal mucosa.

As Calderon-Garciduenas (5) reported: blown secretions, nasal smears, imprints, nasal scrapings, nasal biopsies, and nasal lavage are all methods to obtain material to evaluate structural or biochemical modifications in the nose. Some are more aggressive than others and the methodology selected would depend on the experience of the research group and the output expected.

As we mentioned previously, we have worked on documenting the impact that air pollution has on the respiratory tract. The approach selected was the use of a small brush to obtain nasal cells for cytological and genotoxic evaluation. The technique we used for DNA damage evaluation was the SCGE or comet assay (15–17). The technique allows identifying the severity of the DNA damage (fragmentation) by measuring the length of the "tail" structured by DNA fragments; the larger the tail, the worse the damage [Fig. 1(D)].

SCGE is a very sensitive test, which identifies DNA single-strand breaks, alkali-labile sites, and delayed repaired site, at alkaline pH; or double-strand breaks performed under neutral conditions (2).

We identify the impact that time of exposure to oxidizing atmosphere, gender, or asthma has on nasal cell's DNA. We demonstrated that the exposure to a polluted urban atmosphere with high ozone concentrations induced DNA single-strand breaks in nasal epithelial cells in young adults (17); when

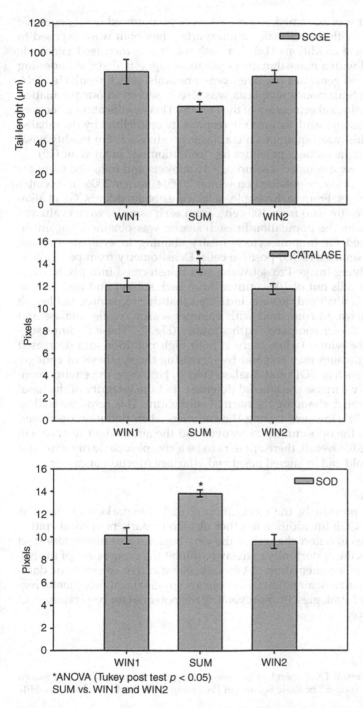

FIGURE 2 An increase in genotoxic damage is observed during winter, while a decrease in superoxide dismutase (SOD) and catalase is evidenced. During summer season, catalase and SOD increase and genotoxicity decrease, suggesting a protective role of these antioxidants.

the effect of gender was examined, males had more pronounced nasal genotoxic damage compared with their female counterparts when both were exposed to the same atmosphere conditions (16). For asthmatics, the increased genotoxic damage compared with a nonasthmatic population suggests that the underling inflammation process generates reactive oxygen radicals, which results in DNA strand breaks (15). Squamous metaplasia was also observed in our population detected by the cytological evaluation of the smears. This modification is the consequence of the constant insult against the respiratory epithelium by the inhaled air pollutants, in this case. Squamous metaplasia, if extensive, can modify nasal function, restricting the tissue capabilities for "conditioning" inhaled air (15).

Additionally, we evaluated a sample of 12 subjects and followed them for one year. Nasal smears were obtained on winter 2005, summer 2006, and winter 2006 from both narines. From each sample, cells were obtained for SCGE at alkaline pH; tail lengths (in μm) from 100 cells, from each sample, were evaluated and the average from the population in each season was obtained. Sacomano fixative was utilized for immunocytochemistry staining to evaluate catalase and superoxide dismutase (SOD)-positive cells. Densitometry from peroxidase-positive cells, applying Image-Pro software, was transformed into pixels; averages from positive cells out of 100 counted from each subject and each season were calculated and analyzed. Results indicated that during winter, tail length (DNA damage) increased compared with summer season, while catalase and SOD decreased in winter compared with summer (Fig. 2). These findings suggest that during the winter (when there is both high pollution and decreased humidity), the epithelium may respond by increasing the synthesis of endogenous antioxidants such as SOD and catalase (18,19). However, the environmental insult apparently surpasses epithelial defenses, and the integrity of the nasal lining cells is disrupted, changing the normal epithelium. This response is illustrated in Figure 1(B). However, during the summer, when pollution decreases, the epithelium has the opportunity to recover and the antioxidant reserve will likewise recover (20). Overall, there appears to be a complex circle of repair and damage, which could end in altered nasal and olfactory function or cancer.

FINAL REMARKS

As we mentioned previously, the recognition of early biomarkers of effect, as illustrated by the SCGE (in addition to other metabolic markers in nasal epithelium), could help us to detect changes at the very beginning of respiratory tract disease and give us the opportunity to delay or inhibit the progression of the respiratory pathology. Documentation of biomarkers indicative of genotoxic damage in response to ambient air pollutants could also motivate environmental regulations and control strategies, thus preventing the potential for respiratory tract damage at its source.

REFERENCES

1. Kendall RJ, Anderson TA, Robert J, et al. Ecotoxicology. In: Klaassen DC, ed. Casarett and Doul's Toxicology: The Basic Science of Poisoning. New York, NY: McGraw-Hill, 2001:1013–1045.
2. Mussali-Galante P, Avila-Costa MR, Piñón-Zarate G, et al. DNA damage as an early biomarker of effect in human health. Toxicol Ind Health 2005; 21:155–166.

3. Cole P. The Respiratory Role of the Upper Airways. St Louis, MO: Mosby Year Book, 1993:122–137.

4. Proctor DF, Andersen IB. The Nose: Upper Airway Physiology and the Atmospheric Environment. Amsterdam: Elsevier Science, 1982:7–25.

5. Calderon-Garciduenas L, Noah TL, Koren HS. Novel Approaches to Study Nasal Responses to Air Pollution. In: Holgate TS, Samet JM, Koren SH, Maynard LR, eds. Air Pollution and Health. San Diego, CA: Academic Press, 1999:199–218 [chap 11].

6. Koening JQ, Pierson WE. Nasal responses to air pollutants. Clin Rev Allergy 1984; 2:255–261.

7. Hilding AC. Phagocytosis, mucous flow and ciliary action. Arch Environ Health 1963; 6:67–79.

8. Harkema JR, Carey SA, Wagner JG. The Nose Revisited: A Brief Review of the Comparative Structure, Function, and Toxicologic Pathology of the Nasal Epithelium. Toxicol Pathol 2006; 34:252–269.

9. Monteiro-Riviere NA, Popp JA. Ultrastructural characterization of the nasal respiratory epithelium in the rat. Am J Anatomy 1984; 169:31–43.

10. Devalia JL, Abdelaziz MM, Davies RJ. Epithelial cells. In: Barnes JP, Rodger WI and Thomson CN, eds. Asthma: Basic Mechanisms and Clinical Management, 3rd ed. San Diego, CA: Academic Press Limited, 1998:189–204 [chap 10].

11. Yamamoto T, Masuda H. Some observations on the fine structure of the goblet cells in the nasal respiratory epithelium of the rat, with special reference to the well-developed agranular endoplasmic reticulum. Okajimas Folia Anat Jpn 1982; 58:583–594.

12. Lewis JL, Nikula KJ, Novak R, et al. Comparative localization of carboxyesterases in F344 rat, beagle dog and human nasal tissue. Fundam Appl Toxicol 1994; 23:510–517.

13. Tsumadikou M, Demedts IK, Brusselle GG, et al. Dendritic cells in chronic obstructive pulmonary disease. Am J Respir Crit Care Med 2008; 177:1180–1186.

14. McDougall CM, Blaylock MG, Douglas JG, et al. Nasal epithelial cells as surrogates for bronchial epithelial cells in airway inflammation studies. Am J Respir Cell Mol Biol. 2008; 39:560–568.

15. Fortoul TI, Valverde M, Lopez MC, et al. Single cell gel electrophoresis assay in nasal epithelium and leukocytes in asthmatic subjects. Arch Environ Health 2003; 58:348–352.

16. Fortoul TI, Valverde M, López MC, et al. Genotoxic differences by sex in nasal epithelium and blood leukocytes in subjects residing in highly polluted areas. Environ Res 2004; 94:243–248.

17. Valverde M, Lopez MC, Lopez I, et al. DNA damage in leukocytes, buccal and nasal epithelial cells of individuals exposed to air pollution in Mexico City. Environ Mol Mutagen 1997; 30:147–152.

18. Feldman Ch, Anderson R, Kanthakumar K, et al. Oxidant-mediated ciliary dysfunction in human respiratory epithelium. Free Radic Biol Med 1994; 17:1–10.

19. Ztto-Knapp R, Yurgovsky K, Schierhorn K, et al. Antioxidative enzymes in human nasal mucosa after exposure to ozone. Possible role of GSTM1 deficiency. Inflamm Res 2003; 52:51–55.

20. Comhair SA, Erzurum SC. Antioxidant responses to oxidant-mediated lung diseases. Am J Physiol Lung Cell Mol Physiol 2002; 283:L246–L255.

12 Nasal Reflexes, Including Alterations in Respiratory Behavior, in Experimental Animals

John B. Morris

Department of Pharmaceutical Sciences, School of Pharmacy,
University of Connecticut, Storrs, Connecticut, U.S.A.

INTRODUCTION

The nasal cavity provides several important respiratory defense mechanisms. It serves to condition the inspired air prior to its entry into the lower respiratory tract. In addition to heating and humidifying the airstream, the nasal cavity scrubs particulates and vapors from the airstream as well (chap. 6). These processes can be quite efficient; in the rodent upper respiratory tract deposition for particles greater than 5 μm can exceed 80%. Essentially complete scrubbing (>95%) of water-soluble reactive vapors such as formaldehyde, acetic acid, or hydrogen fluoride occurs in the rodent nose (chap. 6) (1,2). Efficient scrubbing leads to high delivery of airborne toxicants to nasal tissues, making the nose an ideal site, from a dosimetric view, for detection of noxious airborne substances and initiation of appropriate reflex responses. The nasal cavity also provides important warning functions about the quality of the inspired air. Cranial nerve I, the olfactory nerve, is specialized for the detection of odors. Cranial nerve V, the trigeminal nerve, contains afferent sensory nerves that detect the presence of noxious materials. Several terms have been used to describe the trigeminal nerve detection of such materials, including nociception and chemesthesis (chap. 13).

Materials that stimulate the trigeminal nerve are termed irritants, or more precisely, sensory irritants, and produce the sensation of tickling, burning, and/or pain in the human (3). It should be recognized that the term irritant is also used to describe materials that induce necrosis and a subsequent inflammatory response. Historically much confusion has resulted from the dual meaning of this term. The potential to induce "irritation" as measured by nasal tissue damage is not correlated with the potential to induce "irritation" as measured by sensory nerve effects (4). In this chapter, the term "irritant" is to describe compounds that stimulate sensory nerves. Most irritants stimulate both the olfactory and the trigeminal nerve; typically the threshold for olfactory stimulation is considerably less than that for trigeminal stimulation (5) (chap. 13). Interactions between the olfactory and trigeminal nerve may exist and may be important in the overall perception of irritants. These interactions may be psychological and/or physiological in nature (5,6). In the rat, collateral branches of the trigeminal nerve innervate the olfactory bulb, highlighting the possibility that direct olfactory–trigeminal interactions exist (7). Potential mechanisms for such an interaction are discussed below.

Stimulation of nasal trigeminal sensory nerves elicits a variety of central and local reflex responses, some of which may be species-specific. This chapter will describe these responses and their modulation by disease and/or physiological mediators. The spectra of irritant nerve–mediated responses in rodents have been well characterized in the author's laboratory for two irritants, acrolein and acetic acid. In addition, the responses to capsaicin have been well studied in several laboratories, as this agent, derived from hot peppers, is a widely used experimental tool (8). This chapter will focus on these agents, not because they are necessarily of great environmental importance, but because they can be used to illustrate common response patterns. Many direct analogies exist between upper airway trigeminal and lower airway vagal nerves. A review of vagal nerve responsiveness is beyond the scope of this chapter; however, when highly relevant or insightful, citations to vagal nerve function will be provided. The reader is referred to the work of Bradley Undem or Lu-Yuan Lee for insights into functionality of vagal sensory afferents in health and disease.

INNERVATION

C and Aδ Fibers

The entire nasal cavity is innervated by the trigeminal nerve (chap. 3). Innervation is via two branches of the trigeminal: the ethomoidal and nasopalantine. Nerve endings are present in the lamina propria and may penetrate superficially to the level of epithelial tight junctions. In healthy tissue, it is unlikely that nerve endings protrude to the airspace (9). Many highly reactive vapors (e.g., acrolein, formaldehyde) are potent stimulators of nasal sensory nerves but, due to their reactivity, are likely to be significantly removed by chemical reaction within the mucous and/or epithelial cells, especially at low exposure concentrations. Thus, the concentrations of vapor reaching nerve endings are expected to be quite low. These dosimetric considerations are particularly important in design and interpretation of in vitro studies of sensory neuronal function.

Two types of trigeminal sensory nerves are thought to be important in mediating respiratory reflex mechanisms: C fibers and Aδ fibers (10–12; chap. 3). Differing irritants stimulate differing populations of nerves (see below); therefore, it is highly likely that reflex response patterns to irritant vapors are determined by the integration of signals from both types. Additionally, signals from epithelial cells and/or olfactory neurons may also be important in modulating the neuronal responsiveness (see below). There are few unambiguous functional criteria with which to classify subtypes of these nerves. C fibers are thin nonmyelinated slowly conducting nerves that are thought to be sensitive to chemical stimulation. A large fraction, but not all, C fiber express several neuropeptides including calcitonin gene–related peptide (CGRP), substance P and neurokinin A (11; chap. 3). Neuropeptides exert broad ranging effects in respiratory tract mucosa (13,14). Aδ fibers are thicker, faster conducting, partially myelinated nerves. In healthy conditions, Aδ fibers do not express neuropeptides. From this perspective alone it can be appreciated that the degree to which irritants stimulate Aδ versus C fibers will profoundly influence the magnitude of any neuropeptide-mediated effects. In disease and/or following toxic injury, Aδ nerves undergo a phenotypic switch and express neuropeptides (see below and chap. 26). This is undoubtedly of toxicological importance, although the functional significance of this effect is not fully explored. Finally, it should be noted that myelination of nerves is a relatively late-occurring developmental effect. Thus, it is possible that the role of Aδ and C fibers may differ in neonates, children, and adults. This possibility has received little attention.

Stimulation of sensory nerves may result in generation of an action potential and transmission of impulses to the respiratory centers of the brain stem. The transganglionic projections of the nasal trigeminal nerve terminate in the subnucleus cauldalis and the subnucleus interpolaris (12,15) (chapts. 3 and 4). Considerable integration of signals may occur in the brainstem (12) and is beyond the scope of this chapter. Simulation of sensory nerves also results in a wave of depolarization that extends throughout the branches of the nerve. This is termed antidromal stimulation and/or the axon reflex (10–12,16,17) (chap. 14). Importantly, this results in the relatively widespread release of neuropeptides from those nerves in which they are expressed (e.g., C fibers). From this perspective it can be recognized that local neuropeptide-mediated responses may occur in the absence of central nervous system–mediated responses, especially if suppressing central neuromodulation pathways are present.

Parasympathetic and sympathetic nerves may also be involved in response to irritants via efferent reflex pathways. Their function and integration with trigeminal nerves have been reviewed (10–12). Trigeminal nerve activation can lead to collateral activation of parasympathetic efferents and resultant effects due to acetylcholine and/or vasoactive intestinal peptide release. The importance of these pathways in irritant responsiveness is not fully understood. The cholinergic antagonist atropine is without effect on the sensory nerve responses to chlorine (18,19) (chap. 20) or acrolein (Morris, unpublished observation, 2000), suggesting that collateral parasympathetic stimulation may not be essential in mediating the obstructive response to these two highly reactive irritants.

Receptor Basis for Irritant Detection

The receptor basis for irritant detection is not fully understood, however, recently there have been significant advances in our understanding of this important aspect of irritant toxicology. A variety of receptor systems may be involved, including (but not limited to) the transient receptor potential vanilloid 1 receptor (TRPV1) (20,21), transient receptor potential A1 receptor (TRPA1) (22,23), the acid sensing ion channel (ASIC) (24), a purinergic receptor–P2X (25), and the nicotinic acethylcholine receptor (26). The irritants capsaicin and cylcohexanone act through the TRPV1 receptor (27). Acidic irritants such as acetic acid presumably act, at least in part, through the ASIC receptor (24); data regarding the role of the TRPV1 receptor in acid detection is conflicting (28–30). Oxidants including H_2O_2 and chlorine as well as electrophilic reactive compounds, such as acrolein and allyl isothiocyanate, interact with the TRPA1 receptor (22,23). TRPV1 and TRPA1 receptors are expressed primarily, if not solely, in C fibers. Based on size of the trigeminal neurons expressing the ASIC receptor, this receptor is present in both C and Aδ fibers (24,28,29). Since neuropeptides are expressed predominantly in C fibers, the potential for neuropeptide-induced responses is dependent on the degree to which an irritant stimulates C versus Aδ fibers. The nerve subtype and receptor basis for irritant detection are of fundamental importance and are area in need of much additional research.

TRPV1 and TRPA1 receptors are both nonspecific cation channels that exhibit a preference for calcium. Studies on the role of the TRPV1 and TRPA1 have been greatly aided by the development of knockout strains of mice (21,22). Recent studies have revealed that electrophilic and oxidant irritants stimulate sensory C fibers through the TRPA1 receptor. For example, responsiveness to acrolein (a prototypical electrophilic irritant) is unaltered in TRPV1-/- mice indicating that the TRPV1 receptor is not directly involved in acrolein responsiveness (30). In contrast, the sensory neuronal response to acrolein is absent in TRPA1-/- mice. Acrolein is present in high concentrations in cigarette smoke; interestingly, cigarette smoke extract also stimulates sensory nerves through the TRPA1 receptor (31). The TRPA1 receptor is thought to be expressed exclusively in C fibers that also coexpresses the TRPV1 receptor (23,23,32). Activation of the TRPA1 receptor pathway likely involves irritant interaction with key cysteine or lysine residues (32). This receptor also is critical in mediating sensory nerve activation by chlorine, H_2O_2, and 4-hydroxynonenal, suggesting it may represent an important oxidant-sensitive pathway for sensory nerve activation and may well respond to endogenously generated oxidants (23,33). It is known that chlorine and H_2O_2 stimulate both the rodent and human TRPA1 receptors (23). This species similarity may not be universally true for all irritants because there is relatively low homology between the rodent versus human TRPA1 channels (32). The discovery of the importance of the TRPA1 receptor is an exciting development that may lead to fundamental insights into the mechanisms of action of a variety of irritants.

The most studied agonist of the TRPV1 receptor is capsaicin. Capsaicin is the pharmacologically active constituent of hot peppers. Few are unfamiliar with the reflex responses induced by this agent in everyday life, with upwards of 60% of individuals reporting nasal secretion in response to hot, spicy foods (34). Responsiveness to capsaicin is blocked by TPRV1 antagonists and is absent in TRPV1-/- mice (20,21). This receptor is widely expressed in C fibers. A small subset of Aδ fibers may express this receptor or the vanilloid-like receptor 1 (21), thus capsaicin is a C fiber selective, but not C fiber specific, agent. Acute challenge with capsaicin is often used as a tool to examine sensory nerve responses. Such experiments have provided many insights into sensory nerve effects. It should be recognized, however, that since capsaicin is selective for C fibers, the responses it evokes may not be predictive of the response patterns evoked by all irritants, particularly those that stimulate Aδ fibers. The TRPV1 receptor may not be sensitive to a broad array of irritant air pollutants. Studies with knockout mice have indicated that this receptor pathway may not be involved in the reflex response to such irritants as acrolein, acetic acid, and styrene (30,35).

Systemic administration of high doses of capsaicin has long been used as an experimental tool to cause long-term degeneration of C fibers (36). The C fiber degeneration is likely the result of excitotoxicity due to excess calcium absorption into stimulated nerves (20,36). Many laboratories, including the author's, have used such paradigms to discern the contribution of capsaicin-sensitive (likely C fiber) versus capsaicin-insensitive (likely Aδ fibers and/or the subset of non-TRPV1 expressing C fiber neurons) in mechanistic studies of irritants. In the mouse, capsaicin pretreatment only partially prevented the sensory irritation and obstructive response to acetic acid but totally ablated these responses to acrolein (1), suggesting that sensory responses to acrolein are mediated solely via capsaicin-sensitive nerves whereas acetic acid stimulates additional nerve subtypes as well. This is consistent with the specific expression TRPA1, the receptor through which acrolein acts, in TRPV1 expressing C fibers. The mechanism of sensory nerve activation by acetic acid is not known but may involve the ASIC receptor (11,28,35); the TRPV1 receptor does not appear to be involved in acetic acid–induced trigeminal activation (30). The ASIC receptor is expressed on both C and Aδ fibers (24,28,29). The partial inhibition of acetic acid responsiveness by capsaicin pretreatment suggests that both C and Aδ fibers contribute to the reflex responses, a result consistent with a role of the ASIC receptor in acid detection.

It is often tacitly assumed that irritant vapor molecules directly interact with sensory nerves. Two lines of evidence suggest this may not be the case. First, multiple receptor pathways may contribute to paracrine-mediated initiation (and/or modulation) of sensory nerve irritant responses. For example, the purine ATP is released during epithelial cell stress, and sensory nerves of the mouse are known to be sensitive to a variety of purinergic mediators (37–40). Adenosine aerosols induce a sensory irritation response in the mouse via adenosine receptor 1 (AR1) pathways as evidenced by increased early expiratory pauses that are inhibited by an AR1 antagonist (41). Moreover, the sensory irritation response to acetic acid vapor is partially diminished by the adenosine receptor antagonist theophylline implicating a role for adenosine signaling in sensory nerve activation by this acid. Second, the nasal mucosa is a highly metabolically active (chap. 5). For example, acetaldehyde is metabolized to acetic acid via acetaldehyde dehydrogenases. Acetic acid is considerably a more potent stimulator of nasal vasodilation than acetaldehyde; inhibition of aldehyde dehydrogenase with cyanamide significantly reduces sensory nerve–mediated responses to acetaldehyde (42), suggesting that its acetaldehyde dehydrogenase metabolite is actually responsible for sensory nerve activation. That irritants may stimulate sensory nerves via paracrine pathways and/or after in situ metabolic activation indicates that the process of sensory nerve activation may be considerably more complex than is generally appreciated.

REFLEX RESPONSES
Stimulation of trigeminal nerves causes a variety of reflex responses in animals including characteristic changes in breathing patterns, vasodilation, vascular congestions, mucous secretion, neurogenic edema, and leukocyte recruitment (1,11,12,43). Species differences exist in the spectrum of responses, thus considerable caution should be used in extrapolating experimental animal data to man. As stated by Tai and Baraniuk (11), "perhaps humans are not good models of rodent disease." Conceptually it is useful to think of specific neuronal responses in an animal model merely as biomarkers of sensory nerve activation, recognizing that the spectrum of response to sensory nerve activation in other species (including man) may well differ.

Breathing Patterns
Of the centrally mediated sensory nerve responses, changes in breathing pattern in mice represents the best characterized biomarker for trigeminal sensory nerve activation. While this response is a well-defined biomarker for trigeminal sensory nerve activation, it may not reflect the most sensitive response in all species. The seminal review on breathing pattern responses to irritants was published over 35 years ago (3). The characteristic change in breathing pattern is a decreased breathing frequency due to a prolonged pause and/or slowing of respiration at the *onset* of each expiration (Fig. 1), a process termed braking (3,44). This is due to glottal closure and is followed by rapid expiration when

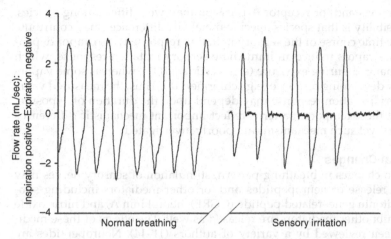

FIGURE 1 Breathing patterns during normal breathing and during exposure to a sensory irritant. The curves show air-flow rate (mL/sec; inspiration up, expiration down) during 1.8 seconds of normal breathing (at a frequency of 250 breaths/min) and 1.8 seconds during sensory irritant exposure (170 breaths/min). The prolonged period of zero flow at the start of expiration is evident. As can be seen the flow rates/duration of the active inspiratory and expiratory phases are not greatly altered, the diminished breathing is dependent on the duration of the zero flow period of each breath.

the glottal resistance is overcome (44). The response is absent following incision of the trigeminal nerve and can be reproduced by electrical stimulation of the trigeminal ganglia (3). Stimulation of vagal nerves can also results in decreased breathing frequencies due to a pause at the *end* of expiration (3,44), thus it is important from a mechanistic sense to determine if breathing frequency changes are the result of pauses at the onset or at the end of expiration. Plethysmography is well suited for this purpose. [It is important to recognize that these changes are not the "enhanced pause" response that is sometimes used as an indirect, but controversial, measure of airway obstruction (45)].

For safety evaluation purposes sensory irritation is quantitated by expressing one minute average breathing frequency during exposure as a percent of the breathing frequency in a preexposure baseline period. The RD50 is the exposure concentration, which results in a 50% reduction in breathing frequency (46). In the absence of additional information, an occupational exposure guideline for sensory irritants is often set at 3% of the RD50 in mice (46,47). Mechanistic studies, including those in the author's laboratory, have relied on measurement of the duration braking to quantitate the response rather than quantitation of breathing frequency. The advantages of this approach are that it is specific for trigeminal activation and is also more sensitive. Increased duration of braking can be documented in the absence of a change in breathing frequency, probably because the rapid expiration following glottal opening can compensate for small increases in the time of braking (48).

Large species differences are often observed in RD50 levels. For some vapors the mouse is often 10 times more sensitive than the rat; however, for other vapors the RD50 values are virtually identical (3,49). The reasons for this are unknown but may reflect the fact that differing vapors stimulate sensory C and Aδ fibers in differing proportions, and/or via differing receptor pathways and nerve populations and/or receptor expression may well differ among species. Another possibility is that species-specific metabolic differences may contribute (chap. 5). The time course of the sensory irritation response is also vapor dependent (3). Some vapors induce an immediate sensory irritation response, which remains unchanged during exposure (e.g., acetic acid) (1), whereas some vapors elicit a slowly developing response (e.g., chlorine) (18). Thus, the measured value for the sensory irritation response may depend upon the duration of exposure. Undoubtedly these temporal patterns reflect important mechanistic differences among vapors, yet such mechanisms are poorly investigated.

Physiological Changes
In addition to changes in breathing pattern, stimulation of sensory nerves may result in the release of neuropeptides and/or other mediators including substance P, calcitonin gene–related peptide (CGRP), neurokinin A, and nitric oxide (NO). No doubt other mediators are released as well. The effects of these mediators have been reviewed by a variety of authors (11–14). Neuropeptides are quickly degraded in nasal tissues by peptidases, the level of which strongly influences the response to these mediators (50).

As noted above, acrolein is a potent sensory irritant that is present in tobacco smoke that acts via stimulation of the TRPA1 receptor (22,31). Its RD50 in the F344 rat is 6 ppm (51). Studies in our laboratory on immediate nasal response to acrolein in the rat serve to highlight the spectrum of responses to inspired irritants. Immediate nasal responses to acrolein include changes in breathing pattern, vasodilation and increased nasal bloodflow, plasma protein extravasation (neurogenic edema), and/or mucus secretion resulting in a steady thickening of the nasal mucous lining layer and nasal airflow obstruction (43). Concentration response studies indicated that marked nasal vasodilation occurred at exposure concentrations as low as 2 ppm, whereas concentrations of 20 ppm were required to produce the other effects. Use of a neurokinin1 (substance P) receptor antagonist, and inhibitors of substance P degradation, strongly suggested a role for substance P in all tissue responses except vasodilation. The role of substance P in nasal neurogenic edema in the rat nasal mucosa is well characterized (52). Subsequent studies revealed a role of CGRP and NO in the vasodilatory response (see below). Thus, it can be appreciated from this example that a wide range of reflex responses can occur during exposure to an irritants, with differing reflex responses being mediating via differing mediators, and exhibiting differing concentration–response relationships.

Several irritant vapors including acrolein, acetaldehyde, acetic acid, and ethyl acrylate induce an immediate nasal vasodilatory response in the F344 rat (42,43,53), suggesting this a common response in this species. In all cases the response was significantly lower in animals pretreated with capsaicin, strongly implicating a role for capsaicin-sensitive sensory nerves (likely C fibers). It should not be assumed that these vapors induce the same pattern of mediator release simply because they all induce a vasodilatory response. This is illustrated by mechanistic studies on acrolein and acetic acid vapors. Shown in Figure 2 is the effect of the short acting CGRP antagonist hCGRP8–37 on the nasal vasodilatory response to acrolein and acetic acid. Nasal vasodilation is assessed by following uptake of the "inert" vapor acetone (43). As can be seen, CGRP has no effect in control (nonirritant exposed) rats, yet produced a profound (but

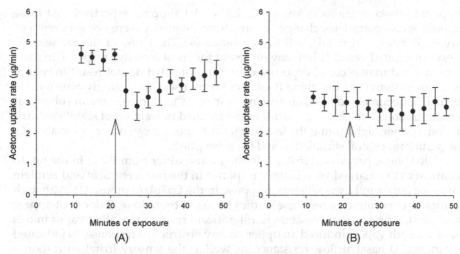

FIGURE 2 The effect of the short acting CGRP antagonist hCGRP8–37 on acrolein-and acetic acid–induced vasodilation [parts (**A**) and (**B**), respectively]. Nasal perfusion was assessed by inert gas (acetone) uptake in the isolated upper respiratory tract of F344 rats at an inspiratory flow rate of 100 mL/min (1) and expressed as steady state acetone uptake rate (μg/min). Acetone uptake rate during both acrolein and acetic acid exposure was significantly elevated over control levels (2.4 μg/min) indicating that both irritants produced a vasodilatory response. The antagonist was injected intravenously at 22 minutes of exposure. As can be seen, the agonist greatly diminished the vasodilatory response in acrolein-exposed rats (as indicated by the decreased acetone uptake rate), but was without effect in acetic acid–exposed rats. The antagonist was without effect on control (air-exposed rats), data not shown. Data are expressed as mean ± standard error.

transient) reduction in the acrolein-induced vasodilatory response. This result strongly suggests that CGRP contributes to the acrolein-induced vasodilatory response. In contrast to acrolein, hCGRP8–37 was without effect on the vasodilatory response to acetic acid. The vasodilatory responses to acrolein and acetic acid (as well as acetaldehyde) are partially inhibited by the general nitric oxide synthase (NOS) inhibitor, L-NAME, and also the neuronal-NOS inhibitor, 7-NI (54). In toto, these results suggest that both acetic acid and acrolein induce neuronal NO release yet only acrolein induces the release of physiologically significant levels of CGRP, highlighting the fact that differing irritants induce reflex responses through differing mechanisms.

Concentration Response Relationships and Species Differences
Of the measured nasal reflex responses of the rat to acrolein, vasodilation was the most sensitive, occurring at concentrations and at least 3-fold lower than the RD50 (43) and 10-fold lower for other effects (e.g., plasma protein extravasation, etc.). Subsequent studies revealed that nasal vasodilation could be demonstrated at exposure concentrations as low as 0.2 ppm, a concentration that did not alter breathing frequency (Morris, unpublished observations, 2001). This relationship appears to hold for other irritants as well. The RD50 (F344 rat) for acetaldehyde, acetic acid, and ethyl acrylate are 3740, 1040, and 3200 ppm, respectively. An immediate nasal vasodilatory response is demonstrable during exposure to these

vapors to concentrations as low as 25, 120, and 100 ppm, respectively. At these exposure concentrations, changes in breathing frequency were not apparent nor were alterations in nasal air flow resistance (42,53). Thus, in the rat sensory nerve–mediated vasodilation may represent the most sensitive response to most irritants and may occur at exposure concentrations that do not result in central nervous system (e.g., breathing frequency) responses. Interestingly, Shusterman et al. (55) documented nasal obstructive responses to chlorine human subjects at concentrations, which do not result in the marked perception of significant irritation, further highlighting the fact that local responses can occur in the absence of significant central stimulation and/or perception.

Response patterns to irritants in the mouse differ from those in the rat. In contrast to the marked vasodilatory response in the rat, acetic acid and acrolein produce only a mild vasodilatory response in the C57Bl/6J mouse (1). Although complete concentration response studies have not been done, it does not appear that plasma protein extravasation is a significant response to either vapor in this species. Both vapors induced an upper airway obstructive response, as indicated by increased nasal airflow resistance, as well as the sensory irritation response as indicated by early expiratory pauses and decreased breathing frequency. The dose–response relationships for these two responses were similar. Thus, the two most sensitive responses in the mouse to these vapors were airway obstruction and sensory irritation. Similarly, sensory irritation and airway obstruction were also sensitive responses to chlorine in the mouse (18) (see chap. 20). Therefore, it appears there are common, but differing, sensory nerve–mediated response patterns to irritants in the mouse (obstruction, sensory irritation) versus rat (vasodilation). Such information should be useful in selecting the most sensitive biomarker for sensory nerve activation for future studies in each species.

MODULATION OF TRIGEMINAL RESPONSIVENESS

Nasal sensory nerve responsiveness is not static but is highly dynamic. Both immediate and long-term modulations in sensory nerve responsiveness have been described. Recent studies indicate that leukotriene LTD4 may enhance C fiber sensitivity (56). Another pathway for modulation of sensory nerve sensitivity is via the purine, adenosine. Elegant studies by Lee (39,57) have shown that adenosine produces an immediate sensitization of vagal afferents, greatly enhancing responsiveness to both capsaicin and lactic acid. Since capsaicin acts predominantly on C fibers this likely reflects effects specific to this class of nerve. Recent studies in the author's laboratory have examined the modulation of sensory nerve responsiveness by the malodorants ethyl sulfide and t-butyl sulfide (48). These studies were initiated because it has been shown that odors result in the release of ATP from olfactory mucosa (58) raising the possibility that odorants may enhance irritant sensitivity through purinergic (adenosine) pathways. [ATP is rapidly degraded extracellularly to adenosine (59)]. Both odorants markedly potentiated the sensory irritation response to capsaicin with the potentiation being blocked by the adenosine receptor antagonist theophylline (48). Subsequent studies revealed that administration of the adenosine precursor adenosine-5-monophosphate also markedly increased the responsiveness to capsaicin (Willis and Morris, unpublished data, 2008). The olfactory bulb receives collateral innervations from the trigeminal nerve (60). It is thought that trigeminal activation may modulate olfactory nerve sensitivity. Perhaps, this interaction is bidirectional, with olfactory stimuli modulating trigeminal sensitivity via purinergic signaling pathways. This might reflect a heretofore unappreciated effect of odorants.

It is widely appreciated that individuals with rhinitis report enhanced sensitivity to irritants (34,61) suggesting a long-term modulation in trigeminal responsiveness. This has been objectively documented in experimental studies of Shusterman, who demonstrate increased obstructive responses to chlorine (chap. 20) and acetic acid (62) in subjects with seasonal rhinitis. Studies using a murine ovalbumin–induced allergic airway disease model have demonstrated increased sensory irritation responses (expiratory pause) to acrolein and acetic acid (1). Interestingly, the nasal obstructive response to acetic acid, but not acrolein, was increased in this disease model (1). This again highlights the fact that sensory neuronal responses are complex and irritant specific. The mechanisms for this irritant-specific enhancement of the nasal obstructive response are not known, but is interesting to note that acrolein stimulates only TRPV1 expressing capsaicin-sensitive cells through the TRPA1 receptor (1), whereas acetic acid stimulates noncapsaicin sensitive nerves, likely Aδ fibers, as well. It is now known that profound changes in Aδ fibers occur in a variety of disease states (see also, chap. 26).

Aδ fibers seldom express neuropeptides, but in animal models of allergic airway disease, these nerves express these potent mediators, a process termed as the phenotypic switch. This is best characterized for vagal nerves (63,64). Modulation of trigeminal nerve substance P expression has been shown in trigeminal nerves as well. In particular, the haptenic moiety toluene diisocyanate has been shown to induce increased expression of substance P in trigeminal nerves (63). For both the vagus and trigeminal, nerve growth factor appears to be a critical mediator relative to enhance substance P expression (63,65). The effect is not limited to allergy; asphalt fume exposure also results in enhanced substance P expression in the nose, in this case accompanied by a neutrophilic inflammation (66). Given the wide-range effects of neuropeptides, it is clear that alterations in neuropeptides expression may exert profound effects on irritant-induced reflex responses. The precise functional significance of altered neuropeptide expression relative to specific environmental irritants and the contribution of C versus Aδ fibers in response elicitation awaits further experimentation.

SUMMARY

In animal models, trigeminal sensory nerve stimulation results in centrally mediated reflex responses (altered breathing patterns) as well as local tissue responses including vasodilation, mucous release, neurogenic edema, and/or nasal airway obstruction. Many of these local responses are mediated via neuropeptides including substance P and CGRP, mediators expressed in C fibers. The pattern of responses is irritant specific, may differ among species, and is likely to be dependent on the degree to which an irritant stimulates C versus Aδ fibers. The receptor pathways through which irritants stimulate trigeminal sensory nerves are currently being unraveled. The TRPA1 receptor appears to be essential for a variety of irritants, including electrophiles and oxidants. While many irritants act directly on sensory nerve receptors, recent evidence highlights the possibility that metabolic activation and/or paracrine signaling pathways may be critical relative to sensory nerve activation. Trigeminal sensory nerves sensitivity may be modulated acutely by paracrine mediators (adenosine, LTD4). Long-term changes in neuropeptides expression can also result from irritant and/or allergen exposure. These phenomena have been demonstrated in a variety of animal models, but it is important to recognize that large differences exist among species. Thus, care is needed in extrapolating animal data to man. Nevertheless, it is clear that animal experimentation has provided critical insights into the complexities of nasal sensory nerve function in health and disease.

REFERENCES

1. Morris JB, Symanowicz PT, Olsen JE, et al. Immediate sensory nerve-mediated respiratory responses to irritants in healthy and allergic airway-diseased mice. J Appl Physiol 2003; 94:1563–1571.
2. Morris JB, Smith FA. Regional deposition and absorption of inhaled hydrogen fluoride in the rat. Toxicol Appl Pharmacol 1982; 62:81–89.
3. Alarie Y. Sensory irritation by airborne chemicals. Crit Rev Toxicol 1973; 3:299–363.
4. Buckley LA, Jiang XZ, James RA, et al. Respiratory tract lesions induced by sensory irritants at the RD50 concentration. Toxicol Appl Pharmacol 1984; 74(3):417–429.
5. Dalton P. Upper airway irritation, odor perception and health risk due to airborne chemicals. Toxicol Lett 2003; 141:239–248.
6. Hummel T, Livermore A. Intranasal chemosensory function of the trigeminal nerve and aspects of its relation to olfaction. Int Arch Occup Environ Health 2002; 75: 305–313.
7. Schaeffer ML, Bottger B, Silver WL, et al. Trigeminal collaterals in the nasal epithelium and olfactory bulb: a potential route for direct modulation of olfactory information by trigeminal stimuli. J Com Neurol 2002; 444:221–226.
8. Geppetti P, Materazzi S, Nicoletti P. The transient receptor potential vanilloid 1: Role in airway inflammation and disease. 2006; 533:207–214.
9. Finger TE, St Jeor VL, Kinnamon JC, et al. Ultrastructure of substance P- and CGRP-immunoreactive nerve fibers in the nasal epithelium of rodents. J Comp Neurol 1990; 294(2):293–305.
10. Undem BJ, McAlexander M, Hunter DD. Neurobiology of the upper and lower airways. Allergy 1999; 54(suppl 57):81–93.
11. Tai C-F, Baraniuk JN. Upper airway neurogenic mechanisms. Curr Opin Allergy Clin Immunol 2002; 2:11–19.
12. Sarin S, Undem B, Sanico A, et al. The role of the nervous system in rhinitis. J Allergy Clin Immunol 2006; 115:999–1014.
13. Barnes PJ, Baraniuk JN, Belvisi MG. Neuropeptides in the respiratory tract. Part I. Am Rev Respir Dis 1991; 144:1187–1198.
14. Barnes PJ, Baraniuk JN, Belvisi MG. Neuropeptides in the respiratory tract. Part II. Am Rev Respir Dis 1991; 144:1391–1399.
15. Anton F, Peppel P. Central projections of trigeminal primary afferents innervating the nasal mucosa: a horseradish peroxidase study in the rat. Neuroscience 1991; 41:617–628.
16. Baraniuk JN. Neural control of the upper respiratory tract. In: Kaliner MA, Barnes PJ, Kunkel GHH, Baraniuk JN, eds. Neuropeptides in Respiratory Medicine. New York, NY: Marcel Dekker, 1994:79–123.
17. Barnes PJ. Role of neural mechanisms in airway defense. In: Cretien J, Dusser D, eds. Environmental Impact on the Airways. New York, NY: Marcel Dekker, 1996:93–121.
18. Morris JB, Wilkie WS, Shusterman DJ. Acute respiratory responses of the mouse to chlorine. Toxicol Sci 2005; 83:380–387.
19. Shusterman D, Murphy MA, Walsh P, et al. Cholinergic blockade does not alter the nasal congestive response to irritant provocation. Rhinology 2002; 40:141–146.
20. Caterina MJ, Schumacher MA, Tominaga M, et al. The capsaicin receptor: a heat-activated ion channel in the pain pathway. Nature 1997; 389:816–824.

21. Caterina MJ, Leffler A, Malmberg AB, et al. Impaired nociception and pain sensation in mice lacking the capsaicin receptor. Science 2000; 288(5464):306–313.
22. Bautista DM, Jordt S-E, Nikai T, et al. TRPA1 mediates the inflammatory actions of environmental irritants and proalgesic agents. Cell 2006; 124:1269–1282.
23. Bessac BF, Sivula M, von Hehn CA, et al. TRPA1 is a major oxidant sensor in murine airway sensory neurons. J Clin Invest 2008; 118:1899–1910.
24. Ichikawa H, Sugimoto T. The co-expression of ASIC3 with calcitonin gene-related peptide and parvalbumin in the rat trigeminal ganglion. Brain Res 2002; 943(2):287–291.
25. Spehr J, Spehr M, Hatt H, et al. Subunit-specific P2X-receptor expression defines chemosensory properties of trigeminal neurons. Eur J Neurosci 2004; 19(9):2497–2510.
26. Alimohammadi H, Silver WL. Evidence for nicotinic acetylcholine receptors on nasal trigeminal nerve endings of the rat. Chem Senses 2000; 25(1):61–66.
27. Silver WL, Clapp TR, Stone LM, et al. TRPV1 receptors and nasal trigeminal chemesthesis. Chem Senses 2006; 31(9):807–812.
28. Kollarik M, Undem BJ. Mechanisms of acid-induced activation of airway afferent nerve fibres in guinea-pig. J Physiol 2002; 543:591–600.
29. Gu Q, Lee L-Y. Characterization of acid signaling in rat vagal pulmonary sensory neurons. Am J Physiol Lung Cell Mol Physiol 2006; 291:L58–L65.
30. Symanowicz PT, Gianutsos G, Morris JB. Lack of role for the vanilloid receptor in response to several inspired irritant air pollutants in the C57Bl/6 J mouse. Neurosci Lett 2004; 362:150–153.
31. Andre E, Campi B, Materazzi S, et al. Cigarette smoke-induced neurogenic inflammation is mediated by a,b-unsaturated aldehydes and the TRPA1 receptor in rodents. J Clin Invest 2008; 118(7):2574–2582.
32. Chen J, Zhang Z-F, Kort ME, et al. Molecular determinants of species-specific activation or blockade of TRPA1 channels. J Neurosci 2008; 28(19):5063–5071.
33. Trevisani M, Siemens J, Materazzi S, et al. 4-Hydroxynonenal, an endogenous aldehyde, causes pain and neurogenic inflammation through activation of the irritant receptor TRPA1. Proc Natl Acad Sci U S A 2007; 104:13519–13524.
34. Shusterman D, Murphy M-A. Nasal hyperreactivity in allergic and nonallergic rhinitis: a potential risk factor for nonspecific building-related illness. Indoor Air 2007; 17:328–333.
35. Kollarik M, Undem BJ. Activation of bronchopulmonary vagal afferent nerves with bradykinin, acid and vanilloid receptor agonists in wild-type and TRPV -/- mice. J Physiol 2004; 555(1):115–123.
36. Holzer P. Capsaicin: cellular targets, mechanisms of action and selectivity for thin sensory neurons. Pharmacol Rev 1991; 43(2):143–201.
37. Schiebert EM, Zsembery A. Extracellular ATP as a signaling molecule for epithelial cells. Biochim Biophys Acta 2003; 1615:7–32.
38. Ahmad S, Ahmad A, McConville G, et al. Lung epithelial cells release ATP during ozone exposure: signaling for cell survival. Free Radic Biol Med 2005; 39:213–226.
39. Hong J-L, Ho C-Y, Kwong, K, et al. Activation of pulmonary C fibres by adenosine in anaesthetized rats: role of adenosine A1 receptors. J Physiol 1998; 508:109–118.
40. Kollarik M, Dinh QT, Fischer A, et al. Capsaicin-sensitive and -insensitive vagal bronchopulmonary C-fibres in the mouse. J Physiol 2003; 551(3):869–879.
41. Vaughan RP, Szewczyk MT, Lanosa MJ, et al. Adenosine sensory transduction pathways contribute to activation of the sensory irritation response to inspired irritant vapors. Toxicol Sci 2006; 93:411–421.
42. Stanek J, Symanowicz PT, Olsen JE, et al. Sensory-nerve mediated nasal vasodilatory response to inspired acetaldehyde and acetic acid vapors. Inhal Toxicol 2001; 13:807–822.
43. Morris JB, Stanek J, Gianutsos G. Sensory nerve-mediated immediate nasal responses to inspired acrolein. J Appl Physiol 1999; 87:1877–1886.
44. Vijayaraghavan R, Schapper M, Thompson R, et al. Characteristic modification of the breathing pattern of mice to evaluate the effects of airborne chemicals on the respiratory tract. Arch Toxicol 1993; 67:478–490.

45. Adler A, Cieslewicz G, Irvin CG. Unrestrained plethysmography is an unreliable measure of airway responsiveness in BALB/c and C57 Bl/6 mice. J Appl Physiol 2004; 97(1):286–292.

46. Alarie Y. Bioassay for evaluating the potency of airborne sensory irritants and predicting acceptable levels of exposure in man. Food Cosmet Toxicol 1981; 19:623–626.

47. Schaper M. Development of a database for sensory irritants and its use in establishing occupational exposure limits. Am Ind Hyg Assoc J 1993; 54:488–544.

48. DeSesa CR, Vaughan RP, Lanosa JM, et al. Sulfur-containing malodorant vapors enhance responsiveness to the sensory irritant capsaicin. Toxicol Sci 2008; 104:198–209.

49. Bos, PM, Zwart, A, Reuzel, PB, et al. Evaluation of the sensory irritation test for the assessment of occupational health risk. CRC Crit Rev Toxicol 1992; 21:423–450.

50. Piedimonte G. Tachykinin peptides. Receptors and peptidases in airway disease. Exp Lung Res 1995; 21:809–834.

51. Babiuk C, Steinhagen WH, Barrow CS. Sensory irritation response to inhaled aldehydes after formaldehyde pretreatment. Toxicol Appl Pharmacol 1985; 79:143–149.

52. Petersson GE, Bacci DM, McDonald M, et al. Neurogenic plasma extravasation in the rat nasal mucosa is potentiated by peptidase inhibitors. J Pharmacol Exp Ther 1993; 264:509–514.

53. Morris JB. Sensory nerve-mediated nasal vasodilatory response to inspired ethyl acrylate. Inhal Toxicol 2002; 14:585–597.

54. Stanek JJ. Characterization of the acute nasal response to inspired acetaldehyde in the F-344 rat [Ph.D. Thesis]. Storrs, CT: University of Connecticut; 1999:166.

55. Shusterman D, Murphy M, Balmes J. Seasonal allergic rhinitic and non-rhinitic subjects react differentially to provocation with chlorine gas. J Allergy Clin Immunol 1998; 101:732–740.

56. Taylor-Clark TE, Nassenstein C, Undem BJ. Leukotriene D4 increases the excitability of capsaicin-sensitive nasal sensory nerves to electrical and chemical stimuli. Br J Pharmacol 2008; 154(6):1359–1368.

57. Gu Q, Ruan T, Hong JL, et al. Hypersensitivity of pulmonary C fibers induced by adenosine in anesthetized rats 2003; 95:1315–1324.

58. Hegg CC, Lucero MT. Purinergic receptor antagonists inhibit odorant-induced heat shock protein 25 induction in mouse olfactory epithelium. Glia 2006; 53:182–190.

59. Schiebert EM, Zsembery A. Extracellular ATP as a signaling molecule for epithelial cells. Biochim Biophys Acta 2003; 1615:7–32.

60. Schaeffer ML, Bottger B, Silver WL, et al. Trigeminal collaterals in the nasal epithelium and olfactory bulb: a potential route for direct modulation of olfactory information by trigeminal stimuli. J Comp Neurol 2002; 444:221–226.

61. Sarin SM, Undem B, Sanico A, et al. The role of the nervous system in rhinitis. J Allergy Clin Immunol 2006; 118:999–1016.

62. Shusterman DJ, Taurin A, Murphy MA, et al. Seasonal allergic rhinitic and normal subjects respond differentially to nasal provocation with acetic acid vapor. Inhal Toxicol 2005; 17(3):147–152.

63. Hunter DD, Satterfield BE, Huang J, et al. Toluene diisocyanate enhances substance P in sensory neurons innervating the nasal mucosa. Am J Respir Crit Care Med 2000; 161(21):543–549.

64. Chuaychoo B, Hunter DD, Myers AC, et al. Allergen-induced substance P synthesis in large-diameter sensory neurons innervating the lungs. J Allergy Clin Immunol 2005; 116(2):325–331

65. Wilfong ER, Dey RD. Nerve growth factor and substance P regulation in nasal sensory neurons after toluene diisocyanate exposure. Am J Respir Cell Mol Biol 2004; 30(6):793–800.

66. Sikora ER, Stone S, Tomblyn S, et al. Asphalt exposure enhances neuropeptide levels in sensory neurons projecting to the rat nasal epithelium. J Toxicol Environ Health A 2003; 66(11):1015–1027.

13 Nasal Chemosensory Irritation in Humans

J. Enrique Cometto-Muñiz and William S. Cain

Chemosensory Perception Laboratory, Department of Surgery (Otolaryngology), University of California, San Diego, La Jolla, California, U.S.A.

Michael H. Abraham, Ricardo Sánchez-Moreno, and Javier Gil-Lostes

Department of Chemistry, University College London, London, U.K.

INTRODUCTION

The detection of external chemicals by humans is accomplished by smell (olfaction), taste (gustation), and chemical sensory irritation (1–4). The latter was originally labeled the common chemical sense (5). More recently, it has been referred to as chemical nociception (6). Nevertheless, at low levels of stimulation, the sensations evoked might not be perceived as painful or even irritating. Since this sensory modality rests on chemically induced somesthesis, or chemical "feel," the quite appropriate and descriptive term "chemesthesis" is now often employed (3,7,8). In the nasal, ocular, and oral mucosae, chemesthesis is principally mediated by the trigeminal nerve (cranial nerve V). Thus, it is common to refer to trigeminal chemosensitivity when addressing this topic (9,10). In the aggregate, eye, nose, and throat irritation has been referred to as "sensory irritation" and, along with secondary reflex symptoms in that anatomical distribution, constitute an important symptom constellation in so-called "problem buildings," as well as being the basis for a substantial fraction of occupational exposure standards (11,12). Nasal chemesthesis, the main focus of this chapter, arises from stimulation with airborne chemicals. Many of these, although not all of them, are volatile organic compounds (VOCs). The chemesthetic sensations that they can evoke in the nose are typically pungent, that is, sharp. They include stinging, freshness, coolness, burning, piquancy, tingling, irritation, prickling, and the like.

Most, if not all, volatile compounds capable of eliciting nasal chemesthesis also elicit olfactory sensations. With few exceptions (e.g., β-phenyl ethyl alcohol), compounds that smell, that is, odorants, can also evoke nasal chemesthesis. As a rule, when the concentration of an airborne chemical rises, it is first noticed by its smell, but as it continues to rise, it also engages a chemesthetic response. It is then important to define the concentration range at which a substance remains undetected, that at which it evokes only an olfactory response, and that at which it evokes an olfactory plus a trigeminal response (13). Carbon dioxide (CO_2) is a compound that, arguably, only elicits chemesthesis with little, if any, smell. On the other hand, some compounds do have a smell but fail to evoke chemesthesis. Such chemicals have been often found by testing homologous series of VOCs, where a certain homolog is reached beyond which chemesthesis cannot be elicited (14) (see the "cut-off" effect section below).

In view of the higher olfactory than trigeminal sensitivity, and the scarcity of compounds that evoke nasal chemesthesis but not smell, it was a challenge to implement bias-controlled (e.g., forced-choice) psychophysical procedures to gauge nasal chemesthesis independently of olfaction. One strategy to achieve this goal entailed testing nasal detection of chemical vapors in subjects lacking a functional olfaction (i.e., anosmics) and, thus, only responding to trigeminal chemesthesis. Another strategy entailed employing subjects with normal olfaction (i.e., normosmics) in a task requiring not the detection but the localization (lateralization) of the vapor to the left or right nostril when air is simultaneously delivered to the contralateral nostril (15–17). Nasal localization rests on trigeminal input (18,19) rather than on olfactory input as originally thought (20). The following sections describe these and other strategies to study nasal chemesthesis in humans.

PSYCHOPHYSICS OF NASAL IRRITATION

Nasal Pungency Thresholds in Anosmics
As mentioned, testing anosmic subjects in tasks of nasal detection of vapors represents one way to eliminate the biasing influence of olfaction in estimating the nasal chemesthetic potency of airborne chemicals (21,22). These early papers suggested that relatively simple physicochemical and molecular structural properties could predict nasal chemesthesis. Nevertheless, the lack of a guiding unit of chemical change among the very wide range of compounds capable of evoking nasal irritation made it difficult to pursue a systematic study of chemesthetic potency across VOCs, for example, by measuring nasal pungency thresholds (NPTs). Along homologous chemical series, carbon chain length constitutes a practical unit of chemical change. This strategy was applied in a series of studies that measured NPTs in anosmics and odor detection thresholds (ODTs) in normosmics for homologous alcohols, acetate esters, 2-ketones, n-alkylbenzenes, aliphatic aldehydes, and carboxylic acids, and for selected terpenes, using a uniform methodology (23–28). The method comprised a two-alternative forced-choice procedure, an ascending concentration approach, and a fixed criterion of five correct choices in a row. The outcome showed that both NPTs and ODTs decline with carbon chain length along n-homologs in each series (Fig. 1). Lower thresholds indicate higher potency of the stimulus. NPTs lay between one and five orders of magnitude above ODTs. The simple delivery system (i.e., "squeeze bottles") used to obtain these thresholds, combined with the stringent criterion chosen to define them, had produced values on the higher end of reported thresholds. Indeed, improved techniques and methodologies did render lower absolute thresholds but left the relative chemesthetic and olfactory potency across VOCs virtually unaltered (29–31). Interestingly, in terms of nasal pungency, many of these series reached a large enough homolog that failed to be detected consistently by one or more anosmics, even at vapor saturation (27). Once such homolog was reached, the failure to detect (i.e., to elicit nasal pungency) became increasingly more evident for all ensuing homologs. In other words, the increasing nasal pungency potency of successive homologs (reflected in decreasing thresholds) ended rather abruptly upon reaching a member that lacked potency altogether. This cut-off effect in chemesthesis is discussed below.

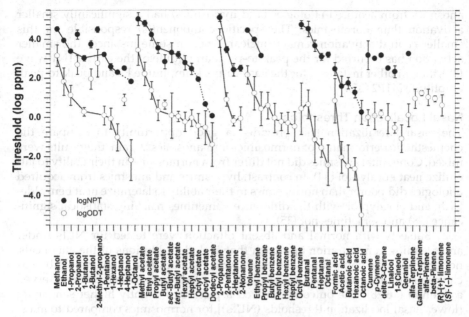

FIGURE 1 NPTs in anosmics and ODTs in normosmics for homologous alcohols, acetate esters, 2-ketones, n-alkylbenzenes, aliphatic aldehydes, and carboxylic acids, for cumene, and for selected terpenes. NPTs could not be measured for some compounds. Only n-homologs are joined by a line. Dashed lines indicate the appearance of a cut-off effect in NPTs such that one or more anosmics consistently failed to achieve the criterion chosen to reach an NPT. Bars indicate SD. *Abbreviations*: NPTs, nasal pungency thresholds; ODTs, odor detection thresholds; SD, standard deviation.

The use of anosmics to gauge nasal trigeminal sensitivity presumes that the lack of olfaction in these subjects does not alter chemesthetic sensitivity to a significant extent, compared to normosmics. A number of investigations have explored this issue but the results have been mixed. In behavioral tests where anosmics showed lower sensitivity and intensity ratings for chemesthesis, normosmics were not "blind" to the odor of the stimuli, which could have affected the results (32,33). Congenital anosmics tested with CO_2 gave lower-intensity ratings than normosmics (34), but anosmics after upper respiratory tract infection and those after head trauma gave intensity ratings no different than those from normosmics (35). Still, anosmics after upper respiratory tract infection and those after head trauma, but not hyposmics, had higher CO_2 detection thresholds than normosmics (36). Unfortunately, these CO_2 thresholds were obtained from yes–no answers (quite prone to criterion-based biases) rather than from a more robust forced-choice procedure. Electrophysiological tests on nasal chemesthesis have included peripheral responses, namely the negative mucosal potential (NMP), and central responses, namely trigeminal event-related potentials (tERP) generated in the cortex. For peripheral responses to CO_2 stimulation, congenital anosmics and those from acquired etiologies had a larger activation than normosmics (34,35). In contrast, for central responses to CO_2 stimulation, congenital anosmics showed no significant differences with normosmics (34), whereas anosmics from acquired etiologies (and hyposmics) had a significantly smaller activation than normosmics. The specific component(s) responsible for this smaller central activation remain unclear, since in some instances the smaller activation has occurred for the peak-to-peak amplitude of the early P1N1 wave (37), and, in other instances, for the base-to-peak amplitude P2 and peak-to-peak amplitude N1P2 (35).

Nasal Localization Thresholds

The nasal lateralization task provided a good opportunity to compare the chemesthetic performance of normosmics and anosmics. Again, the results were mixed. Congenital anosmics did not differ from normosmics in their ability to lateralize neat eucalyptol (34). In contrast, hyposmics and anosmics from acquired etiologies did poorer than normosmics in their ability to lateralize neat benzaldehyde and eucalyptol, with the difference sometimes reaching statistical significance (38) and sometimes not (35).

Subjects with normal and absent olfaction were tested for NLTs under an ascending concentration approach that included dilutions of the chemicals, not only the neat compound, and that involved quantification of vapors by gas chromatography. Homologous n-alcohols, selected terpenes, and cumene served as stimuli (16,28). The outcome showed a picture of slightly higher sensitivity [lower nasal localization thresholds (NLTs)] for normosmics compared to anosmics, but the difference did not achieve statistical significance (Fig. 2). Overall, the three estimates of nasal chemesthesis (NLTs in normosmics, NLTs in anosmics, and NPTs) and even ocular chemesthesis (i.e., eye irritation thresholds, EITs) often provided a similar picture of trigeminal sensitivity for each VOC (Fig. 2). We note that some compounds could not reach the specific criterion for nasal and/or ocular threshold in all participants on all repetitions (even when presented neat) (see details in Refs. 16,28). Among the n-alcohols, 1-octanol did not achieve the NLT criterion in either normosmics or anosmics, and very often failed to achieve it for NPTs; nevertheless, octanol did reach EIT in both groups (16) (Fig. 2). Among the terpenes, delta-3-carene and 1,8-cineole were the most reliable chemesthetic stimuli, reaching NPT, NLT, and EIT (the last two in normosmics and anosmics) in virtually all subjects and repetitions, whereas geraniol very commonly failed to reach criterion for these same thresholds (28) (Fig. 2). On the basis of the results presented in this and the previous section, we conclude that any advantage in nasal chemesthetic sensitivity that normosmics might have over anosmics, if indeed real, appears relatively small.

Cut-Off Point for Eliciting Chemesthesis Along Vapors from Homologous series

As noted above, thresholds for nasal (and ocular) chemesthesis decline with carbon chain length along homologous series, but only until reaching a large enough homolog that begins to fail to evoke chemesthetic detection altogether, even at vapor saturation at room temperature ($\approx 23°C$). The point where this happens within each series has been labeled the cut-off point for chemesthesis, and the shortest homolog failing detection, the cut-off homolog (13). A number of recent studies, particularly in the ocular mucosa, have defined the cut-off homolog for various series and have provided indications on the likely basis for the effect in terms of a structure–activity context (14,39–45). This is discussed next.

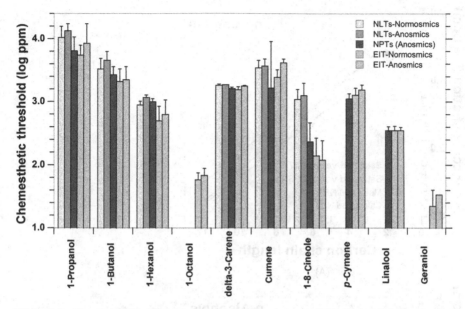

FIGURE 2 Showing NLTs in normosmics and anosmics, and NPTs (by definition, always in anosmics) for homologous alcohols, selected terpenes, and cumene. For comparison, EITs in normosmics and anosmics are also shown for the same VOCs. For some compounds, one or more types of thresholds could not be measured at all; for other compounds, some participants (normosmics and anosmics) could not achieve the criterion for chemesthetic threshold on all repetitions (see details in Refs. 16,28). Bars indicate SD. *Abbreviations*: NLTs, nasal localization thresholds; NPTs, nasal pungency thresholds; EITs, eye irritation thresholds; SD, standard deviation.

The appearance of a cut-off effect in the detection of chemesthesis from homologous vapors can rest on at least two possibilities. One is that the series has reached a member whose vapor pressure is too low to reach the necessary threshold concentration. Another is that the series has reached a member who lacks a key property to trigger transduction. For example, the homolog might be too large to interact effectively with a target site or a binding pocket in a receptive macromolecule. To look into these issues, we devised three approaches. The first approach consisted in applying a successful quantitative structure–activity relationship (QSAR), based on a solvation equation (46), to calculate and model predicted NPTs along homologous series (47–49). (This quantitative structure–activity relationship for nasal chemesthesis is described in detail in chapter 25, "Physicochemical Modeling of Sensory Irritation in Humans and Experimental Animals".) Then, trends in measured and calculated NPTs were compared with trends in saturated vapor concentrations at room temperature (23°C) and at body temperature (37°C) across homologs. The outcome showed that an extrapolation of the trend depicted by measured and calculated NPTs to the point of the cut-off homolog produced a predicted NPT concentration lower than the saturated vapor at 23°C (acetate's case) or lower than that at 37°C (alcohol's case) (Fig. 3). We conclude that (*i*) for the acetates, the cut-off was not likely the result of a low vapor concentration, and (*ii*) for the alcohols, if indeed the cut-off resulted from

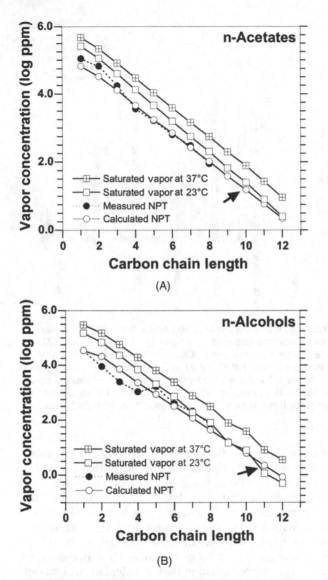

(A)

(B)

FIGURE 3 (**A**) Trends of measured and calculated NPTs, and of saturated vapor at 23 and 37°C, as a function of the variable carbon chain length of n-acetates. The arrow points to the cut-off homolog (decyl acetate, C10) and shows that the saturated vapor concentration at 23°C is higher than the calculated (i.e., expected) threshold. Thus, the observed nasal pungency cut-off is unlikely to arise from a concentration restriction (14). (**B**) Analogous data for n-alcohols. The arrow points to the likely cut-off homolog (from experiments on ocular chemesthesis) (undecyl alcohol, C11) and shows that, if the cut-off indeed arises from a low vapor concentration, raising the concentration to vapor saturation at 37°C should make undecyl alcohol detectable. This did not occur, at least for ocular chemesthesis (39). *Abbreviation*: NPTs, nasal pungency thresholds.

a low vapor concentration, raising the concentration of the cut-off homolog to vapor saturation at 37°C should overcome the effect and precipitate detection. The outcome of this strategy is discussed next.

The second approach consisted in testing the chemesthetic detection of each cut-off homolog at vapor saturation at 37°C, where the concentration for both compounds (decyl acetate and 1-undecanol) will be clearly above the predicted threshold (Fig. 3). For decyl acetate, the test has been performed for both nasal and ocular chemesthesis. The results for nasal chemesthesis revealed no significant increase in its detection at the higher vapor saturation, arguing against a cut-off based on a concentration restriction (14) (Fig. 4). The results for ocular chemesthesis did find a slight, but significant, increase, for the group data, but analysis of the individual data revealed that the increase was only seen for half of the 12 participants, whereas the other half did not show it at all (39). It was suggested that the exact cut-off point might vary slightly among subjects, perhaps due to genetic variability in the receptors involved. For 1-undecanol, the test involving increased vapor saturation has been, so far, only performed for ocular chemesthesis. The results mimicked those obtained for decyl acetate in the eye, including the sharp contrast between the performance of two subgroups of subjects (39).

The third approach to probe into the basis for the cut-off effect has been only explored in the ocular mucosa, up to the moment. It consists in measuring concentration-detection functions for chemesthesis, instead of only a threshold value, and establishing whether the function for the cut-off homolog reaches a plateau at some low level of detection such that further increases in concentration fail to increase detectability. The results from studies of homologous alcohols, acetates, alkylbenzenes, and 2-ketones have supported the existence of this kind of plateau for the respective cut-off homolog: 1-undecanol, decyl acetate, heptylbenzene, and 2-tridecanone (44,45). The outcome provides additional support to the idea that the cut-off often emerges from a restriction other than vapor concentration; for example, it could emerge from the homolog exceeding a critical molecular dimension to activate chemesthesis.

Nasal Chemesthesis and Exposure Time

Nasal irritation from chemicals often increases with time of exposure until reaching a plateau and, sometimes, declining thereafter (50). At the perithreshold level and for short exposures (<10 sec), this temporal effect produces lower chemesthetic thresholds as stimulus duration increases. Whether measured with a trigeminally induced reflex apnea (51) or with the nasal lateralization technique, temporal integration of threshold nasal chemesthesis has been described for ammonia, CO_2, and the n-alcohols ethanol, butanol, and octanol (52–56). The outcome suggests that, within a short time frame ($\approx$4 sec) detection of nasal chemesthesis relies on total mass rather than on concentration of the stimulus. However, the relationship between time and concentration falls, to a smaller or larger degree, short of a perfect trading: it takes somewhat more than doubling the exposure duration to compensate for a twofold decrease in concentration.

Suprathreshold ratings of nasal chemesthesis (i.e., perceived intensity) increase with stimulus duration, at least for up to about four seconds, as shown for ammonia (53,56) and CO_2 (57). In addition, the amplitude of the P3 peak from the tERP response to CO_2 also increases with stimulus duration in the very short

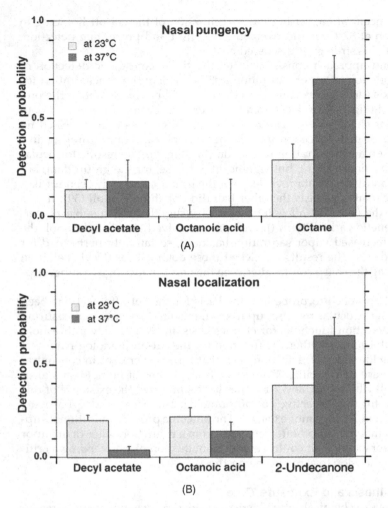

FIGURE 4 (**A**) Homologs located at (decyl acetate) or beyond (octanoic acid) the cut-off point fail to increase their detection via nasal pungency (in anosmics) despite the increase in concentration achieved from vapor saturation at 23°C to vapor saturation at 37°C. In contrast, homologs located before the cut-off (octane) do increase their detection at the higher concentration. (**B**) The same homologs also fail to increase their detection when nasal chemesthesis is gauged via nasal localization (lateralization) in normosmics. In contrast, 2-undecanone, a homolog located just before the cut-off (which occurs with 2-tridecanone in the ketones series), does increase its detection. Bars indicate SE. *Abbreviation*: SE, standard error.

range of 100 to 300 msec (58). Chamber studies with formaldehyde at a fixed concentration showed that the perceived intensity of (nasal and ocular) chemesthesis grows steadily with time during the 29-minute exposure (59). Daily home or occupational exposure to an airborne chemical that might be causing mild irritation can increase the nasal chemesthetic threshold to that chemical (i.e., produce desensitization), but the effect does not seem to generalize to other irritants (15,60,61). Experiments entailing two to three hours exposures to mixtures of

VOCs (62) and to environmental tobacco smoke (63) showed that the perceived intensity of chemesthesis clearly increases with time.

Nasal Chemesthesis and Chemical Mixtures

Most, if not all, studies we have discussed have dealt with the production of chemesthesis in humans from exposures to single chemicals. In contrast, environmental exposures at work and at home, indoors and outdoors, typically involve the presence of complex mixtures of compounds (64). It is, then, very relevant to explore and understand how the human nasal chemesthetic system processes mixtures of irritants. For convenience, we can distinguish those investigations that focused on the perithreshold level, that is, detection of nasal irritation by mixtures, from those that focused on the suprathreshold level, that is, magnitude (or intensity) of nasal irritation by mixtures. Under both strategies, one important goal is to uncover the rules that govern the chemesthetic impact of the mixture as compared to that of the individual components.

Human investigations addressing the nasal chemesthetic intensity of chemical mixtures are relatively few compared to those in olfaction. The total nasal perceived intensity of mixtures of the pungent odorants ammonia and formaldehyde grew with concentration in a way indicating that the perception of the mixture switched from being significantly lower to being significantly higher than the sum of the perceived intensities of its components (65). This suggested that an increase of the relative contribution of nasal chemesthesis over olfaction underlays the effect, an interpretation later confirmed in additional experiments (66). The complex stimuli environmental tobacco smoke elicited both nasal and ocular chemesthesis, although the ocular mucosa seemed the most sensitive (63). These studies have relied on instructing the subjects to assess separately the odorous and the chemesthetic (pungent) component of the nasal sensation, a task quite prone to different biases across subjects (67,68).

Regarding the perithreshold detection of mixtures, an intensive study of nine single chemicals and their mixtures, including two three-component, two six-component, and one nine-component mixtures, observed various degree of agonism (additive effects) among the constituents (69). The outcome comprised not only NPTs, but EITs and ODTs as well. The use of short-chain and longer-chain homologs from four different series facilitated the interpretation of the results in physicochemical terms. The degree of chemosensory agonism became larger as the number of components and their lipophilicity increased. The chemesthetic thresholds showed stronger agonism than the olfactory ones, with eye irritation having stronger agonism than nasal pungency. While physicochemical properties were shown to play an important role in the chemosensory detection of single chemicals (Fig. 1), they seem to have relevance for mixtures too. Later studies took a more comprehensive look at the study of mixtures by measuring not just threshold values but complete concentration-detection, that is, psychometric or detectability, functions (see Ref. 30). This quite intensive and detailed approach geared the effort to the simpler case of binary mixtures. Again, components of the mixtures were selected from homologous series and included 1-butanol and 2-heptanone (70), butyl acetate and toluene (71), and ethyl propanoate and ethyl heptanoate (72). The idea behind this selection was to explore a pair from different series but with similar chemical functionality, a pair from different series with dissimilar functionality, and a pair from the same series and functionality. The results were analyzed in terms of response-addition (sum of detectabilities) and dose-addition (sum of doses within a selected psychometric function). The overall results indicated relative agreement between response- and dose-addition, and, irrespective of the structural or chemical similarity between components, a stronger degree of addition of detection at low than at high levels of detectability (73). (We clarify that in these studies low levels of detectability were still above chance detection, and high levels were still below perfect detection.) Interestingly, nasal pungency showed a stronger degree of addition than eye irritation, particularly at high levels of detectability.

Complex mixtures of airborne compounds, for example, environmental tobacco smoke (74), fragranced household products (75), indoor air (62,76–78), and outdoor air surrounding composting facilities (79) or animal production facilities (80,81), have been employed to assess their potential to elicit human mucosal sensory irritation. In these cases, the focus has been exclusively on the total effect of the mixture, at one or another dilution level, on a healthy sample of subjects, or on two or more groups of subjects with or without a certain condition (for example, asthmatics). No attempts were made to relate the chemesthetic impact of the mixture to that of the individual components. In fact, in some cases, the very complex mixtures could not be fully defined from a chemical standpoint.

RESPIRATORY BEHAVIOR AND NASAL IRRITATION

Reflex Nasal Transitory Apnea
When a chemical irritant entering the nose reaches a certain level, a momentary and reflex interruption of breathing (apnea) occurs. This physiological effect has been employed to understand functional features of nasal chemesthesis and to assess the comparative chemesthetic sensitivity of groups of subjects by gender, age, and smoking status (51,56,82–85). The results indicated that females, young individuals, and nonsmokers are more sensitive than males, older individuals, and smokers, respectively, as probed by the reflex transitory apnea. The differences have been confirmed by psychophysical procedures (38,82,86–91). In addition, physiological techniques (e.g., rhinomanometry, negative mucosal potential, tERP) have also been employed to study these and other group differences, for example, those due to nasal disease or condition (see reviews in Refs. 90,92). These methods and outcomes are described in detail in chapter 9, "Exposure and Recording Systems in Human Studies."

Other Respiratory Behavior Alterations
Airborne chemicals producing odor and irritation can change breathing parameters, for example, they can reduce tidal volume (93). The odorants acetic acid, amyl acetate, and phenyl ethyl alcohol were selected as representative of compounds possessing strong, medium, and null (or minimal) nasal trigeminal impact in an investigation that combined estimates of odor and nasal irritation with measurements of nasal patency and respiratory behavior (94). As expected, the compounds showed greater differences in nasal irritation than in odor strength. It was pointed out that tidal volume seemed to have a close and inverse relationship to nasal irritation.

Using a visual analog scale, normosmics and anosmics estimated the magnitudes of odor and irritation evoked by the presentation of a wide range of concentrations of propionic acid while their breathing characteristics were recorded (95,96). As concentration increased, inhalation volume began to decline in normosmics at a lower concentration and at a faster rate than it did in anosmics. When normosmics first began to show this decline, their estimates of both odor and nasal irritation were already significantly higher than in the clean air condition (control). In contrast, inhalation duration began to decline at the same concentration in normosmics and anosmics, and it did it at a concentration where anosmics first reported nasal irritation. Still, as observed for inhalation volume, inhalation duration declined with concentration faster in normosmics than in anosmics. The authors favored the interpretation that the higher sensitivity of normosmics to a decline in inhalation volume validated their reports of nasal irritation at concentrations undetected by anosmics.

In an environmental chamber study, phlebotomists occupationally exposed to isopropanol and unexposed controls were challenged with a 400-ppm concentration of isopropanol (its time-weighted average threshold limit value) during four hours (97). Exposures to phenyl ethyl alcohol and to clean air served as a negative control for irritation, and as a negative control for both odor and irritation, respectively. The only physiological end point that changed exclusively in the isopropanol condition was respiratory frequency. It increased, relative to baseline, in both subject groups, with no differences among them. Since reports of irritation and odor intensity declined with time, the authors attributed the increase in respiration frequency to a voluntary change in breathing in response to an unpleasant, solvent-like odor (i.e., a cognitive mediation) rather than to a reflexive change due to sensory irritation (i.e., an autonomic event).

The significance of the phenomenon of irritant-induced respiratory alterations lies, at least in part, in establishing an analogy with (and basis for extrapolation from) animal studies of a similar end point. Several decades of work have documented the effects of irritants of different anatomical specificities (i.e., "sensory," "pulmonary," "respiratory") on breathing patterns in experimental animals. Alarie (98) coined the term "RD_{50}" to describe the end point of 50% reduction in respiratory rate that occurs upon exposure to (generally) water-soluble irritants with predominant impact on the upper respiratory tract (see chap. 12). Several reviews have since supported the relevance of the RD_{50} in predicting human sensory irritation (99–101), whereas other reviewers remain skeptical regarding the use of the RD_{50} model for human risk assessment purposes (102,103).

SUMMARY

Nasal chemosensory irritation (i.e., chemesthesis) in humans results from stimulation of the trigeminal nerve. Almost all chemical vapors that produce odor can evoke nasal chemesthesis at higher concentrations, although there is a cut-off point along homologous series beyond which larger homologs fail to be detected by chemesthesis. The failure seems to rest on some aspect of molecular structure or dimensions rather than on a low vapor concentration. In turn, almost all irritants can also elicit an odor with the arguable exception of CO_2. To separate the trigeminal from the olfactory response of the nose, investigators have tested subjects lacking olfaction (i.e., anosmics) and have measured nasal lateralization thresholds, that is, the ability to localize whether a vapor entered the right or the left nostril when air enters the contralateral nostril. Such ability rests on trigeminal, not olfactory input. Detection of nasal chemesthesis from chemical mixtures reveals additive effects among constituents, particularly at low levels of detectability (but still above chance detection). As a rule, increases in time of exposure decrease chemesthetic thresholds (i.e., enhances sensitivity) and produce higher ratings of irritation intensity. Nasal chemesthesis can produce alterations in respiration, including a reflex, transitory apnea, and reductions in the duration and volume of nasal inhalations. Relative consistency has been found for irritation thresholds among two of three anatomical structures subsumed within "sensory irritation"—that is, the nose and eye. Thus for predictive toxicology and risk assessment purposes, an argument can be made that measurements using one system can often be extrapolated to the other.

ACKNOWLEDGMENTS

Preparation of this chapter was supported by grants number R01 DC 002741 and R01 DC 005003 from the National Institute on Deafness and Other Communication Disorders (NIDCD), National Institutes of Health (NIH).

REFERENCES

1. Doty R, ed. Handbook of Olfaction and Gustation, 2nd ed. New York, NY: Marcel Dekker, Inc., 2003.
2. Green B, Mason J, Kare M, eds. Chemical Senses, Vol 2: Irritation. New York, NY: Marcel Dekker, Inc., 1990.
3. Bryant B, Silver WL. Chemesthesis: the common chemical sense. In: Finger TE, Silver WL, Restrepo D, eds. The Neurobiology of Taste and Smell, 2nd ed. New York, NY: Wiley-Liss, 2000:73–100.
4. Doty RL, Cometto-Muñiz JE, Jalowayski AA, et al. Assessment of upper respiratory tract and ocular irritative effects of volatile chemicals in humans. Crit Rev Toxicol 2004; 34(2):85–142.
5. Parker GH. The relation of smell, taste, and the common chemical sense in vertebrates. J Acad Nat Sci Phila 1912; 15:219–234.
6. Lee Y, Lee CH, Oh U. Painful channels in sensory neurons. Mol Cells 2005; 20(3):315–324.
7. Green BG, Mason JR, Kare MR. Preface. In: Green BG, Mason JR, Kare MR, eds. Chemical Senses, Vol 2: Irritation. New York, NY: Marcel Dekker, Inc., 1990:v–vii.
8. Green BG, Lawless HT. The psychophysics of somatosensory chemoreception in the nose and mouth. In: Getchell TV, Doty RL, Bartoshuk LM, et al., eds. Smell and Taste in Health and Disease. New York, NY: Raven Press, 1991:235–253.
9. Finger T, Silver W, Bryant B. Trigeminal nerve. In: Adelman G, Smith B, eds. Encyclopedia of Neuroscience, Vol II. Amsterdam: Elsevier, 1999:2069–2071.
10. Doty RL, Cometto-Muñiz JE. Trigeminal chemosensation. In: Doty RL, ed. Handbook of Olfaction and Gustation, 2nd ed. New York, NY: Marcel Dekker, 2003:981–1000.
11. Cometto-Muñiz JE, Cain WS. Sensory irritation. Relation to indoor air pollution. Ann N Y Acad Sci 1992; 641:137–151.
12. Meldrum M. Setting occupational exposure limits for sensory irritants: the approach in the European Union. AIHAJ 2001; 62(6):730–732.
13. Cometto-Muñiz JE. Physicochemical basis for odor and irritation potency of VOCs. In: Spengler JD, Samet J, McCarthy JF, eds. Indoor Air Quality Handbook. New York, NY: McGraw-Hill, 2001:20.1–20.21.
14. Cometto-Muñiz JE, Cain WS, Abraham MH. Determinants for nasal trigeminal detection of volatile organic compounds. Chem Senses 2005; 30(8):627–642.

15. Wysocki CJ, Dalton P, Brody MJ, et al. Acetone odor and irritation thresholds obtained from acetone-exposed factory workers and from control (occupationally unexposed) subjects. Am Ind Hyg Assoc J 1997; 58(10):704–712.

16. Cometto-Muñiz JE, Cain WS. Trigeminal and olfactory sensitivity: comparison of modalities and methods of measurement. Int Arch Occup Environ Health 1998; 71:105–110.

17. Dalton PH, Dilks DD, Banton MI. Evaluation of odor and sensory irritation thresholds for methyl isobutyl ketone in humans. Am Ind Hyg Assoc J 2000; 61(3):340–350.

18. Schneider RA, Schmidt CE. Dependency of olfactory localization on non-olfactory cues. Physiol Behav 1967; 2:305–309.

19. Kobal G, Van Toller S, Hummel T. Is there directional smelling? Experientia 1989; 45:130–132.

20. von Bèkesy G. Olfactory analogue to directional hearing. J Appl Physiol 1964; 19:369–373.

21. Doty RL. Intranasal trigeminal detection of chemical vapors by humans. Physiol Behav 1975; 14:855–859.

22. Doty RL, Brugger WE, Jurs PC, et al. Intranasal trigeminal stimulation from odorous volatiles: psychometric responses from anosmic and normal humans. Physiol Behav 1978; 20:175–185.

23. Cometto-Muñiz JE, Cain WS. Thresholds for odor and nasal pungency. Physiol Behav 1990; 48:719–725.

24. Cometto-Muñiz JE, Cain WS. Nasal pungency, odor, and eye irritation thresholds for homologous acetates. Pharmacol Biochem Behav 1991; 39:983–989.

25. Cometto-Muñiz JE, Cain WS. Efficacy of volatile organic compounds in evoking nasal pungency and odor. Arch Environ Health 1993; 48:309–314.

26. Cometto-Muñiz JE, Cain WS. Sensory reactions of odor and nasal pungency to volatile organic compounds: the alkylbenzenes. Am Ind Hyg Assoc J 1994; 55: 811–817.

27. Cometto-Muñiz JE, Cain WS, Abraham MH. Nasal pungency and odor of homologous aldehydes and carboxylic acids. Exp Brain Res 1998; 118:180–188.

28. Cometto-Muñiz JE, Cain WS, Abraham MH, et al. Trigeminal and olfactory chemosensory impact of selected terpenes. Pharmacol Biochem Behav 1998; 60(3):765–770.

29. Cometto-Muñiz JE, Cain WS, Hiraishi T, et al. Comparison of two stimulus-delivery systems for measurement of nasal pungency thresholds. Chem Senses 2000; 25(3):285–291.

30. Cometto-Muñiz JE, Cain WS, Abraham MH, et al. Psychometric functions for the olfactory and trigeminal detectability of butyl acetate and toluene. J Appl Toxicol 2002; 22(1):25–30.

31. Cometto-Muñiz JE, Abraham MH. Human olfactory detection of homologous n-alcohols measured via concentration-response functions. Pharmacol Biochem Behav 2008; 89(3):279–291.

32. Kendal-Reed M, Walker JC, Morgan WT, et al. Human responses to propionic acid. I. Quantification of within- and between-participant variation in perception by normosmics and anosmics. Chem Senses 1998; 23(1):71–82.

33. Gudziol H, Schubert M, Hummel T. Decreased trigeminal sensitivity in anosmia. ORL J Otorhinolaryngol Relat Spec 2001; 63(2):72–75.

34. Frasnelli J, Schuster B, Hummel T. Subjects with congenital anosmia have larger peripheral but similar central trigeminal responses. Cereb Cortex 2007; 17(2):370–377.

35. Frasnelli J, Schuster B, Hummel T. Interactions between olfaction and the trigeminal system: what can be learned from olfactory loss. Cereb Cortex 2007; 17(10):2268–2275.

36. Frasnelli J, Schuster B, Zahnert T, et al. Chemosensory specific reduction of trigeminal sensitivity in subjects with olfactory dysfunction. Neuroscience 2006; 142(2): 541–546.

37. Hummel T, Barz S, Lotsch J, et al. Loss of olfactory function leads to a decrease of trigeminal sensitivity. Chem Senses 1996; 21(1):75–79.
38. Hummel T, Futschik T, Frasnelli J, et al. Effects of olfactory function, age, and gender on trigeminally mediated sensations: a study based on the lateralization of chemosensory stimuli. Toxicol Lett 2003; 140/141:273–280.
39. Cometto-Muñiz JE, Cain WS, Abraham MH. Molecular restrictions for human eye irritation by chemical vapors. Toxicol Appl Pharmacol 2005; 207(3):232–243.
40. Cain WS, Lee NS, Wise PM, et al. Chemesthesis from volatile organic compounds: psychophysical and neural responses. Physiol Behav 2006; 88(4/5):317–324.
41. Cometto-Muñiz JE, Cain WS, Abraham MH, et al. Chemical boundaries for detection of eye irritation in humans from homologous vapors. Toxicol Sci 2006; 91(2): 600–609.
42. Abraham MH, Sánchez-Moreno R, Cometto-Muñiz JE, et al. A quantitative structure–activity analysis on the relative sensitivity of the olfactory and the nasal trigeminal chemosensory systems. Chem Senses 2007; 32:711–719.
43. Cometto-Muñiz JE, Cain WS, Abraham MH, et al. Cut-off in detection of eye irritation from vapors of homologous carboxylic acids and aliphatic aldehydes. Neuroscience 2007; 145:1130–1137.
44. Cometto-Muñiz JE, Cain WS, Abraham MH, et al. Concentration-detection functions for eye irritation evoked by homologous n-alcohols and acetates approaching a cut-off point. Exp Brain Res 2007; 182:71–79.
45. Cometto-Muñiz JE, Abraham MH. A cut-off in ocular chemesthesis from vapors of homologous alkylbenzenes and 2-ketones as revealed by concentration-detection functions. Toxicol Appl Pharmacol 2008; 230(3):298–303.
46. Abraham MH. The potency of gases and vapors: QSARs—anesthesia, sensory irritation, and odor. In: Gammage RB, Berven BA, eds. Indoor Air and Human Health, 2nd ed. Boca Raton, FL: CRC Lewis Publishers, 1996:67–91.
47. Abraham MH, Andonian-Haftvan J, Cometto-Muñiz JE, et al. An analysis of nasal irritation thresholds using a new solvation equation. Fundam Appl Toxicol 1996; 31(1):71–76.
48. Abraham MH, Kumarsingh R, Cometto-Muñiz JE, et al. An algorithm for nasal pungency thresholds in man. Arch Toxicol 1998; 72:227–232.
49. Abraham MH, Gola JMR, Cometto-Muñiz JE, et al. The correlation and prediction of VOC thresholds for nasal pungency, eye irritation and odour in humans. Indoor Built Environ 2001; 10(3/4):252–257.
50. Shusterman D, Matovinovic E, Salmon A. Does Haber's law apply to human sensory irritation? Inhal Toxicol 2006; 18(7):457–471.
51. Dunn JD, Cometto-Muñiz JE, Cain WS. Nasal reflexes: reduced sensitivity to CO_2 irritation in cigarette smokers. J Appl Toxicol 1982; 2:176–178.
52. Wise PM, Radil T, Wysocki CJ. Temporal integration in nasal lateralization and nasal detection of carbon dioxide. Chem Senses 2004; 29(2):137–142.
53. Wise PM, Canty TM, Wysocki CJ. Temporal integration of nasal irritation from ammonia at threshold and supra-threshold levels. Toxicol Sci 2005; 87(1):223–231.
54. Wise PM, Canty TM, Wysocki CJ. Temporal integration in nasal lateralization of ethanol. Chem Senses 2006; 31(3):227–235.
55. Wise PM, Toczydlowski SE, Wysocki CJ. Temporal integration in nasal lateralization of homologous alcohols. Toxicol Sci 2007; 99(1):254–259.
56. Cometto-Muñiz JE, Cain WS. Temporal integration of pungency. Chem Senses 1984; 8:315–327.
57. Wise PM, Wysocki CJ, Radil T. Time-intensity ratings of nasal irritation from carbon dioxide. Chem Senses 2003; 28(9):751–760.
58. Frasnelli J, Lotsch J, Hummel T. Event-related potentials to intranasal trigeminal stimuli change in relation to stimulus concentration and stimulus duration. J Clin Neurophysiol 2003; 20(1):80–86.
59. Cain WS, See LC, Tosun T. Irritation and odor from formaldehyde: chamber studies. In: IAQ'86 Managing Indoor Air for Health and Energy Conservation. Atlanta,

Georgia, USA: American Society of Heating, Refrigerating and Air-Conditioning Engineers, Inc., 1986:126–137.
60. Dalton P, Dilks D, Hummel T. Effects of long-term exposure to volatile irritants on sensory thresholds, negative mucosal potentials, and event-related potentials. Behav Neurosci 2006; 120(1):180–187.
61. Smeets M, Dalton P. Perceived odor and irritation of isopropanol: a comparison between naive controls and occupationally exposed workers. Int Arch Occup Environ Health 2002; 75(8):541–548.
62. Hudnell HK, Otto DA, House DE, et al. Exposure of humans to a volatile organic mixture. II. Sensory. Arch Environ Health 1992; 47(1):31–38.
63. Cain WS, Tosun T, See LC, et al. Environmental tobacco-smoke—sensory reactions of occupants. Atmos Environ 1987; 21(2):347–353.
64. Feron VJ, Arts JH, Kuper CF, et al. Health risks associated with inhaled nasal toxicants. Crit Rev Toxicol 2001; 31(3):313–347.
65. Cometto-Muñiz JE, García-Medina MR, Calviño AM. Perception of pungent odorants alone and in binary-mixtures. Chem Senses 1989; 14(1):163–173.
66. Cometto-Muñiz JE, Hernandez SM. Odorous and pungent attributes of mixed and unmixed odorants. Percept Psychophys 1990; 47(4):391–399.
67. Dalton P. Odor, irritation and perception of health risk. Int Arch Occup Environ Health 2002; 75(5):283–290.
68. Dalton P. Upper airway irritation, odor perception and health risk due to airborne chemicals. Toxicol Lett 2003; 140/141:239–248.
69. Cometto-Muñiz JE, Cain WS, Hudnell HK. Agonistic sensory effects of airborne chemicals in mixtures: odor, nasal pungency, and eye irritation. Percept Psychophys 1997; 59(5):665–674.
70. Cometto-Muñiz JE, Cain WS, Abraham MH, et al. Chemosensory detectability of 1-butanol and 2-heptanone singly and in binary mixtures. Physiol Behav 1999; 67(2):269–276.
71. Cometto-Muñiz JE, Cain WS, Abraham MH, et al. Ocular and nasal trigeminal detection of butyl acetate and toluene presented singly and in mixtures. Toxicol Sci 2001; 63(2):233–244.
72. Cometto-Muñiz JE, Cain WS, Abraham MH. Chemosensory additivity in trigeminal chemoreception as reflected by detection of mixtures. Exp Brain Res 2004; 158(2):196–206.
73. Cometto-Muñiz JE, Cain WS, Abraham MH. Detection of single and mixed VOCs by smell and by sensory irritation. Indoor Air 2004; 14(suppl 8):108–117.
74. Walker JC, Kendal-Reed M, Utell MJ, et al. Human breathing and eye blink rate responses to airborne chemicals. Environ Health Perspect 2001; 109(suppl 4):507–512.
75. Opiekun RE, Smeets M, Sulewski M, et al. Assessment of ocular and nasal irritation in asthmatics resulting from fragrance exposure. Clin Exp Allergy 2003; 33(9):1256–1265.
76. Meininghaus R, Kouniali A, Mandin C, et al. Risk assessment of sensory irritants in indoor air—a case study in a French school. Environ Int 2003; 28(7):553–557.
77. Otto D, Molhave L, Rose G, et al. Neurobehavioral and sensory irritant effects of controlled exposure to a complex mixture of volatile organic compounds. Neurotoxicol Teratol 1990; 12(6):649–652.
78. Laumbach RJ, Fiedler N, Gardner CR, et al. Nasal effects of a mixture of volatile organic compounds and their ozone oxidation products. J Occup Environ Med 2005; 47(11):1182–1189.
79. Müller T, Thissen R, Braun S, et al. (M)VOC and composting facilities. Part 2: (M)VOC dispersal in the environment. Environ Sci Pollut Res Int 2004; 11(3):152–157.
80. Schiffman SS, Walker JM, Dalton P, et al. Potential health effects of odor from animal operations, wastewater treatment, and recycling of byproducts. J Agromed 2000; 7:7–81.

81. Schiffman SS, Williams CM. Science of odor as a potential health issue. J Environ Qual 2005; 34(1):129–138.
82. Garcia-Medina MR, Cain WS. Bilateral integration in the common chemical sense. Physiol Behav 1982; 29(2):349–353.
83. Cometto-Muñiz JE, Cain WS. Perception of nasal pungency in smokers and non-smokers. Physiol Behav 1982; 29(4):727–731.
84. Stevens JC, Cain WS. Aging and the perception of nasal irritation. Physiol Behav 1986; 37(2):323–328.
85. Shusterman DJ, Balmes JR. A comparison of two methods for determining nasal irritant sensitivity. Am J Rhinol 1997; 11(5):371–378.
86. Cometto-Muñiz JE, Noriega G. Gender differences in the perception of pungency. Physiol Behav 1985; 34(3):385–389.
87. Stevens JC, Plantinga A, Cain WS. Reduction of odor and nasal pungency associated with aging. Neurobiol Aging 1982; 3(2):125–132.
88. Shusterman D, Murphy MA, Balmes J. Differences in nasal irritant sensitivity by age, gender, and allergic rhinitis status. Int Arch Occup Environ Health 2003; 76(8): 577–583.
89. Shusterman D, Murphy MA, Balmes J. Influence of age, gender, and allergy status on nasal reactivity to inhaled chlorine. Inhal Toxicol 2003; 15(12):1179–1189.
90. Shusterman D. Trigeminally-mediated health effects of air pollutants: sources of inter-individual variability. Hum Exp Toxicol 2007; 26(3):149–157.
91. Wysocki CJ, Cowart BJ, Radil T. Nasal trigeminal chemosensitivity across the adult life span. Percept Psychophys 2003; 65(1):115–122.
92. Shusterman D. Individual factors in nasal chemesthesis. Chem Senses 2002; 27(6):551–564.
93. Warren DW, Walker JC, Drake AF, et al. Assessing the effects of odorants on nasal airway size and breathing. Physiol Behav 1992; 51(2):425–430.
94. Warren DW, Walker JC, Drake AF, et al. Effects of odorants and irritants on respiratory behavior. Laryngoscope 1994; 104(5 pt 1):623–626.
95. Walker JC, Kendal-Reed M, Hall SB, et al. Human responses to propionic acid. II. Quantification of breathing responses and their relationship to perception. Chem Senses 2001; 26(4):351–358.
96. Kendal-Reed M, Walker JC, Morgan WT. Investigating sources of response variability and neural mediation in human nasal irritation. Indoor Air 2001; 11(3):185–191.
97. Smeets MA, Maute C, Dalton PH. Acute sensory irritation from exposure to isopropanol (2-propanol) at TLV in workers and controls: objective versus subjective effects. Ann Occup Hyg 2002; 46(4):359–373.
98. Alarie Y. Sensory irritation by airborne chemicals. CRC Crit Rev Toxicol 1973; 2(3):299–363.
99. de Ceaurriz JC, Micillino JC, Bonnet P, et al. Sensory irritation caused by various industrial airborne chemicals. Toxicol Lett 1981; 9(2):137–143.
100. Schaper M. Development of a database for sensory irritants and its use in establishing occupational exposure limits. Am Ind Hyg Assoc J 1993; 54(9):488–544.
101. Kuwabara Y, Alexeeff GV, Broadwin R, et al. Evaluation and application of the RD50 for determining acceptable exposure levels of airborne sensory irritants for the general public. Environ Health Perspect 2007; 115(11):1609–1616.
102. Bos PM, Zwart A, Reuzel PG, et al. Evaluation of the sensory irritation test for the assessment of occupational health risk. Crit Rev Toxicol 1991; 21(6):423–450.
103. Bos PM, Busschers M, Arts JH. Evaluation of the sensory irritation test (Alarie test) for the assessment of respiratory tract irritation. J Occup Environ Med 2002; 44(10):968–976.

14 Human Nasal Reflexes

Kathryn Sowerwine, Samantha Jean Merck, and James N. Baraniuk

Division of Rheumatology, Immunology and Allergy, Georgetown University Medical Center, Washington, D.C., U.S.A.

INTRODUCTION

The conditions of inhaled air activate trigeminal type Aδ and C afferent neurons. Distinct subpopulations of afferent neurons are presumed to have unique combinations of afferent sensor proteins, sodium, and other ion channels. These are involved in neural depolarization and the spread of the action potential throughout the extensively branched nasal ramification of nerve endings, and transmission to the dorsal root ganglion cell bodies and central nervous system. Some of the branched neurons release neurotransmitters locally in the nasal mucosa to generate the axon response of "neurogenic inflammation." The axon response in human nasal mucosa leads to glandular secretion without any change in vascular processes. This activation–exocytosis loop is a very rapid onset nasal defense mechanism. Depolarization of dorsal root ganglion cells may alter the amount and identity of sensor proteins and neurotransmitters and so alter the functions of these neurons ("neural plasticity").

Central afferent endings synapse with interneurons in the first two layers of the substantia gelantinosa. These secondary neurons include (*i*) local interneurons that regulate central transmission of peripheral sensations, and (*ii*) the long tract neurons that cross the midline and ascend within several lateral spinal cord sensation-restricted spinothalamic pathways. Connections with brain stem nuclei induce parasympathetic, sympathetic, and systemic motor neuron responses. These include the nasobronchial, Hering–Breuer, and nasosystemic reflexes. Cortical connections lead to perceptions of nasal itch, other irritation, cooling, rhinorrhea, fullness/congestion, headache, nasal patency, and obstruction to airflow. Perceptions of these inputs may be modified by previous experience and emotional memories through connections with the limbic system, amygdala, anterior cingulate gyrus, and other centers. Hyperactive nasal reflexes may play roles in autonomic reflexes such as the gustatory reflex, recurrent midfacial pain syndromes, and perhaps even so-called "chemical sensitivity syndromes."

NASAL CYCLE

An intrinsic, subliminal nasal reflex is the normal nasal cycle. This reciprocating bilateral cycle of congestion and decongestion has been observed for over a century (1) and has been confirmed by manometry (2), magnetic resonance imaging (3), and acoustic rhinometry (4). The cycle occurs in a majority of adults and in children as young as three years (4). Normal individuals are generally unaware of this phenomenon (5,6), probably because total nasal airflow resistance remains stable (6,7). Subjective sensations of patency and "congestion" do not correlate with objective measures of left and right nostril "obstruction" to airflow (8). This suggests that central monitoring of total nasal airflow occurs with integration of afferent inputs from both nostrils.

The nasal cycle consists of periodic vasodilation and engorgement of the nasal venous sinusoids leading to airflow obstruction, followed by increased nasal patency due to vasoconstriction, sinusoid collapse, and an increase in nasal cavity air space volume. The cycle occurs in laryngectomized subjects, indicating that it is not driven by airflow.

The obstructive phase of mucosal and turbinate swelling may begin with the unilateral cessation of sympathetic discharge. Vasodilation and mucosal thickening would occur by default. The absence of sympathetic discharge is supported by the ability of the α_1–adrenergic agonist, pseudoephedrine, to constrict vessels and so prevent maximum mucosal thickening and airflow obstruction. Pseudoephedrine was redundant and had no effect during the active, sympathetic, vasoconstrictor, decongestion phase (9). The vascular congestion may permit plasma extravasation through the fenestrated capillaries of the superficial lamina propria and the accumulation of interstitial fluid with edema evident by microstereometry. Airflow during this phase is very slow and predominantly laminar (10). Nitric oxide (NO) may also play a role in initiating and maintaining the obstructive phase (11). Levels in the nasal lumen were as high as 1100 parts per billion during the obstructive phase (11). Levels correlated with the smallest cross-sectional areas for airflow and highest airflow resistance measurements. The high NO levels may promote vasodilation, mucosal thickening, and nostril occlusion. The elevation may be due to trapping of NO escaping from the paranasal sinuses. Alternatively, stimulation of parasympathetic NO/vasoactive intestinal polypeptide neurons may occur at the onset of the nasal obstruction. Simultaneous parasympathetic cholinergic discharge replenishes epithelial-lining fluid mucus (12,13). This scenario suggests brain stem cycling of unilateral parasympathetic and sympathetic discharges in the nasal cycle.

The onset of nasal patency begins with an increase in the cross-sectional area for airflow in the anterior cavum between the anterior tip of the inferior turbinate and septal tuberculum. The required vasoconstriction is likely provided by sympathetic release of norepinephrine and neuropeptide Y. Sympathetic constriction of the nasal venous sinusoids during decongestion and the airway patency phase may "wring" out the turbinate lamina propria and promote exudation of the accumulated plasma-rich interstitial fluid onto the epithelial surface (14). Small vasoconstrictor-induced changes in nasal airspace volume can accelerate flow velocities sufficiently to induce turbulent air motion. Turbulent airflow permits inspired particulate material to contact the mucosa and become firmly adsorbed onto mucus. Mucociliary clearance was 2.5-fold faster in the patent nostril compared to the obstructed side (15). Patent nostrils had low nasal airflow resistance (<6 cm $H_2O/l/sec$) and so increased airflow. This likely explained the reduction in nitric oxide levels to <80 parts per billion. However, if local tissue NO production fell during the transition from nasal obstruction to nasal patency, then the absence of this vasodilator combined with sympathetic vasoconstriction would increase nasal patency. Therefore, sympathetic vasoconstriction, turbulent airflow, and more efficient ciliary activity are present in the more patent nostril.

Eccles proposed that the nasal cycle acts in respiratory defense by (*i*) alternating the work of air conditioning between the two nasal passages, (*ii*) generating a plasma exudate which physically cleanses the epithelium and provides a source of antibodies and inflammatory mediators, and (*iii*) maintaining the patency of the airway during the inflammatory response to infection (16). The synthesis and secretion of innate immune antimicrobial glandular proteins and mucins would be another beneficial effect.

The brain stem autonomic responses may be regulated by a hypothalamic center. All subjects with Kallmann's syndrome of hypothalamic and olfactory center hypoplasia and loss of gonadal and smell functions have abnormal nasal cycles (17). Electrical stimulation of hypothalamic nuclei leads to pronounced bilateral sympathetic activation and nasal vasoconstriction (18). Other centers may also participate, as the nasal cycle can become synchronized to the sleep cycle (19). The switch in patency from one nostril to the other may occur during rapid eye movement sleep.

Nasal cycles can be overridden, modulated, or sensed in many environmental, pathological, and positional (lateral recumbent) situations. It is important to recognize the cycle as a normal phenomenon and to differentiate it from pathological causes of nasal obstruction (20).

NASONASAL REFLEXES

Axon Response
Stimulation of the fine nerve endings of nociceptive neurons in the nasal epithelium depolarizes the entire nerve including all of the branched neural endings throughout the nasal mucosa. These neural processes have swellings or varicosities that contain multiple colocalized neurotransmitters. Stimulation of subpopulations of these nerves with nocifers such as hypertonic saline leads to pain, local substance P release within three minutes, and glandular exocytosis within five minutes (21,22). This local "efferent" function of "afferent" neurons provides a rapidly mobilized defense mechanism for airway mucosa.

Nasonasal Parasympathetic Reflexes
Nasonasal reflexes generally refer to unilateral afferent stimulation that leads to bilateral brain stem efferent reflexes. These can be identified by secretion into the contralateral nostril. Unilateral histamine provocations lead to contralateral secretion that is about 60% of the mass of the challenged side. The afferent limb of the reflex arc is broken by cocaine anesthesia to the ipsilateral trigeminal nerves, while the efferent parasympathetic limb is blocked by contralateral topical anticholinergics or Vidian neurectomy (23).

The majority of the glandular secretion that follows allergen challenge in atopic rhinitic subjects is due to histamine activation of H1 receptors on a subpopulation of nociceptive "itch nerves" that recruit bilateral parasympathetic cholinergic reflexes (24). These very narrow diameter, slowly conducting neurons travel in a separate section of the lateral spinothalamic tract to synapse in a unique region of the thalamus compared to type C pain fibers (25). These nerves offer a specific target for drug development for reducing pruritus and related neural responses.

Sensors on trigeminal neurons of the nasal and soft palate mucosa respond to capsaicin and other piquant stimuli by recruiting parasympathetic cholinergic reflexes. Hot, spicy foods can be a strong stimulus for this reflex, and in their extreme form cause the disability of gustatory rhinitis (26). Responses to these foods can range from congestion and blockage to nasal airflow to copious anterior and posterior nasal discharge. Nasal capsaicin treatment can block the afferent limb of this reflex, while topical nasal anticholinergic drugs effectively block the parasympathetic cholinergic efferent limb (27).

An important issue is why there are ranges of sensory afferent sensitivity and efferent reflex responses within the human population. Cold dry air provides another example. Breathing very cold air leads to rhinorrhea in virtually all subjects. However, a smaller proportion develops copious mucus discharge akin to gustatory rhinitis. This "skiier's rhinitis" (28) ("ski bunny rhinitis") is effectively prevented by topical anticholinergic drugs, indicating the importance of the efferent reflex arc and glandular secretion. Cold dry air inhalation has another important aspect, since it causes dose-dependent nasal obstruction only in subjects with idiopathic nonallergic, noninfectious, noninflammatory rhinitis (iNAR) (also referred to as "nonallergic, noninfectious perennial rhinitis" or "noninfectious, nonallergic rhinitis") and distinguishes this group from control and atopic subjects who do not respond (Table 1).

The iNAR group may be further subdivided into those who have excessive, cholinergic glandular secretion in response to a wide range of stimuli including capsaicin and nicotine. These are the so-called "runners." They have more exaggerated reflex arcs as indicated by dose responses to these nociceptive agents and greater efficacy of topical anticholinergic drugs (29). The contrasting group is the "blockers" who have greater complaints of nasal congestion and minimal responses to anticholinergics. The incidence and prevalence of these subsets have not been established because no criteria have been established to successfully classify these subtypes.

iNAR subjects also have decreased superficial nasal blood flow measured by laser Doppler signals ($14 \pm 4\%$; $p < 0.05$) after breathing nebulized saline

TABLE 1 Nasal Reflexes

Nasal afferents	Systemic afferents
Nociceptive nerve axon responses	Postural reflex
Nasal cycle	Crutch reflex
Nasonasal reflexes	Exercise reflex
Nasopharyngeal reflex	Hot and cold cutaneous temperature
Nasolaryngeal reflex	reflexes
Nasolacrimal reflex	Visible and infrared light reflexes
Nasosalivary reflex	Bronchonasal reflex
Gustatory rhinitis ("salsa sniffles")	Ovulatory rhinitis
Cold dry air–induced rhinitis ("skiier's	Sexual reflexes
nose" "ski bunny rhinitis")	Alcohol reflex
Diving reflex	Psychogenic syndromes
Naso-broncho-cardiac reflex	Crutch reflex
Sneeze	Exercise reflex
Nasal tracheobronchial vasodilation reflex	

at 22°C (30). The effect began after 30 seconds and lasted 60 to 90 seconds. Nasal nitric oxide was decreased 8.0 ± 0.6% after 60 seconds ($p < 0.001$). One potential explanation is that activation of nasal "cold" receptors on afferent nerves recruited transient sympathetic vasoconstriction that reduced the delivery of arginine to the mucosa and so decreased nasal NO production, superficial vasodilation, and blood flow. Measures of tissue NO are required to identify the mechanism(s) involved (31). Activation of cold receptors may play a major role in the symptomatic relief of some nasonasal reflex effects in chronic rhinitis subjects.

Mechanosensitive receptors also stimulate nasonasal reflexes. Rubbing a cotton tipped swab soaked in saline in the middle meatus caused an increase in nasal airflow resistance (32). This increase was not blocked by local application of cocaine, a vasoconstrictor and local anesthetic drug. This suggests that mechanosensitive neurons may represent another distinct population of nasal type C neurons that induce protective, obstructive vasodilatory responses in vivo.

Reflex arcs display neural plasticity. One month of untreated allergic rhinitis induces afferent neural sensitivity to bradykinin and endothelin 1 and the recruitment of bilateral parasympathetic reflexes (33–36). Nerve growth factor may play a critical role in upregulation of the afferent neuron sensitivity (37). Efferent neural reflex responses are also increased with greater glandular exocytosis (38,39) and potentially vasodilation (40,41). Reflex-mediated vascular hypersensitivity detected by increased mucosal blood flow and reduced nasal cavity volume (acoustic rhinometry) was found in about half of allergic subjects who had unilateral nasal allergen challenge (42). The nasonasal reflex effects lasted 15 minutes on the contralateral side compared to over 20 minutes on the allergen-challenged side and had only 45% of the magnitude of the ipsilateral vascular changes. These effects were blocked by atropine sulfate, suggesting that bilateral cholinergic vasodilator responses were recruited. The allergen-induced hyperresponsiveness is only on the order of two- to eightfold. This is much less than the 50-fold changes in the tracheobronchial tree in asthma (43).

It remains to be seen if similar effects occur after chronic occupational, toxic inhalant, and other nasal stimuli in nonallergic rhinitis syndromes (44). Capsaicin, for example, had greater effects in untreated severe allergic rhinitis subjects compared to a group of nonallergic rhinitis subjects (41). Investigations of other provocations and in additional subtypes of nonallergic rhinitis subjects may reveal differences in pathological mechanisms such a neural reflex modulation.

SYSTEMIC REFLEXES ORIGINATING FROM NASAL MUCOSAL STIMULATION

The trigemino-cardiac reflex is a component of the diving reflex (45). It is also a well-recognized surgical phenomenon consisting of bradycardia, arterial hypotension, apnea, and gastric hypermotility. It occurs during surgery of the orbit, nose, sinuses, and the cerebellopontine angle near the central part of the trigeminal nerve, and with nasal packing. Ten percent of subjects having transsphenoidal surgery for pituitary adenomas developed bradycardia of 45%

and drops in mean arterial blood pressure of 54% (46). Electrical stimulation of the trigeminal nerve leads to bradycardia and hypotension that can be prevented with intravenous atropine or transection of the afferent maxillary nerve (47). The bradycardia may be mediated by parasympathetic cholinergic reflexes to the atrioventricular node, while the sudden loss of sympathetic discharge may contribute to both the bradycardia and hypotension.

Irritants also activate this reflex. Significant bradycardia developed in essentially all of 80 healthy volunteers who had 25% ammonia placed on their middle turbinate mucosa (48). Eleven had apnea before bradycardia. These responses were blocked by pretreatment of the nasal mucosa with 2% lidocaine, suggesting the involvement of trigeminal afferents. "Sudden sniffing death" can occur upon sniffing organic solvents such as glue, cleaning fluid, and gasoline. Insertion of nasogastric tubes and inhalation of water into the nose have both been associated with bradycardia and cardiac arrest (49).

Cooling of the facial skin, rather than inhalation of cold air, is predominantly responsible for bronchoconstriction during cold weather in both chronic obstructive pulmonary disease and healthy subjects (50). Placing bags of 23°C water on the face did not cause bronchoconstriction. However, bags of 4°C water placed over areas of trigeminal innervation produced a 14% drop in SGaw (51). These results reinforce the importance of cold receptors in triggering the diving reflex (52). Covering the face is effective at preventing this facio-bronchoconstrictor reflex. Other nasobronchial reflexes will be discussed elsewhere in this book.

SNEEZE

The sneeze reflex has been experienced by everyone, and is an important airway defense response for expelling inhaled irritant materials (53) (Fig. 1). Normal subjects have an average of four sneezes with nose blowing per day (54). While generally benign, paroxysms of sneezing have induced an acute aortic dissection in one hypertensive patient (55). Acute orbital emphysema occurred after sneezing in a chronic rhinosinusitis subject who had had multiple surgeries and potential weakening of their medial orbital wall (56). Mild head trauma, such as jumping from a 1 m height or sneezing, may precipitate cavernous sinus thrombosis (57). This combined with other risk factors such as use of birth control pills and procoagulant states may help explain the 20% of unresolved causes of cavernous sinus thrombosis. Intractable psychogenic sneezing has been described and resolves after appropriate psychotherapy (58–60).

The best-defined afferent pathway involves histamine-mediated depolarization of H1 receptor–bearing type C trigeminal neurons. Other stimuli include allergen, chemical irritants, electrical stimulation of nociceptive afferent neurons in the trigeminal ethmoid and maxillary nerves and potentially sympathetic afferents associated with the Vidian and greater petrosal nerves, sudden exposure to bright lights, and cooling of the skin of various parts of the body (61,62). These stimuli activate a stereotyped series of actions that are choreographed by activation of a complex array of central pathways and nuclei leading to systemic muscle coordination. Intercostal and accessory respiratory muscle contractions provide a rapid oral inspiration to hyperinflated volumes, followed by closure of the eustachian tubes, eyes, glottic, and nasopharyngeal structures when at the maximum lung volume. Abdominal, neck, and other muscles contract in

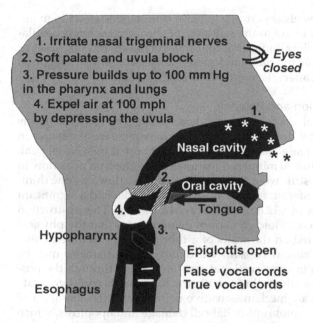

1. Irritate nasal trigeminal nerves
2. Soft palate and uvula block
3. Pressure builds up to 100 mm Hg in the pharynx and lungs
4. Expel air at 100 mph by depressing the uvula

Eyes closed

1.

Nasal cavity

2.

Oral cavity

4.

Tongue

3.

Hypopharynx

Epiglottis open

False vocal cords

Esophagus

True vocal cords

FIGURE 1 The sneeze reflex. Nasal mucosal irritation (1) activates nociceptive afferent nerves. After a deep hyperinflation through the mouth, the soft palate, uvula, posterior tongue, and lips close to occlude the airway. A Valsalva maneuver increases the intrathoracic and glottic pressure to ~100 mm Hg. The uvula is suddenly pulled anteriorly. The pressurized air escapes through the nose at speeds of 100 to 200 miles/hr expelling nasal secretions and inhaled irritants.

a forceful Valsalva maneuver that compresses the thoracic air to pressures of >100 mm Hg. Sudden anterior flexion of the soft palate opens the nasopharyn-geal space so that the pressurized air column can rush through the nose at speeds of over 100 miles/hr (45 m/sec). The shearing forces remove mucus strands and any attached particulate or other irritant from the epithelial surfaces and expel them from the nostrils. The pressure differential may introduce high-pressure waves into sinus cavities, up the nasolacrimal and potentially into the middle ear if the maneuver is not properly coordinated. This process can be rapidly repeated in staccato fashion. Cholinergic nasal, lacrimal, salivary, and posterior pharyngeal gland exocytosis follows to resurface the expelled epithelial-lining fluids. The sneeze reflex may be coordinated by a latero-medulary sneeze center near the spinal trigeminal tract and nucleus. This center appears to be bilateral and functionally independent on both sides based on its unilateral loss in strokes affecting this region (63).

PERIPHERAL STIMULI LEADING TO NASAL REFLEXES

Exercise
Exercise promotes a drop in total nasal airway resistance within 30 seconds that is maximal at five minutes, and may persist for up to 30 minutes after completing aerobic efforts (64,65). Nasal airway resistance drops in proportion to exertion, with a 39% reduction at a workload of 75 W and 49% after 100 W. Sympathetic

vasoconstriction of nasal vessels is part of a general sympathetic effect to maintain the flow of oxygenated blood to the muscles (66). Isocapneic hyperventilation does not alter nasal airflow, indicating that the workload, and not nasal or oral airflow, is the trigger for the nasal and systemic vasoconstrictor response. Body position also does not affect the nasal changes of exercise.

Positional Nasal Obstruction and Patency
Positional regulation of nasal airflow has been demonstrated by having subjects lay in the right and left lateral decubitus positions to maximize nasal patency in the superior nostril and decrease patency in the inferior nostril (67). Nasal peak flow rates were measured after 30 minutes. Position had no effect on peak flow in the superior nostrils, or if nostrils were 100% obstructed (zero flow in some rhinitis subjects). However, the inferior, reflexly obstructed nostrils had a significant reduction in mean peak flow of -12.8 L/min (SD = 4.1 L/min). This obstruction may have clinical implications. Fluid dynamics demonstrates that the physical force (frictional stress) exerted on the walls of a tube increases as tube diameter is decreased (increased airflow resistance). Reduced tube diameter may be equated with the reduction in cross-sectional area for airflow through the nostrils during rhinitis. If so, breathing through obstructed nostrils could generate mechanical forces that activate mechanosensitive neurons and the sensation of nasal obstruction, or even promote epithelial cell damage and apoptosis, which may worsen nasal inflammation. This full hypothesis remains to be tested.

Peripheral Cutaneous Temperature Exposures
Cold water immersion of one upper limb leads to unilateral nasal airflow obstruction in normal nonrhinitic subjects. Both the afferent and efferent arms of the reflex were limited to the chilled side (68). Chilling one foot in water also increases nasal airflow resistance (69).

Immersion of both feet in warm water (42°C) increased the temperature of the nasal mucosa from $\sim$30°C toward core body temperature (70). Lidocaine prevented this nasal mucosal temperature rise. Topical mucosal application of phenoxybenzamine, an α-adrenergic receptor antagonist also increased the mucosal temperature. These data suggested that foot warming led to a decrease in systemic sympathetic activity, decreased norepinephrine release, default vasodilation, and so an increase in arterial blood flow through the superficial nasal vascular plexus. A transient, organ-specific parasympathetic vasodilator effect may also have been induced. Acetylcholine was not involved, although it is conceivable that small-diameter vasoactive intestinal polypeptide/NO sphenopalatine vasodilator neurons were activated.

Crutch Reflex
Five minutes of unilateral axillary pressure decreased the ipsilateral minimum nasal cross-sectional area (median change = 0.09 cm^2, $p < 0.01$) (71). This demonstrated that axillary pressure caused either a loss of sympathetic vasoconstriction in the anterior nasal valve, or increased parasympathetic tone. In either event, ipsilateral nasal airflow obstruction was induced. The contralateral nasal minimum cross-sectional area was significantly increased (median change = 0.35 cm^2, $p = 0.01$), suggesting a contralateral increase in sympathetic vasoconstriction. Systemic sympathetic effects were suggested by increases in heart rate

and diastolic blood pressure, but systolic blood pressure was unaltered. The loss of parasympathetic cholinergic inhibition of the sinoatrial node may also have contributed to the increased heart rate.

CONCLUSION

A wide variety of reflexes have nasal afferent and either nasal, tracheobronchial, or systemic efferent sympathetic or parasympathetic connections. Our understanding of these nasal afferents will improve as we apply new information about subsets of type C and Aδ nerve fibers that are characterized by specific combinations of ion channels, G protein–coupled excitatory and inhibitory autoreceptors, other sensors, and their combinations of neurotransmitters. Recognition that these afferents may be regulated by neurotropic cytokines, pollution, pollen grains, and allergic and nonallergic inflammation is important in order to determine optimal, individualized treatment plans. Unfortunately, there are few drugs for blocking these afferent nasal nerves short of surgical extirpation. Afferent nerve axon responses appear to stimulate the immediate defense mechanism of glandular secretion when induced by relatively low-intensity pain of hypertonic saline. Other sets of afferents may be activated by histamine (itch), capsaicin, and other agents, and lead to different local mucosal responses. These actions are very different in humans from the widely studied rat neurogenic inflammation mechanisms. Nasonasal cholinergic, parasympathetic reflexes also lead to glandular exocytosis and can be effectively blocked by topical anticholinergic agents. The naso-broncho-cardiac or diving reflex is an ancient one that persists in humans. It may have been retained to function at the time of birth. Systemic temperature, pressure (e.g., crutch reflex), and other stimuli can affect the nose by stimulating or suppressing autonomic sympathetic or parasympathetic tone. There is still considerable work to be done to define the neural pathways of these and the other reflexes that have been discussed.

ACKNOWLEDGMENTS

Supported by U.S. Public Health Service Awards RO1-ES015382, M01-RR13297, P50 DC000214, and Department of the Army Award W81XWH-07–1-0618.

REFERENCES

1. Kayser R. Die exacte Messung der Luftdurchgangigtreir der Nase. Arch Laryngol 1895; 3:101.
2. Stoksted P. The physiological cycle of the nose under normal and pathological conditions. Acta Otolaryngol 1952; 42:175.
3. Kennedy DW, Zinreich SJ, Kumar AJ, et al. Physiologic mucosal changes within the nose and ethmoid sinus: imaging of the nasal cycle by MRI. Laryngoscope 1988; 98:928–933.
4. Lund VJ. Nasal physiology: neurochemical receptors, nasal cycle, and ciliary action. Allergy Asthma Proc 1996; 17:179–184.
5. Hallen H, Geisler C, Haeggstrom A, et al. Variations in congestion of the nasal mucosa in man. Clin Otolaryngol Allied Sci 1996; 21:396–399.
6. Gungor A, Moinuddin R, Nelson RH, et al. Detection of the nasal cycle with acoustic rhinometry: techniques and applications. Otolaryngol Head Neck Surg 1999; 120:238–247.
7. Corey JP, Kemker BJ, Nelson R, et al. Evaluation of the nasal cavity by acoustic rhinometry in normal and allergic subjects. Otolaryngol Head Neck Surg 1997; 117: 22–28.

8. Davis SS, Eccles R. Nasal congestion: mechanisms, measurement and medications. Core information for the clinician. Clin Otolaryngol Allied Sci 2004; 29:659–666.
9. Jawad SS, Eccles R. Effect of pseudoephedrine on nasal airflow in patients with nasal congestion associated with common cold. Rhinology 1998; 36:73–76.
10. Lang C, Grutzenmacher S, Mlynski B, et al. Investigating the nasal cycle using endoscopy, rhinoresistometry, and acoustic rhinometry. Laryngoscope 2003; 113:284–289.
11. Qian W, Sabo R, Ohm M, et al. Nasal nitric oxide and the nasal cycle. Laryngoscope 2001; 111:1603–1607.
12. Widdicombe J. The physiology of the nose. Clin Chest Med 1986; 7:159–170
13. Malekzadeh S, Druce HM, Baraniuk JN. Neuroregulation of mucosal vasculature. In: Busse WW, Holgate ST, eds. Asthma & Rhintis, 2nd ed. Oxford, UK: Blackwell Science, 2000:945–960.
14. Eccles R. A role for the nasal cycle in respiratory defense. Eur Respir J 1996; 9:371–376.
15. Soane RJ, Carney AS, Jones NS, et al. The effect of the nasal cycle on mucociliary clearance. Clin Otolaryngol Allied Sci 2001; 26:9–15.
16. Eccles RB. The nasal cycle in respiratory defence. Acta Otorhinolaryngol Belg 2000; 54:281–286.
17. Galioto G, Mevio E, Galioto P, et al. Modifications of the nasal cycle in patients with hypothalamic disorders: Kallmann's syndrome. Ann Otol Rhinol Laryngol 1991; 100:559–562.
18. Eccles R, Lee RL. The influence of the hypothalamus on the sympathetic innervation of the nasal vasculature of the cat. Acta Otolaryngol (Stockh) 1981; 91:127–134.
19. Atanasov AT, Dimov PD. Nasal and sleep cycle—possible synchronization during night sleep. Med Hypotheses 2003; 61:275–257.
20. Kern EB. Symposium. ENT for nonspecialists. The nose: structure and function. Postgrad Med 1975; 57:101–103.
21. Baraniuk JN, Ali M, Yuta A, et al. Hypertonic saline nasal provocation stimulates nociceptive nerves, substance P release, and glandular mucous exocytosis in normal humans. Am J Respir Crit Care Med 1999; 160:655–662.
22. Baraniuk JN, Ali M, Naranch K. Hypertonic saline nasal provocation and acoustic rhinometry. Clin Exp Allergy 2002; 32:543–550.
23. Konno A, Togawa K. Role of the vidian nerve in nasal allergy. Ann Otol Rhinol Laryngol 1979; 88:259–263.
24. Mossiman BL, White MV, Hohman RJ, et al. Substance P, calcitonin-gene related peptide, and vasoactive intestinal peptide increase in nasal secretions after allergen challenge in atopic patients. J Allergy Clin Immunol 1993; 92:95–104.
25. Andrew D, Craig AD. Spinothalamic lamina 1 neurons selectively sensitive to histamine: a central neural pathway for itch. Nat Neurosci 2001; 4:72–77.
26. Raphael G D, Haupstein-Raphael M, Kaliner MA. Gustatory rhinitis: a syndrome of food-induced rhinorrhea. J Allergy Clin Immunol 1983; 83:110–115.
27. Blom HM, van Rijwijk JB, Garrelds IM, et al. Intranasal capsaicin is efficacious in non-allergic, non-infectious perennial rhinitis. Clin Exp Allergy 1997; 27:796–801.
28. Silvers WS. The skier's nose: a model of cold-induced rhinorrhea. Ann Allergy 1991; 67:32–36.
29. Stjarne P, Lundblad L, Lundberg JM, et al. Capsaicin and nicotine sensitive afferent neurones and nasal secretion in healthy human volunteers and in patients with vasomotor rhinitis. Br J Pharmacol 1989; 96:693–701.
30. Landis BN, Beghetti M, Morel DR, et al. Somato-sympathetic vasoconstriction to intranasal fluid administration with consecutive decrease in nasal nitric oxide. Acta Physiol Scand 2003; 177:507–515.
31. Silkoff PE, Roth Y, McClean P, et al. Nasal nitric oxide does not control basal nasal patency or acute congestion following allergen challenge in allergic rhinitis. Ann Otol Rhinol Laryngol 1999; 108(4):368–372.
32. Milicic D, Mladina R, Djanic D, et al. Influence of nasal fontanel receptors on the regulation of tracheobronchal vagal tone. Croat Med J 1998; 39:426–429.

33. Baraniuk JN, Silver PB, Kaliner MA, et al. Perennial rhinitis subjects have altered vascular, glandular, and neural responses to bradykinin nasal provocation. Int Arch Allergy Immunol 1994; 103:202–208.
34. Baraniuk JN, Silver PB, Kaliner MA, et al. Effects of ipratropium bromide on bradykinin nasal provocation in humans. Clin Exp Allergy 1994; 14:724–729.
35. Riccio MM, Proud D. Evidence that enhanced nasal reactivity to bradykinin in patients with symptomatic allergy is mediated by neural reflexes. J Allergy Clin Immunol 1996; 97:1252–1263.
36. Riccio MM, Reynolds CJ, Hay DW, et al. Effects of intranasal administration of endothelin-1 to allergic and nonallergic individuals. Am J Respir Crit Care Med 1995; 152:1757–1764.
37. Sanico AM, Koliatsos VE, Stanisz AM, et al. Neural hyperresponsiveness and nerve growth factor in allergic rhinitis. Int Arch Allergy Immunol 1999; 118:154–158.
38. Druce HM, Wright RH, Kossoff D, et al. Cholinergic nasal hyperreactivity in atopic subjects. J Allergy Clin Immunol 1985; 76:445–452.
39. Jeney EMV, Raphael GD, Meredith SD, et al. Abnormal nasal glandular secretion in recurrent sinusitis. J Allergy Clin Immunol 1990; 86:10–18.
40. Sheahan P, Walsh RM, Walsh MA, et al. Induction of nasal hyper-responsiveness by allergen challenge in allergic rhinitis: the role of afferent and efferent nerves. Clin Exp Allergy 2005; 35:45–51.
41. Sanico AM, Philip G, Proud D, et al. Comparison of nasal mucosal responsiveness to neuronal stimulation in non-allergic and allergic rhinitis: effects of capsaicin nasal challenge. Clin Exp Allergy 1998; 28:92–100.
42. Numata T, Konno A, Terada N, et al. Role of vascular reflex in nasal mucosal swelling in nasal allergy. Laryngoscope 2000; 110:297–302.
43. Dahl R, Mygind N. Mechanisms of airflow limitation in the nose and lungs. Clin Exp Allergy 1998; 28(suppl 2):17–25.
44. Meggs WJ. RADS and RUDS—the toxic induction of asthma and rhinitis. J Toxicol Clin Toxicol 1994; 32(5):487–501.
45. Kratschmer F. On reflexes from the nasal mucous membrane on respiration and circulation. Respir Physiol 2001; 127:93–104.
46. Schaller B. Trigemino-cardiac reflex during transsphenoidal surgery for pituitary adenomas. Clin Neurol Neurosurg 2005; 107:468–474.
47. Patow CA, Kaliner M. Nasal and cardiopulmonary reflexes. Ear Nose Throat J 1984; 63:78.
48. Betlejewski S, Betlejewski A, Burduk D, et al. Nasal-cardiac reflex. Otolaryngol Pol 2003; 57:613–618.
49. James JEA, Daly MdeB. Nasal reflexes. Proc R Soc Med 1969; 1287–1293.
50. Koskela HO, Koskela AK, Tukiaineu HO. Bronchoconstriction due to cold weather in COPD. The roles of direct airway effects and cutaneous reflex mechanisms. Chest 1996; 110:632–636.
51. Johans WT, Melville GN, Ulmer WT. The effect of facial cold stimulation on airway conductance in healthy man. Can J Physiol Pharmacol 1969; 47:453–457.
52. Kawakami T, Natelson BH, DuBois AB. Cardiovascular effects of face immersion and factors affecting diving reflex in man. J Appl Physiol 1967; 23:964–970.
53. Leung AKC, Robson WLM. Sneezing. J Otolaryngol 1994; 23:125–129.
54. Hanes B, Mygind N. How often do normal persons sneeze and blow the nose? Rhinology 2002; 40:10–12.
55. Baydin A, Nural MS, Guven H, et al. Acute aortic dissection provoked by sneeze: a case report. Emerg Med J 2005; 22:756–757.
56. Gonzalez F, Cal V, Elhendi W. Orbital emphysema after sneezing. Ophthal Plast Reconstr Surg 2005; 21:309–311.
57. Rottger C, Trittmacher S, Gerriets T, et al. Sinus thrombosis after a jump from a small rock and a sneezing attack: minor endothelial trauma as a precipitating factor for cerebral venous thrombosis? Headache 2004; 44:812–815.
58. Bhatia MS, Khandpal M, Srivastava S, et al. Intractable psychogenic sneezing: two case reports. Indian Pediatr 2004; 41:503–505.

59. Lin TJ, Maccia CA, Turnier CG. Psychogenic intractable sneezing: case reports and a review of treatment options. Ann Allergy Asthma Immunol 2003; 91:575–578.
60. Gopalan P, Browning ST. Intractable paroxysmal sneezing. J Laryngol Otol 2002; 116:958–959.
61. Widdicombe JG. Reflexes from the upper respiratory tract. In: Fishman AP, Cherniak NS, Widdicombe JG, et al., eds. Handbook of Physiology. Section 3. The Respiratory System. Vol II, Control of Breathing, Part 1. Washington, DC: American Physiological Society, 1986:363–394.
62. Garcia-Moreno JM. Sneezing. A review of its causation and pathophysiology. Rev Neurol 2005; 41:615–621.
63. Seijo-Martinez M, Varela-Freijanes A, Grandes J, et al. Sneeze-related area in the medulla: localisation of the human sneezing centre? J Neurol Neurosurg Psychiatry 2006; 77(4):559–561.
64. Forsyth RD, Cole P, Shephard RJ. Exercise and nasal patency. J Appl Physiol 1983; 55:860–865.
65. Syabbalo NC, Bundgaard A, Widdicombe JG. Effects of exercise on nasal airflow resistance in healthy subjects and in patients with asthma and rhinitis. Bull Eur Physiolpathol Respir 1985; 21:507–513.
66. Olson LC, Strohl KP. The response of the nasal airway to exercise. Am Rev Respir Dis 1987; 135:356–359.
67. Singh V, Chowdhary R, Chowdhary N. Does nasal breathing cause frictional trauma in allergic rhinitis? J Assoc Physicians India 2000; 48:501–504.
68. Wilde AD. The effect of cold water immersion on the nasal mucosa. Clin Otolaryngol Allied Sci 1999; 24:411–413.
69. Mygind N. Non-immunological factors. In: Mygind N, ed. Nasal Allergy. Oxford, UK: Blackwell Scientific, 1978:140–154.
70. Assanasen P, Baroody FM, Haney L, et al. Elevation of the nasal mucosal surface temperature after warming of the feet occurs via a neural reflex. Acta Otolaryngol 2003; 123:627–636.
71. Wilde AD, Jones AS. The nasal response to axillary pressure. Clin Otolaryngol Allied Sci 1996; 21:442–444.

15 Olfactory Toxicity in Humans and Experimental Animals

Pamela Dalton

Monell Chemical Senses Center, Philadelphia, Pennsylvania, U.S.A.

INTRODUCTION

In both humans and other animals, the nose is an important sensory organ, containing receptors capable of detecting a wide variety of airborne chemicals through the perception of odor. Among humans, impairment of our olfactory function can have serious consequences for the detection of many volatile warning signals (e.g., smoke, spoiled food, and gas leaks) and can impact food choice, nutritional status, social relationships, and many other issues related to quality of life.

In addition to the foregoing, however, the nose works as a filter, humidifier, and thermoregulator of inspired air. To accomplish this it uses multiple mechanisms for removing volatiles and particulates from inspired airstreams—through mucociliary clearance, vascular uptake and diffusion, and extensive metabolic capacity to remove and detoxify soluble particulates and vapors. Both the "scrubbing" ability of the nose and the chemosensory reflexes and responses to airborne chemicals initiated in the nose are the primary means of protecting the lower airways from toxicants. Impairments of these functional chemo-detecting systems render individuals susceptible to a wide range of health impacts. Yet, in the service of protecting the lower airways, the peripheral structures of the olfactory system can be exposed to numerous xenobiotics that have the potential to cause olfactory dysfunction.

In this chapter, the goal is to clarify the importance and functionality of the olfactory system in humans as well as to summarize the latest knowledge about the adverse effects associated with exposure to occupational and environmental toxicants on this sensory system. Because the majority of data on olfactory toxicity from occupational and environmental toxicants derives from subchronic and chronic exposure studies in animals, the evidence from those studies will be reviewed, but the chapter will primarily focus on documented adverse effects of exposure in humans. The chapter also discusses the mechanisms underlying olfactory toxicity from chemicals and particulates as well as the anatomical and metabolic factors that complicate the ability to understand the risks to the human olfactory system from data derived in animal studies.

OLFACTORY SENSORY INNERVATION

The nasal cavity is innervated by the olfactory nerve (CN I) whose primary function is odorant detection. The olfactory epithelium contains cells of three types: olfactory receptor neurons (ORNs), their precursors (basal cells), and sustentacular cells (serving glia-like, supportive functions). The receptor cells (ORNs) are located beneath a watery, mucus layer in this epithelium; on one end of each receptor, projections of the hair-like, olfactory cilia extend down into the watery layer covering the membrane. The receptor sites for odorant molecules are on the cilia and as such, the cilia are the structures involved in the initial stages of olfactory signal transduction (1). Odor information is transmitted via the bundles of axons that form the olfactory nerve (CN I), which extend from the olfactory receptor cells in the olfactory epithelium through the cribriform plate to synapse, unbranched, within the olfactory bulb, a small structure in the base of the forebrain where receptor input is integrated. The potential for some metals to reach the brain via the olfactory nerve can produce impairment in olfactory ability via damage to central olfactory structures such as the olfactory bulb (2).

One of the primary differences in the nasal passages of humans and other animals is the percentage of the nasal cavity that is covered by olfactory epithelium. In humans, the olfactory mucosa occupies approximately 3% of the nasal cavity (3), while in rats (the most common inhalation toxicology model) the olfactory epithelium comprises nearly 50% of the nasal cavity (4). Mice, rabbits, and dogs are much closer to rats than humans or monkeys in the relative amount of OE within the nasal passages.

The olfactory neuroepithelium (OE) in humans is located high in the superior region of the nasal vault, although recent studies have indicated a more extensive distribution, extending to the anterior middle turbinate and the body of the middle turbinate itself (5). Because of this relatively protected location of the OE in humans, less than 15% of the air inhaled through the nose actually reaches the olfactory epithelium during a normal breath (6–10). In contrast, the rodent OE receives a far greater concentration of inhaled pollutants than does the human OE. However, because of the more complex nasal structure, narrower channels, and lower ventilation rates in the rat, airborne chemicals are more efficiently removed by the nasal airway in rats than in humans, thus resulting in significant differences in dosimetry in the posterior and superior regions of the nasal cavity between the two species.

OLFACTORY DYSFUNCTION

Because the function of the nasal chemosensory systems is to detect ambient chemicals, the olfactory sensory neurons must constantly interact with chemical stimuli. Although relatively protected by virtue of their location in the nasal passages, olfactory receptors are located on specialized neurons that extend into the nasal cavity and are therefore uniquely and consistently exposed to the external environment. Numerous animal studies have shown that OSNs are continuously replaced throughout the adult vertebrate lifespan, even in healthy animals housed in clean environments (11). Halpern (12) has suggested that the cost of this continuous turnover implies that normal olfactory function is inherently damaging to the neurons. Perhaps not surprisingly, then, acute exposure to high levels of pollutants or chronic, lower level exposure might exceed the regenerative capacity of the olfactory system and lead to dysfunction. For example, damage to the progenitor cells in the basal cell layer can result in loss of OSN regeneration and subsequent olfactory impairment (13,14).

Awareness of the potential for olfactory loss from occupational exposure to volatiles has been noted in the medical literature for more than 100 years (15). However, until recently, the incidence of morbid or lethal outcomes from occupational exposures, such as lung disease or cancer, likely overshadowed concerns about the impact of occupational exposure on olfactory function. With the understanding that the olfactory system may represent a sentinel sensory system for exposure and damage to lower airways, the importance of olfactory evaluations in at-risk populations has increased.

Impairment of olfactory function due to acute or chronic exposure to airborne toxicants can be temporary, long-lasting, or permanent (16,17) and of different levels of severity. While a total loss of olfactory sensitivity, *anosmia*, is a relatively rare outcome following chronic low-level occupational exposure, *hyposmia*, manifested as varying gradations of sensitivity loss to multiple odorants, is a far more common sequelae and can render an individual susceptible to environmental dangers and produce substantial decrements in their quality of life. Additionally, the distortion of odorant quality or *dysosmia* is a disturbingly common outcome following chemical damage and regeneration of the olfactory epithelium in both humans and animal models (13,18) and may be a more disconcerting experience than other forms of chemosensory dysfunction for patients (19).

INHALED AGENTS AND OLFACTORY TOXICITY

Although estimates of the prevalence of olfactory dysfunction following inhalation chemical exposure are imprecise, approximately 22% of the patients who have presented to the Monell-Jefferson Clinic with olfactory complaints/dysfunction had a history of chronic or serious acute exposure to one or more environmental chemicals (e.g., herbicides, pesticides, formaldehyde, industrial solvents or cleaning products, or wood dust) (Cowart, personal communication, 2000). Based on evidence from a wide variety of epidemiological (human) and experimental (animal) studies, it is likely that this figure represents a significant underestimate of the individuals who may incur alterations in olfactory perception following chemical exposure, but who fail to seek attention for their condition.

Unfortunately, most evidence of human olfactory dysfunction following xenobiotic exposures has accrued from retrospective assessments of accidental exposures or other circumstances in which exposures could not be quantified. Although Amoore (18) identified more than 100 airborne substances that were reported to disrupt olfactory function following either acute or chronic exposures, including organic solvents, metals, inorganic nonmetallic compounds, and dusts, much of the evidence he compiled was based on single case studies or anecdotal observations in occupational environments (18,20,21) and relied on subjective reports of olfactory function or limited olfactory testing.

Despite the paucity of systematic human data, two lines of evidence provide consistent support for the association between upper airway exposure to inhaled toxicants and impairment of olfactory function. The first body of evidence comes from assessments of workers with exposure to chemical vapors and particulates. Although few in number, studies of occupationally exposed individuals and/or populations have reported problems with their sense of smell or have exhibited decrements on certain measures of olfactory function. For example, an analysis of work history and olfactory ability among the 712,000 (20–79 years old) U.S. and Canadian respondents to the National Geographic Smell Survey revealed that factory workers of all ages reported poorer senses of smell and demonstrated objective evidence of poorer odor detection ability, although this effect was most pronounced among elderly individuals (22). Notably, among this sample, factory workers reported the highest rates of smell loss secondary to chemical exposure and head injury. More recently, a study of 50 adults aged 50 to 96 years (M = 70.4 years) found that environmental risk, defined as having worked in a wood-processing plant with common and long-term exposure to irritant or corrosive chemicals (e.g., formaldehyde, toluene) or particles (e.g., wood dust), was a significant predictor (along with age) of a decrement in olfactory sensitivity (R^2 = 0.97) (23). A comprehensive cross-sectional study of workers exposed to styrene vapor found a significant interaction between age and exposure to styrene on olfactory sensitivity to styrene (24). Additionally, decrements in performance on the University of Pennsylvania Smell Identification Test (UPSIT), a 40-item test of olfactory identification ability, have been observed among workers exposed to organic solvents (25–27) and acrylate and methacrylate vapors (28). In perhaps the most at-risk population studied to date, workers and volunteers exposed to the fumes and dust cloud at the World Trade Center site have shown significantly greater risk of olfactory dysfunction than individuals having the same occupations, but without exposure at the WTC site (Dalton, unpublished data, 2009).

Because of ethical issues concerning controlled human exposures to chemical toxicants, the most extensive and systematic body of evidence linking airborne chemical exposure to damage the olfactory system comes from animal toxicological studies. As shown in Table 1, a large number of inhalation studies in rodents have demonstrated selective and dose-dependent histopathologic alterations in the olfactory mucosa from controlled exposures to a diverse range of chemical substances (e.g., formaldehyde, acrylates, methyl bromide, chlorine, and isobutyraldehyde) (29–32). Although the precise nature and distribution of these chemically induced nasal lesions varies as a function of regional deposition of the inhaled substance and the local susceptibility of the nasal tissue, the available evidence suggests that the rat olfactory epithelium is particularly vulnerable to damage by inhaled compounds (31,33), with a variety of nonneoplastic lesions of the OE associated with chronic exposure to a variety of chemicals (34–37).

Although animal studies can highlight the potential for damage to the human olfactory system, the data must be interpreted cautiously. With a relatively much larger area of the nasal passages composed of olfactory epithelium (OE) in rodents than in humans, it is more likely that the OE will exhibit evidence of damage, but less likely that functional aspects will be compromised. For this reason, it is not surprising to find little evidence of olfactory dysfunction or loss in animals following chemical ablation of the olfactory epithelium by inhalation of compounds such as methyl bromide or zinc sulfate (82,83). Yet, for several reasons, these findings may represent an underestimation of the impact on humans from comparable exposures. First, the majority of functional olfactory assessments employed in animal studies are fairly gross evaluations, capable only of discriminating total loss from some minor level of preserved olfactory function; studies that have used more sophisticated assays have indeed found significant functional deficits and persistent quality distortions following chemical damage to the epithelium (13). Second, the relatively smaller amount of OE in humans than in rodents raises the possibility that similar levels of chemical damage may compromise human olfactory function to a much greater degree. Finally, the protective and repair processes that occur following chemical exposure may differ substantially between animals and humans, and there is evidence

TABLE 1 Agents Associated with Olfactory Toxicity in Experimental Animals

Compound, concentration	Duration	Effects on histology and function	References
Acetaldehyde, 400–5000 ppm	6 hr/5 day/wk for 1–28 mo	Degeneration, metaplasia, loss of Bowman's glands and nerve bundles, adenomas, squamous cell carcinoma	(38–41)
Acrolein, 1.7 ppm	6 hr/day 5 days	Hypertrophy, hyperplasia, erosion, ulceration, necrosis inflammation	(41)
Acrylic Acid, 5–75 ppm	6 hr/day, 5 day/wk for 13 wk	Degeneration, replacement with respiratory epithelium, inflammation, hyperplasia of Bowman's glands	(42)
Benomyl, 50–200 mg/m3	6 hr/6 day/wk	Degeneration	(43)
Bromobenzene, 25 umol/kg ip	[5 min–3 day]	Degeneration of olfactory epithelium and Bowman's glands	(44)
Cadmium, 250–500 ug/m3	5 hr/day, 5 day/wk for 20 wk	Degeneration of olfactory epithelium	(45)
Chlorine gas, 0.4–11 ppm	6 hr/day, 5 day/wk for 16 wk	Degeneration, septal perforations, intracellular deposits of eosinophilic material, mucus cell hypertrophy	(46,47)
Chloroform, 300 ppm	6 hr/day for 7 days	Degeneration of bowman's glands, cell proliferation in periosteum and bone	(48)
Chloropicrin, 8 ppm	6 hr/day for 5 days	Hypertrophy, hyperplasia, ulceration, necrosis, inflammation	(41)
Coumarin, 50 mg/kg ip	[48 hr]	Necrosis, cell loss, and basal cell metaplasia in the olfactory mucosa	(49)
Chlorthiamid, 6–50 mg/kg ip	[8 hr–7 day]	Degeneration of olfactory epithelium and Bowman's glands, replacement with respiratory epithelium fibrosis in lamina propria	(50)
Dibasic esters, 20–900 mb/m3	4 hr/day for 7–13 wk	Degeneration, sustentacular cells injured initially, cell proliferation	(51,52)
1,2-Dibromo-3-chloropropane, 5–60 ppm	6 hr, 5 day/wk for 13 wk	Degeneration, metaplasia, hyperplasia	(53)
1,2-Dibromo-ethane, 3–75 ppm	6 hr, 5 day/wk for 13 wk	Degeneration, metaplasia, hyperplasia	(53)
1,3-Dichloropropene, 30–150 ppm	6 hr/day, 5 day/wk for 6–24 mo	Degeneration and/or metaplasia	(54,55)
Dimethylamine, 10–511 ppm	6 hr/day, 5 day/wk for 6–12 mo	Degeneration, loss of nerve bundles, hypertrophy of Bowman's glands	(41,56)

(Continued)

TABLE 1 Agents Associated with Olfactory Toxicity in Experimental Animals (Continued)

Compound, concentration	Duration	Effects on histology and function	References
Ferrocene, 3–30 mg/m3	6 hr/day, 5 day/wk for 13 wk	Iron accumulation, necrotizing inflammation, metaplasia	(57)
Formaldehyde, 0.25–15 ppm	6 hr/day, 5 day for 4 mo	Decrease number of bipolar cells, increased number of basal cells, degeneration of nerve bundles, reduced odor discrimination	(58)
Furfural, 250–400 ppm	7 hr/day, 5 day/wk for 52 wk	Disorientation of sensory cells, degeneration of Bowman's glands, cyst-like structures in lamina propria	(59)
Furfural alcohol, 2–250 ppm	13 wk	Squamous and respiratory metaplasia of OE, inflammation, hyaline droplets, squamous metaplasia of ducts	(60)
Hexamethylene diisocyanate, 0.005–0.175 ppm	6 hr/day, 5 day/wk for 12 mo	Degeneration, mucus hyperplasia	(61)
B,B'-Iminodi-propionitrile, 200–400 mg/kg ip	[6 hr–56 day]	Degeneration of axon bundles, increase of glial fibrillary acidic protein	(62)
Methyl bromide, 200 ppm	4 hr/day, 4 day/wk for	Degeneration, decreased carnosine, behavioral deficits	(63)
3-Mthylfuran, 148–322 umol/1	1 hr	Degeneration, more severe in rats than hamsters	(64)
3-Methylindole, 100–400 mg/kg ip	[7–90 day]	Degeneration, fibrous adhesions, osseous remodeling, Bowman's gland hypertrophy, behavioral deficits	(65,66)
Methyl isocyanate, 10, 30 ppm	2 hr	Degeneration of the respiratory and olfactory epithelia	(67)
Napthalene, 400–1600 mg/kg ip	[24 hr]	Cytotoxicity (mice and hamsters), necrosis (rats)	(68)
Nickel subsulfide, 0.11–1.8 mg/m3	6 hr/day, 5 day/wk for 13 wk	Atrophy of OE	(69)
Nickel sulfate, 3.5–635 mg/m3	6 hr/day for 12–16 days	Atrophy of OE, degeneration	(70)
N-Nitroso-dimethylamine, 20–80 mg/kg ip	[6 hr–30 day]	Degeneration of olfactory epithelium and Bowman's glands	(71)

N-Nitroso-pyrrolidine, 30–100 mg/kg ip	[6 hr–30 day]	Degeneration of olfactory epithelium and Bowman's glands	(72)
Propylene glycol monomethyl ether acetate, 3000 ppm	2 wk	Slight to moderate degeneration of olfactory epithelium	(73)
Propylene oxide, 0–525 ppm	4 wk	Degeneration of the olfactory epithelium	(74)
Pyridine, 5–444 ppm	6 hr–4 day	Degeneration of olfactory epithelium	(75)
RP 73401, 1 mg/kg/day	1 hr–5 day	Degeneration of olfactory epithelium and Bowman's glands	(76)
Sulfur dioxide, 10–117 ppm	72 hr, or 6 hr/day for 5 days	Necrosis, edema, destruction, hyperplasia, hypertrophy	(41,77)
Sulfuryl flouride, 0–600 ppm	6 hr/day, 5 day for 2 wk	Inflammation	(78)
Tetramethoxy-silane, 1–45 ppm	6 hr/day, 5 day for 28 days	Ulceration, inflammation, and necrosis of olfactory and respiratory epithelia	(79)
Toluene, 1000 ppm	5 hr/day, 5 day for 4 wk	Transient hyposmia, inflammation OE	(80)
2,4-Toluene diisocyanate, 0.4 ppm	6 hr/day for 5 days	Ulceration, necrosis, inflammation, degeneration of OE	(41)
3-Trifluoromethyl pyridine, 0.1–329 ppm	6 hr/day for 10–90 days	Degeneration of OE, reduced Bowman's activity	(81)

that these regenerative changes themselves may have the greatest detrimental impact on chemosensory function (84).

INHALED AGENTS AND OLFACTORY TOXICITY—SYSTEMATIC STUDIES

Table 2 lists a variety of inhaled agents and the associated adverse olfactory effects in humans. This list is primarily based on single-case studies or clinical reports (18,85) and less often on epidemiological investigations with appropriately matched control groups. In contrast, here the author reviews only those effects that have been documented in epidemiological studies conducted in the workplace. While the value of single-case studies to identify potential agents of concern is acknowledged, the impoverished database on documented effects on olfactory toxicity in humans and the associated exposure concentrations complicates the interpretation of many clinical reports and highlights the need for longitudinal studies of olfactory toxicity in occupational cohorts and well-matched controls.

Metals

A number of metals have been investigated for their ability to cause olfactory dysfunction. Occupational exposure to inhalation of metal dusts or fumes can lead to loss of olfactory acuity among other adverse effects (2). **Nickel** is often considered the classic example of a compound with olfactory toxicity, ranging from relatively innocuous effects such as rhinorrhea to more detrimental outcomes such as anosmia, polypoid disease, or squamous cell carcinoma (111,112). However, exposure to nickel in occupational settings frequently occurs in combination with other putative olfactory toxicants, such as cadmium or chromium, thus complicating the ability to establish clear dose–response outcomes in humans. For this reason, a review of epidemiological studies of nickel-exposed workers specifically excluded cohorts that were not free of other known or suspected carcinogenic metals (i.e., chromium and cadmium) (112). The review provided evidence that occupational exposure to nickel dust was associated with an extraordinarily high incidence of nasal sinus cancer (which is quite rare in the general population) with a strong association between level of risk and duration of employment in refinery operations with the highest levels of airborne nickel compounds. The nasal neoplasms of nickel refinery workers generally involve the turbinates and the ethmoid sinuses, are locally aggressive, and metastasize widely, thus leading to poor prognosis. Rhinoscopic examination of refinery workers has revealed hyperplastic rhinitis of the middle turbinates, with advanced polypoid disease. Biopsies of the nasal mucosa revealed dysplastic and preneoplastic lesions and evidence of keratinization, suggesting that prolonged assault by nickel fumes leads to alterations that render the epithelium more resistant to further damage. The effect of nickel in rodents is somewhat different; two-year bioassays conducted by the National Toxicology Program (NTP) revealed distinctly different carcinogenic risks for three nickel compounds, with nickel subsulfide showing substantially greater potential for promoting tumor incidence than nickel oxide and nickel sulfate hexahydrate eliciting no carcinogenic activity whatsoever (113–115).

Olfactory effects from exposure to **cadmium** have also been documented. As early as 1948, Friberg reported that 37% of workers in an alkaline battery

TABLE 2 Agents Associated with Olfactory Toxicity in Humans

Category	Symptom	Reference
Metals		
Cadmium	Rhinorrhea, decreased mucociliary function	(85–90)
	Anosmia or hyposmia	
Cadmium compounds	Anosmia or hyposmia	(18)
	Nasal lesions/hyposmia	(91,92)
Chromate salts	Anosmia or hyposmia	(18)
Chromium and chromium salts Chromium (IV)	Anosmia or hyposmia	(2,85,93–95)
	Nasal irritation	
Iron carboxyl	Unspecified olfactory deficit	(96)
Lead	Anosmia or hyposmia	(85)
	Decreased sensitivity; no effect on odor identification	(97)
Mercury	Anosmia or hyposmia	(85)
Manganese	Anosmia or hyposmia	(98)
Nickel	Rhinitis, polyps, carcinoma, congestion, decreased mucociliary function, benign and malignant histopathologic change	(89)
	Anosmia or hyposmia	(85)
Nickel hydroxide	Anosmia or hyposmia	(18,85)
Zinc	Anosmia or hyposmia	(85)
Zinc chromate	Anosmia or hyposmia	(18,85)
Zinc sulfate	Temporary hyposmia	(18)
Welding fumes	Decreased odor ID	(99)
Nonmetals/Inorganics		
Ammonia	Anosmia or hyposmia	(18,85)
	Upper respiratory tract irritation	(94)
Carbon disulfide	Anosmia or hyposmia	(18,85)
Carbon monoxide	Anosmia or hyposmia	(18,85)
Chlorine	Anosmia or hyposmia	(18,85)
	Rhinitis	(100)
Fluorides	Anosmia or hyposmia	(85)
Hydrazine	Anosmia or hyposmia	(18,85)
Hydrogen selenide	Anosmia or hyposmia	(85)
Hydrogen sulfide	Temporary hyposmia	(18)
Methyl bromide	Dysosmia	(101)
Nitrogen oxides	Anosmia or hyposmia	(18)
Phosphorous oxychloride	Unspecified olfactory deficit	(96)
Silicone dioxide	Unspecified olfactory deficit	(96)
Sulfur dioxide	Nasal irritation, rhinorrhea, xenobiotic metabolism	(89)
	Permanent hyposmia	(18)
Sulfur oxides	Anosmia or hyposmia	(18)
Sulfuric acid	Temporary hyposmia	(18)
Organics		
Acetate	Anosmia or hyposmia	(85)
Acetone	Temporary/permanent hyposmia/anosmia	(18)
	Sensory irritation	(102)

(Continued)

TABLE 2 Agents Associated with Olfactory Toxicity in Humans (*Continued*)

Category	Symptom	Reference
Acetophenone	Anosmia or hyposmia	(18)
Acrylates	Anosmia or hyposmia	(85)
	Increase in NAL ECP	(103)
Benzene	Permanent hyposmia or anosmia; sensory irritation	(18,104)
Benzol	Unspecified olfactory deficit	(96)
Butyl acetate	Unspecified olfactory deficit	(96)
Chloromethanes	Anosmia or hyposmia	(85)
Cyclohexanone	Temporary hyposmia	(18)
Dipropylene glycol methyl ether	Nasal irritation	(105)
Ethyl acetate	Unspecified olfactory deficit	(96)
Formaldehyde	Nasal irritation, rhinorrhea, benign histopathologic change	(89)
	Temporary hyposmia	(18)
	Rhinorrhea and crusting	(106)
	Nasal irritation	(107)
	Increase in NAL total protein, albumin and eosinophils	(108)
	Nasal lesions, cancer	(101)
	Rhinnorhea, crusting, cilia loss, goblet cell hyperplasia, squamous cell metaplasia, dysplasia	(106)
	Nasal irritation	(109)
Menthol	Anosmia or hyposmia	(18)
Organophosphates	Anosmia or hyposmia	(85)
Pentachlorophenol	Anosmia or hyposmia	(85)
Petroleum	Anosmia or hyposmia	(85)
Tetrahydrofuran	Hyposmia, phantosmia	(85)
Trichloroethylene	Anosmia or hyposmia	(18)
Vinyl toluene	Nasal irritation	(105,110)
Dusts		
Cement	Unspecified olfactory deficit	(96)
Chalk	Anosmia or hyposmia	(18)
Coke	Unspecified olfactory deficit	(96)
Grain	Anosmia or hyposmia	(85)
Hardwood	Anosmia or hyposmia	(85)
Lime	Anosmia or hyposmia	(18,85)
Potash	Unspecified olfactory deficit	(96)
Tobacco	Hyposmia	Ref mgr. 1472

plant in Sweden experienced anosmia (116), a finding that was later supported by studies in Germany (117), the United Kingdom (118), Poland (92), and Italy (86), with deficits found in 27% to 65% of workers with exposures to cadmium and nickel hydroxide. Although most of the investigators attributed the olfactory deficits to cadmium, the mixed exposures found in those worksites preclude determining whether the impairment was due to cadmium, nickel, or their combination.

Chromium exposure often occurs in combination with nickel and other metals, particularly in the manufacture of steel alloys. Occupational exposure to chromium has long been associated with perforations of the nasal septum, with accompanying rhinitis, although this seldom resulted in reports of olfactory deficits. Evaluating workers who had a minimum of seven years exposure in a chromate-production factory revealed that 51% had nasal septal perforations, while 54% exhibited increased odor detection thresholds (i.e., impaired olfaction) to five different compounds (119). By comparison, however, with workers having exposure to nickel or cadmium compounds, sinonasal cancers were relatively uncommon in workers exposed to chromium, with only 22 documented cases of nasal cancer in chromate-exposed workers worldwide as of 1992 (120).

Gases and Vapors

The toxicity of vapor interactions with biological receptors has historically been proposed to occur via two primary mechanisms, physical or chemical, leading to these volatiles being classified as nonreactive or reactive, respectively. Alarie et al. evaluated 145 volatile organic chemicals, thus classified in order to determine what properties could be used to predict their respiratory sensory irritant potency (121). Although both categories of vapors caused sensory irritation in the upper respiratory tract, the irritancy of nonreactive chemicals could be estimated using a variety of physicochemical properties (e.g., solubility, lipophilicity). In contrast, the irritant potential of reactive chemicals could best be estimated using five different mechanisms of chemical reactivity, and so doing led to potency estimates that were higher than what would be predicted from physicochemical parameters alone (121). This classification has also proven useful for generating predictions of the deposition, clearance, metabolism, and toxic impact of gases in the nasal passages (122) and will be used to organize the discussion of effects that follow.

Nonreactive Volatiles

There are numerous case studies to document that acute and accidental exposures to high concentrations of many nonreactive volatile organics can have an adverse affect on nasal health and function. However, the vast majority of occupational exposures occur at far lower concentrations over much longer periods of time and the long-term impact on the health and function of the human nose is relatively unknown. Animal studies, primarily conducted on rodents, provide some evidence that chronic exposure to a number of nonreactive gases and solvents can damage the olfactory epithelium and its supporting structures when exposure occurs at sufficiently high concentrations (refer to Table 1).

Because of their ubiquitous presence in many occupational environments, the nasal-toxicity of solvents has probably been investigated more than any other class of inhaled agents. Occupational exposure to solvents or solvent mixtures appears to be associated with (*i*) some degree of impairment of olfactory sensitivity (20,123,124) and (*ii*) nasal irritation as measured using psychophysical techniques (125–127) or assays of inflammatory biomarkers (128). Although solvent effects on olfactory function have been shown to be transient and reversible following cessation of exposure (20,127), the ability to draw any firm conclusions regarding the long-term effects of solvent exposure on nasal health and function are limited by the small number of studies which have been conducted.

Reactive Volatiles

Animal studies provide ample evidence that inhalation exposure to reactive gases and vapors can cause a variety of toxic effects in all nasal regions, but the degree and the dominant site of injury will be determined by either the dose to the tissue, the cellular susceptibility, or both. Gases that are highly water soluble and highly reactive will be rapidly extracted from air streams and thus cause a predictable pattern of lesions based on sites of the highest local concentrations (129). In rodents, acute and chronic exposure to a variety of reactive compounds (e.g., diethylamine, chlorine, glutaraldehyde, formaldehyde) can induce inflammation coupled with loss of olfactory cilia, sensory cells, atrophy, or necrosis in the olfactory epithelium and Bowman's glands with ultimate replacement of OE by respiratory epithelium. Although many reactive compounds such as acrolein, furfural, dimethylamine, chlorine, and ethyl acrylate are cytotoxic to nasal mucosa and can induce hyperproliferative changes, they do not typically lead to the formation of nasal tumors in rodents (84). Thus, whether an inhaled toxicant that is capable of inducing proliferative changes in nasal tissue (with consequent olfactory loss) will lead to tumor formation is highly dependent on a number of factors, including the parent compound, its metabolism in nasal tissue, the local tissue dose, and tissue-specific differences in sensitivity.

Acute high-level exposures to a reactive and soluble gas such as formaldehyde, chlorine, or ammonia can directly damage the respiratory and sensory epithelia in the nose and produce lesions and necrosis that may lead to permanent changes in physiologic functions such as olfaction (84). Biopsies of the nasal mucosa of individuals acutely exposed to such irritants have found evidence of epithelial desquamation. The response of the respiratory epithelium to irritants is first manifest by the loss of cilia, which is then followed by degeneration of epithelial cells, cell separation, and exfoliation. Ongoing repeated exposures may lead to squamous cell metaplasia, where respiratory epithelium is replaced by squamous epithelium, which is presumed to be more resistant to different challenges. Although less is known about the changes in the olfactory epithelium following chronic irritant challenge, lesions there can range from slight loss of olfactory cilia to complete replacement of olfactory epithelium with respiratory or (following more severe damage) squamous epithelium (130). Thus, repetitive exposure to even low concentrations of chemicals and/or particulates can lead to a sequence of modifications to the nasal mucosa that begins with an inflammatory response and may culminate in total desquamation of the sensory epithelium or the development of polyps and consequent olfactory dysfunction (130).

However, there is a paucity of data from cross-sectional or longitudinal studies of exposure to reactive volatiles on human nasal toxicity. While all reactive compounds at sufficiently high doses appear to elicit intranasal sensory irritation, their ability to produce inflammation, necrosis, or metaplastic changes in the olfactory epithelium may differ dramatically from the effects seen in rodents due to anatomical influences on airflow-driven dosimetry and metabolic differences in the underlying tissue (see chaps. 5–7).

Several epidemiological studies addressing the effects of exposure to reactive gases on olfaction have produced mixed results. Schwartz et al. (28) evaluated the olfactory function of 731 workers exposed to a variety of acrylates and methacrylates using a 40-item smell identification test. Workers having the highest level of cumulative exposures showed the greatest deficit in olfactory function, although the impairment did not appear to be clinically significant. Among workers exposed to styrene vapors in the reinforced-plastics industry, Dalton et al. (24) found that, contrary to findings in the animal literature, there was no evidence of any long-term alteration in olfaction due to styrene exposure. Exposed workers did, however, exhibit a specific loss in sensitivity to styrene itself, which has been observed in other chemically exposed populations and likely represents a transient, reversible form of sensory adaptation (127,131,132).

Dusts and Particulates

In comparison with research on exposure to metals and volatiles, there have been relatively few studies evaluating the toxicity of particulates and dusts in human nasal health. Much of the evidence for adverse effects on nasal health and function in animal assays, moreover, has examined the health impact of metal dusts (i.e., nickel, cadmium) (2) or mixed exposures (i.e., wood dust and formaldehyde) (133,134), and there is considerable uncertainty as to whether the physical or chemical interactions produce the greatest impact.

Both fine and coarse particles can enter the nasal cavity and can induce nasal irritation (135). Epidemiologic studies of particulate exposure have reported associations between concentrations of ambient particulate matter and nasal effects. It should be noted that not only can the particles themselves produce irritation via physical mechanisms, but, depending upon their composition, particles can also be intrinsically toxic and/or act as carriers and may concentrate odorants such as organic acids and ammonia on their surfaces (135). Recent studies evaluating exposure to calcium carbonate dust on various assays of human nasal function have found a dose-dependent decrease in nasal patency and mucociliary clearance as well as increases in symptoms such as dryness and perceived obstruction (136).

Mixed Exposures

In populations exposed to multicomponent mixtures such as fire combustion products (137) or urban air pollution (138), the detrimental effects on olfactory function may be apparent, but cannot easily be attributed to a single agent. There is the possibility, however, that such mixed exposures may increase the potential for olfactory dysfunction, as damage to the nasal metabolic capacity can render the olfactory mucosa more susceptible to damage from other compounds (139).

MECHANISMS OF OLFACTORY TOXICITY

The mechanisms underlying deficits in olfactory function have traditionally been described as either "conductive" or "neurological" in origin. In the former case, inhaled volatiles or dusts affect olfactory function *indirectly*, through inflammation-induced modifications to the airway geometry thereby impeding airflow and odorant transport to the olfactory epithelium (OE). In the latter case, the peripheral olfactory epithelium is at *direct* risk of damage from exposure to certain classes of inhaled toxicants (e.g., reactive chemicals such as chlorine or formaldehyde), but may also be compromised by the cascade of inflammatory mediators that can follow chemical exposure as a protective response, but which can also damage the olfactory epithelium.

Neurological Defects

In the category of neurologic injury, the following types of histologic changes are observed in the olfactory epithelium following exposure to inhaled toxicants: (*i*) *degeneration,* manifested by the loss of sensory and sustentacular cells, which results in thinning of the affected mucosal surface or with more extensive lesions, the loss of Bowman's glands and nerve bundles, and necrosis of individual cells. (*ii*) *Regeneration* of the damaged epithelium from the basal cells, which, due to the highly regenerative capacity of olfactory neurons, nearly always occurs. However, regenerative epithelium is distinguished by disorganized proliferation of undifferentiated basal cells undergoing organization. (*iii*) *Repair* processes characterized by the replacement of olfactory epithelium by epithelium that resembles squamous or respiratory epithelia, with little evidence of active regeneration. (*iv*) *Inflammation,* both acute and chronic, that may be observed following either short-term or long-term insults to the mucosa, respectively. (*v*) *Keratinization* of the epithelium, a protective response of epithelia produced by structural proteins of the intermediate filament family in response to chronic inflammation, which may actually impair recovery of sensory or respiratory epithelium after injury. In addition, exposure to reactive chemicals that lead to the loss of sustentacular cells, the source of P450 biotransformation enzymes in the nasal mucosa, can render the nasal mucosa and the entire respiratory tract infinitely more susceptible to damage from a host of environmental pollutants, which would otherwise be metabolized to less toxic metabolites.

Overall, the responses of olfactory epithelium (OE) to toxic insult can include degeneration, necrosis, atrophy, hyperplasia, metaplasia, and neoplasia, with the specific type of response due to the mechanism of action of the toxicant. For example, nonspecific necrosis of cells in OE may be a result of reactive irritants such as chlorine or sulfur dioxide, while cell-specific toxicity (e.g., sustentacular or sensory cells) may occur following exposure to toxicants such as dibasic esters and methyl bromide (62). It is also important to recognize that certain changes in the OE (i.e., regeneration of OE to respiratory epithelium) may reflect an adaptation that enhances the resistance of the epithelial barrier to further toxic insult, while inadvertently reducing the sensory acuity of the olfactory system.

Physiologically, the influx of cells that respond to chemical irritant stimulation of the mucosa themselves release a host of inflammatory mediators, which directly or indirectly affect the structure or function of the neuroepithelium. Inflammatory mediators can lead to ORN damage or inhibition, as well as fibrosis, gland hyperplasia, keratinization, and edema, which interfere with axonal growth and impair successful regeneration of the neuroepithelium.

As noted above, inflammation induced by pollutant exposure can also be cytotoxic to sustentacular cells, which are the source of microsomal cytochrome P450 (CYP) proteins—the key detoxification enzymes catalyzing the biotransformation of environmental contaminants. The highest levels of CYP expression occur in the liver, which is the primary locus of first-pass clearance and metabolism of ingested compounds. However, extrahepatic tissues, particularly those that are the initial sites of exposure to contaminants, also express these enzymes (see also chap. 5). Importantly, the mammalian olfactory mucosa appears unique in this regard, having high levels of tissue-specific forms of cytochrome P450 enzymes (CYP2A13). These enzymes are produced in the sustentacular cells of the olfactory mucosa and thus, any pathological conditions (i.e., chronic rhinosinusitis) or exogenous or endogenous agents, which elicit nasal inflammation can damage or erode these cells and inhibit metabolic activity. In addition to inflammatory damage, genetic polymorphisms (140,141), tobacco smoke (142), and other exogenous agents [diallyl sulfide (143), *m*-xylene (144,145), 8-methoxypsoralen (146)] have all been shown to inhibit respiratory CYP 450 isozymes.

The degree to which these enzymes are induced or inhibited can significantly alter the toxicological impact of inhaled airborne chemicals on the respiratory and olfactory mucosa. Upon contact with these enzymes in the nasal mucosa, many toxic compounds are rapidly metabolized to a less toxic intermediate, whereas benign compounds can be metabolized to more toxic forms. Interestingly, many xenobiotic compounds become toxicants only upon metabolism to a reactive intermediate. A well-validated example of this is the induction of carcinogenic nitrosamines from tobacco smoke only when catalyzed by CYP2A13 (147,148). Although CYP2A13 has overlapping substrate specificity with several other enzymes, it has recently been shown to have the highest efficiency for metabolism of several airborne pollutants, such as naphthalene, styrene, and toluene (139). This provides compelling support for its localization and high levels of expression in the nasal mucosa—the sentinel portal of entry for these pollutants.

These metabolic processes have been speculated to underlie the seemingly disparate results from studies of olfactory acuity in smokers who are coexposed to other occupational pollutants (149,150). In one study, smokers were at higher risk of experiencing exposure-related deficits in olfactory function, for example Ref. 92, while in another study, for example Ref. 28, the adverse effects on olfactory function were found only among nonsmokers. Either directly, or through effects on inflammation, exposure to tobacco smoke could damage the sustentacular cells. If enzymatic biotransformation metabolizes a pollutant to a more toxic form, then inhibition of this process will protect nasal function from exposure to the toxic metabolite. Alternatively, however, if the parent molecule is a toxicant, inhibition of metabolic activity could reduce detoxification and result in more damage to the olfactory mucosa.

The nature of the changes in olfactory epithelium and alterations in olfactory function will depend on the relationship between the timing of the assessment and the stage of chemical exposure. For example, during the early days or weeks of exposure to a chemical, one might find evidence of acute inflammation such as elevations in IL-6 in nasal lavage fluid (NLF), while histological examination of tissue biopsied from the high middle turbinate might show early evidence of loss of sensory cells. Such early stage changes will be correlated with self-reports and observations of rhinitis symptoms and irritation of the nasal passages. At a sensory level, however, there is likely scant evidence of olfactory alterations during this stage. After weeks of chronic exposure, however, when self-reports of acute irritation often appear to decrease, histological evidence of repair or evidence of protective processes such as erosion in the sensory epithelium can be noted (151,152), which could signal the beginning of olfactory deficits. At either stage, however, obstructive changes due to increased vascularization or inflammation-induced edema might compromise olfactory perception by altering airflow and the transport of odorant molecules to the olfactory epithelium, as described below.

Conductive Deficits in Odorant Transport

Intermittent or continuous obstruction of airflow through the nasal cavities is a frequently reported symptom among individuals exposed to airborne irritants. The early phase of the inflammatory response to an irritant chemical elicits the expression of chemokines and cytokines, which subsequently leads to the recruitment of inflammatory cells and increased glandular secretions (153). Along with irritant-induced increases in nasal microcirculation, these factors promote mucosal swelling and alteration of the internal geometry of the nasal passages. Even without treatment, the degree of nasal obstruction (and therefore impairment of olfactory sensitivity) will often fluctuate spontaneously due perhaps to physical, environmental, or internal factors.

Reductions in nasal airflow and patency stemming from mucosal inflammation have frequently been implicated as a causal factor in chemosensory loss following exposure to chemicals or irritants. However, the nonuniformity in the relationship between nasal patency and chemosensory acuity has led to efforts to determine the precise parameters of congestive disease that produce sensitivity impairments. While it is widely accepted that shape and volume of the nasal cavity can influence olfactory function (154,155), obstruction or deviations in certain regions of the nasal passages appear to have more impact on olfactory function in other regions (156,157), which may account for the lack of association generally reported between total nasal airway resistance measured by rhinomanometry and olfactory performance (158).

The most significant advances in understanding the degree to which even minor alterations in airway geometry can lead to olfactory deficits have come from the use of anatomically correct, 3-D, nasal cavity models. Zhao et al. (159) developed a method to rapidly transform an individual CT scan into a geometric mesh that can be used to model airflow and odorant transport. In one study, varying the nasal anatomy in two critical regions (the nasal valve and the olfactory cleft), they showed that although the total nasal airflow through the nostril and nasal resistance did not change significantly, the amount of airflow though the olfactory region could be increased by more than 700%. This increased airflow to the olfactory region was further shown to have even greater impact on the transport of airborne chemicals (odorant) to the olfactory mucosa, especially for chemicals with high mucosal solubility and diffusivity. Results of olfactory evaluations implicating the importance of local airflow at certain critical regions are consistent with these modeling results (160).

More importantly, the study by Zhao et al. revealed that the overall nasal airflow pattern was highly sensitive to even the slightest local nasal geometric configurations: Figure 1(A) shows a smooth streamline pattern for inspiratory steady laminar flow in the right nasal cavity of a normal subject. Figure 1(B) shows the streamline pattern in the left side of the same subject; eddies can be seen in the anterior part of the nose induced by a small airway constriction in the nasal valve region, along with a secondary eddy in the superior and posterior parts of the nasal cavity. If the constriction was artificially removed, the streamline pattern became smooth again, as in the right side. However, if the airway was further constricted in the nasal valve region [Fig. 1(C)], the anterior eddy

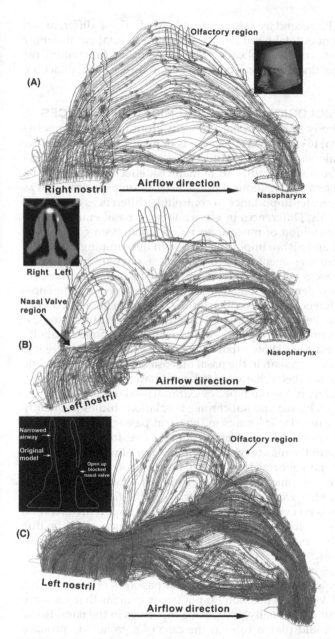

FIGURE 1 Nasal airflow patterns and odorant transport can be highly sensitive to local airway geometry. (**A**) Smooth airflow streamline pattern for inspiratory steady laminar flow in the smooth right nasal cavity. (**B**) In the left nasal cavity of the same subject, airflow eddies in the anterior part are induced by a small airway constriction in the nasal valve region (CT scan), along with a secondary eddy in the superior and posterior parts. If the constriction was artificially removed, the streamline pattern became smooth again (not shown) as in the right side. (**C**) However, if the airway was further constricted in the nasal valve region, the anterior eddy was suppressed, while the secondary eddy was enhanced. *Source*: From Ref. 161.

was suppressed, while the secondary eddy was enhanced. These different air-flow patterns, independent of total nasal airway resistance or total nasal airflow rate, are likely to have enormous implications for the sensory function of the nose: pungency (i.e., trigeminal function), patency, and in this context, olfaction.

CROSS-SPECIES PHYSIOLOGICAL AND BIOCHEMICAL DIFFERENCES

Although a substantial body of data on nasal toxicity from inhaled agents has been established in animal toxicological studies, it is of great importance to take into account the anatomic and metabolic differences among species that may preclude extrapolation from such studies to human risk effects. Perhaps in no other organ or organ system are anatomical differences among species, such as rodents and humans, of greater importance in controlling differences in regional toxicity of inhaled materials. Differences in gross anatomy, nasal epithelia, and the distribution and composition of mucous secretions have been comprehensively documented (162) and all are important factors in determining the actual inhaled dose of any airborne agent. One of the most obvious and significant physiological differences between rodents and humans is that humans are oronasal breathers, while rodents are obligate nose breathers. Species differences in clearance include differences in mucociliary flow, chemotactic attraction of macrophages, and the biochemistry of airway activation, detoxification, and tissue response (163). For example, interspecies comparison of metabolic capability in the upper respiratory tract has shown that cytochrome P450 activities are less efficient in human nasal mucosa than in the nasal mucosa of rodents (164). While some of these differences have been characterized quantitatively, many have not, and this presents uncertainty in the interspecies extrapolation of risk.

Although rodents in chronic and subchronic inhalation studies frequently develop nonneoplastic or neoplastic lesions in the nasal passages, the evidence for similar effects in humans is far from clear. As noted above, for example, findings of extensive damage to the olfactory epithelium in mice and rats following chronic exposure to styrene vapor at levels comparable to occupational exposures prompted a human epidemiological investigation of olfactory function in a group of reinforced-plastics workers (24). Contrary to the findings from the animal literature, there was no evidence of any dose-related long-term alteration in olfactory function due to styrene exposure. This discrepancy could be due to a number of factors. For example, the cross-species differences in the distribution and location of olfactory epithelium and the airflow and deposition patterns in the nasal passages can greatly alter the dose at any site (see chaps. 6–8). Thus, inhaled styrene vapor may deposit in the nasal passages of humans, but unlike in rodents, high levels of deposition do not occur in areas that subserve olfaction. Alternatively (or additionally), metabolic capacities in the nasal tissue of rodents and humans could play a role. In the case of styrene, the primary metabolic pathway is the oxidation by cytochromes P450 to two enantiomeric forms of styrene oxide (165). When this metabolic process is prevented by pre-exposure to the cytochrome P450 enzyme inhibitor (5-phenyl-1-pentyne), the development of olfactory lesions in rodents following exposure to styrene does not occur (166). This strongly suggests that the lesions found in rodent olfactory tissue are induced by the primary metabolite of styrene, styrene oxide, and not by exposure to styrene per se.

Evidence that this metabolite may not be present in human nasal epithelium exposed to styrene comes from an investigation that compared metabolic activity of styrene in vitro in rat, mouse, and human nasal respiratory tissue and found important differences in the metabolic activity of styrene across these species (167). Specifically, rat and mouse nasal respiratory fractions were found to contain high concentrations of the two cytochrome P450 isoforms necessary for the conversion of styrene into styrene oxide, whereas human nasal fractions did not. Species differences in the nasal metabolism of other chemicals that produce rat nasal tumors have been recently reported as well (168), suggesting that the relevance of nasal tumorigenesis studies in rodents for human risk prediction may need to be evaluated on a chemical-by-chemical basis. Both anatomical and metabolic differences could explain differences in the toxic effects of styrene on the olfactory epithelium in humans and rodents.

SUMMARY

The toxicity of inhaled chemicals, particulates, and dusts on olfactory function has historically been neglected in occupational and environmental health fields. Certainly, areas of the human nose that subserve olfaction are refractory to easy examination during routine physical examinations; endoscopic inspections capable of visualizing the OE require both special skills and equipment. Furthermore, evaluations of olfactory function have also been restricted to specialized clinics or research laboratories and are typically not routinely evaluated unless a problem is identified. Due to the often gradual nature of sensory loss, moreover, exposed individuals may not recognize their impairment until the damage has progressed to an extreme degree.

Ethical issues that preclude experimental studies of nasal toxicity from inhaled chemicals in humans necessitate reliance on data from controlled animal assays. However, significant differences in nasal anatomy and metabolism lend uncertainty to our understanding of the relevance of nasal effects in rodents for human health. As Morgan has suggested, however, a unified strategy consisting of mapping of nasal lesions in rodents combined with computational modeling of airflow-driven dosimetry and knowledge of local metabolism can be used as a sound basis for extrapolation of animal data to human risk assessment (169). However, the availability of this approach should not replace evaluations of nasal health and olfactory function in at-risk populations. This is important not only to validate interspecies extrapolation, but also to develop a substantive database of human olfactory toxicity from environmental and occupational inhalants, which can be used to promote better surveillance among occupational medicine practitioners and heightened attention to protection from adverse effects on olfactory function in the field of toxicology and industrial hygiene.

REFERENCES

1. Lowe G, Gold GH. The spatial distributions of odorant sensitivity and odorant-induced currents in salamander olfactory receptor cells. J Physiol 1991; 442:147–168.
2. Sunderman FW Jr. Nasal toxicity, carcinogenicity, and olfactory uptake of metals. Ann Clin Lab Sci 2001; 31(1):3–24.
3. Sorokin SP. The respiratory system. In: Weiss L. ed., Cell and Tissue Biology: A Textbook of Histology 6th ed. Baltimore: Urban & Schwarzenberg, 1988:753–814.

4. Gross EA, Swenberg JA, Fields S. Comparative morphometry of the nasal cavity in rats and mice. J Anatomy 1982; 135:83–88.
5. Clerico DM, To WC, Lanza DC. Anatomy of the human nasal passages. In: Doty RL, ed. Handbook of Olfaction and Gustation, 2nd ed. New York, NY: Marcel Dekker, Inc., 2003:1–16.
6. Stuiver M. Biophysics of the Sense of Smell. Groningen, The Netherlands: Rijks University, 1958.
7. Hahn I, Scherer PW, Mozell MM. Velocity profiles measured for airflow through a large-scale model of the human nasal cavity. J Appl Physiol 1993; 75:2273–2287.
8. Keyhani K, Scherer PW, Mozell MM. Numerical simulation of airflow in the human nasal cavity. J Biomech Eng 1995; 117(4):429–441.
9. Subramaniam RP, Richardson RB, Morgan KT, et al. Computational fluid dynamics simulations of inspiratory airflow in the human nose and nasopharynx. Inhal Toxicol 1999; 10:91–120.
10. Kelly JT, Prasad AK, Wexler AS. Detailed flow patterns in the nasal cavity. J Appl Physiol 2000; 89:323–337.
11. Loo AT, Youngentob SL, Kent PF, et al. The aging olfactory epithelium: neurogenesis, response to damage, and odorant-induced activity. Dev Neurosci 1996; 14(7/8):881–900.
12. Halpern BP. Environmental factors affecting chemoreceptors: an overview. Environ Health Perspect 1982; 44:101–105.
13. Schwob JE, Youngentob SL, Mezza RC. Reconstitution of the rat olfactory epithelium after methyl bromide-induced lesion. J Comp Neurol 1995; 359(1):15–37.
14. Schwob JE, Youngentob SL, Ring G, et al. Reinnervation of the rat olfactory bulb after methyl bromide-induced lesion: timing and extent of reinnervation. J Comp Neurol 1999; 412:439–457.
15. Mackenzie M. A manual of diseases of the throat and nose: Vol. II: Diseases of the oesophagus, nose, and naso-pharynx. New York, NY: Wood, 1884.
16. Doty RL, Bartoshuk LM, Snow JB Jr. Causes of olfactory and gustatory disorders. In: Getchell TV, Doty RL, Bartoshuk LM, Snow JB Jr., eds. Smell and Taste in Health and Disease. New York, NY: Raven Press, 1991:449–462.
17. Shusterman D. Toxicology of nasal irritants. Curr Allergy Asthma Rep 2003; 3:258–265.
18. Amoore JE. Effects of chemical exposure on olfaction in humans. In: Barrow CS, ed. Toxicology of the Nasal Passages. Washington, D.C.: Hemisphere Publishing Corporation, 1986:155–190.
19. Cowart BJ, Rawson NE. Olfaction. In: Goldstein EB, ed. The Blackwell Handbook of Perception. Oxford, U.K.: Blackwell Publishers Ltd., 2001:567–600.
20. Emmett EA. Parosmia and hyposmia induced by solvent exposure. Br J Ind Med 1976; 33:196–198.
21. Prudhomme JC, Shusterman D, Blanc PD. Acute-onset persistent olfactory deficit resulting from multiple overexposures to ammonia vapor at work. J Am Board Fam Pract 1998; 11(1):66–69.
22. Corwin J, Loury M, Gilbert AN. Workplace, age and sex as mediators of olfactory function: data from the National Geographic Smell Survey. J Gerontol B Psychol Sci Soc Sci 1995; 50B(4):179–186.
23. Elsner RJ. Environment and medication use influence olfactory abilities of older adults. J Nutr Health Aging 2001; 5(1):5–10.
24. Dalton P, Lees PSJ, Cowart BJ, et al. Olfactory function in workers exposed to styrene in the reinforced-plastics industry. Am J Ind Med 2003; 44:1–11.
25. Sandmark B, Broms I, Löfgren L, et al. Olfactory function in painters exposed to organic solvents. Scand J Work Environ Health 1989; 15:60–63.
26. Schwartz BS, Ford DP, Bolla KI, et al. Solvent-associated decrements in olfactory function in paint manufacturing workers. Am J Ind Med 1990; 18:697–706.
27. Schwartz BS, Ford DP, Bolla KI, et al. Solvent-associated olfactory dysfunction: not a predictor of deficits in learning and memory. Am J Psychiatry 1991; 148(6):751–756.

28. Schwartz BS, Doty RL, Monroe C, et al. Olfactory function in chemical workers exposed to acrylate and methacrylate vapors. Am J Public Health 1989; 79(5):613–618.

29. Abdo KM, Haseman JK, Nyska A. Isobutyraldehyde administered by inhalation (whole body exposure) for up to thirteen weeks or two years was a respiratory tract toxicant but was not carcinogenic in F344/N rats and B6C3F1 mice. Toxicol Sci 1998; 42(2):136–151.

30. Hurtt ME, Thomas DA, Working PK, et al. Degeneration and regeneration of the olfactory epithelium following inhalation exposure to methyl bromide: pathology, cell kinetics, and olfactory function1. Toxicol Appl Pharmacol 1988; 94: 311–328.

31. Jiang XZ, Buckley LA, Morgan KT. Pathology of toxic responses to the RD^{50} concentration of chlorine gas in the nasal passages of rats and mice. Toxicol Appl Pharmacol 1983; 71:225–236.

32. Monticello TM, Miller FJ, Morgan KT. Regional increases in rat nasal epithelial cell proliferation following acute and subchronic inhalation of formaldehyde. Toxicol Appl Pharmacol 1991; 111:409–421.

33. Genter MB, Deamer-Melia NJ, Wermore BA, et al. Herbicides and olfactory/neurotoxicity responses. Rev Toxicol 1998; 2:93–112.

34. Feron VJ, Woutersen RA, Spit BJ. Pathology of chronic nasal toxic responses including cancer. In: Barrow CS, ed. Toxicology of the Nasal Passages. Washington, D.C.: Hemisphere Publishing Corporation, 1986:67—89.

35. Ekblom A, Flock A, Hansson P, et al. Ultrastructural and electrophysiological changes in the olfactory epithelium following exposure to organic solvents. Acta Otolaryngol 1984; 98:351–361.

36. Rose CS, Heywood PG, Costanzo RM. Olfactory impairment after chronic occupational cadmium exposure. J Occup Med 1992; 34:600–605.

37. Odkvist LM, Edling C, Hellquist H. Influence of vapors on the nasal mucosa among industry workers. Rhinology 1985; 23:121–127.

38. Appelman LM, Woutersen RA, Feron VJ. Inhalation toxicity of acetaldehyde in rats. I. Acute and subacute studies. Toxicology 1982; 23:293–307.

39. Woutersen RA, Appelman LM, Van Garderen-Hoetmer A, et al. Inhalation toxicity of acetaldehyde in rats. Toxicology 1986; 41:213–231.

40. Woutersen RA, Appelman LM, Feron VJ, et al. Inhalation toxicity of acetaldehyde in rats. II. Carcinogenicity study: interim results after 15 months. Toxicology 1984; 31:123–133.

41. Buckley LA, Jiang XZ, James RA, et al. Respiratory tract lesions induced by sensory irritants at the RD 50 concentration. Toxicol Appl Pharmacol 1984; 74: 417–429.

42. Miller RR, Ayres JA, Jersey GC, et al. Inhalation toxicity of acrylic acid. Fundam Appl Toxicol 1981; 1:271–277.

43. Warheit DB, Kelly DP, Carakostas MC, et al. A 90-day inhalation toxicity study with benomyl in rats. Fundam Appl Toxicicol 1989; 12:333–345.

44. Brittebo EB, Eriksson C, Brandt I. Activation and toxicity of bromobenzene in nasal tissues in mice. Arch Toxicol 1990; 64:54–60.

45. Sun TJ, Miller ML, Hastings L. Effects of inhalation of cadmium on the rat olfactory system: behavior and morphology. Nuerotoxicol Teratol 1996; 18:89–98.

46. Wolf DC, Morgan KT, Gross EA, et al. A 2-year Inhalation exposure of female and male B6C3F1 mice and F344 rats to chlorine gas induces lesions confined to the nose. Fundam Appl Toxicol 1995; 24(1):111–131.

47. Ibanes JD, Leininger JR, Jarabek AM, et al. Reexamination of respiratory tract responses in rats, mice, and rhesus monkeys chronically exposed to inhaled chlorine. Inhal Toxicol 1996; 8(9):859–876.

48. Mery S, Larson JL, Butterworth BE, et al. Nasal toxicity of chloroform in male F344 rats and female B6C3F(1) mice following a 1-week inhalation exposure. Toxicol Appl Pharmacol 1994; 125(2):214–227.

49. Gu J, Walker VE, Lipinskas TW, et al. Intraperitoneal administration of coumarin causes tissue-selective depletion of cytochromes P450 and cytotoxicity in the olfactory mucosa. Toxicol Appl Pharmacol 1997; 146:133–143.

50. Brittebo EB, Eriksson C, Feil V, et al. Toxicity of 2,6-dichlorothiobenzamide (chlorothiamid) and 2,6-dicholorobenzamide in the olfactory nasal mucosa of mice. Fundam Appl Toxicol 1991; 17:92–102.

51. Keenan CM, Kelly DP, Bogdanffy MS. Degeneration and recovery of rat olfactory epithelium following inhalation of dibasic esters. Fundam Appl Toxicol 1990; 15:381–393.

52. Bogdanffy MS, Frame SR. Olfactory mucosal toxicity—integration of morphological and biochemical data in mechanistic studies—dibasic esters as an example. Inhal Toxicol 1994; 6:205–219.

53. Reznik G, Stinson SF, Ward JM. Respiratory pathology in rats and mice after inhalation of 1,2-dibromo-3-chloropropane or 1,2-dibromoethane for 13 weeks. Arch Toxocol 1980; 46:233–240.

54. Stott WT, Young JT, Calhoun LL, et al. Subchronic toxicity of inhaled technical grade 1,3-dichloropropene in rats and mice. Fundam Appl Toxicol 1988; 11:207–220.

55. Lomax LG, Stott WT, Johnson KA, et al. The chronic toxicity and oncogenicity of inhaled technical-grade 1,3-dichloropropene in rats and mice. Fundam Appl Toxicol 1989; 12:418–431.

56. Buckley LA, Morgan KT, Swenberg JA, et al. The toxicity of dimethylamine in F-344 rats and B6C3P1 mice following a 1-year inhalation exposure. Fundam Appl Toxicol 1985; 5:341–352.

57. Nikula KJ, Sun JD, Barr EB, et al. 13 Week repeated inhalation exposure of F344/N rats and B6C3F1 mice to ferrocene. Fundam Appl Toxicol 1993; 21:127–139.

58. Apfelbach R, Weiler E. Sensitivity to odors in Wistar rats is reduced after low-level formaldehyde-gas exposure. Naturwissenschaften 1991; 78(5):221–223.

59. Feron VJ, Kruysse A. Effects of exposure to furfural vapour in hamsters simultaneously treated with benzo(alpha) pyreme or diethyllnitrosamine. Toxicology 1978; 11:127–144.

60. Miller RA, Mellick PW, Leach CL, et al. Nasal toxicity in B6C3F1 mice inhaling furfuryl alcohol for 2 to 13 weeks. Toxicologist 1991; 11:669a.

61. Foureman GL, Greenberg MM, Sangha GK, et al. Evaluation of nasal tract lesions in derivation of the inhalation reference concentration for hexamethylene diisocyanite. Inhal Toxicol 1999; 6(suppl):341–355.

62. Genter MB, Llorens J, O'Callaghan JP, et al. Olfactory toxicity of beta,beta'-iminodipropionitrile in the rat. J Pharmacol Exp Ther 1992; 263(3):1432–1439.

63. Hastings L, Miller ML, Minnema D, et al. Effects of methyl bromide on the rat olfactory system. Chem Senses 2002; 16:43–55.

64. Morse CC, Boyd MR, Witschi H. The effect of 3-methylfuran inhalation exposure on the rat nasal cavity. Toxicology 1984; 30:195–204.

65. Turk MA, Henk WG, Flory W. 3-Methylindole-induced nasal mucosa damage in mice. Vet Pathol 1987; 24:400–403.

66. Peele DB, Allison SD, Bolon B, et al. Functional deficits produced by 3 methylindole-induced olfactory mucosal damage revealed by a simple olfactory learning task. Toxicol Appl Pharmacol 1991; 107:191–202.

67. Uraih LC, Talley FA, Mitsumori K, et al. Ultrastructural changes in the nasal mucosa of Fischer 344 rats and B6C3F1 mice following an acute exposure to methyl isocyanate. Environ Health Perspect 1987; 72:77–88.

68. Plopper CG, Suverkropp C, Morin D, et al. Relationship of cytochrome P-450 activity to Clara cell cytoxicity. Histopathologic comparison of the respiratory tract of mice, rats and hamsters after parenternal administration of napthalene. J Pharmacol Exp Ther 1992; 261:353–363.

69. Dunnick JK, Elwell MR, Benson JM, et al. Lung toxicity after 13-week inhalation exposure to nickel oxide, nickel subsulfide, or nickel sulfate hexahydrate in F344/N rats and B6C3F1 mice. Fundam Appl Toxicol 1989; 12:584–594.

70. Evans JE, Miller ML, Andrings A, et al. Behavioral, histological and neurochemical effects of nickel (II) on the rat olfactory system. Toxicol Appl Pharmacol 1995; 130:209–220.
71. Ranggatabbu C, Sleight SD. Development of preneoplastic lesions in the liver and nasal epithelium of rats initiated with N-nitrosodimethylamine or N-nitrosopyrrolidine and promoted with polybrominated biphenyls. Food Chem Toxicol 1992; 30(11):921–926.
72. Ranggatabbu C, Sleight SD. Sequential study in rats of nasal and hepatic lesions induced by N-nitrosopyrrolidine. Fundam Appl Toxicol 1992; 19:147–156.
73. Miller RR, Hermann EA, Young, JT, et al. Propylene glycol monomethyl ether acetate metabolism, disposition, and short-term vapor inhalation toxicity studies. Toxicol Appl Pharmacol 1984; 75:521–530.
74. Eldridge SR, Bogdanffy MS, Jokinen MP, et al. Effects of propylene oxide on nasal epithelial cell proliferation in F344 rats. Fundam Appl Toxicol 1995; 27:25–32.
75. Nikula KJ, Novak RF, Chang IY, et al. Induction of nasal carboxylesterase in F344 rats following inhalation exposure to pyridine. Drug Metab Dispos 1995; 23(5):529–535.
76. Pino M, Valerio MG, Miller GK, et al. Toxicologic and carcinogenic effects of the type iv phosphodiesterase inhibitor RP 73401 on the nasal olfactory tissue in rats. Toxicol Pathol 1999; 27:383–394.
77. Giddens WE, Fairchild GA. Effects of sulfur dioxide on the nasal mucosa of mice. Arch Environ Health 1972; 25:166–173.
78. Eisenbrandt DL, Nitschke KD. Inhalation toxicity of sulfuryl fluoride in rats and rabbits. Fundam Appl Toxicol 1989; 12:540–557.
79. Kolesar GB, Siddiqui WH, Geil RG, et al. Subchronic inhalation toxicity of tetramethoxysilane in rats. Fundam Appl Toxicol 1989; 13:285–295.
80. Jacquot L, Pourie G, Buron G, et al. Effects of toluene inhalation exposure on olfactory functioning: behavioral and histological assessment. Toxicol Lett 2006; 165(1):57–65.
81. Gaskell BA, Hext PM, Pigott GH, et al. Olfactory and hepatic-changes following inhalation of 3-trifluoromethyl pyridine in rats. Toxicology 1988; 50(1):57–68.
82. Youngentob SL, Schwob JE, Sheehe PE, et al. Odorant threshold following methyl bromide-induced lesions of the olfactory epithelium. Physiol Behav 1997; 62(6):1241–1252.
83. Slotnick B, Glover P, Bodyak N. Does intranasal application of zinc sulfate produce anosmia in the rat? Behav Neurosci 2000; 114(4):814–829.
84. Feron VJ, Arts JH, Kuper CF, et al. Health risks associated with inhaled nasal toxicants. Crit Rev Toxicol 2001; 31(3):313–347.
85. Schiffman SS, Nagle HT. Effect of environmental pollutants on taste and smell. Otolaryngol Head Neck Surg 1992; 106:693–700.
86. Mascagni P, Consonni D, Bregante G, et al. Olfactory function in workers exposed to moderate airborne cadmium levels. Neurotoxicology 2003; 24(4–5):717–724.
87. Davis JM, Dorman D. Health risk assessments of manganese—differing perspectives: session VIII summary and research needs. Neurotoxicology 1998; 19(3):488–489.
88. Friberg L. Health hazards in the manufacture of alkaline accumulators with special reference to chronic cadmium poisoning. Acta Med Scand 1950; 138(suppl 240):1–124.
89. Leopold DA. Nasal toxicity: end point of concern in humans. Inhal Toxicol 1994; 6:23–39.
90. Liu YZ, Huang JX, Luo CM, et al. Effects of cadmium on cadmium smelter workers. Scand J Work Environ Health 1985; 11:29–32.
91. Rydzewski B, Sulkowski W, Miarzynska M. Olfactory disorders induced by cadmium exposure: a clinical study. Int J Occup Med Environ Health 1998; 11(3):235–245.
92. Sulkowski WJ, Rydzewski B, Miarzynska M. Smell impairment in workers occupationally exposed to cadmium. Acta Otolaryngol 2000; 120(2):316–318.

93. National Library of Medicine Hazardous substances data bank; Chromium II chloride, chromium II dioxide, chromium II oxalate; National Library of Medicine: 95.
94. Amdur MO, Doull J, Klaassen CD. Toxicology: The Basic Science of Poisons, 4th ed. New York, NY: Pergamon Press, 1991.
95. Kitamura F, Yokoyama K, Araki S, et al. Increase of olfactory threshold in plating factory workers exposed to chromium in Korea. Ind Health 2003; 41(3): 279–285.
96. Murphy C, Doty RL, Duncan HJ. Clinical disorders of olfaction. In: Doty RL, ed. Handbook of Olfaction and Gustation, 2nd ed. New York, NY: Marcel Dekker, 2003:461–478.
97. Caruso A, Lucchini R, Toffoletto F, et al. Study of the olfactory function of a group of workers with significant lead exposure. G Ital Med Lav Ergon 2007; 29(3):460–463.
98. Mergler D, Huel G, Bowler R, et al. Nervous system dysfunction among workers with long-term exposure to manganese. Environ Res 1994; 64:151–180.
99. Antunes MB, Bowler R, Doty RL. San-Francisco/Oakland Bay Bridge welder study. Neurology 2007; 69(12):1278–1284.
100. Leroyer C, Malo J, Girard D,et al. Chronic rhinitis in workers at risk of reactive airways dysfunction syndrome due to exposure to chlorine. Occup Environ Med 1999; 56:334–338.
101. Morgan KT. Nasal dosimetry, lesion distribution, and the toxicologic pathologist: a brief review. Inhal Toxicol 1994; (suppl 6):41–57.
102. Morgott DA. Acetone. In: Clayton GD, Clayton FE, eds. Patty's Industrial Hygiene and Toxicology, 4th ed. New York, NY: John Wiley & Sons, Inc., 1993:149–281.
103. Granstrand P, Nylander-French L, Holmstrom M. Biomarkers of nasal inflammation in wood-surface coating industry workers. Am J Ind Med 1998; 33:392–399.
104. Callaghan-Rose M, Nickola TJ, Voynow JA. Airway mucus obstruction: Mucin glycoproteins, MUC gene regulation and goblet cell hyperplasia. Am J Respir Cell Mol Biol 2001; 25:533–537.
105. Clayton G, Clayton F. Patty's industrial hygiene and toxicology, 3rd ed. New York, NY: John Wiley & Sons, 1981.
106. Edling C, Hellquist H, Odkvist LM. Occupational exposure to formaldehyde and histopathological changes in the nasal mucosa. Br J Ind Med 1988; 45:761–765.
107. Baradana EJ, Montanaro A. Formaldehyde: an analysis of its respiratory, cutaneous, and immunologic effects. Ann Allergy 1991; 66:441–452.
108. Pazdrak K, Gorski P, Krakowiak A, et al. Changes in nasal lavage fluid due to formaldehyde inhalation. Int Arch Occup Environ Health 1993; 64:515–519.
109. Cain WS, See LC, Tosun T. Irritation and odor from formaldehyde: chamber studies. ASHRAE Proc 1986; 14:126–137.
110. American Conference of Governmental Industrial Hygienists (Ed.). Threshold Limit Values for Chemical Substances and Physical Agents and Biological Exposure Indices; 1993–1994; Cincinnati, OH: ACGIH, 1993.
111. Boysen M, Solberg LA, Torjussen W, et al. Histological changes, rhinoscopical findings and nickel concentration in plasma and urine in retired nickel workers. Acta Otolaryngol 1984; 97(1–2):105–115.
112. Seilkop SK, Oller AR. Respiratory cancer risks associated with low-level nickel exposure: an integrated assessment based on animal, epidemiological, and mechanistic data. Regul Toxicol Pharmacol 2003; 37(2):173–190.
113. National Toxicology Program. Toxicological and carcinogenesis studies of nickel subsulfide in F344/N rats and B6C3F1 mice; NPT TR 453; NIH Publication Series No. 96–3369: 96.
114. National Toxicology Program. Toxicological and carcinogenesis studies of nickel oxide in F344/N rats and B6C3F1 mice; NTP TR 451; NIH Publication Series No. 96–3363: 96.
115. National Toxicology Program. Toxicological and carcinogenesis studies of nickel sulfate hexahydrate—in F344/N rats and B6C3F1 mice; NTP TR 454; NIH Publication Series No. 96–3370: 96.

116. Friberg L. Proteinuria and kidney injury among workmen exposed to cadmium and nickel dust. J Ind Hyg Toxicol 1948; 30:32–36.
117. Baader EW. Chronic cadmium poisoning. Ind Med Surg 1952; 21:427–430.
118. Adams RG, Crabtree N. Anosmia in alkaline battery workers. Br J Ind Med 1961; 18:216–221.
119. Watanabe S, Fukuchi Y. Occupational impairment of the olfactory sense of chromate producing workers. J Ind Health 2000; 23:606–611.
120. Dingle AF. Nasal disease in chrome workers. Clin Otolaryngol 1992; 17:287–288.
121. Alarie Y, Schaper M, Nielsen GD, et al. Estimating the sensory irritating potency of airborne nonreactive volatile organic chemicals and their mixtures. SAR QSAR Environ Res 1996; 5(3):151–165.
122. Morris JB, Hassett DN, Blanchard KT. A physiologically based pharmacokinetic model for nasal uptake and metabolism of nonreactive vapors. Toxicol Appl Pharmacol 1993; 123:120–129.
123. Schwartz BS, Bolla KI, Ford DP, et al. Solvent-associated decrements in olfactory function in paint manufacturing workers. Am J Ind Med 1990; 18:697–706.
124. Sandmark B, Broms I, Lofgren L, et al. Olfactory function in painters exposed to organic solvents. Scand J Work Environ Health 1989; 11:703–714.
125. Dalton P, Wysocki CJ, Brody MJ, et al. Perceived odor, irritation and health symptoms following short-term exposure to acetone. Am J Ind Med 1997; 31:558–569.
126. Seeber A, Kiesswetter RR, Vangala M, et al. Combined exposure to organic solvents: an experimental approach using acetone and ethyl acetate. Appl Psychol Int Rev 1992; 41(3):281–292.
127. Åhlstrom R, Berglund B, Berglund U, et al. Impaired odor perception in tank cleaners. Scand J Work Environ Health 1986; 12:574–581.
128. Mann WJ, Muttray A, Schaefer D, et al. Exposure to 200 ppm of methanol increases the concentrations of interleukin-1beta and interleukin-8 in nasal secretions of healthy volunteers. Ann Otol Rhinol Laryngol 2002; 111(7 Pt 1):633–638.
129. Kimbell JS, Gross EA, Joyner DR, et al. Application of computational fluid dynamics to regional dosimetry of inhaled chemicals in the upper respiratory tract of the rat. Toxicol Appl Pharmacol 1993; 121(2):253–263.
130. Jiang X-Z, Morgan KT, Beauchamp RO Jr. Histopathology of acute and subacute nasal toxicity. In: Barrow CS, ed. Toxicology of the Nasal Passages. Washington, D.C.: Hemisphere Publishing Corp., 1986:51–66.
131. Wysocki CJ, Dalton P, Brody MJ, et al. Acetone odor and irritation thresholds obtained from acetone-exposed factory workers and from control (occupationally non-exposed) subjects. Am Ind Hyg Assoc J 1997; 58:704–712.
132. Smeets MA, Maute CM, Dalton P. Acute sensory irritation from exposure to isopropanol in workers and controls: objective versus subjective effects. Ann Occup Hyg 2002; (46):359–373.
133. Gosselin NH, Brunet RC, Carrier G. Comparative occupational exposures to formaldehyde released from inhaled wood product dusts versus that in vapor form. Appl Occup Environ Hyg 2003; 18(5):384–393.
134. Vaughan TL, Stewart PA, Teschke K, et al. Occupational exposure to formaldehyde and wood dust and nasopharyngeal carcinoma. Occup Environ Med 2000; 57(6):376–384.
135. Schiffman SS, Walker JM, Dalton P, et al. Potential health effects of odor from animal operations, wastewater treatment, and recycling of byproducts. J Agromedicine 2001; 7(1):7–81.
136. Riechelmann H, Rettinger G, Weschta M, et al. Effects of low-toxicity particulate matter on human nasal function. J Occup Environ Med 2003; 45(1):54–60.
137. McDermott RD, Wilson TL, Dalton PH. Chemosensory function in firefighters: a longitudinal and cross-sectional analysis. Chem Senses 2008; 33:S82–S83.
138. Hudson R, Arriola A, Martinez-Gomez M, et al. Effect of air pollution on olfactory function in residents of Mexico City. Chem Senses 2006; 31(1):79–85.

139. Fukami T, Katoh M, Yamazaki H, et al. Human cytochrome p450 2A13 efficiently metabolizes chemicals in air pollutants: naphthalene, styrene, and toluene. Chem Res Toxicol 2008; 21(3):720–725.

140. Zhang XL, Chen Y, Liu YQ, et al. Single nucleotide polymorphisms of the human CYP2A13 gene: evidence for a null allele. Drug Metab Dispos 2003; 31(9): 1081–1085.

141. Zhang XL, Su T, Zhang QY, et al. Genetic polymorphisms of the human CYP2A13 gene: identification of single-nucleotide polymorphisms and functional characterization of an Arg257Cys variant. J Pharmacol Exp Ther 2002; 302(2):416–423.

142. von Weymarn LB, Brown KM, Murphy SE. Inactivation of CYP2A6 and CYP2A13 during nicotine metabolism. J Pharmacol Exp Ther 2006; 316(1):295–303.

143. Hong JY, Smith T, Lee MJ, et al. Metabolism of carcinogenic nitrosamines by rat nasal-mucosa and the effect of diallyl sulfide. Cancer Res 1991; 51(5):1509–1514.

144. Foy JWD, Schatz RA. Inhibition of rat respiratory-tract cytochrome P-450 activity after acute low-level *m*-xylene inhalation: role in 1-nitronaphthalene toxicity. Inhal Toxicol 2004; 16(3):125–132.

145. Vaidyanathan A, Foy JWD, Schatz RA. Inhibition of rat respiratory-tract cytochrome P-450 isozymes following inhalation of *m*-xylene: possible role of metabolites. J Toxicol Environ Health A 2003; 66(12):1133–1143.

146. Visoni S, Meireles N, Monteiro L, et al. Different modes of inhibition of mouse Cyp2a5 and rat CYP2A3 by the food-derived 8-methoxypsoralen. Food Chem Toxicol 2008; 46(3):1190–1195.

147. Su T, Bao ZP, Zhang QY, et al. Human cytochrome p450 CYP2A13: predominant expression in the respiratory tract and its high efficiency metabolic activation of a tobacco-specific carcinogen, 4-(methylnitrosamino)-1-(3-pyridyl)-1-butanone. Cancer Res 2000; 60(18):5074–5079.

148. Wong HL, Zhang XL, Zhang QY, et al. Metabolic activation of the tobacco carcinogen 4-(methylnitrosamino)-(3-pyridyl)-1-butanone by cytochrome P450 2A13 in human fetal nasal microsomes. Chem Res Toxicol 2005; 18(6):913–918.

149. Ding XX, Kaminsky LS. Human extrahepatic cytochromes P450: function in xenobiotic metabolism and tissue-selective chemical toxicity in the respiratory and gastrointestinal tracts. Annu Rev Pharmacol Toxicol 2003; 43:149–173.

150. Ding XX, Spink DC, Bhama JK, et al. Metabolic activation of 2,6-dichlorobenzonitrile, an olfactory-specific toxicant, by rat, rabbit, and human cytochromes p450. Mol Pharmacol 1996; 49(6):1113–1121.

151. Wilmer JW, Woutersen RA, Appelman LM, et al. Subchronic (13-week) inhalation toxicity study of formaldehyde in male rats: 8-hour intermittent versus 8-hour continuous exposures. Toxicol Lett 1989; 47(3):287–293.

152. Schlage WK, Bulles H, Friedrichs D, et al. Cytokeratin patterns of epithelial cells of the rat nasal cavity in vivo and in vitro. Toxicol Lett 1996; 88(1–3):65–73.

153. Klimek L, Hundorf I, Delank KW, et al. Assessment of rhinological parameters for evaluating the effects of airborne irritants to the nasal epithelium. Int Arch Occup Environ Health 2002; 75(5):291–297.

154. Keyhani K, Scherer PW, Mozell MM. A numerical model of nasal odorant transport for the analysis of human olfaction. J Theor Biol 1997; 186:279–301.

155. Youngentob SL, Stern NM, Mozell M, et al. Effect of airway resistance on perceived odor intensity. Am J Otolaryngol 1986; 7:187–193.

156. Leopold DA. The relationship between nasal anatomy and human olfaction. Laryngoscope 1988; 98(11):1232–1238.

157. Hornung DE, Leopold DA. Relationship between uninasal anatomy and uninasal olfactory ability. Arch Otolaryngol Head Neck Surg 1999; 125:53–58.

158. Doty RL, Mishra A. Olfaction and its alteration by nasal obstruction, rhinitis, and rhinosinusitis. Laryngoscope 2001; 111(3):409–423.

159. Zhao K, Scherer PW, Hajiloo SA, et al. Effect of anatomy on human nasal air flow and odorant transport patterns: implications for olfaction. Chem Senses. 2004; 29:365–379.

160. Trotier D, Bensimon JL, Herman P, et al. Inflammatory obstruction of the olfactory clefts and olfactory loss in humans: a new syndrome? Chem Senses 2007; 32(3):285–292.
161. Zhao K, Dalton P. The way the wind blows: implications of modeling nasal airflow. Curr Allergy Asthma Rep 2007; 7:117–125.
162. Harkema, 1999.
163. Bogdanffy and Jarabek, 1995.
164. Dahl and Hadley, 1991.
165. Bond JA. Review of the toxicology of styrene. CRC Crit Rev Toxicol 1989; 19(3):227–249.
166. Green T. The metabolism of styrene by rat, mouse and human nasal cytochrome P-450's; CTL/R/1412; Cheshire, U.K., 99.
167. Green T, Lee R, Toghill A, et al. The toxicity of styrene to the nasal epithelium of mice and rats: studies on the mode of action and relevance to humans. Chem Biol Interact 2001; 137(2):185–202.
168. Green T, Lee R, Moore RB, et al. Acetochlor-induced rat nasal tumors: further studies on the mode of action and relevance to humans. Regul Toxicol Pharmacol 2000; 32(1):127–133.
169. Morgan KT. Nasal dosimetry, lesion distribution, and the toxicologic pathologist—a brief review. Inhal Toxicol 1994; 6:41–57.

16 Inflammatory and Epithelial Responses in the Nose and Paranasal Sinuses of Experimental Animals

James G. Wagner and Jack R. Harkema

Department of Pathobiology and Diagnostic Investigation, College of Veterinary Medicine, Michigan State University, East Lansing, Michigan, U.S.A.

INTRODUCTION

Inflammatory responses and toxicant-induced nasal lesions in laboratory animals generally exhibit characteristic, site-specific, distribution patterns (1–4). For example, formaldehyde induces lesions in rats that are essentially confined to the anterior nose, in regions lined by transitional and respiratory epithelia. In contrast, the nasal damage induced by methyl bromide is confined to the olfactory epithelium, and the transitional and respiratory epithelia are not affected. Each nasal toxicant appears to exhibit its own characteristic pattern of lesion distribution.

Site specificity of nasal lesions has been reported for each of the four principal epithelial types lining the nasal airways of rodents, including the squamous (5), transitional (6), respiratory (7), and olfactory (8) regions. This chapter provides a broad survey of inflammatory responses that are typically found in these specific epithelial populations in the nasal cavity (rhinitis), as well as a brief discussion of the paranasal sinuses (sinusitis). In addition, the special case of allergic rhinitis is discussed in the context of toxicological interaction with airborne pollutants and chemical-induced allergic rhinitis associated with workplace exposures.

Regional tissue susceptibility is generally a consequence of local metabolism (9), whereas local dosimetry may be influenced by airflow, mucous flow, blood flow, physicochemical properties of the chemical, or other factors in addition to metabolism (4). The intranasal metabolism of chlothiamid (10), acetaminophen (11), and beta–beta'-iminodipropionitrile (8) by cytochromes P-450 is a principal factor for the site-specific nature of acute and chronic nasal lesions induced by these compounds in laboratory rodents. In addition, the metabolism of dibasic esters by nasal carboxylesterase activity appears to be involved in the site-specific injury to olfactory epithelial sustentacular cells (12,13).

Reed et al. combined nasal mapping with immunohistochemistry to illustrate the distribution of the cellular antioxidant capacity in the nasal mucosa of normal laboratory rats (14). Superoxide dismutase, catalase, DT-diaphorase, glutathione reductase, and peroxidase are minimally detected in squamous and transitional epithelium, with much higher occurrence in respiratory and olfactory epithelia. Enzymes were especially concentrated in the dorsal meatus and the tips of turbinates projecting into the medial meatus. These sites are coincident with the regions of estimated high air flow in computational models of rats (3).

Analysis of nasal lavage provides an estimate in epithelial lining fluid of secreted, non-region-specific antioxidants, which are predominated by urate and ascorbic acid (approximately 10:1 ratio) (15). Nasal lavage concentrations of urate and ascorbate fluctuate in response to exposures to air pollutants, with depletion during inhalation to ozone (16) or diesel exhaust (17). Postexposure repletion of these antioxidants often exceeds preexposure levels, suggesting an adaptive response to offset the effects of future exposures (16–19). Regional sources of these secreted antioxidants have not been described; however, olfactory epithelium contains up to fourfold more ascorbate than respiratory epithelium (14). These differences may reflect active production and secretion by cells in the respiratory epithelium.

INFLAMMATORY RESPONSES IN THE SQUAMOUS EPITHELIUM

Though many inhaled nasal toxicants cause lesions in the proximal nasal passages, only a few of these irritants induce structural damage to the squamous epithelium lining the nasal vestibule or the ventral meatus in rodents. Therefore, the squamous epithelium is believed to be more resistant to injury than transitional or respiratory epithelium. A few irritants, like dimethylamine (20), glutaraldehyde (5), ammonia (21), and hydrogen chloride (22) do cause lesions to this nasal epithelium. The caustic nature of these chemicals and the airflow driven, locally high dose to this tissue are the probable reasons for these chemical-induced lesions in the squamous epithelium, rather than cellular susceptibility. Acute alterations of squamous epithelium are usually erosion or ulceration, with or without accompanying inflammation. Lesions induced after long-term exposures may include hyperplasia or hyperkeratosis. These latter changes may represent defensive or adaptive responses to the prolonged exposure to the irritant, and/or early indicators of a subsequent neoplastic response.

INFLAMMATORY RESPONSES IN THE TRANSITIONAL EPITHELIUM

The transitional epithelium is thought to be more sensitive than squamous epithelium to certain toxicants. Exposure to less irritating oxidants, like ozone (6) and chlorine gas (23), causes hyperplastic and metaplastic changes in the transitional epithelium of laboratory animals. These changes are often preceded by acute inflammation with an influx of neutrophils into the lamina propria, luminal epithelium, and airway lumen.

Ozone, the principal oxidant air pollutant in photochemical smog, has been shown to cause nasal epithelial and inflammatory responses in laboratory animals and humans (24). This irritating, oxidant gas induces epithelial hyperplasia and mucous cell metaplasia in the transitional epithelium of both rats (25) and monkeys (26) after short-term exposures. In Fischer rats (F344/N) exposed for seven days to 0.8 ppm ozone, 6 hr/day, mucous cell metaplasia in the nasal transitional epithelium (NTE) lining the maxilloturbinates, lateral wall, and lateral aspects of the nasoturbinates develops in the proximal nasal passages of these rodents (25). As much as 15% of the hyperplastic and metaplastic epithelium consisted of mucous cells compared to a normal mucous cell density of

0% to 1% in the NTE of control rats exposed to filtered air (6,25). These ozone-induced responses of hyperplasia and mucous cell metaplasia, as well as a robust influx of inflammatory cells in rat maxilloturbinates are depicted in Figure 1. The ozone-induced lesions in the F344/N resembled those previously observed in the nasal cavity of bonnet monkeys repeatedly exposed to 0.15 or 0.3 ppm ozone for 6 or 90 days (27).

The reason(s) for the rapid induction of mucous cell metaplasia in transitional epithelium after ozone exposure is not fully known. With further investigation of this rat model, it was shown that ozone-induced mucous cell metaplasia and epithelial hyperplasia could be induced with only three consecutive, 6 hr/day exposures to 0.5 ppm ozone (28). Seven days after the start of the exposures, rats exposed to ozone for three days had mucous cell metaplasia that was indistinguishable from that in rats exposed to the same concentration of ozone for seven consecutive days. Thus, once initiated the development of ozone-induced phenotypic changes within the epithelium are not dependent on additional ozone exposure. Influx of neutrophils, an increase in epithelial DNA synthesis, and an overexpression of a mucin-specific gene (MUC5AC) precede the onset of mucous cell metaplasia induced by ozone (29,30). Immunodepletion of neutrophils prior to ozone exposure demonstrated that mucous cell metaplasia is dependant in part on the influx of neutrophils in the nasal airways (31). Similar inhibition was achieved with intervention with fluticasone topical steroid, which blocked neutrophil influx by 60% and metaplastic responses by 85% (32).

Neutrophils are a potent source of inflammatory mediators that can induce mucin gene expression and epithelial metaplasia. For example, intraairway instillation of neutrophil elastase induces mucous cell metaplasia in mouse and hamster airways (33,34), and elastase inhibits mucus hypersecretion caused by ozone (35). Epithelial studies in vitro suggest an oxidative stress mechanism underlies elastase-induced expression of mucin-specific mRNA and protein expression (36,37). In addition to proteases, neutrophils can release a number of other inflammatory mediators, including TNFα, platelet-activating factor, and interleukins-1 and -6 (IL-1 and IL-6), that can induce mucus secretion and/or overexpression of mucin genes (e.g., MUC5AC) in airway epithelial cells (38). A common pathway for mucous production in airway epithelial cells may be via the epidermal growth factor receptor (EGFR) system, a protease-activated receptor with multiple ligands (39–41). Exposure of humans to ozone induces expression of EGFR and production of EGFR ligands in the nasal mucosa (42), suggesting a similar process may occur during ozone-induced mucous cell metaplasia that occurs in the rat. Furthermore, the increased EGFR response is correlated with neutrophil influx.

Neutrophil-dependence of ozone-induced mucous metaplasia suggests that other inhaled inflammatory stimuli might augment the epithelial responses to ozone. Endotoxin is a potent inflammagen derived from Gram-negative bacteria, inhalation of which provokes a rapid and robust influx of neutrophils into nasal airways and mucous cell metaplasia in respiratory epithelium of nasal septum (43). The presence of endotoxin-containing bioaerosols in the industrial workplace (44,45) and office buildings (46) is associated with nasal symptoms in workers. To determine the effects of combinations of ozone and endotoxin exposures on nasal epithelial responses, Fanucchi et al. (47) devised a regimen of three

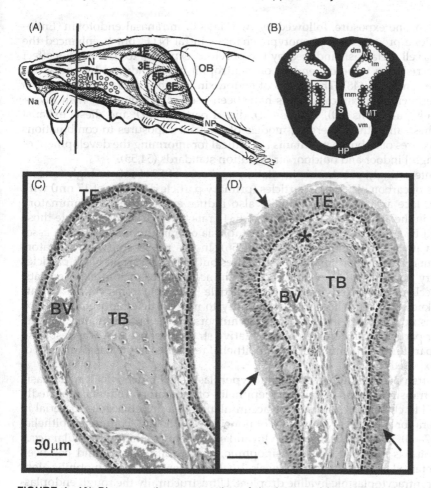

FIGURE 1 (**A**) Diagrammatic representation of the intranasal location of inflammatory and epithelial lesions (dots distributed along the maxilloturbinate, MT) in the nasal mucosa lining the lateral wall of the right passage in a laboratory rat repeatedly exposed to ozone. Verticle line through the proximal nasal passage represents the location of the transverse section of the nasal cavity illustrated in part (**B**). Dots in (**B**) represent the site-specific, bilateral location of ozone-induced mucosal lesions along the MT, N, and lateral wall (transitional epithelium lining the lateral meatus, l m) at this proximal level of the nasal passages. Rectangular box highlights the intranasal site from which the light photomicrographs in parts (**C**) and (**D**) were taken. Light photomicrographs of the MTs (**C** and **D**) are from the proximal nasal airways of rats that were exposed to either filtered air alone (0 ppm ozone) (**C**) or to a continuous exposure of 0.5 ppm ozone for 13 weeks (**D**). Turbinate tissue sections were stained with hematoxylin and eosin, and alcian blue (AB; pH 2.4). The transitional epithelium (TE) lining the maxilloturbinate in part (**D**) is markedly thickened due to epithelial hyperplasia and mucous cell metaplasia (arrows identify mucous cells with AB-stained mucosubstances). Numerous inflammatory cells are also present in the underlying lamina propria (*asterisk*) in part (**D**), but not in part (**C**). The turbinate bone (TB) is also markedly atrophic in the MT of the rat that was exposed to ozone (**D**) compared to the normal maxilloturbinate in the air-exposed animal (**C**). Dotted lines in parts (**C**) and (**D**) highlight the basal lamina separating the TE from the underlying lamina propria of the nasal mucosa. *Abbreviations*: Na, naris; N, nasoturbinate; 1–6E, ethmoturbinates; OB, olfactory bulb; NP, nasopharynx. S, septum; HP, hard palate; dm, dorsal medial meatus (airway); mm, middle meatus; vm, ventral meatus.7

days of ozone exposure, followed by two days of intranasal endotoxin. Endotoxin alone produced no phenotypic changes in rat NTE, but it enhanced the mucous cell metaplasia induced by the previous exposure to ozone. This effect was later shown to be dependent on neutrophil influx induced by endotoxin into nasal mucosa (48). The effects of endotoxin or ozone as a co-inflammagen in experimental or clinical rhinitis have been used in combination with mycotoxins (49), allergens (50), viruses (51), diesel particles (52), and diesel exhaust (53). These and other emerging models of relevant exposures to combinations and mixtures of airborne pollutants are critical for informing the development of meaningful indoor and outdoor air pollution standards (54,55).

Interestingly, subchronic inhalation exposure of F344 rats to high concentrations of carbon black nanoparticles (primary particle diameter of 17 nm) with high surface area and low solubility also induce epithelial and inflammatory lesions in the anterior nasal passages of F344 rats that markedly resemble those induced by ozone (e.g., mucous cell metaplasia of transitional epithelium associated with neutrophilic inflammation) (56). In this recent study, investigators also found that similar inhalation exposures but to larger carbon black particles (primary particle size of 70 nm) did not induce nasal lesions in exposed rats. The mechanisms underlying this nanoparticle-induced chronic rhinitis in rats are unknown and the human susceptibility to nanoparticle-induced nasal toxicity has not yet been investigated. The authors, however, speculated that the smaller particles induced a greater oxidative stress response that was responsible in part for the inflammatory and epithelial responses in the nasal airways of these exposed rodents.

Often accompanying epithelial hyperplasia and mucous cell metaplasia in the transitional and respiratory epithelia of laboratory rodents repeatedly exposed to chemical irritants is the accumulation of a proteinaceous material in the supra- or subnuclear cytoplasm of nonciliated cuboidal/columnar epithelial cells (57–60). With routine hematoxylin and eosin staining, this intracellular accumulation has a homogeneous eosinophilic, hyaline, appearance and has been often referred to as epithelial hyalinosis, hyaline degeneration, eosinophilic globules, or intracytoplasmic hyaline droplets. Ultrastructurally the rough endoplasmic reticulum of the affected epithelial cells is markedly dilated with this proteinaceous material. This epithelial change may also be associated with secreted droplets of this proteinaceous material in the nasal airway lumen and/or the accumulation of eosinophilic crystals either in the airway lumen or within the altered epithelium. This alteration is a commonly observed nonspecific epithelial change in the nasal epithelium of both mice and rats (61), but it has also been reported in the nasal transitional and respiratory epithelia of laboratory monkeys exposed to ozone (27,61). In rodents, intracytoplasmic accumulation of proteinaceous material may also occur in the subepithelial glands in the lamina and often at the junction of olfactory and respiratory epithelia in the nasal mucosa of rodents. This epithelial alteration has also been commonly observed in the sustentacular cells of nasal olfactory epithelium of normal aging mice or in younger mice exposed to certain inhaled toxicants (57,58,61–63).

Ward et al. recently reported that this intracytoplasmic proteinaceous material in the nasal epithelium of mice contains Ym1/2 chitinase proteins (64). Ym proteins (isotypes Ym1 and Ym2) are members of the chitinase-like gene family and have been identified in various murine tissues, besides the nose, including the lung, oral cavity, esophagus, glandular stomach, bile duct, gall bladder, and in peritoneal and bronchoalveolar lavage fluid associated with tissue injury and inflammation (63–67). Though the function(s) of mammalian Ym proteins are not fully known, these novel lectin-binding proteins may play important roles in the pathogenesis of allergen-induced airway inflammation and remodeling (65) and the regeneration of olfactory epithelium after toxicant-induced injury (63). More research is needed to fully understand the function and importance of these unique proteins in nasal epithelium after exposure to inhaled toxicants.

INFLAMMATORY RESPONSES IN THE RESPIRATORY EPITHELIUM

As in transitional epithelium, lesions in respiratory epithelium may be superficial or extend to the underlying lamina propria. A common superficial, and often reversible, effect of irritants on respiratory epithelium involves attenuation and/or loss of cilia along the luminal surface in the proximal nasal cavity. This effect is frequently seen in mice and rats exposed to chlorine gas (68) and is a common alteration observed in monkeys exposed to 0.15 and 0.30 ppm ozone for 6 or 90 days (8 hr/day) (27). Ciliated cell necrosis, mucous cell hyperplasia, and inflammatory cell influx were additional features in the nasal respiratory epithelium of monkeys exposed to ozone.

Another common response of respiratory epithelium, which already contains secretory cells, is an increase in the synthesis and storage of mucosubstances after exposure to such noxious stimuli such as cigarette smoke (69), allergens (70,71), and endotoxin (43). Prior to exposure, secretory cells in these populations do not necessarily have mucin-containing globules, but rather contain serous fluids with various proteins, lipids, and low-molecular-weight antioxidants. Recently, alterations in the secretory apparatus in pulmonary conducting airways have been shown to proceed as phenotypic switch by Clara cells, which decrease production of Clara cell secretory protein (CCSP), and begin to produce and secrete mucin glycoproteins (72–74). It is possible that a similar diversion of protein production in secretory cells occurs during mucous (goblet) cell metaplasia in the nose. CCSP and uteroglobin are among at least a dozen members of the secretoglobin family of disulfide-bridged dimeric proteins found throughout the body. Uteroglobin, but not CCSP, has been detected in nasal tissue. Although its function in the normal nose is unknown, nasal uteroglobin gene expression and protein production are decreased after intranasal endotoxin challenge (75), and in symptomatic allergic rhinitis (76). Although not reported, one would assume that mucin gene expression and increased mucus production was associated with this downregulation of uteroglobin. Further studies are needed to confirm this possible relationship in the nose.

INFLAMMATORY RESPONSES IN THE PARANASAL SINUSES

Like humans, the maxillary sinus of mice and rats is lined with respiratory epithelium, with distribution of ciliated and secretory cells similar to that found on the lateral meatus of the nasal cavity, and is surrounded with glands of Steno and maxillary glands (77,78). There are few detailed descriptions of inflammatory responses to toxicant exposure within rodent paranasal sinuses.

Either the inhaled or instilled materials do not easily gain access to sinus spaces, or any pathological changes that do occur are not identified and reported. Exposure to ozone for 20 months induced a dose-dependent increase in intraepithelial mucous in maxillary sinus epithelium, though at levels 30% to 50% less than responses in the lateral meatus (79). After acute cigarette smoke inhalation (1–4 days), a robust and persistent influx of neutrophils, the formation of abscesses and epithelial hyperplasia was evident in sinuses (80). Increased staining for acidic mucosubstances occurred in glands, but not in epithelium after four days of inhalation.

During human sinusitis, the gland density surrounding the maxillary sinus increases up to sixfold, while surface goblet cells decrease by 20% (81). Similar relationships have not been reported in animal models of nasal allergy. Unfortunately, a number of reports using rodent models with the terminology "chronic rhinosinusitis" or "chronic sinusitis" have misidentified the airspaces between turbinates and lateral walls as sinuses (82). Hence, inflammatory lesions in these regions that should be termed as rhinitis were reported as a sinusitis (83–85). As such these animal models bear little relevance to understanding pathological responses in the maxillary sinus in humans. Indeed some of the regions discussed in these models contain olfactory, not respiratory epithelium. Caution should therefore be used in interpreting experimental results that describe sinusitis. For the toxicologist, strategies for nasal histopathologic analyses should begin by consulting nasal diagrams (maps) of the intranasal location of epithelial cell populations generated by Mery et al. (86), and using a modified approach proposed originally by Young (77).

ALLERGIC INFLAMMATION; EXPERIMENTAL ALLERGIC RHINITIS
The study of allergic rhinitis (AR) by the toxicologist is important to understand the pathophysiology and health risk of (i) occupational exposure and airway hypersensitivity to chemicals, and (ii) exacerbation of existing allergic airways disease by environmental airborne pollutants. In general, animal models of AR are less reported and lack the diversity and understanding of experimental animal models of asthma. Most experimental AR protocols range from hours to weeks of allergen challenge, with at most 12 exposures to allergen before measuring endpoints. These brief treatment regimens are most often designed to test the efficacy of pharmaceutical agents against acute exacerbations, leaving relatively few animal studies that model chronic AR of humans.

The guinea pig and, to a lesser extent, the Brown Norway rat and various susceptible mouse strains (e.g., BALB/c) have provided most of the experimental descriptions for nasal inflammatory and physiological responses to allergic stimuli. Occlusion of nasal passages, the most common complaint in AR after acute provocation with allergen, is characterized by early and late phases of inflammation (87,88). An immediate and transient episode of itching and sneezing begins within seconds of exposure and lasts for 5 to 30 minutes. A secondary (late) phase is characterized by rhinorrhea and airway obstruction that can last for hours. Evidence from both animal and human clinical studies demonstrate that preformed mediators released from mast cells and basophils—specifically histamines, tryptase, cysteinyl leukotrienes (cysLTs), and platelet-activating factor (PAF)—promote the initial irritation and sneeze reflex. Mucus hypersecretion with airway obstruction during the secondary phase is accompanied by a progression in mucosal swelling, tissue infiltration of eosinophils and neutrophils, and the synthesis and release of prostaglandins, interleukins, and reactive oxygen species (ROS).

In the allergic guinea pig, enumerating the frequency of nasal rubbing and sneezes is a subjective but useful measure, especially for testing the immediate and early mechanisms and therapies involving histamine- and leukotriene-dependent pathways (88–90). For example, observers will count between three and six sneezes and six and 10 rubbings per minute after acute exposure (89,91). Recently, these observational data have been corroborated in part with the direct physiological measurements of airway resistance in the isolated upper respiratory tract of allergic guinea pigs (89,92). While acute tissue infiltration of eosinophils and their appearance in nasal lavage fluid accompanies airway obstruction, the associated histopathology and tissue inflammation as they relate to nasal obstruction in both early and late phases is not fully understood. This is in stark contrast to the airway remodeling and pathology that drive analogous responses in lower airways, i.e., early and late bronchoconstriction, which are well studied in mice (93,94).

Like asthma, AR is a chronic disease marked by episodic rounds of inflammation, yet few rodent AR models have been designed to examine long-term alterations and potential airway remodeling of the nasal mucosa. This limitation might easily have been filled, in part, by examining the nose from mice used in a number of well-designed, chronic experimental asthma models (95–99). Repeated challenge with allergen for weeks or months produces many of the pathological hallmarks of human asthma, including subepithelial fibrosis, smooth muscle and mucus cell hyperplasia, and epithelial exfoliation. In the few chronic experimental AR where histopathological changes are reported, some epithelial and inflammatory responses are consistent with human AR.

Using a three-month protocol of multiple intranasal ovalbumin challenges in BALB/mice, Lim et al. (100) documented time- and challenge-dependent development of subepithelial fibrosis and goblet cell hyperplasia in the proximal aspects of nasoturbinates. Immunohistochemical detection of matrix metalloproteinase and tissue inhibitors of metalloproteinase was localized to the fibrotic lesions. Transient tissue infiltration of eosinophils occurred at early (1 week), but not later timepoints (1–3 months). Similar associations of decreasing inflammatory cell recruitment with repeated allergen provocation was found in C57BL/6 mice, where airway mucosal remodeling was evident only after four to eight weeks of challenges, and eosinophil influx peaked after two weeks (101). In allergic BALB/c mice that were challenged three times a week, goblet cell hyperplasia in lateral walls occurred after five, but not two weeks, and persists through four months; by 10 weeks of multiple challenges collagen deposition was evident (102). Despite the brevity in reports on experimental chronic AR, these studies nonetheless suggest that chronic remodeling of nasal mucosa after repeated exposures is preceded by a transient inflammatory response.

Relevance to Human AR
As described above, it is not difficult to induce allergic inflammation and airway resistance in upper airways of laboratory animals. However, the

translation of any new knowledge from these experimental models to human AR is not straightforward. The distinct gross structural differences and distribution of epithelium of the rodent and human are important considerations. From a review of the literature, further examples of the limitations of histopathologic comparisons across and within human and experimental AR in animals include (i) inconsistencies in site-specific selection for evaluation, (ii) misidentification of nasal anatomy, and (iii) the use of subjective quantitative and qualitative analyses (e.g., number of goblet cells vs. amounts of stored mucosubstances).

Interspecies variability in nasal gross anatomy has been emphasized in Chapter 1 and in previous reviews (103,104). In the proximal nasal airway, the complex turbinate structures of small laboratory rodents likely provide more effective protection of the lower respiratory tract than the simple middle and inferior turbinates of the human nose. Mucosal swelling in turbinates, especially where they are in close opposition to the septum and lateral wall, can impede both airflow and mucus drainage through the nasal cavity. As such changes in airflow and nasal resistance in rodents may be more easily or rapidly achieved compared to those in humans. Most of the histopathologic analyses in both humans and rodents have been in regions populated with respiratory epithelium. The anterior portion of the middle and inferior turbinates are common sampling sites for biopsies in humans, partly because of their accessibility. In rodents, by comparison, sites of analyses are usually the nasal septum and the lateral wall, and less commonly the turbinate structures favored in human studies. The septal mucosa overlies cartilage, whereas the mucosa of turbinates overlies bone. Thus, when responses in respiratory epithelium of rodents and humans are compared, the surface epithelium may be similar, but the cellularity and vascularization of the underlying mucosa may be quite different and belie inaccurate conclusions with regard to structure/function relationships and its impact on the pathophysiology.

Most analyses of human AR have focused on mucus goblet cell enumeration, where modest increases during seasonal AR were not statistically significant (105). Similar modest changes in the epithelial hyperplasia lining the nasal septum after acute allergen challenge has been reported in BALB/c mice (106) and Brown Norway rats (71). However in the rat model, there was a profound increase in the amount of intraepithelial mucosubstances (71), suggesting that hypertrophy and hyperproduction of mucosubstances within individual cells, rather than an increase in mucous cells (hyperplasia), may underlie the hypersecretory mucosa associated with human AR. Supporting this notion are reports of secreted mucosubstances within the nasal lumen of allergic rats (71,107), which parallels the findings of Berger et al. (108), who found more actively secreting goblet cells in AR patients than in healthy controls.

Infiltrations of eosinophils, neutrophils, and mast cells are commonly reported in both experimental AR and in clinical studies (71,106,109). Eosinophils in biopsies or in nasal lavage fluid are highly correlated with most symptoms in AR patients (110,111). Nitric oxide, leukotrienes, and interleukins derived from eosinophils are potential mediators of nasal obstruction (mucosal swelling) and goblet cell secretory responses during AR. However, at least two animal studies have found no causative role for eosinophils in AR responses. Blockade of IL-5 in guinea pigs inhibits eosinophil accumulation in nasal mucosa, but mucus secretion and nasal airway obstruction are unaffected after chronic allergen exposure (112). Furthermore, in IgE-receptor–deficient mice, nasal obstruction is independent of eosinophil recruitment into nasal tissues (113). By comparison, eosinophils are strongly suggested, but not clinically proven to mediate late responses that lead to obstruction in human AR (110,111). Furthermore, eosinophil-independent pathways of airway hyperreactivity and mucus cell metaplasia have also been demonstrated in murine asthma models (114,115).

Histamine-dependent physiological responses initiated by activated mast cells are well defined in AR. Histologically, increased numbers of degranulated mast cells are detected in turbinate biopsies from patients with AR (116,117). In kinetic studies of the response to allergen provocation in human AR, investigators have reported mast cell migration from the lamina propria into nasal epithelium where degranulation occurs (118). In allergic guinea pigs by comparison, mast cell migration, but not increased numbers or degranulation, was detected in the subepithelial mucosa (119). Beyond this example, comparative descriptions of nasal mast cell histopathology are rarely reported in experimental AR models.

Chemical-Induced AR

A variety of airborne agents in the workplace have been implicated in allergic airway disease, ranging broadly from hairdresser exposures to persulfates (120,121) to the volatile organic hydrocarbon exposures associated with yacht making (122). Inhalation of low molecular weight (LMW) chemicals, which include anhydrides and isocyanates, and in particular toluene diisocyanate (TDI), is the leading cause of occupational asthma and allergic rhinitis. While TDI is well documented as a respiratory irritant (123,124), repeated exposures can also induce respiratory sensitization, involving the same T-cell and immunoglobulin responses as classic environmental allergens (e.g., pollens, pet dander). Elicitation of nasal eosinophilic inflammation and Th2-cytokine gene expression (IL-4, -5, -13) after subchronic inhalation exposure to TDI is associated with serum IgE and IgG responses in mice (125). Nasal sensitization and provocation in mice with trimellitic anhydride, a LMW chemical used to make plastics, resins, and industrial polymers, produces eosinophilic inflammation, IgE, and cytokine expression similar to TDI, but also creates mucous cell metaplasia, regenerative hyperplasia in nasal transitional epithelium, and moderate lymphoid hyperplasia in NALT (70). These lesions are identical to those elicited by nonirritant, allergens in rodents such as ovalbumin (71,107,126). Many LMW and isocyanates can induce dermal sensitivity, but not all can create allergic airways responses and AR. Farraj et al. (127) demonstrated a disparity in the ability of several isocyanate compounds to induce local (dermal) expression of Th2 cytokines and serum IgE responses, but not to elicit allergic airway reactivity. While all isocyanates they tested caused nasal epithelial cytotoxicity (epithelial necrosis and exfoliation), the structural antigenic determinants of isocyanate and anhydrides to induce specific immune responses were not determined. It is thought that LMW chemicals are too small (<10 kDa) to initiate an IgE antibody response without first conjugating to a host protein. Antibodies to several isocyanate/protein conjugates have been identified, including the conjugates with serum albumin and the matrix protein keratin (128,129).

Diesel Exhaust and AR

Exposure to diesel engine (DE) exhaust, a major component of urban air pollution, has been associated with direct effects on nasal mucosal inflammation, as well as to exacerbate allergic rhinitis. Gaseous components of DE emissions include carbon dioxide, nitrogen oxides, sulfur dioxide, and polycyclic aromatic hydrocarbons (PAHs). Elemental carbon makes up the core of DE particles, onto which gaseous components absorb, notably PAHs, sulfates, and trace metals. Inhalation of DE exhaust ($3 \text{ mg}/\text{m}^3$) for up to four weeks produces little physiological response (sneezing, hyperresponsiveness) in guinea pigs (130), but can enhance histamine-induced responses of nasal secretion, vascular permeability, and intranasal pressure (130,131). Interestingly, the inflammatory cell response of guinea pigs to DE exhaust is primarily composed of eosinophils, both in the nose (131) and lung (132). By contrast, acute pulmonary responses to diesel stimuli in mice (133), and rats (134), usually consist of neutrophils, while the nasal responses are comparatively understudied in these laboratory animals. In all species tested however, DE exhaust or particles (DEP) exposures can enhance eosinophil, IgE, or hyperreactive airway responses to an allergen provocation in experimental asthma (133,134) and rhinitis (135–137). In a series of studies in mice by Farraj et al. (138,139), enhancement of airway resistance by DEP inhalation was due primarily to obstruction in nasal flow, and dependent on neutrotrophin receptor function. They found that neurotrophins (including nerve growth factor) were increased during AR (140) and localized to submucosal mast cell and eosinophils (141). Activated eosinophils can produce and secrete neurotrophins (142), suggesting that DEP may act via eosinophils to produce neurotrophin-dependent responses in the nose. It is notable that these effects were induced by DEP and not DE exhaust that contains more gaseous components and VOCs. Because DE emissions vary widely in composition and depend on the fuel, engine, and operating conditions during the generation of DE, the identification of the specific agents responsible for modulating allergic responses requires a more systematic study (143,144).

Ozone and Allergic AR

Inhalation of ozone enhances allergic nasal responses in humans, specifically by increasing eosinophil influx and eosinophil cationic protein in nasal lavage fluid (145,146). In allergic guinea pigs, five weeks of ozone exposure dramatically increased mucosal eosinophil influx and goblet cell hypertrophy in the nasal septum (147). In acute rat studies, only the combination of ozone and allergen induced eosinophil influx and mucous cell metaplasia in nasal transitional epithelium, and mucous hyperplasia in respiratory epithelium lining the septum, lateral wall, and nasoturbinates (71). In a separate study, two days of intranasal ovalbumin challenge induced mild eosinophil accumulation and increased epithelial mucosubstances in the maxillary sinuses of Brown Norway rats (148). Acute exposure to ozone exacerbated these responses, as well inducing an intramural eosinophil influx along the length of the nasolacrimal duct. Interestingly, a significant eosinophil presence was detected as far as 4 mm distal from the duct opening to the proximal nasal cavity, or almost half the distance to the orbital socket. It is tempting, therefore, to speculate that ozone irritation of ocular tissues might contribute to the downstream inflammation found in the nasolacrimal duct. Increases in ambient ozone concentrations are associated with ocular irritation and enhancement of symptoms in allergic rhinoconjunctivitis patients (149,150). Soluble inflammatory mediators such as IL-4 and TNFα are produced in conjunctiva of OVA-induced allergic Brown Norway rats (151), and may drain into the nasolacrimal duct to mediate the recruitment of eosinophils into the mucosal tissues surrounding this duct. It is possible that any whole body exposure of laboratory rodents to gases (e.g., ozone) or aerosols (e.g., particles, ovalbumin) will induce ocular inflammatory responses that modify nasal responses, particularly in the nasolacrimal duct.

SUMMARY

In this chapter, we have presented a brief and broad survey of nasal inflammatory and epithelial responses to inhaled toxicants. In this regard, we have emphasized the specialized structure of the nose and its airway mucosa. Site-specific nasal epithelial and inflammatory responses depend both on airway dosimetry and cell susceptibility. Acute inflammatory cell influx after toxicant exposure can contribute both to the injury and the repair of epithelial lesions, including necrosis, hyperplasia, and metaplasia. The specific example of ozone was used to illustrate the neutrophil-dependent and independent alterations that occur in transitional epithelium of the laboratory rat. The secretory apparatus of the nasal epithelium responds to inhaled toxicants and inflammagens by rapidly remodeling (hyperplasia, metaplasia) and altering the production and secretion of airway mucus. Another important health effect of upper airways to certain toxicant exposures is the initiation or exacerbation of allergic rhinitis that may also lead to epithelial remodeling and overproduction of airway mucus. The example of isocyanates was discussed as an example of occupational induced allergic rhinitis, while ozone and diesel exhaust were highlighted as examples of air pollutants that may exacerbate allergen-induced allergic rhinitis. Lastly, a caveat was presented with regard to the need for correct identification of nasal anatomy and the appreciation for species differences in epithelial responses. Understanding differences and similarities of inflammatory and epithelial responses to inhaled toxicants in the nasal airways of laboratory animals and humans is necessary to adequately translate results from animal studies to human health.

REFERENCES

1. Morgan KT, Monticello TM. Airflow, gas deposition, and lesion distribution in the nasal passages. Environ Health Perspect 1990; 85:209–218.
2. Morgan KT, Jiang XZ, Starr TB, et al. More precise localization of nasal tumors associated with chronic exposure of F-344 rats to formaldehyde gas. Toxicol Appl Pharmacol 1986; 82(2):264–271.
3. Kimbell JS, Godo MN, Gross EA, et al. Computer simulation of inspiratory airflow in all regions of the F344 rat nasal passages. Toxicol Appl Pharmacol 1997; 145(2): 388–398.
4. Morgan KT. Nasal dosimetry, lesion distribution, and the toxicologic patholgist: A brief review. Inhal Toxicol 1994; 6:41–57.
5. Gross EA, Mellick PW, Kari FW, et al. Histopathology and cell replication responses in the respiratory tract of rats and mice exposed by inhalation to glutaraldehyde for up to 13 weeks. Fundam Appl Toxicol 1994; 23(3):348–362.
6. Harkema JR, Hotchkiss JA. Ozone-induced proliferative and metaplastic lesions in nasal transitional and respiratory epithelium: Comparative pathology. Inhal Toxicol 1994; 6:187–204.

7. Morgan KT, Patterson DL, Gross EA. Responses of the nasal mucociliary apparatus of F-344 rats to formaldehyde gas. Toxicol Appl Pharmacol 1986; 82(1):1–13.

8. Genter MB, Llorens J, O'Callaghan JP, et al. Olfactory toxicity of beta,beta'-iminodipropionitrile in the rat. J Pharmacol Exp Ther 1992; 263(3):1432–1439.

9. Vollrath M, Altmannsberger M, Weber K, et al. An ultrastructural and immunohistological study of the rat olfactory epithelium: unique properties of olfactory sensory cells. Differentiation 1985; 29:243–253.

10. Brittebo EB, Eriksson C, Feil V, et al. Toxicity of 2,6-dichlorothiobenzamide (chlorthiamid) and 2,6-dichlorobenzamide in the olfactory nasal mucosa of mice. Fundam Appl Toxicol 1991; 17(1):92–102.

11. Jeffery EH, Haschek WM. Protection by dimethylsulfoxide against acetaminophen-induced hepatic, but not respiratory toxicity in the mouse. Toxicol Appl Pharmacol 1988; 93(3):452–461.

12. Blesa S, Cortijo J, Mata M, et al. Oral N-acetylcysteine attenuates the rat pulmonary inflammatory response to antigen. Eur Respir J 2003; 21(3):394–400.

13. Pottenger LH, Malley LA, Bogdanffy MS, et al. Evaluation of effects from repeated inhalation exposure of F344 rats to high concentrations of propylene. Toxicol Sci 2007; 97(2):336–347.

14. Reed CJ, Robinson DA, Lock EA. Antioxidant status of the rat nasal cavity. Free Radic Biol Med 2003; 34(5):607–615.

15. Van Der Vliet A, O'Neill CA, Cross CE, et al. Determination of low-molecular-mass antioxidant concentrations in human respiratory tract lining fluids. Am J Physiol 1999; 276(2 Pt 1):L289–L296.

16. Mudway IS, Blomberg A, Frew AJ, et al. Antioxidant consumption and repletion kinetics in nasal lavage fluid following exposure of healthy human volunteers to ozone. Eur Respir J 1999; 13(6):1429–1438.

17. Blomberg A, Sainsbury C, Rudell B, et al. Nasal cavity lining fluid ascorbic acid concentration increases in healthy human volunteers following short term exposure to diesel exhaust. Free Radic Res 1998; 28(1):59–67.

18. Tunnicliffe WS, Harrison RM, Kelly FJ, et al. The effect of sulphurous air pollutant exposures on symptoms, lung function, exhaled nitric oxide, and nasal epithelial lining fluid antioxidant concentrations in normal and asthmatic adults. Occup Environ Med 2003; 60(11):e15.

19. Lund K, Refsnes M, Ramis I, et al. Human exposure to hydrogen fluoride induces acute neutrophilic, eicosanoid, and antioxidant changes in nasal lavage fluid. Inhal Toxicol 2002; 14(2):119–132.

20. Buckley LA, Morgan KT, Swenberg JA, et al. The toxicity of dimethylamine in F-344 rats and B6C3F1 mice following a 1-year inhalation exposure. Fundam Appl Toxicol 1985; 5(2):341–352.

21. Bolon B, Bonnefoi MS, Roberts KC, et al. Toxic interactions in the rat nose: pollutants from soiled bedding and methyl bromide. Toxicol Pathol 1991; 19(4 Pt 2):571–579.

22. Jiang XZ, Buckley LA, Morgan KT. Pathology of toxic responses to the RD50 concentration of chlorine gas in the nasal passages of rats and mice. Toxicol Appl Pharmacol 1983; 71(2):225–236.

23. Wolf DC, Morgan KT, Gross EA, et al. Two-year inhalation exposure of female and male B6C3F1 mice and F344 rats to chlorine gas induces lesions confined to the nose. Fundam Appl Toxicol 1995; 24(1):111–131.

24. Nikasinovic L, Momas I, Seta N. Nasal epithelial and inflammatory response to ozone exposure: a review of laboratory-based studies published since 1985. J Toxicol Environ Health B Crit Rev 2003; 6(5):521–568.

25. Harkema JR, Hotchkiss JA, Henderson RF. Effects of 0.12 and 0.80 ppm ozone on rat nasal and nasopharyngeal epithelial mucosubstances: quantitative histochemistry. Toxicol Pathol 1989; 17(3):525–535.

26. Harkema JR, Plopper CG, Hyde DM, et al. Effects of an ambient level of ozone on primate nasal epithelial mucosubstances. Quantitative histochemistry. Am J Pathol 1987; 127(1):90–96.

27. Harkema JR, Plopper CG, Hyde DM, et al. Response of the macaque nasal epithelium to ambient levels of ozone. A morphologic and morphometric study of the transitional and respiratory epithelium. Am J Pathol 1987; 128(1):29–44.
28. Hotchkiss JA, Harkema JR, Henderson RF. Effect of cumulative ozone exposure on ozone-induced nasal epithelial hyperplasia and secretory metaplasia in rats. Exp Lung Res 1991; 17(3):589–600.
29. Cho HY, Hotchkiss JA, Harkema JR. Inflammatory and epithelial responses during the development of ozone-induced mucous cell metaplasia in the nasal epithelium of rats. Toxicol Sci 1999; 51(1):135–145.
30. Hotchkiss JA, Harkema JR, Johnson NF. Kinetics of nasal epithelial cell loss and proliferation in F344 rats following a single exposure to 0.5 ppm ozone. Toxicol Appl Pharmacol 1997; 143(1):75–82.
31. Cho HY, Hotchkiss JA, Bennett CB, et al. Neutrophil-dependent and neutrophil-independent alterations in the nasal epithelium of ozone-exposed rats. Am J Respir Crit Care Med 2000; 162(2 Pt 1):629–636.
32. Hotchkiss JA, Hilaski R, Cho H, et al. Fluticasone propionate attenuates ozone-induced rhinitis and mucous cell metaplasia in rat nasal airway epithelium. Am J Respir Cell Mol Biol 1998; 18(1):91–99.
33. Jamil S, Breuer R, Christensen TG. Abnormal mucous cell phenotype induced by neutrophil elastase in hamster bronchi. Exp Lung Res 1997; 23(4):285–295.
34. Voynow JA, Fischer BM, Malarkey DE, et al. Neutrophil elastase induces mucus cell metaplasia in mouse lung. Am J Physiol Lung Cell Mol Physiol 2004; 287(6):L1293–L1302.
35. Nogami H, Aizawa H, Matsumoto K, et al. Neutrophil elastase inhibitor, ONO-5046 suppresses ozone-induced airway mucus hypersecretion in guinea pigs. Eur J Pharmacol 2000; 390(1–2):197–202.
36. Ghosh S, Janocha AJ, Aronica MA, et al. Nitrotyrosine proteome survey in asthma identifies oxidative mechanism of catalase inactivation. J Immunol 2006; 176(9):5587–5597.
37. Fischer BM, Voynow JA. Neutrophil elastase induces MUC5AC gene expression in airway epithelium via a pathway involving reactive oxygen species. Am J Respir Cell Mol Biol 2002; 26(4):447–452.
38. Voynow JA, Gendler SJ, Rose MC. Regulation of mucin genes in chronic inflammatory airway diseases. Am J Respir Cell Mol Biol 2006; 34(6):661–665.
39. Casalino-Matsuda SM, Monzon ME, Forteza RM. Epidermal growth factor receptor activation by epidermal growth factor mediates oxidant-induced goblet cell metaplasia in human airway epithelium. Am J Respir Cell Mol Biol 2006; 34(5):581–591.
40. Deshmukh HS, Case LM, Wesselkamper SC, et al. Metalloproteinases mediate mucin 5AC expression by epidermal growth factor receptor activation. Am J Respir Crit Care Med 2005; 171(4):305–314.
41. Takeyama K, Dabbagh K, Jeong Shim J, et al. Oxidative stress causes mucin synthesis via transactivation of epidermal growth factor receptor: role of neutrophils. J Immunol 2000; 164(3):1546–1552.
42. Polosa R, Sapsford RJ, Dokic D, et al. Induction of the epidermal growth factor receptor and its ligands in nasal epithelium by ozone. J Allergy Clin Immunol 2004; 113(1):120–126.
43. Harkema JR, Hotchkiss JA. In vivo effects of endotoxin on nasal epithelial mucosubstances: quantitative histochemistry. Exp Lung Res 1991; 17(4):743–761.
44. Rusca S, Charriere N, Droz PO, et al. Effects of bioaerosol exposure on work-related symptoms among Swiss sawmill workers. Int Arch Occup Environ Health 2008; 81(4):415–421.
45. Heldal KK, Halstensen AS, Thorn J, et al. Upper airway inflammation in waste handlers exposed to bioaerosols. Occup Environ Med 2003; 60(6):444–450.
46. Reynolds SJ, Black DW, Borin SS, et al. Indoor environmental quality in six commercial office buildings in the midwest United States. Appl Occup Environ Hyg 2001; 16(11):1065–1077.

47. Fanucchi MV, Hotchkiss JA, Harkema JR. Endotoxin potentiates ozone-induced mucous cell metaplasia in rat nasal epithelium. Toxicol Appl Pharmacol 1998; 152(1):1–9.
48. Wagner JG, Van Dyken SJ, Hotchkiss JA, et al. Endotoxin enhancement of ozone-induced mucous cell metaplasia is neutrophil-dependent in rat nasal epithelium. Toxicol Sci 2001; 60(2):338–347.
49. Islam Z, Amuzie CJ, Harkema JR, et al. Neurotoxicity and inflammation in the nasal airways of mice exposed to the macrocyclic trichothecene mycotoxin roridin a: kinetics and potentiation by bacterial lipopolysaccharide coexposure. Toxicol Sci 2007; 98(2):526–541.
50. Eldridge MW, Peden DB. Allergen provocation augments endotoxin-induced nasal inflammation in subjects with atopic asthma. J Allergy Clin Immunol 2000; 105(3):475–481.
51. Spannhake EW, Reddy SP, Jacoby DB, et al. Synergism between rhinovirus infection and oxidant pollutant exposure enhances airway epithelial cell cytokine production. Environ Health Perspect 2002; 110(7):665–670.
52. Kongerud J, Madden MC, Hazucha M, et al. Nasal responses in asthmatic and nonasthmatic subjects following exposure to diesel exhaust particles. Inhal Toxicol 2006; 18(9):589–594.
53. Cassee FR, Boere AJ, Bos J, et al. Effects of diesel exhaust enriched concentrated PM2.5 in ozone preexposed or monocrotaline-treated rats. Inhal Toxicol 2002; 14(7):721–743.
54. Mauderly JL, Samet JM. Is there evidence for synergy among air pollutants in causing health effects? Environ Health Perspect 2009; 117(1):1–6.
55. Nadadur SS, Miller CA, Hopke PK, et al. The complexities of air pollution regulation: the need for an integrated research and regulatory perspective. Toxicol Sci 2007; 100(2):318–327.
56. Santhanum P, Wagner J, Elder A, et al. Effects of subchronic inhalation exposure to carbon black nanoparticles in the nasal airways of laboratory rats. Int J Nanotech 2008; 5(1):30–54.
57. Herbert RA, Hailey JR, Grumbein S, et al. Two-year and lifetime toxicity and carcinogenicity studies of ozone in B6C3F1 mice. Toxicol Pathol 1996; 24(5): 539–548.
58. Herbert RA, Leininger JR. Nose, Larynx, and Trachea. In: Maronpot R, Boorman GA, Gaul BW, eds. Pathology of the Mouse: Reference and Atlas. Vienna, IL: Cache River Press, 1999:259–292.
59. Boorman GA, Morgan KT, Uraih LC. Nose, larynx and trachea. In: Boorman GA, Eustis SL, Elwell MR, Montgomery CA Jr., Mackenzie WF, eds. Pathology of the Fischer Rat. San Diego: Academic Press, 1990:315–337.
60. Harkema JR. Comparative pathology of the nasal mucosa in laboratory animals exposed to inhaled irritants. Environ Health Perspect 1990; 85:231–238.
61. Nagano K, Katagiri T, Aiso S, et al. Spontaneous lesions of nasal cavity in aging F344 rats and BDF1 mice. Exp Toxicol Pathol 1997; 49(1–2):97–104.
62. Katagiri T, Takeuchi T, Mine T, et al. Chronic inhalation toxicity and carcinogenicity studies of 3-chloro-2-methylpropene in BDF1 mice. Ind Health 2000; 38(3): 309–318.
63. Giannetti N, Moyse E, Ducray A, et al. Accumulation of Ym1/2 protein in the mouse olfactory epithelium during regeneration and aging. Neuroscience 2004; 123(4): 907–917.
64. Ward JM, Yoon M, Anver MR, et al. Hyalinosis and Ym1/Ym2 gene expression in the stomach and respiratory tract of 129S4/SvJae and wild-type and CYP1A2-null B6, 129 mice. Am J Pathol 2001; 158(1):323–332.
65. Zhu Z, Zheng T, Homer RJ, et al. Acidic mammalian chitinase in asthmatic Th2 inflammation and IL-13 pathway activation. Science 2004; 304(5677):1678–1682.
66. Boot RG, Bussink AP, Verhoek M, et al. Marked Differences in Tissue-specific Expression of Chitinases in Mouse and Man. J Histochem Cytochem 2005; 53(10):1283–1292.

67. Hele DJ, Birrell MA, Webber SE, et al. Mediator involvement in antigen-induced bronchospasm and microvascular leakage in the airways of ovalbumin sensitized Brown Norway rats. Br J Pharmacol 2001; 132(2):481–488.
68. Jiang XZ, Morgan KT, Beauchamp RO. Histopathology of acute and subacute nasal toxicity. In: Barrow CS, ed. Toxicology of the nasal passages. New York, NY: Hemisphere, 1986:51–66.
69. Hotchkiss JA, Evans WA, Chen BT, et al. Regional differences in the effects of mainstream cigarette smoke on stored mucosubstances and DNA synthesis in F344 rat nasal respiratory epithelium. Toxicol Appl Pharmacol 1995; 131(2):316–324.
70. Farraj AK, Harkema JR, Kaminski NE. Allergic rhinitis induced by intranasal sensitization and challenge with trimellitic anhydride but not with dinitrochlorobenzene or oxazolone in A/J mice. Toxicol Sci 2004; 79(2):315–325.
71. Wagner JG, Hotchkiss JA, Harkema JR. Enhancement of nasal inflammatory and epithelial responses after ozone and allergen coexposure in Brown Norway rats. Toxicol Sci 2002; 67(2):284–294.
72. Evans CM, Williams OW, Tuvim MJ, et al. Mucin is produced by clara cells in the proximal airways of antigen-challenged mice. Am J Respir Cell Mol Biol 2004; 31(4):382–394.
73. Young HW, Williams OW, Chandra D, et al. Central role of Muc5ac expression in mucous metaplasia and its regulation by conserved 5′ elements. Am J Respir Cell Mol Biol 2007; 37(3):273–290.
74. Zhu Y, Ehre C, Abdullah LH, et al. Munc13-2-/- baseline secretion defect reveals source of oligomeric mucins in mouse airways. J Physiol 2008; 586(7):1977–1992.
75. Fransson M, Adner M, Uddman R, et al. Lipopolysaccharide-induced downregulation of uteroglobin in the human nose. Acta Otolaryngol 2007; 127(3):285–291.
76. Benson M, Jansson L, Adner M, et al. Gene profiling reveals decreased expression of uteroglobin and other anti-inflammatory genes in nasal fluid cells from patients with intermittent allergic rhinitis. Clin Exp Allergy 2005; 35(4):473–478.
77. Young JT. Histopathologic examination of the rat nasal cavity. Fundam Appl Toxicol 1981; 1(4):309–312.
78. Edranov SS, Kovalyeva IV, Kryukov KI, et al. Anatomic and histological study of maxillary sinus in albino rat. Bull Exp Biol Med 2004; 138(6):603–606.
79. Harkema JR, Hotchkiss JA, Griffith WC. Mucous cell metaplasia in rat nasal epithelium after a 20-month exposure to ozone: a morphometric study of epithelial differentiation. Am J Respir Cell Mol Biol 1997; 16(5):521–530.
80. Vidic B, Rana MW, Bhagat BD. Reversible damage of rat upper respiratory tract caused by cigarette smoke. Arch Otolaryngol 1974; 99(2):110–113.
81. Tos M, Mogensen C. Mucus production in chronic maxillary sinusitis. A quantitative histopathological study. Acta Otolaryngol 1984; 97(1–2):151–159.
82. Jacob A, Chole RA. Survey anatomy of the paranasal sinuses in the normal mouse. Laryngoscope 2006; 116(4):558–563.
83. Jacob A, Faddis BT, Chole RA. Chronic bacterial rhinosinusitis: description of a mouse model. Arch Otolaryngol Head Neck Surg 2001; 127(6):657–664.
84. Blair C, Nelson M, Thompson K, et al. Allergic inflammation enhances bacterial sinusitis in mice. J Allergy Clin Immunol 2001; 108(3):424–429.
85. Bomer K, Brichta A, Baroody F, et al. A mouse model of acute bacterial rhinosinusitis. Arch Otolaryngol Head Neck Surg 1998; 124(11):1227–1232.
86. Mery S, Gross EA, Joyner DR, et al. Nasal diagrams: a tool for recording the distribution of nasal lesions in rats and mice. Toxicol Pathol 1994; 22(4):353–372.
87. Patou J, De Smedt H, van Cauwenberge P, et al. Pathophysiology of nasal obstruction and meta-analysis of early and late effects of levocetirizine. Clin Exp Allergy 2006; 36(8):972–981.
88. Widdicombe JG. Nasal pathophysiology. Respir Med 1990; 84(suppl A):3–9; discussion 9–10.
89. Al Suleimani M, Ying D, Walker MJ. A comprehensive model of allergic rhinitis in guinea pigs. J Pharmacol Toxicol Methods 2006; 55(2):127–134.

90. Szelenyi I, Marx D, Jahn W. Animal models of allergic rhinitis. Arzneimittelforschung 2000; 50(11):1037–1042.
91. Tsunematsu M, Yamaji T, Kozutsumi D, et al. Establishment of an allergic rhinitis model in mice for the evaluation of nasal symptoms. Life Sci 2007; 80(15):1388–1394.
92. Al Suleimani YM, Dong Y, Walker MJ. Differential responses to various classes of drugs in a model of allergic rhinitis in guinea pigs. Pulm Pharmacol Ther 2008; 21(2):340–348.
93. Zosky GR, Larcombe AN, White OJ, et al. Ovalbumin-sensitized mice are good models for airway hyperresponsiveness but not acute physiological responses to allergen inhalation. Clin Exp Allergy 2008; 38(5):829–838.
94. Glaab T, Ziegert M, Baelder R, et al. Invasive versus noninvasive measurement of allergic and cholinergic airway responsiveness in mice. Respir Res 2005; 6:139.
95. Yu M, Tsai M, Tam SY, et al. Mast cells can promote the development of multiple features of chronic asthma in mice. J Clin Invest 2006; 116(6):1633–1641.
96. McMillan SJ, Lloyd CM. Prolonged allergen challenge in mice leads to persistent airway remodelling. Clin Exp Allergy 2004; 34(3):497–507.
97. Wegmann M, Fehrenbach H, Fehrenbach A, et al. Involvement of distal airways in a chronic model of experimental asthma. Clin Exp Allergy 2005; 35(10):1263–1271.
98. Ikeda RK, Miller M, Nayar J, et al. Accumulation of peribronchial mast cells in a mouse model of ovalbumin allergen induced chronic airway inflammation: modulation by immunostimulatory DNA sequences. J Immunol 2003; 171(9):4860–4867.
99. Hirota JA, Ask K, Fritz D, et al. Role of STAT6 and SMAD2 in a model of chronic allergen exposure: a mouse strain comparison study. Clin Exp Allergy 2009; 39(1):147–158.
100. Lim YS, Won TB, Shim WS, et al. Induction of airway remodeling of nasal mucosa by repetitive allergen challenge in a murine model of allergic rhinitis. Ann Allergy Asthma Immunol 2007; 98(1):22–31.
101. Wang H, Lu X, Cao PP, et al. Histological and immunological observations of bacterial and allergic chronic rhinosinusitis in the mouse. Am J Rhinol 2008; 22(4):343–348.
102. Nakaya M, Dohi M, Okunishi K, et al. Prolonged allergen challenge in murine nasal allergic rhinitis: nasal airway remodeling and adaptation of nasal airway responsiveness. Laryngoscope 2007; 117(5):881–885.
103. Harkema JR. Comparative aspects of nasal airway anatomy: relevance to inhalation toxicology. Toxicol Pathol 1991; 19(4 Pt 1):321–336.
104. Harkema JR, Carey SA, Wagner JG. The nose revisited: a brief review of the comparative structure, function, and toxicologic pathology of the nasal epithelium. Toxicol Pathol 2006; 34(3):252–269.
105. Berger G, Marom Z, Ophir D. Goblet cell density of the inferior turbinates in patients with perennial allergic and nonallergic rhinitis. Am J Rhinol 1997; 11(3):233–236.
106. Miyahara S, Miyahara N, Matsubara S, et al. IL-13 is essential to the late-phase response in allergic rhinitis. J Allergy Clin Immunol 2006; 118(5):1110–1116.
107. Wagner JG, Jiang Q, Harkema JR, et al. gamma-Tocopherol prevents airway eosinophilia and mucous cell hyperplasia in experimentally induced allergic rhinitis and asthma. Clin Exp Allergy 2008; 38(3):501–511.
108. Berger G, Moroz A, Marom Z, et al. Inferior turbinate goblet cell secretion in patients with perennial allergic and nonallergic rhinitis. Am J Rhinol 1999; 13(6):473–477.
109. Nakaya M, Dohi M, Okunishi K, et al. Noninvasive system for evaluating allergen-induced nasal hypersensitivity in murine allergic rhinitis. Lab Invest 2006; 86(9):917–926.
110. Ciprandi G, Cirillo I, Vizzaccaro A, et al. Airway function and nasal inflammation in seasonal allergic rhinitis and asthma. Clin Exp Allergy 2004; 34(6):891–896.
111. Ciprandi G, Cirillo I, Vizzaccaro A, et al. Correlation of nasal inflammation and nasal airflow with forced expiratory volume in 1 second in patients with perennial allergic rhinitis and asthma. Ann Allergy Asthma Immunol 2004; 93(6):575–580.
112. Yamasaki M, Mizutani N, Sasaki K, et al. No involvement of interleukin-5 or eosinophils in experimental allergic rhinitis in guinea pigs. Eur J Pharmacol 2002; 439(1–3):159–169.

113. Miyahara S, Miyahara N, Takeda K, et al. Physiologic assessment of allergic rhinitis in mice: role of the high-affinity IgE receptor (FcepsilonRI). J Allergy Clin Immunol 2005; 116(5):1020–1027.
114. Humbles AA, Lloyd CM, McMillan SJ, et al. A critical role for eosinophils in allergic airways remodeling. Science 2004; 305(5691):1776–1779.
115. Singer M, Lefort J, Vargaftig BB. Granulocyte depletion and dexamethasone differentially modulate airways hyperreactivity, inflammation, mucus accumulation, and secretion induced by rmIL-13 or antigen. Am J Respir Cell Mol Biol 2002; 26(1): 74–84.
116. Amin K, Rinne J, Haahtela T, et al. Inflammatory cell and epithelial characteristics of perennial allergic and nonallergic rhinitis with a symptom history of 1 to 3 years' duration. J Allergy Clin Immunol 2001; 107(2):249–257.
117. Berger G, Goldberg A, Ophir D. The inferior turbinate mast cell population of patients with perennial allergic and nonallergic rhinitis. Am J Rhinol 1997; 11(1): 63–66.
118. Fokkens WJ, Godthelp T, Holm AF, et al. Dynamics of mast cells in the nasal mucosa of patients with allergic rhinitis and non-allergic controls: a biopsy study. Clin Exp Allergy 1992; 22(7):701–710.
119. Kawaguchi S, Majima Y, Sakakura Y. Nasal mast cells in experimentally induced allergic rhinitis in guinea-pigs. Clin Exp Allergy 1994; 24(3):238–244.
120. Moscato G, Pignatti P, Yacoub MR, et al. Occupational asthma and occupational rhinitis in hairdressers. Chest 2005; 128(5):3590–3598.
121. Aalto-Korte K, Makinen-Kiljunen S. Specific immunoglobulin E in patients with immediate persulfate hypersensitivity. Contact Dermatitis 2003; 49(1):22–25.
122. Volkman KK, Merrick JG, Zacharisen MC. Yacht-maker's lung: A case of hypersensitivity pneumonitis in yacht manufacturing. WMJ 2006; 105(7):47–50.
123. Sangha GK, Alarie Y. Sensory irritation by toluene diisocyanate in single and repeated exposures. Toxicol Appl Pharmacol 1979; 50(3):533–547.
124. Weyel DA, Rodney BS, Alarie Y. Sensory irritation, pulmonary irritation, and acute lethality of a polymeric isocyanate and sensory irritation of 2,6-toleune diisocyanate. Toxicol Appl Pharmacol 1982; 64(3):423–430.
125. Johnson VJ, Yucesoy B, Reynolds JS, et al. Inhalation of toluene diisocyanate vapor induces allergic rhinitis in mice. J Immunol 2007; 179(3):1864–1871.
126. Farraj AK, Harkema JR, Jan TR, et al. Immune responses in the lung and local lymph node of A/J mice to intranasal sensitization and challenge with adjuvant-free ovalbumin. Toxicol Pathol 2003; 31(4):432–447.
127. Farraj AK, Boykin E, Haykal-Coates N, et al. Th2 Cytokines in Skin Draining Lymph Nodes and Serum IgE Do Not Predict Airway Hypersensitivity to Intranasal Isocyanate Exposure in Mice. Toxicol Sci 2007; 100(1):99–108.
128. Choi JH, Nahm DH, Kim SH, et al. Increased levels of IgG to cytokeratin 19 in sera of patients with toluene diisocyanate-induced asthma. Ann Allergy Asthma Immunol 2004; 93(3):293–298.
129. Wisnewski AV. Developments in laboratory diagnostics for isocyanate asthma. Curr Opin Allergy Clin Immunol 2007; 7(2):138–145.
130. Kobayashi T, Ikeue T, Ikeda A. Four-week exposure to diesel exhaust induces nasal mucosal hyperresponsiveness to histamine in guinea pigs. Toxicol Sci 1998; 45(1):106–112.
131. Hiruma K, Terada N, Hanazawa T, et al. Effect of diesel exhaust on guinea pig nasal mucosa. Ann Otol Rhinol Laryngol 1999; 108(6):582–588.
132. Ishihara Y, Kagawa J. Dose-response assessment and effect of particles in guinea pigs exposed chronically to diesel exhaust: analysis of various biological markers in pulmonary alveolar lavage fluid and circulating blood. Inhal Toxicol 2002; 14(10): 1049–1067.
133. Gowdy K, Krantz QT, Daniels M, et al. Modulation of pulmonary inflammatory responses and antimicrobial defenses in mice exposed to diesel exhaust. Toxicol Appl Pharmacol 2008; 229(3):310–319.

134. Dong CC, Yin XJ, Ma JY, et al. Effect of diesel exhaust particles on allergic reactions and airway responsiveness in ovalbumin-sensitized brown Norway rats. Toxicol Sci 2005; 88(1):202–212.
135. Kobayashi T. Exposure to diesel exhaust aggravates nasal allergic reaction in guinea pigs. Am J Respir Crit Care Med 2000; 162(2 Pt 1):352–356.
136. Takafuji S, Suzuki S, Koizumi K, et al. Diesel-exhaust particulates inoculated by the intranasal route have an adjuvant activity for IgE production in mice. J Allergy Clin Immunol 1987; 79(4):639–645.
137. Maejima K, Tamura K, Taniguchi Y, et al. Comparison of the effects of various fine particles on IgE antibody production in mice inhaling Japanese cedar pollen allergens. J Toxicol Environ Health 1997; 52(3):231–248.
138. Farraj AK, Haykal-Coates N, Ledbetter AD, et al. Neurotrophin mediation of allergic airways responses to inhaled diesel particles in mice. Toxicol Sci 2006; 94(1):183–192.
139. Farraj AK, Haykal-Coates N, Ledbetter AD, et al. Inhibition of pan neurotrophin receptor p75 attenuates diesel particulate-induced enhancement of allergic airway responses in C57/B16 J mice. Inhal Toxicol 2006; 18(7):483–491.
140. Sanico AM, Stanisz AM, Gleeson TD, et al. Nerve growth factor expression and release in allergic inflammatory disease of the upper airways. Am J Respir Crit Care Med 2000; 161(5):1631–1635.
141. Bresciani M, Laliberte F, Laliberte MF, et al. Nerve growth factor localization in the nasal mucosa of patients with persistent allergic rhinitis. Allergy 2009; 64(1):112–117.
142. Kobayashi H, Gleich GJ, Butterfield JH, et al. Human eosinophils produce neurotrophins and secrete nerve growth factor on immunologic stimuli. Blood 2002; 99(6):2214–2220.
143. Ris C. U.S. EPA health assessment for diesel engine exhaust: a review. Inhal Toxicol 2007; 19(suppl 1):229–239.
144. Singh P, DeMarini DM, Dick CA, et al. Sample characterization of automobile and forklift diesel exhaust particles and comparative pulmonary toxicity in mice. Environ Health Perspect 2004; 112(8):820–825.
145. Hiltermann TJ, de Bruijne CR, Stolk J, et al. Effects of photochemical air pollution and allergen exposure on upper respiratory tract inflammation in asthmatics. Am J Respir Crit Care Med 1997; 156(6):1765–1772.
146. Peden DB, Setzer RW Jr., Devlin RB. Ozone exposure has both a priming effect on allergen-induced responses and an intrinsic inflammatory action in the nasal airways of perennially allergic asthmatics. Am J Respir Crit Care Med 1995; 151(5):1336–1345.
147. Iijima MK, Kobayashi T. Nasal allergy-like symptoms aggravated by ozone exposure in a concentration-dependent manner in guinea pigs. Toxicology 2004; 199(1):73–83.
148. Wagner JG, Harkema JR, Jiang Q, et al. g- tocopherol attenuates ozone-induced exacerbation of allergic rhinosinusitis in rats. Toxicol Pathol 2009; 37:481–491.
149. Sanchez-Carrillo CI, Ceron-Mireles P, Rojas-Martinez MR, et al. Surveillance of acute health effects of air pollution in Mexico City. Epidemiology 2003; 14(5):536–544.
150. Riediker M, Monn C, Koller T, et al. Air pollutants enhance rhinoconjunctivitis symptoms in pollen-allergic individuals. Ann Allergy Asthma Immunol 2001; 87(4): 311–318.
151. Fukushima A, Ozaki A, Fukata K, et al. Differential expression and signaling of IFN-gamma in the conjunctiva between Lewis and Brown Norway rats. Microbiol Immunol 2003; 47(10):785–796.

17 Inflammatory Conditions of the Nose and Paranasal Sinuses in Humans

Ricardo Tan and Jonathan Corren

Allergy Research Foundation, Los Angeles, California, U.S.A.

INTRODUCTION

Chronic inflammatory conditions of the nose and paranasal sinuses include rhinitis, sinusitis, and nasal polyposis. This chapter reviews the pathophysiology, epidemiology, diagnosis, and treatment of these conditions.

ALLERGIC RHINITIS

Pathophysiology

Following the nasal inhalation of airborne allergens, allergenic proteins become bound to specific IgE antibodies on the surface of mast cells in the nasal mucosa. Cross-linkage of allergen-bound IgE causes activation of key intracellular proteins, which leads to fusion of cytoplasmic granules with the cell membrane (1). This step of mast cell activation results in the release of preformed mediators, including histamine, into the extracellular fluid. Somewhat later, newly formed mediators, such as sulfidopeptide leukotrienes (LTC4, D4, E4) and prostaglandins (primarily PGD2), are synthesized and subsequently released into the milieu surrounding mucosa. These mediators have a number of acute effects in the nose, including the elicitation of neural reflexes, which eventuate in sneezing and itching; extravasation of fluid into the mucosa and submucosa, which causes edema; and stimulation of mucus glands, which results in watery discharge (2). During the hours following nasal allergen exposure, a number of blood cells, primarily lymphocytes and eosinophils, influx into the mucosa at the site of the allergic reaction. Eosinophils are currently thought to be the primary effector cell in allergic rhinitis, releasing a number of cationic proteins (e.g., eosinophil cationic protein), which cause tissue damage, as well as sulfidopeptide leukotrienes (2,3). This inflammatory cell infiltrate is thought to be largely responsible for persistent symptoms in allergic rhinitis as well the nonspecific mucosal hyperreactivity to nonallergic triggers.

Epidemiology and Classification

The prevalence of allergic rhinitis has increased over the past several decades and the condition is currently estimated to affect 30% of the American population (4). Approximately 80% of allergic rhinitis patients develop the disease before the age of 30, and the prevalence diminishes gradually as the population ages (4).

Historically, allergic rhinitis has been categorized as being either perennial with year-around symptoms, or seasonal, occurring in the spring, summer, or fall seasons. Allergic rhinitis can alternatively be classified according to duration or severity. Duration is divided into "intermittent" (<4 wk/yr) or "persistent" (>4 wk/yr) categories, while severity is classified as mild, moderate, or severe.

Clinical Presentation

Patients with allergic rhinitis most commonly present with symptoms of nasal congestion, pruritus of the nose, eyes, or palate, sneezing, and watery rhinorrhea. These symptoms may be present throughout the year or may be primarily seasonal, and patients may be able to identify specific triggers, such as animal danders, dust, or mold. Patients often have a preceding history of other atopic problems, such as atopic dermatitis or food allergy. Similarly, most have a positive family history of atopy in first-degree relatives. Other symptoms that are frequently reported by patients with allergic rhinitis include headache, cough, muffled hearing, and clicking of the ears (5).

The physical examination frequently reveals "allergic shiners" (periorbital darkening) and pale, boggy nasal turbinates with clear secretions. In children, it is not uncommon to see a transverse crease across the bridge of the nose, often associated with repetitive pushing of the nasal alae (allergic salute) (5).

Diagnostic Tests

The gold standard for confirming the diagnosis of allergic rhinitis is measurement of specific IgE. This may be done by either skin testing or analysis of serum.

The preferred method of allergy skin testing is the percutaneous technique, in which an allergenic extract is placed onto the skin and is followed by a prick or puncture of the skin. The skin is examined 15 to 20 minutes later and a positive reaction is denoted by the presence of a wheal, defined as a bump, and flare, defined as an area of erythema, which are larger than a predetermined standard. The sensitivity and specificity of percutaneous allergy testing is on the order of 90%. Another method of skin testing, which is frequently used, is the intradermal method. While this technique is extremely sensitive (95%+), the specificity of the test is on the order of 50%, resulting in a low positive predictive value (6).

While skin testing is considered to be the best method for assessing specific IgE, there are several situations in which it cannot be used, including diffuse dermatitis, dermagraphia, and an inability to withhold medications which have antihistaminic effects. There are a number of in vitro methods that measure specific IgE in serum, and the most recent and popular of these is the Immuno-Cap ELISA assay. This test is somewhat less sensitive than allergy skin testing (approximately 70%), but is extremely specific (6).

Treatment

In general, there are three main modalities used to treat allergic rhinitis: allergen avoidance strategies, medications, and specific allergen immunotherapy.

Allergen Avoidance Strategies

Indoor allergens, including house dust mites, animal danders, cockroach, and molds, have proven to be amenable to environmental control measures. Outdoor sources of allergens, such as seasonal plant pollens and molds, are much more difficult to control effectively other than to limit outdoor activity during specific times of the year.

Allergen control strategies for the microscopic dust mites are aimed chiefly at: (*i*) removing mite proteins from the indoor environment, by frequent washing of bedding in hot water, vacuum cleaning the carpeting, and dusting, and; (*ii*) preventing mite allergen from reaching the patient, by using impermeable encasings on the pillows, mattress, and box springs (7).

Animal danders are most effectively reduced by removing the animal from the inside of the patient's dwelling (8). Keeping the animal out of the bedroom, use of high-efficiency particulate air (HEPA) filters, frequent washing of the pet, and regular and thorough indoor cleaning are also effective.

Pharmacotherapy

Antihistamines are the oldest class of medications for treating allergic rhinitis (9). The first-generation agents (e.g., diphenhydramine) primarily reduce sneezing, rhinorrhea, and nasal and ocular pruritus and have minimal effects on nasal congestion and can have side effects including anticholinergic (e.g., dryness of the mouth and eyes) and central nervous system depression (e.g., sedation). Newer agents (e.g., loratadine, fexofenadine, cetirizine, desloratadine, levocetirizine) are equally effective but lack the anticholinergic effects. Two very potent antihistamines, azelastine and olopatadine, are also available as intranasal formulations and have been shown to be at least as effective as oral preparations (10).

Oral and intranasal decongestants (e.g., pseduoephedrine, phenylpropanolamine) are commonly employed for the relief of nasal congestion and rhinorrhea in allergic rhinitis. Intranasal decongestant sprays should be used only for short-term relief (less than seven days at a time). Prolonged topical decongestant use can result in rebound congestion, also known as drug-induced rhinitis or rhinitis medicamentosa, which is characterized by worsening congestion caused by vasodilation in the mucosa when the decongestant is stopped (11).

Intranasal steroids are considered the most potent form of treatment for rhinitis and are effective in controlling most symptoms of allergic rhinitis (12). They are well-tolerated and recommended for regular daily maintenance use to control and prevent symptoms. As-needed use has been shown in studies to be effective but regular daily use provides better control (13). The available preparations include triamcinolone, mometasone, budesonide, fluticasone, ciclesonide, and beclomethasone. These preparations have differences in binding affinity, topical potency, and lipid solubility but all have comparable clinical efficacy (14).

Leukotriene inhibitors (montelukast), intranasal cromolyn, and intranasal anticholinergics (e.g., ipatropium bromide) have also been shown to be effective in controlling allergic rhinitis symptoms (15–18).

Immunotherapy

Immunotherapy is the process of exposure, either sublingually or through subcutaneous injections (allergy shots), to gradually increasing doses of specific allergen extracts to which a patient is allergic in order to decrease sensitivity and improve symptoms. It is indicated in patients who show specific IgE responses by skin or RAST testing. Patients who have responded poorly to other therapy are unable to avoid allergens or who would prefer not to use medications are all candidates for immunotherapy (19–21).

NONALLERGIC RHINITIS

Nonallergic rhinitis refers to a group of rhinitis syndromes that are characterized by the absence of evidence of IgE sensitivity on skin and/or serum testing.
These include vasomotor rhinitis, NARES (nonallergic rhinitis with eosinophilia
syndrome), atrophic rhinitis, hormonal rhinitis, and environmental nonallergic
rhinitis.

Vasomotor rhinitis presents with nasal symptoms induced by nonspecific
triggers such changes in temperature, strong smells, irritant vapors, and smokes
(e.g., perfume, cigarette smoke), as well as changes in humidity or baromet-
ric pressure. Intranasal steroids and the intranasal antihistamine, azelastine, are
effective in the overall treatment of vasomotor rhinitis, and intranasal ipratropium bromide is effective for the control of hypersecretion (4).

NARES is characterized by the presence of nasal eosinophils but no other
evidence of specific IgE sensitivity. Symptoms are perennial and may include
hyposmia (reduced sense of smell) (22).

Atrophic rhinitis is often seen in elderly patients and presents with crusting and foul odor in the nasal passages. Treatment consists of nasal lavage
and removal of crusting. Infection is treated with topical or systemic antibiotics
(23) Hormonal rhinitis is associated with rhinitis that is triggered or worsened by
menstrual periods or pregnancy. (24)

Environmental nonallergic rhinitis refers to rhinitis due to physical and
chemical factors in the atmosphere. If triggered by chemical agents, the term
"irritant rhinitis" is often used. The term "occupational nonallergic rhinitis" is
also used if the etiologic trigger occurs in the workplace. Irritants in ambient and
indoor air include, among others: tobacco smoke, volatile organic compounds
(VOCs), ozone, sulfur dioxide, ammonia, chlorine, acetic acid, carbonless copy
paper, and paper dust (25).

CHRONIC SINUSITIS

Introduction

Sinusitis refers to inflammation in the paranasal sinuses, which is most commonly infectious in origin. The infectious etiology is usually bacterial but may
also be fungal. Thirty one million Americans suffer from sinusitis every year and
lose an average of four days of work per episode. It is a frequent cause of lost
school days as well.

Acute sinusitis is commonly defined as sinus inflammation lasting for up to
three to four weeks. Subacute sinusitis lasts four to eight weeks. Chronic sinusitis
is defined as persisting beyond eight weeks. Recurrent sinusitis is defined as
three or more separate episodes of acute sinusitis per year (26).

Pathophysiology

The sinuses are air cavities within the facial and skull bones with small openings or ostia which when obstructed cause changes in oxygenation and acidity
that lead to bacterial growth and infection. Visualization of purulent drainage
from these ostia confirms the presence and location of sinusitis. The maxillary,
frontal, anterior, and middle ethmoidal sinuses all drain into the middle meatus, the sphenoid sinuses open above the superior turbinate, and the posterior
portion of the ethmoids drain into the superior meatus (27). The most important

factors in the development of sinusitis are (*i*) the function of the cilia, (*ii*) the patency of the ostia, and (*iii*) the nature of the sinus secretions. Allergic, nonallergic and viral rhinitis, immotile cilia syndrome, and other disorders of ciliary movement, immune deficiencies, mechanical obstruction (e.g., nasal polyps, septal deviation), and cystic fibrosis can predispose to sinusitis. In both acute and chronic bacterial sinusitis, the organisms most commonly responsible are *Streptococcus pneumoniae*, *Hemophilus influenzae*, and *Moraxella catarrhalis*. In chronic sinusitis, infection may also be commonly due to *Staphylococcus aureus*, Group A streptococcus, *Pseudomonas aeruginosa*, and anaerobic bacteria such as *Fusobacterium* and *Bacteroides* (28). In cystic fibrosis, the most common organism is *P aeruginosa*. Fungal sinusitis, although less common than bacterial sinusitis, is no longer seen only in immunocompromised individuals (29).

Clinical Presentation

Acute sinusitis most commonly presents as a "common cold" that persists for more than five to seven days. Patients complain of headache and facial pressure, which is aggravated by bending over. Pain may be referred to the teeth embedded in the floor of the maxillary sinuses or felt between the eyes when the ethmoids are involved. Fever and generalized malaise may be present. Purulent nasal discharge may be the first sign of bacterial infection as secretions change from clear to thick and dark yellow or green.Chronic sinusitis is less obvious by history and indeed is often missed and symptoms attributed to other conditions. Patients may present with only one long-standing and refractory symptom such as cough, postnasal drip, headache, nasal congestion, or halitosis with no other associated symptoms (30).

Underlying conditions that predispose to chronic sinusitis include allergic or vasomotor rhinitis, congenital immunodeficiency, AIDS, cystic fibrosis, Wegener's granulomatosis, Kartagener's syndrome (with sinusitis, bronchiectasis, and situs inversus), or aspirin sensitivity syndrome (with aspirin sensitivity, nasal polyps, sinusitis and asthma) (30).

Diagnostic Tests

Nasal Smear

The nasal smear is an inexpensive, useful test that is easy to perform in the office. In clinical studies, the specificity of nasal smear neutrophils in sinusitis ranges from 40% to 90% and the sensitivity ranges from 67% to 80% (31).

Sinus Culture

Cultures of nasal secretions are not reflective of the true bacterial picture in the sinuses because of contamination in the nasal passages. Only a specimen obtained from inside the sinuses through an invasive puncture procedure can give an accurate identification of the microorganisms involved. An antral puncture or middle meatal aspiration can be performed but are not routinely done and are indicated only in complicated and refractory cases (32).

Ultrasound

The A-mode ultrasound is used in many physician's offices as an inexpensive, noninvasive procedure for screening for sinusitis. Characteristic patterns of sound waves are produced by the presence of fluid in the sinuses. It can be used only for the maxillary and frontal sinuses. When compared with radiographic imaging, ultrasound showed false-positive rates of 39% to 45% and false-negative rates of 42% to 56%. Presently, ultrasound is not considered sensitive enough to substitute for radiographic imaging and is recommended mainly for pregnant women to avoid radiation (33).

x-Ray

The Waters view is frequently used to screen for air–fluid levels in acute maxillary and frontal sinusitis. The Caldwell or anterior–posterior view is also useful for determining frontal and maxillary involvement. Standard x-rays are not considered as useful in chronic sinusitis as mild-moderate thickening may be missed. The ethmoid sinuses and the osteomeatal complex are also not visualized well.

CT Scan

The CT scan is considered the gold standard for diagnosing sinus disease (34). It can show the ethmoid sinuses and the osteomeatal complex in detail and detect subtle thickening in all the sinuses in chronic sinusitis. The limited-cut, four-slice CT scan of the sinuses is a very cost-effective and adequate alternative to a full CT scan. High-resolution CT scans are an essential part of the pre-operative assessment for sinus surgery. Although an MRI is more sensitive than the CT scan for detecting soft tissue inflammation, the CT scan is sensitive enough for diagnosis in most cases (35).

Treatment

Medical Management

Adjunctive measures may be helpful in relieving the symptoms of acute and chronic sinusitis. Although studies supporting their efficacy are inconclusive or not available, clinical experience has shown agents and measures including saline irrigation, mucolytics, expectorants, topical and oral decongestants, and intranasal and systemic corticosteroids to be helpful.

Antibiotic therapy is the mainstay of the treatment of acute and chronic sinusitis. Several guidelines for the use of antibiotic therapy in sinusitis have been published (30,36–39). Acute sinusitis should be treated with 10 to 14 days of antibiotics. Chronic sinusitis needs at least three to four weeks of therapy or longer. Antibiotic therapy for at least one week after the patient with chronic sinusitis reports resolution of symptoms is recommended.

Surgical Treatment

Functional endoscopic sinus surgery (FESS) is the procedure of choice in patients who do not respond to maximal medical therapy (40). This approach aims to improve drainage from the sinuses by removing inflamed or dead tissue obstructing the ostia in the osteomeatal complex. FESS is markedly less invasive than prior sinus surgical procedures and has been very helpful in refractory cases of chronic sinusitis.

NASAL POLYPS

Pathophysiology

Nasal polyps are inflammatory masses that originate in the paranasal sinuses and cause blockage in the nasal passages. Nasal polyps are the result of chronic hyperplastic inflammation of the nasal and sinus mucous membranes but the etiology and pathogenetic mechanism are still unclear (41,42). The majority of patients with nasal polyps, up to 70%, have no history of allergic rhinitis (43). The polyp surface has pseudostratified respiratory epithelium. Epithelial damage and inflammation from various factors with eventual goblet cell hyperplasia and mucus hypersecretion are seen in histologic examination of polyps (44). Several inflammatory factors have been isolated from nasal polyps including endothelial vascular cell adhesion molecule (VCAM)-1, nitric oxide synthase, granulocyte-macrophage colony–stimulating factor (GM–CSF), eosinophil survival enhancing activity (ESEA), cys-leukotrienes (Cys-LT), and many other cytokines (45,46). Free radical damage resulting from environmental factors such as air pollution has also been implicated in polyp formation (47). Nasal polyps are not malignant or premalignant but may cause destruction of nasal bones if growth is uncontrolled (48). Polyps can originate in any of the paranasal sinuses but most commonly develop on the middle meatus and osteomeatal complex. Polyps are also often seen to originate in the ethmoid region, from the maxillary sinus ostium (antral choanal polyps), the turbinates, or the septum (49). Inverted papilloma, a malignant lesion, should be distinguished from nasal polyps (50).

Epidemiology

Nasal polyps occur equally among men and women and there has been no observed racial predilection. Nasal polyps have been observed in 5% of the non-allergic population and 1.5% of persons with allergic rhinitis. Nasal polyps can be seen in up to 40% of patients with cystic fibrosis. It is also part of the Samter triad with aspirin sensitivity and asthma (51).

Clinical Presentation

The presence of nasal polyps is associated with diminished or absent sense of smell (hyposmia or anosmia), diminished or absent of sense of taste (hypogeusia or ageusia), rhinorrhea, and postnasal drip. The degree of nasal obstruction depends on the size and extent of nasal polyposis (52). Complications from blockage of sinus ostia include acute and chronic sinusitis. Rarely, pressure or bony destruction can lead to proptosis, diplopia or intracranial infections such as meningitis or encephalitis (48). Physical examination shows the characteristic grape-like, translucent masses in the nasal passages.

Diagnostic Tests

Cystic fibrosis should be ruled out in children with the sweat chloride test.

Coronal CT scan of the sinuses is the imaging procedure of choice for identifying the extent and origin of nasal polyps (53). It is also essential in delineating the anatomy of the sinuses in case of surgical removal. MRI is not considered useful in nasal polyps.

Treatment

Systemic and intranasal steroids are the treatment of choice for medical treatment of nasal polyps (54). Oral steroids provide the most significant reduction of polyp size and relief of nasal obstruction (55). A dose of prednisone, starting at 30–60 mg/day and tapered over 7 to 10 days may be effective. Intranasal steroids are effective in reducing the size of smaller polyps and preventing recurrence of smaller polyps (56). Leukotriene inhibitors, intranasal cromolyn, topical diuretics, macrolide antibiotics, and intranasal lysine-acetylsalicylic acid have been shown in studies to have verying efficacy for control of nasal polyps (57).

Surgical removal of polyps may be indicated if medical treatment is not effective, especially if there is severe obstruction and recurrent infection (58). Endoscopic sinus surgery is the procedure of choice and is considered safe and effective. The recurrence rate after surgery is high and has been reported to be up to 60% in studies with many patients requiring repeat surgical polyp removal (59,60).

SUMMARY

The human upper airway is susceptible to a variety of inflammatory conditions of allergic or nonallergic etiology. Nasal hyperreactivity to physical and/or chemical stimuli is reported by approximately 40% of individuals with allergic rhinitis, and is a defining characteristic of the most common variety of nonallergic rhinitis (61). As documented in Chapters 26 and 27 of this volume, a reciprocal relationship exists between nasal allergy and nasal irritation, with allergic rhinitis exhibiting heightened sensitivity to chemical irritants, and with selected chemical irritants promoting allergic sensitization and/or potentiating the nasal allergic response.

REFERENCES

1. Pearlman DS. Pathophysiology of the inflammatory response. 1999; 104: S132–S137.
2. Gelfand EW. Inflammatory mediators in allergic rhinitis. J Allergy Clin Immunol 2004; 114:S135–S138.
3. Bascom R, Pipkorn U, Lichtenstein LM, et al. The influx of inflammatory cells into nasal washings during the late response to antigen challenge: Effect of systemic steroid pretreatment. Am Rev Respir Dis 1988; 138:406–412.
4. Settipane RA, Charnock DR. Epidemiology of rhinitis: Allergic and nonallergic. Clin Allergy Immunol 2007; 19:23–34.
5. Joint Task Force on Practice Parameters. The diagnosis and management of rhinitis: An updated practice parameter. J Allergy Clin Immunol 2008; 122(suppl 2):S1–S84.
6. Bernstein L, Li J, Bernstein D, et al. Allergy diagnostic testing: An updated practice parameter. Ann Allergy Asthma Immunol 2008; 100:S1–S148.
7. Sheikh A, Hurwitz B, Shehata Y. House dust mite avoidance measures for perennial allergic rhinitis. Cochrane Database Syst Rev 2007; (1):CD001563.
8. Wood RA, Chapman ME, Adkinson NF, et al. The effect of cat removal on allergen content in household-dust samples. J Allergy Clin Immunol 1989; 83:730–734.
9. Meltzer EO. Evaluation of the optimal oral antihistamine for patients with allergic rhinitis. Mayo Clin Proc 2005; 80:1170–1176.
10. Lee C, Corren J. Review of azelastine nasal spray in the treatment of allergic and nonallergic rhinitis. Expert Opin Pharmacother 2007; 8(5):701–709.

11. Morris S, Eccles R, Martez SJ, et al. An evaluation of nasal response following different treatment regimes of oxymetazoline with reference to rebound congestion. Am J Rhinol 1997; 11:109–115.
12. LaForce C. Use of nasal steroids in managing allergic rhinitis. J Allergy Clin Immunol 1999; 103:S388–S394.
13. Jen A, Baroody F, de Tineo M, et al. As needed-use of fluticasone propionate nasal spray reduces symptoms of seasonal allergic rhinitis. J Allergy Clin Immunol 2000; 105:732–738.
14. Corren J. Intranasal corticosteroids for allergic rhinitis: How do different agents compare? J Allergy Clin Immunol 1999; 104:S144–S149.
15. Patel P, Philip G, Yang W, et al. Randomized, double-blind, placebo-controlled study of montelukast for treating perennial allergic rhinitis. Ann Allergy Clin Immunol 2005; 95:551–557.
16. Philip G, Malmstrom K, Hampel FC, et al. Montelukast Spring Rhinitis Study Group. Montelukast for treating seasonal allergic rhinitis: A randomized, double-blind, placebo-controlled trial performed in the spring. Clin Exp Allergy 2002; 32(7):1020–1028.
17. Birchall MA, Henderson JS, Studham JM, et al. The effect of sodium cromoglycate on intranasal histamine challenge in allergic rhinitis. Clin Otolaryngol Allied Sci 1994; 19:521–525.
18. Meltzer EO. Intranasal anticholinergic therapy of rhinorrhea. J Allergy Clin Immunol 1992; 90(6 Pt 2):1055–1064.
19. Joint Task Force on Practice Parameters. Allergen Immunotherapy: a practice parameter second update. J Allergy Clin Immunol 2007; 120:S25–S85.
20. Durham SR. Sublingual immunotherapy: What have we learnt from the 'big trials'? Curr Opin Allergy Clin Immunol 2008; 8(6):577–584.
21. Niggeman B, Jacobsen L, Dreborg S, et al. Five year follow-up on the PAT study: Specific immunotherapy and long-term prevention of asthma in children. Allergy 2006; 61:855–859.
22. Jacobs RL, Freedman PM, Boswell RN. Nonallergic rhinitis with nasal eosinophilia (NARES syndrome). J Allergy Clin Immunol 1981; 67:253–262.
23. Moore EJ, Kern EB. Atrophic rhinitis: a review of 242 cases. Am J Rhinol 2001; 15:355–361.
24. Ellegard EK. Pregnancy rhinitis. Immunol Allergy Clin North Am 2006; 26:119–135.
25. Shusterman D. Environmental nonallergic rhinitis. Clin Allergy Immunol 2007; 19:249–266.
26. Meltzer EO, Hamilos DL, Hadley JA, et al. Rhinosinusitis: Establishing definitions for clinical research and patient care. J Allergy Clin Immunol 2004; 114(suppl): S156–S212.
27. Rohr AS, Spector SL. Paranasal sinus anatomy and pathophysiology. Clin Rev Allergy 1984; 2:387.
28. Hamilos DL. Chronic sinusitis. J Allergy Clin Immunol 2000; 106:213–227.
29. Schubert MS, Goetz DW. Evaluation and treatment of allergic fungal sinusitis. Demographics and diagnosis. J Allergy Clin Immunol 1998; 102:387–394.
30. Joint Task Force on Practice Parameters. The diagnosis and management of sinusitis: A practice parameter update. J Allergy Clin Immunol 2005; 116:S13–S47.
31. Giff FF, Neiburger JB. The role of nasal cytology in the diagnosis of chronic sinusitis. Am J Rhinol 1989; 3:13–15.
32. Gold SM, Tami TA. Role of middle meatus aspiration culture in the diagnosis of chronic sinusitis. Laryngoscope 1997; 107:1586–1589.
33. Rohr AS, Spector SL, Siegel SC, et al. Correlation between A-mode ultrasound and radiography in the diagnosis of maxillary sinusitis. J Allergy Clin Immunol 1986; 78:58–61.
34. Mudgil SP, Wise SW, Hopper KD, et al. Correlation between presumed sinusitis-induced pain and paranasal sinus computed tomographic findings. Ann Allergy Asthma Immunol 2002; 88:223–226.

35. Zinreich SJ. Imaging of chronic sinusitis in adults: X-ray, computed tomography and magnetic resonance imaging. J Allergy Clin Immunol 1992; 90(suppl):445–451.
36. Sinus and Allergy Health Partnership. Antimicrobial treatment guidelines for acute bacterial rhinosinusitis. Otolaryngol Head Neck Surg 2000; 123(suppl):S1–S32.
37. American Academy of Pediatrics. Subcommittee on the Management of Sinusitis and Committee on Quality Improvement. Clinical practice guideline: Management of sinusitis. Pediatrics 2001; 108:798–808.
38. Hickner JM, Bartlett JG, Besser RE, et al. Principles of appropriate antibiotic use for acute rhinosinusitis in adults. Ann Intern Med 2001; 134:495–497.
39. Benninger MS, Sedory-Holzer SE, Lau J. Diagnosis and treatment of uncomplicated acute bacterial rhinosinusitis: Summary of the Agency for Health Care Policy and Research evidence-based report. Otolaryngol Head Neck Surg 2000; 122:1–7.
40. Kennedy DW. Functional endoscopic sinus surgery: Technique. Arch Otorhinolaryngol 1985; 111:643–649.
41. Tos M. The pathogenic theories on the formation of nasal polyps. Am J Rhinol 1990; 4:51–56.
42. Fokkens W, Lund V, Mullol J. European position paper on rhinosinusitis and nasal polyps 2007. Rhinol Suppl 2007; (20):1–136.
43. Bernstein JM, Gorfien J, Noble B. Role of allergy in nasal polyposis: A review. Otolaryngol Head Neck Surg 1995; 113(6):724–732.
44. Burgel PR, Escudier E, Coste A, et al. Relation of epidermal growth factor receptor expression to goblet cell hyperplasia in nasal polyps. J Allergy Clin Immunol 2000; 106(4):705–712.
45. Singh H, Ballow M. Role of cytokines in nasal polyposis. J Investig Allergol Clin Immunol 2003; 13(1):6–11.
46. Steinke JW, Bradley D, Arango P, et al. Cysteinyl leukotriene expression in chronic hyperplastic sinusitis-nasal polyposis: importance to eosinophilia and asthma. J Allergy Clin Immunol 2003; 111(2):342–349.
47. Dagli M, Eryilmaz A, Besler T, et al. Role of free radicals and antioxidants in nasal polyps. Laryngoscope 2004; 114(7):1200–1203.
48. Winestock DP, Bartlett PC, Sondheimer FK. Benign nasal polyps causing bone destruction in the nasal cavity and paranasal sinuses. Laryngoscope 1978; 88(4):675–679.
49. Norlander T, Fukami M, Westrin KM, et al. Formation of mucosal polyps in the nasal and maxillary sinus cavities by infection. Otolaryngol Head Neck Surg 1993; 109(3 Pt 1):522–529.
50. Garavello W, Gaini RM. Incidence of inverted papilloma in recurrent nasal polyposis. Laryngoscope 2006; 116(2):221–223.
51. Bachert C, Gevaert P, van Cauwenberge P. Nasal polyps and rhinosinusitis. In: Adkinson, ed. Middleton's Allergy: Principles and Practice, 7th edn. 2008:991–997.
52. Radenne F, Lamblin C, Vandezande LM, et al. Quality of life in nasal polyposis. J Allergy Clin Immunol 1999; 104(1):79–84.
53. Kingdom TT, Orlandi RR. Image-guided surgery of the sinuses: Current technology and applications. Otolaryngol Clin North Am 2004; 37(2):381–400.
54. Nores JM, Avan P, Bonfils P. Medical management of nasal polyposis: A study in a series of 152 consecutive patients. Rhinology 2003; 41(2):97–102.
55. Patiar S, Reece P. Oral steroids for nasal polyps. Cochrane Database Syst Rev 2007:CD005232.
56. Hamilos DL, Thawley SE, Kramper MA, et al. Effect of intranasal fluticasone on cellular infiltration, endothelial adhesion molecule expression, and proinflammatory cytokine mRNA in nasal polyp disease. J Allergy Clin Immunol 1999; 103(1 Pt 1):79–87.
57. Parnes SM. Targeting cysteinyl leukotrienes in patients with rhinitis, sinusitis and paranasal polyps. Am J Respir Med 2002; 1(6):403–408.
58. Bhattacharyya N. Progress in surgical management of chronic rhinosinusitis and nasal polyposis. Curr Allergy Asthma Rep 2007; 3:216–220.

59. Bikhazi NB. Contemporary management of nasal polyps. Otolaryngol Clin North Am 2004; 37(2):327–337.
60. Garrel R, Gardiner Q, Khudjadze M, et al. Endoscopic surgical treatment of sinonasal polyposis-medium term outcomes (mean follow-up of 5 years). Rhinology 2003; 41(2):91–96.
61. Shusterman D, Murphy M-A. Nasal hyperreactivity in allergic and nonallergic rhinitis: A potential risk factor for nonspecific building-related illness. Indoor Air 2007; 17:328–333.

18 Chronic Tissue Changes and Carcinogenesis in the Upper Airway

Ruud A. Woutersen and C. Frieke Kuper

Department of Toxicology and Applied Pharmacology, Business Unit Quality and Safety, TNO Quality of Life, Zeist, The Netherlands

Piet J. Slootweg

Department of Pathology, Radboud University Medical Center, HB Nijmegen, The Netherlands

INTRODUCTION

Inhalation toxicity testing is becoming increasingly important, among other things because of the increasing number of drugs that are administered through the intranasal route to avoid enzymatic and acid breakdown and first pass metabolism in the gastrointestinal mucosa and the liver. The nature and site of damage to the nasal mucosa are influenced by a series of factors related to the airborne substance (dose, exposure concentration, solubility, polarity, diffusion rate, particle size, shape, and density), the exposure pattern (continuous, interrupted, peak loads), the subject (genetic constitution, anatomy of the respiratory tract, respiration rate, detoxification systems), and coexposures (lifestyle, occupation). A great number of chemicals are capable of inducing upper respiratory tract tumors in rodents. Nasal tumors are encountered most frequently, whereas laryngeal tumors are relatively rare (1–5). A large number of nitrosamines are capable of inducing nasal tumors in rats and hamsters (6–9). Furthermore, tumors of the nose develop in experimental animals after exposure to a variety of industrial chemicals such as ethylene dibromide (10), dimethyl sulfate (11), formaldehyde (12), acetaldehyde (13), bis(chloromethyl) ether (14), hexamethylphosphoramide (15), propylene oxide (16,17), phenylglycidyl ether (18), dimethylcarbamoyl chloride (19), epichlorohydrin (20), and vinyl chloride (21).

Although exposure to several of these chemicals is common in industrial as well as domestic environments, epidemiological studies are scarce and have not provided convincing evidence that exposure to these chemicals individually is associated with nasal cancer in humans (21). The striking differences in size and internal shapes of the nose and the possible differences in physiology of the nose between rodents, which in contrast to humans are obligatory nose breathers, and humans is a factor that may be responsible for the absence of a carcinogenic response in humans (22).

On the other hand, the evidence for nasal carcinogenicity of inhaled chemical mixtures in experimental animals is very limited, while there is ample evidence in humans that occupational exposure to certain chemical mixtures is associated with increased risk of nasal cancer. Examples of such (complex) chemical mixtures are wood dust, textile dust, chromium- and nickel-containing materials, and leather dust. It is remarkable that these mixtures are aerosols, suggesting that a "dusty working environment" may increase nasal cancer risk (23). Recently, Albertini and Sweeney (17) published a review paper on the genotoxicity of the rodent nasal carcinogen propylene oxide. They concluded that apart from the DNA reactivity of propylene oxide, associated toxicities such as increased cell proliferation and depletion of detoxification mechanisms may be necessary for its carcinogenicity and suggested that this is probably a much more general mechanism for nasal carcinogens.

The present review concentrates on the role of sustained tissue damage and repair in the development of upper respiratory tumors induced in experimental animals and humans by a variety of compounds with emphasis on the effects of formaldehyde, acetaldehyde, and propylene oxide. The histopathology of upper respiratory tract tumors in humans and rodents have been described in detail in a previous paper (22) and will only be summarized in the present paper.

HISTOPATHOLOGY OF UPPER RESPIRATORY TRACT TUMORS IN EXPERIMENTAL ANIMALS

Tumors of the Squamous Epithelium and Respiratory Mucosa of the Nose

Polypoid Adenoma

Polypoid adenomas (Fig. 1) (24) were observed in male rats exposed to 20 ppm formaldehyde (25,26), and in one male rat exposed to 1.5/2.0 ppm trichlorobutene for 83 weeks (27). There seems to be little doubt that polypoid adenomas originate from respiratory epithelium in view of the tumors' localization, histology, and cytology (28). Polypoid adenomas occur only very rarely in an untreated control rat. Therefore, they are considered most probably to be related to treatment and in the case of formaldehyde as distinctively so.

Carcinoma In Situ

In several rats exposed to high concentrations of acetaldehyde (1500 or 3000 ppm) carcinomas in situ developed in metaplastic stratified squamous respiratory epithelium (Fig. 2) (13). Clearly, these tumors were part of the neoplastic response of the nasal respiratory epithelium to acetaldehyde and may be considered precursors of squamous cell carcinomas.

Squamous Cell Carcinomas

Squamous cell carcinomas (Fig. 3) varied from large tumors, filling one or both sides of the nasal cavity, destroying turbinates and bones, and extending into the subcutis and brain, to small neoplasms invading the submucosa of the nasal epithelium (29). The large tumors often showed extensive keratinization, and their origin could not be determined with certainty. Most small squamous cell carcinomas were seen to originate from metaplastic, keratinized, stratified squamous respiratory epithelium, and they occurred in the anterior part of the nose. A few squamous cell carcinomas appeared to be derived from metaplastic, keratinized, squamous olfactory epithelium located in the dorsomedial and posterior part of the nasal cavity (Fig. 4).

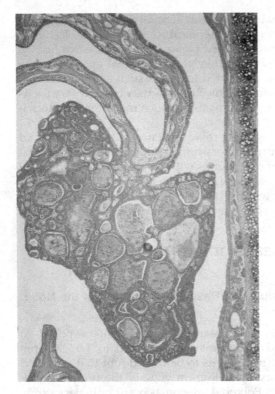

FIGURE 1 Polypoid adenoma originating from a nasoturbinate. Male rat exposed to 20 ppm formaldehyde for 4 weeks. H&E, x40.

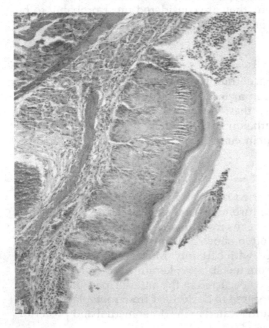

FIGURE 2 Carcinoma-in-situ originating from the olfactory epithelium of the organ of Masera on the nasal septum. Male rat exposed to 1500 ppm acetaldehyde for 28 months. H&E, x40.

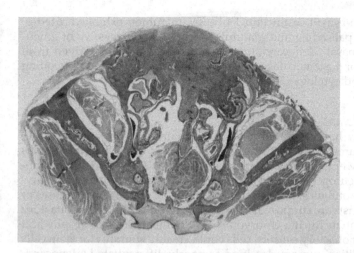

FIGURE 3 Male rat exposed to 3000/1000 ppm acetaldehyde for 28 months showing both an adenocarcinoma originating from the olfactory epithelium dorsomedian in the nose and a squamous cell carcinoma originating from the repiratory epithelium lining the septum ventrally in the nose. H&E, x10

Tumors of the Olfactory Mucosa of the Nose

Adenocarcinomas

These tumors consist of compact sheets and cords of cells separated by strands of fibrous tissue varying widely in thickness (Fig. 3). Tumor cells are pleomorphic, and bizarre mitotic figures are often observed. Tumors often contain both dark and light cells with large hyperchromatic round-to-oval nuclei. Dark cells were small with scanty cytoplasm, sharp nuclear membranes, a fine nuclear

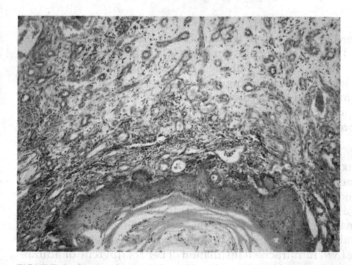

FIGURE 4 Squamous metaplasia (keratinised) of the olfactory epithelium in the dorsomedial part of the nasal cavity. Rat exposed to 3000/1000 ppm acetaldehyde for 28 months. H&E, x40

chromatin pattern, and small, distinct nucleoli. Several tumors exhibited rosettes, pseudorosettes, and palisading and glandular formations, suggestive of a neurogenic origin. The nasal adenocarcinomas are considered to be derived from olfactory stem cells or sustentacular cells based on their localization, prevailing histologic pattern, and cytology (30).

Olfactory Neuroblastomas

Several investigators have reported neuroepitheliomas, esthesioneuroepitheliomas, (esthesio)neuroblastomas, esthesioneuromas, or esthesioneurocytomas of the nasal olfactory epithelium in experimental animals treated with carcinogens (4,31–35). The neurogenic origin of this lesion is apparent from the unequivocal presence of neurotubules, neurosecretory granules, or neuritic processes of tumor cells. Neuroblasts are supposedly the precursor cells from which olfactory sensory cells differentiate during embryonic development (36,37). Atrophy and toxic degeneration of olfactory sensory cells could create a stimulus for stem cells to proliferate. In addition, tumors classified as poorly differentiated adenocarcinomas may originate from neuroblasts that have lost their characteristic morphological markers (e.g., neurotubules, axons), and in this case, they are probably of neurogenic origin and should be classified as olfactory neuroblastomas. On the other hand, poorly differentiated adenocarcinomas exhibiting duct-like spaces resembling rosettes and pseudorosettes may be misdiagnosed as olfactory neuroblastomas.

Carcinosarcoma

One carcinosarcoma was found in a female rat exposed to 5000 ppm vinyl chloride (38). In this tumor sarcomatous structures predominated, although neoplastic epithelial elements were also present.

Schwannoma

Three male rats exposed to 1.5/2.0 ppm trichlorobutene exhibited nasal tumors, which were diagnosed as malignant Schwannomas (4). Two of the tumors were very large osteolytic masses growing outside the nasal cavity, while the third was a limited tumorous process expanding in the olfactory lamina propria. The diagnosis was based mainly on the involvement of mucosal nerve bundles, and the typical histologic picture showing whorls of plump spindle or epitheloid cells and circumscribed fibrous areas.

Tumors of the Larynx (Fig. 5)

In rats, laryngeal tumors included carcinomas (in situ), squamous cell carcinomas, and adenocarcinomas. These tumors were all considered to be related to treatment with either acetaldehyde or propylene oxide. In the larynx of hamsters, papillomas, carcinomas (in situ), and squamous cell carcinomas were found when exposed to acetaldehyde vapor alone or simultaneously to benzo[a]pyrene or diethylnitrosamine (39). Intratracheal instillation of benzo[a]pyrene or administration of carcinogenic nitrosamines through various routes also leads to laryngeal tumors in Syrian golden hamsters (40).

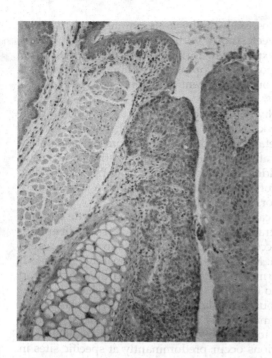

FIGURE 5 Carcinoma-in-situ of the larynx. Female rat exposed to 1500 ppm acetaldehyde for 122 weeks. H&E, x 40.

IRRITATING NASAL CARCINOGENS IN RODENTS

Formaldehyde Carcinogenesis

There is consistent evidence for the genotoxicity of formaldehyde in in vitro systems, laboratory animals, and exposed humans (41). Formaldehyde is weakly genotoxic but capable to react with nucleic acids and proteins. DNA adducts, DNA–protein crosslinks, stand breaks, and the induction of repair mechanisms have been detected in vitro. In V79 Chinese hamster cells, formaldehyde induced DNA–protein crosslinks, sister-chromatid exchanges, and micronuclei, but no gene mutations, in concentrations similar to those inducing cytotoxicity, suggesting that formaldehyde-induced DNA–protein crosslinks are related to cytotoxicity and clastogenicity. In vivo, DNA–protein crosslinks have been detected in tracheal (42) and nasal epithelium (43–45). In monkeys, the levels of DNA–protein crosslinks were highest in the mucosa of the middle turbinates. Very few crosslinks were found in the larynx, trachea, and in the major bronchi (46). In the nasal epithelium of F344 rats, DNA–protein crosslinks were still detected at formaldehyde concentrations as low as 0.3 ppm (47).

In rats exposed to formaldehyde concentrations of 10 ppm, daily for six hours on five days a week, rhinitis, and hyperplasia, and squamous metaplasia of the nasal respiratory epithelium were described in all studies. In rats exposed to 1.0 ppm for two years, no histopathological changes were observed (48). Concentrations of 2 ppm and higher induced rhinitis, epithelial dysplasia and even papillomatous adenomas and squamous metaplasia of the respiratory epithelium of the nose, and 6 ppm and higher induced squamous cell carcinomas (49,50). Uninterrupted exposure of rats for 8 hr/day (continuous) was compared with eight exposures for 30 minutes followed by a 30-minute phase without exposure (intermittent) in two 13-week studies with the same total dose. Effects were seen only after intermittent exposure to formaldehyde concentrations of 4 ppm, but not after continuous exposure to 2 ppm. The authors concluded that the toxicity in the nose depends on the concentration and not on the total dose (51).

In mice exposed to formaldehyde concentrations of 2.0, 5.6, or 14.3 ppm for two years (6 hr/day, 5 day/wk), rhinitis and epithelial hyperplasia was observed. At 5.6 ppm dysplasia, metaplasia and atrophy was present; squamous cell carcinomas were observed only after exposure to a concentration of 14.3 ppm (49). In hamsters exposed to formaldehyde concentrations of 10 ppm (5 hr/day, 5 day/wk) for life, the incidence of hyperplasia and metaplasia was slightly increased, but the incidence of tumors was unaffected (52).

Induction of nasal squamous carcinomas in rats by formaldehyde requires long-term exposure to high concentrations (10–20 ppm), which results in epithelial degeneration and cell death accompanied by rhinitis, followed by regenerative hyperplasia and metaplasia. Moderate-to-severe histopathological nasal changes are observed after long-term exposure to 6 ppm, and they increase in severity in a linear fashion from 10 to 15 ppm. However, more subtle nasal epithelial changes such as slight squamous metaplasia, mild focal hyperplasia with or without squamous metaplasia, and slightly increased cell proliferation have been observed in rats exposed to formaldehyde concentrations ranging from 0.3 to 2 ppm (53–55). These alterations occur predominantly at specific sites in the nasal mucosa lined by transitional or respiratory epithelium. Despite differences in anatomy and physiology of the nose between rats and humans, the respiratory tract defense systems are similar in both species. It is, therefore, reasonable to assume that the response of the respiratory tract to formaldehyde will be qualitatively similar in rats and humans. If in humans exposure to formaldehyde is accompanied by recurrent tissue damage and repair at the site of contact, formaldehyde in cytotoxic concentrations may be assumed to have carcinogenic (and also mutagenic) potential in humans. Correspondingly, if the respiratory tract tissue is not recurrently injured, exposure of humans to relatively low, noncytotoxic levels of formaldehyde can be assumed to represent a negligible, if any, cancer risk. Therefore, in humans, formaldehyde exposure should be controlled to levels below that likely to produce a cytotoxic effect. This threshold is expected to be in the range of 0.3 to 0.5 ppm (56–58).

In conclusion, formaldehyde is a highly irritating, genotoxic carcinogen capable of inducing malignant tumors in the nose of rats and mice after long-term inhalation exposure (49). There is substantial experimental evidence to suggest a crucial role for tissue damage and persistent hyper- and metaplasia of the nasal respiratory epithelium in formaldehyde carcinogenesis (59,60). A very important piece of evidence originates from a long-term inhalation study in which male rats with a severely damaged or an undamaged nasal mucosa were exposed to 0, 0.1, 1.0, or 10.0 ppm formaldehyde for 6 hr/day, 5 day/wk, for either 28 months or for 3 months followed by an observation period of 25 months (48). Treatment-related nasal tumors occurred only in the 10-ppm group of rats with a damaged nose exposed to formaldehyde for the full 28 months. Evidently, "drastic" conditions were required for tumor formation: severe damage plus a high concentration of formaldehyde.

Acetaldehyde Carcinogenesis

Acetaldehyde has been shown to induce chromosomal aberrations, micronuclei, and/or Sister Chromatid Exchanges in cultured mammalian cells including human lymphocytes (61–67). Acetaldehyde may produce similar cytogenetic effects in vivo (68). The production of cytogenetic effects may be related to the ability of acetaldehyde to form DNA–DNA and/or DNA–protein crosslinks (68–72). Acetaldehyde has been found to produce gene mutations (sex-linked recessive lethals) in Drosophila (73), but most tests in bacteria have been negative and no information is available on the ability of acetaldehyde to induce gene mutations in cultured mammalian cells (68). It can be concluded that acetaldehyde is a genotoxic cross-linking agent with limited or no ability to induce gene mutations.

In rats, the nose and larynx appeared to be the main target sites of acetaldehyde vapor (74). From a 28-month carcinogenicity study in rats, using exposure levels of 0, 750, 1500, and 3000/1000 ppm, it was clear that acetaldehyde vapor produced nasal carcinomas at each exposure level tested (13,75), At each exposure concentration, in addition to nasal tumors, degenerative hyperplastic and metaplastic changes of the nasal mucosa were observed. The results of a recovery study suggested that non-neoplastic, hyperproliferative changes of the nasal epithelium seen at the end of the exposure period (52 weeks) may progress to malignant tumors despite discontinuation of treatment (76). Chronic exposure of hamsters to a high concentration of acetaldehyde (2500 ppm decreased to 1650 ppm after 45 weeks) resulted in only a few nasal tumors (one adenoma, one adenocarcinoma, and one anaplastic carcinoma in a total of 53 animals), despite severe and extensive keratinized squamous metaplasia of the respiratory epithelium (39,77).

The distribution of nasal lesions induced by acetaldehyde correlated with regional acetaldehyde dehydrogenase deficiencies, indicating that regional susceptibility to the toxic effects of acetaldehyde may be due, at least in part, to a lack of aldehyde dehydrogenase in the susceptible regions (78). The aforementioned data indicate that acetaldehyde meets the widely accepted criteria to be a complete carcinogen. Since, in analogy to formaldehyde, acetaldehyde also may react preferentially with single-stranded DNA and the proliferation rate in the normal intact nasal epithelium is low, it is reasonable to assume that the initiating potential of acetaldehyde in concentrations not leading to cell damage is very low, whereas concentrations of acetaldehyde causing recurrent tissue damage may be very effective with respect to initiation (13).

In summary, acetaldehyde is a genotoxic carcinogen, of which the irritating cytotoxic potential probably but not certainly plays a crucial role in its carcinogenicity. As long as the role of the cytotoxicity in the carcinogenicity of acetaldehyde has not been demonstrated more clearly, its genotoxicity rather than cytotoxicity should be considered the determining factor in cancer risk assessment.

Propylene Oxide Carcinogenesis

Propylene oxide is a DNA-reactive genotoxic agent, but its potency as a DNA-reactive mutagen is weak. It causes glutathione (GSH) depletion, cell proliferation, and necrosis (17,79,80). In a National Toxicology Program study, papillary adenomas were found in the nasal epithelium in rats and carcinomas, papillomas, hemangiomas, and hemangiosarcomas in mice (16,81).

The development of these tumors was accompanied by an exposure-related increased incidence of inflammation, hyperplasia, and/or squamous metaplasia of the respiratory epithelium lining the upper respiratory tract (rats) and nasal turbinates (mice). A long-term inhalation study with propylene oxide in Wistar rats (82,83) demonstrated nonsignificant increased incidences of carcinomas in the nose, larynx/pharynx, trachea, and lungs. In these studies there was a concentration-related increase in focal hyperplasia and degenerative changes of the nasal epithelium. The incidence and severity of the responses increased with the concentration from slight at 30 ppm to moderate at 100 ppm and marked in the animals exposed to 300 ppm (82,83). Inflammation, degeneration, and hyperplasia of the nasal epithelium occurred in cooperation with tumor formation at the port of entry (17,84). The pattern of tumor induction in rodents provides significant clues as to the role of the genotoxicity of propylene oxide in its carcinogenicity. The authors concluded that the weak DNA-reactivity of propylene oxide will not be sufficient to induce cancer. Enhancement of this weak DNA-reactivity by cell proliferation and glutathione depletion, which is known to result in Reactive Oxygen Species and hence DNA adduct formation, seems to be required for tumor induction.

IRRITATING NASAL NONCARCINOGENS IN RODENTS

There is an increasing volume of information which shows that chemically induced chronic toxicity, with histological evidence of epithelial regeneration (cell proliferation, hyperplasia) is not necessarily associated with carcinogenesis in rodents. Acrolein, a very cytotoxic unsaturated aldehyde, induced rhinitis and hyperplasia and metaplasia of the nasal respiratory and olfactory epithelium in Syrian hamsters, but no (nasal) tumors (85). Chronic exposure of Syrian hamsters to furfural, a heterocyclic aldehyde, resulted in highly characteristic alterations of the olfactory epithelium but did not lead to tumors (86). Two-year exposure of rats to dimethylamine led to concentration-dependent lesions of the nasal olfactory mucosa, including destruction of sensory cells and nerve bundles, but no nasal tumors occurred (87). In a two-year inhalation study with ethyl acrylate in rats, the compound did not induce nasal tumors, but the olfactory epithelium was severely damaged, showing atrophy of sensory cells, basal cell hyperplasia, and intraepithelial glandular structures (88). Despite marked damage to the nasal mucosa, no nasal tumors were found in Syrian hamsters after chronic exposure to formaldehyde vapor (52). Chronic inhalation exposure of rats and mice to chlorine resulted in epithelial degeneration, hyperplasia, and metaplasia but no neoplasia of the nasal passages (89). Ward et al. (90) summarized the results of inhalation studies with 19 chemicals in rats and mice sponsored by the National Toxicology Program. Five chemicals were found to be nasal carcinogens, and 14 chemicals were not carcinogenic to the nasal tissues. The five carcinogens induced inflammatory and proliferative (regenerative) nasal lesions. Of the 14 noncarcinogens, 12 induced similar inflammatory and proliferative nasal lesions. Of the five nasal carcinogens, three also caused tumors at other sites, and 7 of the 14 nasal noncarcinogens caused tumors elsewhere. The nasal carcinogens induced nasal lesions in all experiments, whereas the nasal noncarcinogens induced nasal lesions in part of the experiments (36/59). Compounds such as acrolein, furfural, dimethylamine, chlorine, and ethylacrylate are very cytotoxic to the nasal mucosa and induced hyperproliferative changes but did not cause nasal tumors. Apparently, the local tissue conditions created by these compounds did not meet the requirements for tumor initiation, promotion, and progression.

NASAL EFFECTS OF CHEMICAL MIXTURES IN RODENTS

Inhalation studies addressing nasal injury due to exposure to defined mixtures of chemicals are very scarce. Those found in the literature deal with mixtures of aldehydes (45,91–93) or mixtures consisting of one or more aldehydes and chemicals such as ozone or ammonia (94–97). Teredesai and Stinn (97) carried out three-day inhalation studies (7 hr/day) in male rats with formaldehyde, acrolein, and ammonia alone or as a mixture. Nasal ulceration was found after exposure to acrolein alone (0.9 or 1.8 ppm) and appeared to occur more frequently in rats simultaneously exposed to formaldehyde plus ammonia at levels not leading to nasal damage themselves (0.9 or 1.9 ppm for formaldehyde; 34 or 158 ppm for ammonia). Moreover, basal cell hyperplasia and squamous metaplasia of the nasal respiratory epithelium were seen after exposure to a mixture of 0.4 ppm acrolein, 0.4 ppm formaldehyde, and 34 ppm ammonia, whereas each of these compounds alone at the same concentration did not visibly affect the nasal respiratory epithelium, indicating that these chemicals have the same target site and act according to the concept of (partial) dose addition.

A number of three-day inhalation toxicity studies in male rats with mixtures of aldehydes and with the individual compounds as well revealed that the histopathological changes and cell proliferation of the nasal epithelium induced by mixtures of formaldehyde (3.2 ppm), acetaldehyde (1500 ppm), and/or acrolein (0.67 ppm) were more severe and more extensive than those observed after exposure to the individual aldehydes at comparable exposure levels (93). The combined effect indicated at least summation of the acrolein and formaldehyde effects on the nasal respiratory epithelium and probably potentiation of the acetaldehyde effect on the olfactory epithelium by formaldehyde and/or acrolein. However, neither additivity nor potentiating interaction occurred following exposure to mixtures of these aldehydes at the no-observed-adverse-effect level (NOAEL) for formaldehyde (1.0 ppm) and acetaldehyde (750 ppm) and the lowest-observed-adverse-effect level (LOAEL) for acrolein (0.25 ppm), because the severity of the nasal changes seen after exposure to this mixture appeared to be similar to the severity of the nasal changes induced by exposure to 0.25 ppm acrolein alone.

These findings suggest that combined exposure to these chemicals (aldehydes) with the same target organ (nose) and exerting the same type of adverse effect (nasal cytotoxicity), but with slightly different target sites (different regions of the nasal passages), is not associated with a greater hazard than exposure to the individual chemicals, provided the exposure concentrations are NOAELS.

In a long-term inhalation study, Holström et al. (98) exposed female Sprague-Dawley rats to formaldehyde (15 mg/m^3) or wood dust (25 mg/m^3) or to a mixture of both at comparable concentrations. Pronounced squamous cell metaplasia with or without keratinization of the nasal respiratory epithelium was only found in rats exposed to formaldehyde alone or in combination with wood dust. Squamous metaplasia with dysplasia was observed in 4/15 rats exposed to the combination, and in 1/16 rats exposed to formaldehyde alone. The animals exposed to wood dust alone did not show pronounced metaplasia. One keratinized squamous cell carcinoma was found in the nasal cavity of a rat exposed to formaldehyde alone. A benign fibro-osseous tumor was encountered in a rat exposed to the combination. In a study with Syrian golden hamsters exposed to beech wood dust, one of the animals developed a well-differentiated keratinized squamous cell carcinoma, while another animal exhibited focal cuboidal metaplasia with mild dysplasia of the nasal epithelium (99,100).

NASAL CARCINOGENS IN RODENTS AFTER ORAL ADMINISTRATION

Chronic lesions in the nasal tissues can be induced by quite a number of compounds upon exposure other than by inhalation or nasal application (Table 1; 101–103). In these studies, respiratory tissues may have been exposed through

TABLE 1 Nasal Lesions Upon Chronic Oral Exposure

Nasal lesion	Substances	Route
Respiratory epithelium hyperplasia	Mixture of PCB126 + PCB153	Gavage[a]
	Mixture PCB126 + PCB118	
Respiratory metaplasia, sometimes including hyperplasia with papillary projections, of olfactory epithelium	Mixture of PCB126 + PCB153	Gavage[a]
	Mixture PCB126 + PCB118	
	Monochloroacetic acid	
	Mercuric chloride	
	Benzophenone	Feed[a]
Epithelial dysplasia	2,3-Dibromo-1-propanol	Dermal[a]
Atrophy/degeneration/metaplasia of olfactory epithelium	Methacrylonitrile	Gavage[a]
	Dipropylene glycol	Drinking water[a]
	Cyclohexanone Oxime	
	Butanal oxime	
	Methyl ethyl ketoxime	
	Benzyl acetate	Feed[a]
	o-Nitrotoluene	Feed[a]
Cystic hyperplasia of glands	Benzyl acetate	Feed[a]
Pigmentation (brown intracytoplasmic granules) of olfactory epithelium	Pentachloroanisole	Gavage[a]
	trans-Cinnamaldehyde	Feed[a]
Adenoma	2,3-Dibromo-1-propanol	Dermal[a]
	2,6-Xylidine	Feed[a]
Carcinoma	2,6-Xylidine	Feed[a]
	Pentachlorophenol	
	Dimethylvinylchloride	Gavage[b]
Squamous cell carcinoma	1,4-Dioxane	Drinking
	N-Nitrosodiethylamine	water[a,b,c]
	1,4-Dioxane	
Carcinoma, NOS	Dimethylvinyl chloride	Gavage[a]
Neuroblastoma	p-Cresidine	Feed[a]
	N-Nitrosodiethylamine	Drinking water[c]
Rhabdomyosarcoma	2,6-Xylidine	Feed[a]

[a]Reviewed by Sells et al. (2007) (102).
[b]Reviewed by Jeffrey et al. (2006) (103).
[c]Reviewed by IARC (2000) (159).

the blood, by exhaling the substance or its toxic metabolite(s) (104), by inhalation of volatile substances from drinking water (105) or feed, or from regurgitation/reflux of substances from the gastrointestinal tract, especially after gavage of the substance. Local metabolism of the parent substance could be one of the pivotal mechanisms by which the nasal lesions were induced. This is supported by the association between lesions in the nasal tissues and the liver (101–103). In that case, humans may be protected by their lesser bioactivation of the nasal mucosa (103).

MALOCCLUSION SYNDROME AND NASAL CANCER IN RATS

The rat incisor teeth develop from cells that are derived from the continuously proliferating odontogenic epithelium, which is located at the base of the tooth and which encloses the connective tissue of the primitive pulp. An adult incisor tooth undergoes functional attrition as a normal wearing process. When one of the incisors is broken or when a malocclusion occurs the nonattrition of the opposing incisor and its continuous growth results in an elongation or overgrowth. The term "overgrowth" is misleading as, in reality, this elongation is not related to the rate of growth of the tooth but is the result of the lack of wear. The malocclusion syndrome, which occurs in about 10% of Cpb: WU (Wistar random) rats, is very frequently accompanied by odontitis, periodontitis, sinus maxillary sinusitis, and rhinitis (106). It is reasonable to assume that a condition of severe necrotizing (peri)odontitis that involves the whole area of the incisor tooth, the maxillary turbinate, and the sinus maxillaris may enhance the risk of the development of squamous cell carcinoma or odontoblastoma, originating from metaplastic respiratory or odontogenic epithelium, respectively (5). Occasionally, we have observed a nasal squamous cell carcinoma in untreated controls in which the tumor was seen to be associated with severe necrotizing periodontitis and rhinitis, indicating that this condition of chronic irritation is probably associated with the formation of some of the nasal tumors found in our strain of rats. It is possible that in rats, nasal squamous cell carcinomas could be considered part of the malocclusion syndrome in some instances. Since inflammatory reactions such as odontitis, periodontitis, and rhinitis are accompanied by recurrent tissue damage and repair, rats suffering from the malocclusion syndrome maybe more susceptible to nasal carcinogens than unaffected rats.

UPPER RESPIRATORY TRACT CARCINOGENESIS IN HUMANS

Histopathology of Upper Respiratory Tract Tumors in Humans

Upper respiratory tract tumors arise in the corresponding mucosal lining. Recently, the WHO has updated the classification of these tumors (107). Basically, two different groups can be discerned: squamous cell carcinomas and adenocarcinomas. Within the squamous cell carcinoma group, one discerns between the keratinizing and the nonkeratinizing type.

Keratinizing squamous cell carcinoma shows squamous differentiation, in the form of extracellular keratin or intracellular keratin. The tumor may be arranged in nests, masses, or as small groups of cells or individual cells. There is often a desmoplastic stromal reaction.

Nonkeratinizing (cylindrical cell, transitional) carcinoma is characterized by a ribbon-like growth pattern. It invades into the underlying tissue with a smooth,

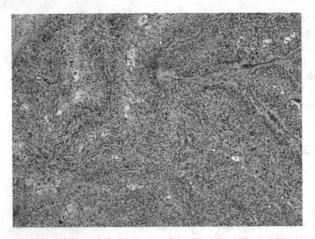

FIGURE 6 Human non-keratinizing sinonasal squamous cell carcinoma consisisting of juxtaposed epithelial ribbons with little intervening stroma.

generally well-delineated border. As its name implies, this tumor does not generally show histologic evidence of keratinization (Fig. 6).

Adenocarcinomas are glandular malignancies of the sinonasal tract, excluding defined types of salivary gland carcinoma. Two main categories are recognized: (1) intestinal-type adenocarcinoma and (2) nonintestinal-type adenocarcinoma.

Intestinal-type adenocarcinoma is a tumor of the nasal cavity and paranasal sinuses that histologically resembles adenocarcinoma of the large bowel. These tumors may show varying growth patterns leading to their subdivision in various subtypes: papillary, colonic, solid, mucinous, and mixed. The most common histologic types seen in association with wood workers as well as in sporadic cases are the papillary and colonic types. The papillary type shows a papillary architecture with occasional tubular glands (Fig. 7). The colonic type shows a

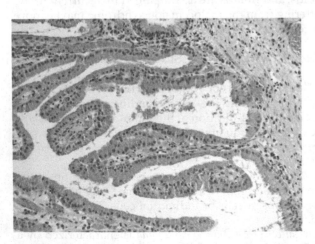

FIGURE 7 Human intestinal type sinonasal adenocarcinoma of the papillary type. High columnar cells are arranged in a papillary fashion.

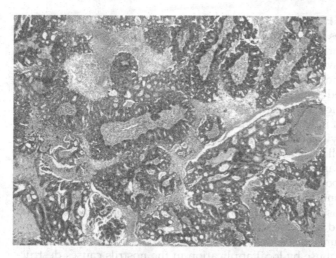

FIGURE 8 Human intestinal type sinonasal adenocarcinoma of the colonic type. Strands composed of atypical columnar cells contain lumina with necrotic content.

tubulo-glandular architecture (Fig. 8). The solid type is characterized by solid and trabecular growth with isolated tubule formation. Analogous to colonic adenocarcinoma, some tumors may contain abundant mucus.

Sinonasal nonintestinal-type adenocarcinoma is defined as a tumor arising in the sinonasal tract that is not of minor salivary gland origin and does not demonstrate histopathologic features of the above-described sinonasal intestinal-type adenocarcinoma. The nonintestinal-type adenocarcinoma has a glandular or papillary growth pattern. Numerous uniform small glands or acini are arranged in a back-to-back or coalescent pattern with little or no intervening stroma (Fig. 9). Occasionally, large, cystic spaces are observed.

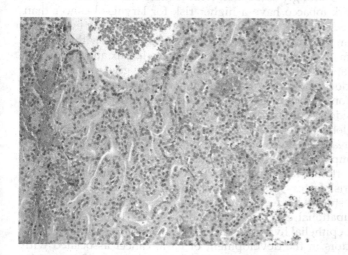

FIGURE 9 Human non-intestinal-type sinonasal adenocarcinoma composed of tightly packed tubuli.

Nasal Carcinogens in Humans

Epidemiological studies have demonstrated an increased risk of upper respiratory tract cancer, primarily nasal tumors, in several occupations, such as nickel refinery workers and workers engaged in the manufacture of wooden furniture or leather boots and shoes (108). In addition, nasopharyngeal cancer in southeastern China and Hong Kong has been associated with the consumption of salted fish, which appeared to contain high concentrations of volatile nitrosamines capable of inducing nasal tumors in experimental animals (7).

Upper respiratory tract cancers in humans have a multifactorial etiology: sinonasal squamous cell cancers may develop from exposure to tobacco smoke, nickel, softwood dust, and mustard gas production, whereas adenocarcinomas may develop from exposure to hardwood, chrome pigment, and leather dust (108–112). Another agent frequently cited as being involved with cancer of the nasal cavity is thorotrast (113). Moreover, an elevated risk of adenocarcinoma among women exposed to textile dust has been observed and there are some suggestions that exposure to asbestos may increase the risk of squamous cell carcinoma (114). Cocaine abuse by local application in the nostrils causes destruction of the nasal septum and chronic inflammatory tissue alterations but no carcinogenic potential has been shown of this compound until now (115).

According to IARC (41), there is sufficient evidence of carcinogenicity of formaldehyde in humans with respect to nasopharyngeal carcinomas (114,116). The induction of human malignancies at this site is likely based on similar mechanisms as the experimental inductions of nasal tumors in rats. On one hand, dosimetry modelings have indicated that human nasal flux patterns shifted distally as inspiratory flow rate increased (117), on the other hand it appears important that the rat is an obligatory nose breather while humans, especially upon physical work, show considerable mouth breathing in addition. In essence, it is expected that the dose–response of human nasopharyngeal tumors elicited by formaldehyde must be nonlinear at low doses, based on the modes of action established experimentally in rodents.

Laryngeal cancer is strongly associated with tobacco use and alcohol consumption. Users of dark tobacco have a higher risk for laryngeal cancer than users of light (flue-cured) tobacco (118). In a large multicenter study evaluating alcohol consumption and tobacco use, the relative risk associated with cigarette smoking was approximately 10 for all subsites within the larynx and hypopharynx (119). Local nasal application of smokeless tobacco may give rise to chronic tissue alterations but does not appear to have any carcinogenic potential (120). A Latin American custom of drinking a nonalcoholic drink, mate, has been associated with an increased risk for laryngeal, oral, oropharyngeal, and esophageal cancer. In this era of global travel, mate has become available in all parts of the world. This drink is a tea-like infusion of the herb *Ilex paraguariensis (yerba mate)*. Possibly a phenolic compound in the drink may act as a promoter (121). The exact mechanism is still uncertain. Mate drinking in the traditional manner should be considered one of the risk factors for cancer of the head and neck (122,123).

Epidemiological studies have convincingly shown an association between nasal cancer and occupational exposure to wood dust (124,125). Reduction in mucociliary transport, epithelial hyperplasia, metaplasia, and dysplasia are considered etiological factors in the development of nasal cancer associated with exposure to wood dust. (110,126–128). Chemicals introduced in wood or wood's natural constituents, and mechanical irritation of the nasal mucosa by wood dust particles, have been suggested as causative or enhancing agents (110,126). However, using scanning electron microscopy for characterizing the morphology of dust particles from a series of wood species, Schmitt et al. (129) concluded that mechanical irritation in the nose cannot be the main principle of action of wood dust carcinogenesis. Moreover, in a recent evaluation of the carcinogenicity of hardwood and softwood dust, the Health Council of the Netherlands (130) concluded that the available data are insufficient to classify wood dust as a direct or indirect genotoxic carcinogen or as a nongenotoxic carcinogen with an essential role for regenerative hyperplasia following recurrent tissue damage.

DISCUSSION AND CONCLUSIONS

Cell proliferation, an essential component of the multistage process of carcinogenesis, is required for both initiation and promotion of neoplasia in certain organs, and it plays an essential role in the later stages of carcinogenesis, including progression of benign lesions to malignancy and metastasis. Each time a cell divides there is a chance, albeit rare, that a mutational event related to the carcinogenic process will occur (131–138). Enhanced cell proliferation may increase the frequency of these spontaneous mutations either by errors in replication or by the conversion of endogenous or exogenous DNA adducts to mutations before DNA repair can occur. Cell division is required to convert a DNA adduct to a permanent mutation. A genotoxic chemical administered at a toxic dose that also induces cell proliferation will be disproportionally more effective as a mutagen and as a carcinogen when administered at a toxic dose that also induce cell proliferation. In a population of proliferating cells, there will be less time for DNA repair processes to remove DNA adducts before cell replication converts adducts to mutations.

There is substantial evidence that following tissue damage, sustained cell proliferation (repair and hyperplasia) in experimental animals and humans may act as a precursor of tumor formation (139). In humans, liver cancer is 10 times more likely in patients with cirrhotic than with noncirrhotic livers (140); colon cancer is frequently seen in persons with chronic colitis (141); squamous cell carcinoma may develop in the hyperplastic epithelium around chronic skin ulcers (142); lung cancer grows in areas of scarring (143). In experimental animals, subcutaneous sarcomas develop at the site of repeated injection of compounds such as common salt or glucose (144); liver cancer in rats often occurs in enlarged livers (139); urinary bladder cancer develops in rats following implantation of solid objects into the bladder (145). Taken collectively, an overall picture emerges that sustained hyperplasia renders various sites in the rodent (139) and in humans vulnerable to tumor development. As far as the nose is concerned, in humans chronic inflammation is not considered a predisposing factor for cancer of the maxillary sinus (146). However, chronic necrotizing (peri)odontitis, rhinitis, and sinusitis may be involved in the development of nasal squamous cell carcinomas in untreated control rats (60). Moreover, rats suffering from such chronic inflammatory changes, such as seen in animals with the malocclusion syndrome, may be more susceptible than normal rats to nasal carcinogens.

Likewise, the role of tissue damage and hyperproliferation is clear in nasal carcinogenesis induced by formaldehyde; its role in acetaldehyde carcinogenesis, however, seems similar but needs further study to justify inclusion in risk estimation. Compounds such as acrolein, furfural, dimethylamine, and ethyl acrylate are very cytotoxic to the nasal mucosa and induced hyperproliferative changes but did not cause nasal tumors. Apparently, the local tissue conditions created by these compounds did not meet the requirements for tumor initiation, promotion, and progression. However, ethyl acrylate, being an irritating nasal noncarcinogen, did cause carcinomas in the forestomach of rats and mice after administration by gavage (147). Such differences in response between organs may be due to differences in local tissue dose and/or tissue-specific differences in sensitivity, leading to metabolic overload and the production of genotoxic metabolites in one tissue but not in another tissue, or with one parent compound and not with another parent compound, or in one species and not in another species (23).

Even if genotoxic metabolites are not involved, it is conceivable that hyperproliferation induced by one compound will lead to tumors on the basis of a certain epigenetic mechanism, whereas hyperproliferation induced by another substance will not lead to tumor formation. For instance, differences in toxifying or detoxifying mechanisms, or in time available for DNA repair, ultimately determine whether or not "background" initiators and promoters will get hold of the tissue and eventually will result in neoplastic transformation. As far as neoplastic transformation is concerned, hyperplastic nasal tissue may easily behave differently from hyperplastic forestomach epithelium even when the hyperplasia is induced by the same compound. Furthermore, hyperplastic nasal epithelium induced by ethyl acrylate may behave differently from hyperplastic nasal epithelium induced by furfural, acrolein or electrocoagulation. Thus species-, tissue-, and compound-specific factors largely, if not entirely determine the role of tissue injury in (nasal) carcinogenesis. Consequently, a status of sustained hyperplasia may, but does not necessarily, lead to (nasal) neoplasia. It is not yet clear, however, how preneoplastic potential of (nasal) hyper- and metaplastic changes can be distinguished from nonpreneoplastic counterparts.

Humans have relatively simple noses as compared to experimental animals like rodents and dogs. Consequently, the anatomy of the nasal cavity in relation to the oral cavity is arranged in such a way that humans can breathe both orally and oronasally, whereas rodents are obligatory nose breathers (148). Despite the differences in complexity and the variations in shape, the noses of most rodents have characteristics similar to those of humans. On the other hand, there are major structural differences between humans and rodents in the nasal cavities that can modify the course of the air currents. The nature and site of damage in the respiratory tract are influenced by a series of factors relating to the airborne substance (solubility, polarity, diffusion rate, particle size, shape, and density), the exposure pattern (e.g., continuous, interrupted, peak loads), or the subject (e.g., anatomy of the respiratory tract, respiration rate, detoxification systems, pathological conditions such as distortions, occlusions, impaired lung, or other functions). Obviously, the combined action of a large number of factors determines the site(s) and type of biological response. For example, a polar, highly soluble gas inhaled through the nose will affect the nasal mucosa but will not reach the pulmonary tissue, whereas an apolar poorly soluble gas inhaled through the mouth penetrates the lungs and may damage the pulmonary tissue. In view of the increasing significance of inhalation toxicity testing, analyses of airborne substances in the different segments of the respiratory system are becoming increasingly important, particularly with respect to their relevance to humans. Acetaldehyde and formaldehyde, for example, show similarities but also considerable differences in site and type of the lesions induced in the upper respiratory tract. Both aldehydes appeared capable of damaging the nasal mucosa and inducing squamous cell carcinomas. Adenocarcinomas of the olfactory epithelium, however, occurred only after exposure to acetaldehyde (3,12,75). Low effective concentrations of acetaldehyde (400–1000 ppm) affected only the olfactory epithelium, while low effective concentrations of formaldehyde (2–3 ppm) damaged only the respiratory epithelium (12), indicating that the impact of acetaldehyde occurred more than that of formaldehyde in the posterior part of the nose. That impact, rather than susceptibility of the two types of epithelium to the cytotoxic action of these aldehydes, determines initially the site of damage is also supported by the observation that alterations of the respiratory epithelium induced by formaldehyde were restricted to the anterior part of the nasal septum and the nasomaxillary turbinates and, as the study progressed, extended to the posterior segments of the nose.

Druckrey et al. (149) found that dinitrosopiperazine and N-nitrosopiperidine given subcutaneously and N-nitrosomorpholine and N-nitrosomethylallylamine given intraveneously produced olfactory neuroblastomas in the nose of rats. When the rats were subjected to exposure by inhalation to nitrosodimethylamine, they also developed olfactory neuroblastomas, but rats exposed to N-nitrosomethylvinylamine by inhalation developed squamous cell carcinomas of the nasal cavity similar to those induced in mice by cutaneous applications of nitrosodiethylamine. Thus, tumor induction in the upper respiratory tract may be influenced by the route of administration and the type of carcinogen. Species differences also play an important role in the effects of chemicals on the upper respiratory tract. Formaldehyde induced nasal carcinomas were observed in numerous rats, in only a few mice, and not at all in hamsters exposed to similar concentrations for comparable periods of time (50,150,151). Exposure of rats to acetaldehyde at concentrations of 750 ppm and higher resulted in nasal carcinomas (13), whereas in hamsters only a few nasal tumors were found following prolonged exposure to the "maximum tolerated concentration" of acetaldehyde (2500 ppm gradually decreased to 1650 ppm). Similar species differences were observed with bis(chloromethyl) ether (BCME). Although BCME produced some pulmonary adenomas and adenocarcinomas in mice and one nasal olfactory neuroblastoma in Syrian golden hamsters, the rat appeared much more sensitive to the carcinogenic effect of this chemical. In 86.5% of the rats exposed to 100 ppb BCME for 6 months, Leong et al. (14) found olfactory neuroblastomas. However, such differences in sensitivity between species should be interpreted with circumspection. For example, it has been concluded that rats are more sensitive than mice to the cytotoxicity of formaldehyde. However, it has been shown that mice are able to minimize the inhalation of irritating concentrations of formaldehyde more effectively than rats (152). Taking into account minute volumes and surface area of the nasal mucosa, it was calculated that the dose of formaldehyde per square centimeter of nasal mucosa during a six-hour exposure to 15 ppm formaldehyde is about 40% lower in mice than in rats (153). This finding is consistent with the lower incidence of nasal tumors in mice than in rats at 15 ppm (154) and also indicates that the nasal mucosa of rats and mice differ much less in sensitivity to formaldehyde than would appear from the results of the long-term inhalation studies. This example also demonstrates that certain differences in response between species may be concentration specific.

Humans are more likely than rodents to inhale through the mouth, because rodents are obligatory nose breathers. Mouth breathing means bypassing filtration by the nasal passages (155). Active inhalation of cigarette smoke is an illustrative example of mouth breathing. The gas phase of cigarette smoke contains high concentrations of aldehydes (156). It is therefore conceivable that smokers who actively inhale cigarette smoke expose their larynx and bronchi to high concentrations of these compounds. Consequently, these chemicals may contribute significantly to the induction of laryngeal and bronchial cancer by cigarette smoke.

In experimental carcinogenesis, one should be aware of the impact that tissue damage and hyperproliferation may have on the different steps of the process of carcinogenesis (139,157). Moreover, the occurrence of tissue damage and hyperproliferation should be taken into account in interpreting the results of carcinogenicity studies, particularly in predicting low-dose effects from high-dose findings, and in estimating human health risk (158). Undoubtedly, chronic tissue damage, often accompanied by hyperproliferation, may play a role in the formation of cancer in both humans and experimental animals. Although many factors may be involved (deposition, clearance, metabolism, DNA repair), sustained hyperproliferation as such seems to be a key factor. Therefore, in humans, recurrent chronic tissue injury should be avoided where possible.

REFERENCES

1. Schouten LJ, Knipschild PG. Epidemiologie van het larynxcarcinoom, een overzicht van de literatuur. Tijdschr Soc Gezondheidsz 1985; 63:375–378.
2. Reznik G, Stinson SF. Nasal tumors in animals and man. In: Experimental Nasal Carcinogenesis, Vol. 3. Boca Raton, FL: CRC Press, 1983.
3. Feron VJ, Woutersen RA, Appelman LM. Epithelial damage and tumours of the nose after exposure to four different aldehydes by inhalation. In: Grosdanoff P, Bass R, Hackenberg U, et al., eds. Problems of inhalatory toxicity studies. Munich: MMV Medizin Verlag, 1984:587–609.
4. Feron VJ, Woutersen RA, Spit BJ. Pathology of chronic nasal toxic responses including cancer. In: Barrow CS, ed. Toxicology of the nasal passages. Washington, DC: Hemisphere, 1986:67–89.
5. Feron VJ, Woutersen RA, van Garderen-Hoetmer A, Dreef-van der Meulen HC. Upper respiratory tract tumors in Cpb:WU (Wistar random) rats. Environ Health Perspect 1990; 85:305–315.
6. Tricker AR, Preussmann R. Carcinogenic N-nitrosamines in the diet: occurrence, formation, mechanisms and carcinogenic potential. Mutat Res 1991; 259:277–289.
7. Reznik-Schüller HM. Nitrosamine-induced nasal cavity carcinogenesis. In: Reznik G, Stinson SF, eds. Experimental nasal carcinogenesis, Vol 3. Boca Raton, FL: CRC Press, 1983:47–78.
8. Lijinsky W, Reuber MD. Transnitrosation by nitrosamines *in vivo*. IARC Sci Publ 1987; 41:625–631.
9. Klein RG, Janowsky I, Pool-Zobel BL, et al. Long-term inhalation of low-dose N-nitrosodimethylamine (NDMA) in rats. Proc Am Assoc Cancer Res 1990; 31:87.
10. NCI. Carcinogenesis bioassay of 1,2-dibromomethane (inhalation study), TR-210 (CAS no 106–93-4). In: *Carcinogenesis Testing Program*. DHHS publication no. (NIH), 81-1766. Bethesda, MD: National Cancer Institute, 1981.

11. Schlögel FA, Bannasch P. Toxicity and cancerogenic properties of inhaled dimethyl sulfate. Arch Pharmacol 1970; 266:441.
12. Battelle A. A chronic inhalation toxicology study in rats and mice exposed to formaldehyde. Final Report, CIIT, 173. Columbus, OH: Battelle Columbus Laboratories, 1981.
13. Woutersen RA, Appelman LM, van Garderen-Hoetmer A, et al. Inhalation toxicity of acetaldehyde in rats. III. Carcinogenicity study. Toxicology 1986; 41:213–231.
14. Leong BKJ, Kociba RJ, Jersey GC. A lifetime study of rats and mice exposed to vapours of bis(chloromethyl) ether. Toxicol Appl Pharmacol 1981; 58:269–281.
15. Lee KP, Trochimowicz HJ. Metaplastic changes of nasal respiratory epithelium in rats exposed to hexamethylphosphoramide (HMPA) by inhalation. Am J Pathol 1982; 106:8–19.
16. Renne RA, Giddens WE, Boorman GA, et al. Nasal cavity neoplasia in F344/N rats and (C57BL/6xC3H)F1 mice inhaling propylene oxide for up to two years. J Natl Cancer Inst 1986; 77:573–582.
17. Albertini RJ, Sweeney LM. Propylene oxide: genotoxicity profile of a rodent nasal carcinogen. Crit Rev Toxicol 2007; 37:489–520.
18. Lee KP, Schneider PW, Trochimowicz HJ. Morphological expression of glandular differentiation in the epidermoid nasal carcinomas induced by phenylglycidylether inhalation. Am J Pathol 1983; 111:140–148.
19. Sellakumar A, Snyder C, Patil G, et al. Inhalation carcinogenesis of dimethylcarbamoyl chloride. A dose response effect and induction of naso-pharyngeal and nasal cavity tumors in rats. Proc Am Assoc Cancer Res 1989; 30:140.
20. Laskin S, Sellakumar AR, Kuschner M, et al. Inhalation carcinogenicity of epichlorohydrin in noninbred Sprague-Dawley rats. J Natl Cancer Inst 1980; 65:751–757.
21. Feron VJ, Kroes R. One year time-sequence inhalation toxicity study of vinyl chloride in rats. II. Morphological changes in the respiratory tract, ceruminous glands, brain, kidneys, heart and spleen. Toxicology 1979; 13:131–141.
22. Woutersen RA, van Garderen-Hoetmer A, Slootweg PJ, et al. Upper respiratory tract carcinogenesis in experimental animals and in humans. Carcinogenesis 1994; 8: 215–264.
23. Kuper CF, Woutersen RA, Slootweg PJ, et al. Carcinogenic response of the nasal cavity to inhaled chemical mixtures. Mutat Res 1997; 380:19–26.
24. Kerns WD. Polypoid adenoma of the nasal cavity in the laboratory rat. In: Jones TC, Mohr U, Hunt RD, eds. Monographs on pathology of laboratory animals: respiratory system. New York, NY: Springer-Verlag, 1985:41–47.
25. Feron VJ, Immel HR, Wilmer JWGM, et al. Nasal tumours in rats after severe injury to the nasal mucosa and exposure to formaldehyde vapour: preliminary results. In: Tyiak E, Gullner G, eds. The role of formaldehyde in biological systems. Budapest, Hungary: Sote Press, 1987:73–77.
26. Feron VJ, Bruijntjes JP, Woutersen RA, et al. Nasal tumours in rats after short-term exposure to a cytotoxic concentration of formaldehyde. Cancer Lett 1988; 39:101–111.
27. Reuzel PGJ, Feron VJ, Immel HR, et al. Chronic (25-month) inhalation toxicity/carcinogenicity study with 2,3,4-trichlorobutene-1 in rats. CIYO/TNO Report V81.l33/267399. Zeist, The Netherlands: CIVO, 1981.
28. Monteiro-Riviere NA, Popp JA. Ultrastructural characterization of the nasal respiratory epithelium in the rat. Am J Anat 1984; 169:31–43.
29. Woutersen RA, van Garderen-Hoetmer A, Slootweg PJ, et al. Upper respiratory tract carcinogenesis in experimental animals and in humans. Carcinogenesis 1984; 8: 215–264.
30. Javek BW. Ultrastructure of human nasal mucosa. Laryngoscope 1983; 93:1576–1599.
31. Pepelko WE. Experimental respiratory carcinogenesis in small laboratory animals. Environ Res 1984; 33:144–188.
32. Reznik C, Reznik-Schüller HM, Hayden DW, et al. Morphology of nasal cavity neoplasms in F-344 rats after chronic feeding of *p*-cresidine, an intermediate of dyes and pigments. Anticancer Res 1981; 1:279–286.

33. Reznik G, Reznik-Schüller HM, Ward JM, et al. Morphology of nasal cavity tumours in rats after chronic inhalation of 1,2-dibromo-3-chloropropane. Br J Cancer 1980; 42:772–781.

34. Cardesa A, Pour P, Haas H, et al. Histogenesis of tumours from the nasal cavities induced by diethylnitrosamine. Cancer 1976; 31:346–358.

35. Reznik-Schüller HM. Pathogenesis of tumours induced with N-nitrosomethylpiperazine in the olfactory region of the rat nasal cavity. J Natl Cancer Inst 1983; 71:165–172.

36. Graziadei PPC, Monti Graziadei GA. Neurogenesis and neuron regeneration in the olfactory system of mammals. I. Morphological aspects of differentiation and structural organization of the olfactory sensory neurons. J Neurocytol 1979; 8: 1–18.

37. Monti Graziadei GA, Graziadei PPC. Neurogenesis and neuron regeneration in the olfactory system of mammals. II. Degeneration and reconstitution of the olfactory sensory neurons after axotomy. J Neurocytol 1979; 8:191–213.

38. IARC (International Agency for Research on Cancer). In: IARC Monographs on the Evaluation of Carcinogenic Risks to Humans, Vol. 38. Tobacco Smoking. IARC: Lyon, France, 1986.

39. Feron VJ, Kruysse A, Woutersen RA. Respiratory tract tumours in hamsters exposed to acetaldehyde vapour alone or simultaneously to benzo(a)pyrene or diethylnitrosamine. Eur J Cancer Clin Oncol 1982; 18:13–31.

40. Saffiotri U, Stinson SF, Keenan KP, et al. Tumor enhancement factors and mechanisms in the hamster respiratory tract carcinogenesis model. Carcinogenesis 1985; 8:63–92.

41. IARC (International Agency for Research on Cancer). In: Monographs on the Evaluation of Carcinogenic Risks to Human, Formaldehyde, 2-Butoxyethanol and 1-ter-Butoxy-2-Propanol, Vol. 88. Lyon, France: IARC, 2005.

42. Cosma GN, Wilhite AS, Marchok AC. The detection of DNA-protein cross-links in rat tracheal implants exposured *in vivo* to benzo[*a*]pyrene and formaldehyde. Cancer Lett 1988; 42:13–21.

43. Casanova M, Heck H D'A. Further studies of the metabolic incorporation and covalent binding of inhaled [^{3}H]- and [^{14}C]-formaldehyde in Fischer 344 rats: effects of glutathione depletion. Toxicol Appl Pharmacol 1987; 89:105–121.

44. Heck H D'A, Casanova M. Nasal dosimetry of formaldehyde: modeling site specificity and the effects of preexposure. In: Dugger EL, Schweiter C, eds. Nasal toxicity and dosimetry of inhaled xenobiotics: implications for human health. Washington DC, USA: Taylor & Francis, 1995:159–175.

45. Lam CW, Casanova M, Heck H D'A. Depletion of nasal glutathione by acrolein and enhancement of formaldehyde induced DNA-protein cross-linking by simultaneous exposure to acrolein. Arch Toxicol 1985; 58:67–71.

46. Casanova M, Morgan KT, Steinhagen WH, et al. Covalent binding of inhaled formaldehyde to DNA in the respiratory tract of rhesus monkeys: pharmacokinetics, rat-to-monkey interspecies scaling, and extrapolation to man. Fundam Appl Toxicol 1991; 77:409–428.

47. Casanova M, Morgan KT, Gross EA, et al. DNA-protein cross-links and cell replication at specific sites in the nodse of F344 rats exposed subchronically to formaldehyde. Fundam Appl Toxicol 1994; 23:252–536.

48. Woutersen RA, van Garderen-Hoetmer A, Bruijntjes JP, et al. Nasal tumours in rats after severe injury to the nasal mucosa and prolonged exposure to 10 ppm formaldehyde. J Appl Toxicol 1989; 9:39–46.

49. Kerns WD, Pavkov KL, Donofrio DJ, et al. Carcinogenicity of formaldehyde in rats and mice after long-term inhalation exposure. Cancer Res 1983; 43: 4382–4392.

50. Swenberg JA. In: Twenty four month final report: inhalation toxicity of dimethylamine in F-344 rats and B6C3F1 mice (Docket 11957). Research Triangle Park, NC: Chemical Industry Institute of Toxicology.

51. Wilmer JWGM, Woutersen RA, Appelman LM, et al. Subchronic (13-week) inhalation toxicity study of formaldehyde in male rats: 8-h hour intermittent versus 8-hour continuous exposures. Toxicol. Lett 1989; 47:287–293.
52. Dalbey WE. Formaldehyde and tumours in hamster respiratory tract. Toxicology 1982;24:9–14.
53. Kamata E, Nakadate M, Uchida O, et al. Results of a 28-month chronic inhalation toxicity study of formaldehyde in male Fisher-344 rats. J Toxicol Sci 1997; 22: 239–254.
54. Woutersen RA, Appelman LM, Wilmer JWGM, et al. Subchronic (13-week) inhalation toxicity study of formaldehyde in rats. J Appl Toxicol 1987; 7:43–49.
55. Zwart A, Woutersen RA, Wilmer JWGM, et al. Cytotoxic and adaptive effects in rat nasal epithelium after 3-day and 13-week exposure to low concentrations of formaldehyde vapour. Toxicology 1988; 51:87–99.
56. Bolt HM. Experimental toxicology of formaldehyde. J Cancer Res Clin Oncol 1987; 113:305–309.
57. DFG. Formaldehyde. Occupational Toxicants 2000; 17:163–201.
58. Conolly RB, Kimbell JS, Janszen D, et al. Human respiratory tract cancer risks of inhaled formaldehyde: dose-response predictions derived from biologically-motivated computational modeling of a combined rodent and human dataset. Toxicol Sci 2004; 82:279–296.
59. Morgan KT, Jiang X-Z, Starr JB, et al. More precise localization of nasal tumours associated with chronic exposure of F-3M rats to formaldehyde gas. Toxicol Appl Pharmacol 1986; 82:264–271.
60. Feron VJ, Wilmer JWGM, Woutersen RA, et al. Inhalation toxicity and carcinogenicity of formaldehyde in animals: significance for assessment of human health risk. In: Mohr U, Bates DV, Dungworth DL, et al. Assessment of inhalation hazards. ILSI monographs. Berlin: Springer-Verlag, 1989:131–138.
61. Obe G, Beek B. Mutagenic activity of aldehydes. Drug Alcohol Depend 1979; 4:91–94.
62. Bird RP, Draper HH, Basrur PK. Effect of malonaldehyde and acetaldehyde on cultured mammalian cells, Production of micronuclei and chromosomal aberrations. Mutat Res 1982; 101:237–246.
63. Böhlke JU, Singh S, Goedde HW. Cytogcnetic effects of acetaldehyde in lymphocytes of Germans and Japanese: SCE clastogenic activity, and cell cycle delay. Hum Genet 1983; 63:285–289.
64. Jansson T. The frequency of sister chromatid exchanges in human lymphocytes treated with ethanol and acetaldehyde. Hereditas 1982; 97:301–303.
65. De Raat, WK, Davis PB, Bakker GL. Induction of sister-chromatid exchanges by alcohol and alcoholic beverages after metabolic activation by rat-liver homogenate. Mutat Res 1983; 124:85–90.
66. He SM, Lambert B. Induction and persistence of SCE-inducing damage in human lymphocytes exposed to vinyl acetate and acetaldehyde *in vitro.* Mutat Res 1985; 158:201–208.
67. Norppa H, Tursi F, Pfafti P, et al. Chromosome damage induced by vinyl acetate through in vitro formation of acetaldehyde in human lymphocytes and Chinese hamster ovary cells. Cancer Res 1985; 45:4816–4821.
68. Dellarco VL. A mutagenicity assessment of acetaldehyde. Mutat Res 1988; 195:1–20.
69. Ristov H, Obe G. Acetaldehyde induces cross-links in DNA and causes sister-chromatid exchanges in human cells. Mutat Res 1978; 58:115–119.
70. Lambert B, Chen Y, He SM. DNA cross-links in human leukocytes treated with vinyl acetate and acetaldehyde in vitro. Mutat Res 1985; 146:301–303.
71. Minini U. Untersuchung von DNS-Schäden durch Formaldehyd, Acetaldehyd, Tetrachlorkohlenstoff und Hydrazin [Thesis] Zurich, 1985. Diss. Natuurwiss. ETH Zurich nr. 7763,0000. DOI 103929/ethz-a-000342608.
72. Lam CW, Casanova M, Heck H D'A. Decreased extractability of DNA from proteins in the rat nasal mucosa after acetaldehyde exposure. Fund Appl Toxicol 1986; 6: 541–550.

73. Woodruff RC, Mason JM, Valencia R, et al. Chemical mutagenesis testing in Drosophila. V. Results of 53 coded compounds tested for the National Toxicology Program. Environ Mutagen 1985; 7:677–702.

74. Appelman LM, Woutersen RA, Feron VJ. Inhalation toxicity of acetaldehyde in rats. I. Acute and subacute studies. Toxicology 1982; 23:293–307.

75. Woutersen RA, Feron VJ, Appelman LM, et al. Inhalation toxicity of acetaldehyde in rats. II. Carcinogenicity study: interim results after 15 months. Toxicology 1984; 31:123–133.

76. Woutersen RA, Feron VJ. Inhalation toxicity of acetaldehyde in rats. IV. Progression and regression of nasal lesions after discontinuation of exposure. Toxicology 1987; 47:295–305.

77. Kruysse A, Feron VJ, Til HP. Repeated exposure to acetaldehyde vapor. Studies in Syrian golden hamsters. Arch Environ Health 1975; 31:449–452.

78. Bogdanffy MS, Randall HW, Morgan KT. Histochemical localization of aldehyde dehydrogenase in the respiratory tract of the Fischer-344 rat. Toxicol Appl Pharmacol 1986; 82:560–567.

79. Eldridge SR, Bogdanffy MS, Jokinen MP, et al. Effects of propylene oxide on nasal epithelial cell proliferation in F-344 rats. Fundam Appl Toxicol 1995; 27:25–32.

80. Rios-Blanco MN, Yamaguchi S, Dhawan-Robl M, et al. Effects of propylene oxide exposure on rat nasal respiratory cell proliferation. Toxicol Sci 2003; 75: 279–288.

81. National Toxicology Program. In: Technical Report on the Toxicology and Carcinogenesis Studies of Propylene Oxide (CAS no. 75–56-9) in F344/N Rats and B6C3F1 Mice (Inhalation Studies). NIH publication 85–2527, NTP-TR 267. Research Triangle Park, NC: U.S. Department of Health and Human Services, 1985.

82. Reuzel PG, Kuper CF. Chronic (28-month) inhalation toxicity/carcinogenicity study of 1,2-propylene oxide in rats. In: Final Report V82.215/280853, Zeist, The Netherlands, 1983.

83. Kuper CF, Reuzel PG, Feron VJ, et al. Chronic inhalation toxicity and carcinogenicity study of propylene oxide in Wistar rats. Food Chem Toxicol 1988; 26:159–167.

84. Csanády GA, Filser JG. A physiological toxicokinetic model for inhaled propylene oxide in rat and human with special emphasis on the nose. Toxicol Sci 2007; 95: 37–62.

85. Feron VJ, Kruysse A. Effects of exposure to acrolein vapour in hamsters simultaneously treated with benzo(a)pyrene or diethylnitrosamine. J Toxicol Environ Health 1977; 3:379–394.

86. Feron VJ, Kruysse A. Effects of exposure to furfural vapour in hamsters simultaneously treated with benzo(a)pyrene or diethylnitrosamine. Toxicol 1978; 11:127–144.

87. Swenberg JA, Kerns WD, Mitchell RI, et al. Induction of squamous cell carcinomas of the rat nasal cavity by inhalation exposure to formaldehyde vapour. Cancer Res 1980; 40:3398–3402.

88. Miller RR, Young JT, Kociba R, et al. Chronic toxicity and oncogenicity bioassay of inhaled ethyl acrylate in Fischer 344 rats and B6C3F1 mice. Drug Chem Toxicol 1985; 8:1–42.

89. Wolf DC, Morgan KT, Gross EA, et al. Two-year inhalation exposure of female and male B6C3F1 mice and F344 rats to chlorine gas induces lesions confined to the nose. Fund Appl Toxicol 1995; 24:111–131.

90. Ward JM, Uno H, Kurata Y, et al. Cell proliferation not associated with carcinogenesis in rodents and humans. Environ Health Perspect 1993; 10:125–136.

91. Cassee FR, Groten JP, Feron VJ. Combined exposure of rat upper respiratory tract epithelium to aldehydes. Human Exp Toxicol 1994; 13:726.

92. Cassee FR. Upper Respiratory Tract Toxicity of Mixtures of Aldehydes; in vivo and in vitro studies [Thesis]. Utrecht, The Netherlands: Utrecht University; 1995:175.

93. Cassee FR, Groten JP, Feron VJ. Changes in the nasal epithelium of rats exposed by inhalation to mixtures of formaldehyde, acetaldehyde and acrolein. Fund Appl Toxicol 1996; 29:208–218.

94. Mautz WJ, Kleinman MT, Phalen RF, et al. Effects of exercise exposure on toxic inter-actions between inhaled oxidants and aldehyde air pollutants. J Toxicol Environ Health 1988; 25:165–177.
95. Reuzel PGJ, Wilmer JWGM, Woutersen RA, et al. Interactive effects of ozone and formaldehyde on the nasal respiratory lining epithelium in rats. J Toxicol Environ Health 1990; 29:279–292.
96. Cassee FR, Feron VJ. Biochemical and histopathological changes in nasal epithelium of rats after 3-day intermittent exposure to formaldehyde and ozone alone or in com-bination. Toxicol Lett 1994; 72:257–268.
97. Teredesai A, Stinn W. Histopathological effects observed in rat nasal epithelium in two 3-day inhalation studies with formaldehyde, acetaldehyde, acrolein, ammonia, and a mixture of formaldehyde, acrolein, and ammonia, respectively. In: Feron VJ, Bosland MC, eds. Nasal Carcinogenesis in Rodents: Relevance to Human Health Risk. Wageningen, The Netherlands: Pudoc Press, 1989:215.
98. Holmström M, Wilhelmsson B, Hellequist H. Histological changes in the nasal mucosa in rats after long-term exposure to formaldehyde and wood dust. Acta Oto-laryngol 1989; 108:274–283.
99. Drettner B, Wilhelmsson B, Lundh B. Experimental studies on carcinogenesis in the nasal mucosa. Acta Otolaryngol 1985; 99:205–207.
100. Wilhelmsson B, Lundh B, Drettner B, et al. Effects of wood dust exposure and diethylnitrosamine. A pilot study in Syrian golden hamsters. Acta Otolaryngol 1985; 99:160–171.
101. Nyska A, Yoshizawa K, Jokinen MP, et al. Olfactory epithelial metaplasia in female Harlan Sprague-Dawley rats following chronic treatment with polychlorinated biphenyls. Toxicol Pathol 2005; 33:371–377.
102. Sells DM, Brix AE, Nyska A, et al. Respiratory tract lesions in noninhalation studies. Toxicol Pathol 2007; 35:170–177.
103. Jeffrey AM, Iatropoulos MJ, Williams GM. Nasal cytotoxic and carcinogenic activities of systemically distributed organic chemicals. Toxicol Pathol 2006; 34: 827–852.
104. Ghanayem BI, Sanchez IM, Burka LT. Effects of dose, strain, and dosing vehicle on metacrylonitrile disposition in rats and identification of a novel-exhaled metabolite. Drug Metab. Dispos 1992; 20:643–652.
105. Goldsworthy TL, Monticello TM, Morgan KT, et al. Examination of potential mech-anisms of carcinogenicity of 1,4-dioxane in rat nasal epithelial cells and hepatocytes. Arch Toxicol 1991; 65:1–9.
106. Kuijpers MHM, van de Kooij AJ, Slootweg PJ. The Rat Incisor in Toxicologic Pathol-ogy. Toxicol Pathol 1996; 24:346–361
107. Barnes L, Tse LLY, Hunt J, et al. Tumours of the nasal cavity and paranasal sinuses. In: Barnes L, Eveson JW, Reichart P, Sidransky D, eds. World Health Organization Clas-sification of Tumours. Pathology and Genetics of Head and Neck Tumours. Lyon, France: IARC Press, 2005:9–80.
108. Doll R, Morgan IG, Speizer FE. Cancer of the lung and sinuses in nickel workers. Br J Cancer 1970; 24:623–632.
109. Vaughan TL, Davis S. Wood dust exposure and squamous cell cancers of the upper respiratory tract. Am J Epidemiol 1991; 133:560–564.
110. Mohtashamipur E, Norpoth K. Zur Frage beruflich bedingter Tumoren in der holzverarbeitenden Industrie. Arbeitsmed Sozialmed Praeventivmed 1983; 18:49–52.
111. Davies JM, Easton DF, Birdstrup PL. Mortality from respiratory cancer and other causes in United Kingdom chromate production workers. Brit J Ind Med 1991; 48:299–313.
112. Wada S, Miyanishi P, Nishimoto Y. Mustard gas as a cause of neoplasia in man. Lancet 1968; 1:1161–1163.
113. Osguthorpe JD. Sinus neoplasia. Arch Otolaryngol Head Neck Surg 1994; 120:19–25.
114. Luce D, Leclerc A, Bégin D, et al. Sinonasal cancer and occupational exposures: a pooled analysis of 12 case-control studies. Cancer Causes Control 2002; 13:147–157.

115. Trimarchi M, Miluzio A, Nicolai P, et al. Massive apoptosis erodes nasal mucosa of cocaine abusers. Am J Rhinol 2006; 20:160–164.
116. Cogliano VJ, Grosse Y, Baan RA, et al. Meeting report: summary of IARC monographs on formaldehyde, 2-butoxyethanol, and 1-tert-butoxy-2-propanol. Environ Health Perspect 2005; 113:1205–1208.
117. Kimbell JS, Overton JH, Subramaniam RP, et al. Dosimetry modeling of inhaled formaldehyde: binning nasal flux predictions for quantitative risk assessment. Toxicol Sci 2001; 64:111–121.
118. Spitz MR. Epidemiology and risk factors for head and neck cancer. Semin Oncol 1994; 21:281–288.
119. Tuyns AJ, Esteve J, Raymond L, et al. Cancer of the larynx/hypopharynx, tobacco and alcohol. IARC International Case Control Study in Turin and Varese (Italy), Zaragoza and Navarra (Spain), Geneva (Switzerland) and Calvados (France). Int J Cancer 1988; 41:483–491.
120. Sreedharan S. Hegde MC, Pai R, et al. Snuff-induced malignancy of the nasal vestibule: a case report. Am J Otolaryngol 2007; 28:353–356.
121. DeStefani E, Correa D, Oreggia F. Risk factors for laryngeal cancer. Cancer 1987; 60:308–309.
122. Goldenberg D, Golz A, Joachims HZ. The beverage mate: a risk factor for cancer of the head and neck. Head Neck 2003; 25:595–601.
123. Goldenberg D, Lee J, Koch WM, et al. Habitual risk factors for head and neck cancer. Otolaryngol Head Neck Surg 2004; 131:986–993.
124. Acheson ED. Nasal cancer in the furniture and boot and shoe manufacturing industries. Prev Med 1976; 5:295–315.
125. Acheson ED. Epidemiology of nasal cancer. In: Barrow CS, ed. Toxicology of the nasal passages. Washington, DC: Hemisphere, 1986:135–141.
126. Boysen M. Histopathology of the nasal mucosa in furniture workers. Rhinology 1985; 23:109–113.
127. Boysen M, Voss R, Solberg LA. The nasal mucosa in softwood exposed furniture workers. Acta Otolaryngol 1986; 101:501–508.
128. Wilhelmsson B, Hellquist H, Olofsson J, et al. Nasal cuboidal metaplasia with dysplasia. Acta Otolaryngol 1985; 99:641–648.
129. Schmitt U, Peek R-D, Rapp AO. 1977. In: SEM of wood dust particles, no. IRG/WP/97–50084, 28th Annual Meeting of the International Research Group on Wood Preservatives; May 25–30, 1997; Whistler, Canada.
130. Health Council of the Netherlands. Hardwood and softwood dust: evaluation of the carcinogenicity and genotoxicity. In: Draft report of the Committee on the Evaluation of the Carcinogenicity of Chemical Substances, The Hague; 2000.
131. Marquardt H. Chemical carcinogenesis. In: Marquardt V, Schäfer SG, McClellan RO, Welsch F, eds. Toxicology. San Diego, CA: Academic Press, 1999: 151–178.
132. Loeb LA. Endogenous carcinogenesis: molecular oncology into the twenty-first century. Cancer Res 1989; 49:5489–5496.
133. Ames BN, Gold LS. Too many rodent carcinogens: Mitogenesis increases mutagenesis. Science 1990; 249:970–971.
134. Butterworth BE, Goldsworthy TL. The role of cell proliferation in multistage carcinogenesis. Proc Soc Exp Biol Med 1991; 198:683–687.
135. Craddock VM. Cell proliferation and experimental liver cancer. In: Cameron HM, Linsell CA, Warwick GP, eds. Liver Cell Cancer. Amsterdam: Elsevier North-Holland, 1976:153–201.
136. Columbano A, Rajalakshmi S, Sarma DSR. Requirement of cell proliferation for the initiation of liver carcinogenesis as assayed by three different procedures. Cancer Res 1981; 41:2079–2083.
137. Richardson FC, Boucheron JA, Dyroff M, et al. Biochemical and morphological studies of heterogenous lobe responses in hepatocarcinogenesis. Carcinogenesis 1986; 7:247–251.

138. Cohen SM, Ellwein LB. Genetic errors, cell proliferation, and carcinogenesis. Cancer Res 1991; 51:6493–6505.

139. Grasso P, Sharrat M, Cohen AJ. Role of persistent, non-genotoxic tissue damage in rodent cancer and relevance to humans. Ann Rev Pharmacol Toxicol 1991; 31:253–287.

140. Johnson P, Williams R. Cirrhosis and the aetiology of hepatocellular carcinoma. J Hepatol 1987; 4:140–147.

141. Laroye GJ. How efficient is immunological surveillance against cancer and why does it fail? Lancet 1974; 1:1097–1100.

142. Haber H, Milne JA, Symmers WSC. The skin. In: Symmers WSC, ED. Systemic Pathology, Vol 6. 2nd ed. London: Churchill Livingstone, 1980:2575–2604.

143. Bennett DE, Sasser WF, Ferguson TB. Adenocarcinoma of the lung in men. Cancer 1969; 23;431–439.

144. Grasso P. Persistent organ damage and cancer production in rats and mice. Mechanisms and models in toxicology. Arch Toxicol 1987; 11:75–83.

145. Roe FJC. An illustrated classification of the proliferative and neoplastic changes in mouse bladder epithelium in response to prolonged irritation. Br J Urol 1964; 36: 253–328.

146. Higginson J, Bolt HM, Bosland MC. Concluding remarks. In: Feron VJ, Bosland MC, eds. Nasal Carcinogenesis in Rodents: Relevance to Human Health Risk. Wageningen, The Netherlands: Pudoc Press, 1989:205–209.

147. Maronpot RR. NTP Technical Report on the Carcinogenesis Bioassay of Ethyl Acrylate in Rats and Mice. In: National Toxicology Program. NTP Technical Report 259. Research Triangle Park, NC; 1983.

148. Reznik GK. Comparative anatomy, physiology, and function of the upper respiratory tract. Environ Health Perspect 1990; 85:171–176.

149. Druckrey H, Ivankovic S, Mennel HD, et al. Selektive Erzeugung von Carcinomen der Nasenhöhle bei Ratten durch *N,N'*-Dinitrosopiperazin, Nitrosopiperidin und Methyl-vinyl-nitrosamin. Z Krebsforsch 1964; 66:138–150.

150. Pour P, Gingell R, Langenbach R, et al. Carcinogenicity of *N*-nitrosomethyl(2-oxopropyl)amine in Syrian hamsters. Cancer Res 1980; 40:3585–3590.

151. Albert RE, Sellakumar AR, Laskin S, et al. Gaseous formaldehyde and hydrogen chloride induction of nasal cancer in the rat. J Natl Cancer Inst 1982; 68:597–603

152. Chang JCF, Steinhagen WH, Barrow CS. Effect of single or repeated formaldehyde exposure on minute volume of B6C9FI mice and F-344 rats. Toxicol Appl Pharmacol 1981; 61:451–459.

153. Chang JCF, Gross EA, Swenberg JA, et al. Nasal cavity deposition, histopathology, and cell proliferation after single or repeated formaldehyde exposures in B6C9FI mice and F-344 rats. Toxicol Appl Pharmacol 1983; 68:161–176.

154. Heck H D'A, Casanova-Schmitz M. Biochemical toxicology of formaldehyde. Rev Biochem Toxicol 1984; 6:155–189.

155. Page NP. Concepts of a bioassay program in environmental carcinogenesis. In: Kraybill H, Mehlman H, eds. Environmental carcinogenesis. Washington, DC: Hemisphere, 1987:87–171.

156. Groenen PJ. Bestanddelen van tabaksrook. Aard en hoeveelheid, potentiële invloed op de gezondheid. CIVO/TNO report 5787. Zeist, The Netherlands: CIVO, 1978.

157. Roe FJC. Non-genotoxic carcinogenesis: implications for testing and extrapolation to man. Mutagenesis 1989; 4:407–411.

158. Feron VJ, Woutersen RA. Role of tissue damage in nasal carcinogenesis. In: Feron VJ, Bosland MC, eds. Nasal carcinogenesis in rodents: relevance to human health risk. The Netherlands: Pudoc, Wageningen, 1989:76–84.

159. IARC (International Agency for Research on Cancer). In: Monographs on the evaluation of carcinogenic risks to humans (2000), Vol. 77. Some Industrial Chemicals. Lyon, France: IARC, 2000:414–416.

19 Secondhand Tobacco Smoke Exposure in Humans

Suzaynn Schick and Dennis J. Shusterman

Division of Occupational and Environmental Medicine, University of California, San Francisco, California, U.S.A.

INTRODUCTION

Secondhand cigarette smoke (SHS) is one of the most important air pollutants in the world today. In the developed world, exposure to SHS is experienced more frequently and at higher concentrations than exposure to any other air pollutant. In the developing world, exposure to smoke from cooking and heating fires is greater than SHS for many populations, but sales and consumption of cigarettes are growing explosively (1). The upper respiratory tract is an important site of SHS action. Sensory irritation, nasal airway obstruction, inflammatory diseases, and nasal cancer are all causally related to SHS exposure.

SECONDHAND SMOKE CHEMISTRY

Secondhand cigarette smoke, also known as environmental tobacco smoke or tobacco smoke pollution (2), is composed primarily of sidestream smoke which is the smoke that comes from the burning tip of a cigarette as it smolders between puffs (3). Sidestream smoke contains higher concentrations of toxic and carcinogenic compounds than the smoke the smoker inhales (mainstream smoke): the burn temperature is lower when the cigarette is smoldering than when it is being puffed on, combustion is less complete and there are more complex, toxic compounds in the smoke.

The differences between mainstream and secondhand cigarette smoke do not end with the differences in combustion temperature and initial toxicity. Most of what we encounter as secondhand smoke is not freshly emitted sidestream smoke. The chemical and physical composition of sidestream smoke changes rapidly as the smoke is diluted into room air. As the plume of smoke mixes with room air, many chemicals evaporate from the particulate phase into the vapor phase and mean particle size decreases (4). Some portion of the smoke is typically lost to ventilation. However, many of the compounds found in sidestream smoke stick or "sorb" to room surfaces before they can be removed by ventilation. Semivolatile compounds, like nicotine, are highly sorbtive. In the absence of ventilation, 80% to 98% of the nicotine in sidestream smoke sorbs to surfaces in a furnished room within two hours (5–7). Typical air exchange rates are quite low (0.5–2 air changes per hour) (8), so surfaces in indoor spaces where people smoke often are covered with smoke compounds. These compounds desorb over time and thus extend the period of potential smoke exposure by hours and days (9).

Finally, there is a new and growing body of evidence showing that the chemical compounds in secondhand smoke can react to create new chemical compounds. Unpublished research from Philip Morris Co shows that concentrations of the tobacco-specific nitrosamine NNK in sidestream smoke in an unventilated furnished room can increase two- to threefold over two hours (10). Also, sorbed nicotine on surfaces can interact with concentrations of ozone and water that are normally present, to form formaldehyde (11). Research on the chemical reactions occurring in secondhand smoke is just beginning.

Relative Toxicity of Mainstream, Sidestream, and Aged Sidestream Cigarette Smoke

A series of animal toxicology experiments by Philip Morris Co. provide a measure of how the chemical and physical differences between mainstream cigarette smoke and SHS affect toxicity. Studies comparing effects of inhaled smoke on rat respiratory epithelium show that fresh sidestream cigarette smoke is two to six times more toxic than fresh mainstream smoke (12). The effects of the chemical changes that occur as secondhand smoke ages were revealed by another series of experiments done at Philip Morris comparing the toxicity of fresh and 30- to 90-minute-old sidestream smoke. They found that aging reduces the toxicity of sidestream smoke because of the loss of smoke constituents to sorbtion (13). However, when equal quantities of fresh and aged smoke were compared on the basis of total particulate material mass, the aged smoke was two to four times more toxic to respiratory tract cells (14). Combining results from the experiments comparing inhaled fresh mainstream and sidestream smoke and those comparing fresh and aged sidestream, it appears that aged sidestream smoke is 4 to 16 times more toxic than mainstream smoke, per gram particulate material, to the rat respiratory epithelium (12,14). The markedly higher toxicity of SHS helps explain the fact that even though smokers are exposed to much higher concentrations of cigarette smoke, one nonsmoker dies of exposure to SHS for every eight smokers who die of smoking.

Dosimetry

Studies with human subjects have shown that the odor and irritation associated with SHS appear to derive more from the vapor phase than the particulate phase (15–17). The chemical constituents of SHS thought responsible for sensory irritation include organic acids (acetic, propionic), aldehydes (formaldehyde and acrolein), nicotine, ammonia, pyridine, toluene, sulfur dioxide, and nitrogen oxides (3,18,19).

The deposition of a given smoke component in the upper respiratory tract depends on: (*i*) particle size (for irritants adsorbed to particulates), and (*ii*) water solubility (for gaseous organics and vapor-phase inorganics). Generally speaking, the larger the particle and the more water soluble the compound, the higher the proportion of the inhaled dose that will be deposited in the upper respiratory tract during nasal breathing. Because of the changes that occur as SHS ages, more smoke components of SHS than mainstream smoke are in the gas phase. Many of these gas and vapor phase irritants are water soluble enough to be active on the upper respiratory tract (i.e., nasal cavity, nasopharynx, and hypopharynx) (3).

SECONDHAND SMOKE-INDUCED EFFECTS IN THE UPPER AIRWAYS

Pathophysiology

SHS stimulates the upper respiratory tract through three neural pathways: the olfactory, trigeminal, and glossopharyngeal nerves (cranial nerves I, V, and IX). The olfactory nerve is responsible for the sense of smell, and projects directly to areas of the primitive forebrain responsible for emotional arousal, including the amygdala and portions of the frontal and temporal lobes. The nasal and oral cavities are innervated by the trigeminal nerve, and the naso-, oro-, and hypopharynx by the glossopharyngeal. These nerves project to the primary somatosensory cortex via the brainstem and thalamus. The trigeminal and glossopharyngeal nerves are responsible for the perception of touch, temperature, and sensory irritation (or what has been termed "chemesthesis") in all head and neck mucosae. The two nasal senses—olfaction and irritant chemoreception—as well as the related sense of taste, functionally interact to produce an integrated impression of the chemical environment (20–22).

Combined across the two sides, the olfactory epithelium occupies a total area of between 4 and 8 cm^2 in the upper reaches of the nasal cavity. The olfactory apparatus is intermittently stimulated during normal relaxed nasal breathing. "Sniffing," or attentive smelling, creates eddy currents that facilitate the delivery of odorant molecules to the olfactory epithelium. Odorant molecules diffuse through the nasal mucus layer, probably aided by an odorant binding protein, to make contact with receptor sites on olfactory receptor cells (23). The olfactory receptor cells are the only neurons known to regenerate on a regular basis; this may constitute a functional response to the fact that olfactory receptor cells are directly exposed to a variety of environmental insults (22).

As noted above, in contrast to the limited distribution of olfactory receptor cells, trigeminal and glossopharyngeal nerve endings are located throughout the nasal and oral cavities, as well as the nasopharynx, oropharynx, and hypopharynx. At the inferior border of the hypopharynx, the vagus nerve innervates the epiglottis and larynx, and is involved in a variety of airway-protective reflexes. The nerve fiber types of greatest importance in the nociceptive roles of the trigeminal and vagal nerves are the small-diameter, unmyelinated, capsaicin-sensitive C fiber (24–26), and Aδ fibers (27).

Exposure to intense irritants anywhere within the nasal or oral cavities (for example, ingestion of horseradish) produces reflex rhinorrhea (and sometimes lacrimation) via a long reflex arc, an autonomic (cholinergic) response (28). This response is mimicked by local instillation of methacholine (an acetylcholine analog), and is blocked by pretreatment with atropine (an acetylcholine antagonist). For intense irritant stimuli, local reflexes may also contribute to the response. The so-called "axon" reflex is a local reaction in which neuropeptides are released near the mucosal surface. This phenomenon is reviewed in detail in Chapter 4 (24,26,29,30).

Newer studies of airway responses to SHS should be viewed against a background of more traditional methods in inhalation toxicology. Animals exposed to highly water-soluble upper respiratory tract irritants reveal predictable changes in respiratory pattern, including slowing of respiration, sneezing, coughing, and increased secretions (31). In humans, the analogous respiratory pattern is an involuntary pause during inspiration or frank breath-holding

(32). Exposure levels necessary to produce these responses, however, are generally high, and researchers in the field of indoor air quality and SHS have searched for more sensitive indices of respiratory tract irritation. As noted below, evidence of true allergic (IgE-mediated) upper airway reactions to SHS is quite limited.

Specific Health Effects

Definitions

"Annoyance" is a subjective state of displeasure resulting from a defined environmental stimulus. SHS contains a number of odorant compounds (e.g., pyridine) that are typically described as unpleasant. Thus, it is not surprising that even in the absence of clinically detectable eye, nose, or throat irritation, non-smokers often complain of annoyance from the odor of SHS. This has been discussed extensively in the National Research Council report (33) and by Samet et al. (16). In addition to annoyance, indoor air quality researchers have shown that unpleasant odors detract from the sense of well-being and, at times, interfere with concentration and productivity (34,35).

"Sensory irritation" refers to subjectively reported tingling, stinging, burning, or pain involving the mucous membranes of the upper respiratory tract and/or cornea (in humans), or to unconditioned aversive responses to an airborne chemical agent in experimental animals. When reflex *functional* alterations are associated with SHS exposure (e.g., changes in respiratory behavior or blink rate), these reflect sensory irritation. "Pathological irritation" refers to irritant-related changes in tissue structure or biochemistry, including necrosis, mucosal desquamation, cellular infiltration, and release of inflammatory mediators.

This chapter defines the "upper airway" as the area from the external nares ("nostrils") to the glottis (larynx). It includes the nasal cavity, naso-, oro-, and hypopharynx. It also includes the paranasal sinuses (which open into the nasal cavity) and Eustachian tubes (which open into the nasopharynx). In light of this anatomy, SHS exposure may, in addition to producing sensory irritation symptoms, contribute to the genesis of inflammatory conditions—including rhinitis, sinusitis, and otitis media.

Sensory Irritation and Odor "Annoyance"

Sensory irritation, odor "annoyance," and reflex changes in nasal patency are among the most common problems associated with SHS exposure. They are acute effects. These health endpoints have been studied both epidemiologically and experimentally.

Epidemiologic Studies

Epidemiologic studies of the upper airway effects of SHS have been of various designs, including cross-sectional and longitudinal. Among the latter are symptom prevalence surveys performed before and after the institution of smoking restrictions (e.g., in restaurants, bars, and taverns). These longitudinal studies are, in effect, "natural experiments," and their findings are particularly persuasive.

Mizoue et al. (36) examined data from a 1998 cross-sectional survey of 1281 municipal employees, who worked in a variety of buildings in a Japanese city. The authors were interested in overtime work and ETS exposure as determinants of symptoms consistent with nonspecific building-related illness or "sick building syndrome" (SBS). Potential confounders, which were adjusted for in a logistic regression model, included age, gender, hierarchical position, use of video display terminal >4 hr/day, psychological stress at work, and lifestyle factors. Using workers exposed to ETS for less than 1 hr/day as the reference group, the odds ratio for the SBS symptom constellation among nonsmokers exposed to ETS >4 hr/day was 2.7 (95% CI: 1.6, 4.8). For symptoms referable to the eyes, nose, throat, and skin, odds ratios increased with increasing hours of ETS exposure. These relationships persisted after adjustment for all covariates, including overtime, which was an independent predictor of SBS symptoms.

Jones et al. (37) surveyed bar staff, waiters, and restaurant managers and owners in New Zealand to determine attitudes and beliefs regarding the health consequences of ETS exposure. A minor component of the questionnaire also dealt with ETS-related symptoms and annoyance. The investigators were able to complete 435 interviews at 364 locations. The self-reported ETS exposure prevalence among respondents was 59%. More than half of those exposed to ETS reported irritation from secondhand smoke to their "throat or lungs," and three-quarters of interviewees indicated that they wanted some sort of smoking restriction in bars.

Wieslander et al. (38) surveyed 80 commercial aircraft crew members on smoking-permitted and smoking-prohibited international flights of long (11–12 hours) duration. Interviews and physical examinations were conducted, including 39 performed in-flight and 41 post-flight. Half of the flights permitted smoking, and the other half occurred soon after a smoking ban. Endpoints included cabin air quality (CAQ—both measured and perceived), upper respiratory tract/mucous membrane symptoms, tear-film stability, nasal patency (by acoustic rhinometry), and biomarkers in nasal lavage fluid (eosinophilic cationic protein, myeloperoxidase, lysozyme, and albumin). Cabin air was found to be of low relative air humidity (2–10%) although carbon dioxide concentrations—a surrogate for the adequacy of ventilation relative to occupancy—were in an acceptable range. Total respirable particles were reduced dramatically by the smoking ban, with the mean falling from 66 to 3 $\mu g/m^3$. The perceived CAQ was improved, and symptoms—particularly ocular—were less prevalent on nonsmoking flights. Tear-film stability increased after the smoking ban but, although there was a trend toward increased nasal patency, it was not consistent by study subgroup. The authors concluded that in-flight ETS exposure is associated with poor perceived air quality, as well as with symptomatic and objective indices of upper respiratory tract/mucous membrane irritation.

Eisner et al. (39) obtained a random sample of bars and taverns and surveyed bartenders before and after a statewide prohibition on smoking in such establishments. Interviewers assessed lower respiratory tract symptoms, sensory irritation symptoms (eye, nose, or throat irritation), ETS exposure, personal smoking, and recent upper respiratory tract infections. Spirometry was also performed. Fifty-three of 67 eligible bartenders were interviewed; all reported workplace ETS exposure at baseline. Respondents reported a reduction in median weekly workplace ETS exposure from 28 hours pre-intervention to two hours post-intervention ($p < 0.001$). One-quarter of bartenders were active smokers, a number that was unchanged post-intervention. Of the 41 (77%) respondents who initially reported sensory irritation symptoms, 32 (78%) reported resolution of symptoms post-intervention ($p < 0.001$). The authors concluded that "... establishment of smoke-free bars and taverns was associated with a rapid improvement of respiratory health."

Raynal et al. (40) studied 375 office employees in a large, open-plan smoking-permitted building and 26 individuals from a building in which no smoking was permitted. Participants were asked to report prevalence of mucous membrane (eye, nose, throat) irritation, lethargy, flu-like illness, chest tightness, and "difficulty breathing" which had improved outside away from work. A composite score was constructed for each individual, with adjustment for demographic variables. Active smoking histories were taken, and both exhaled breath carbon monoxide (CO) and salivary cotinine levels measured for validation purposes. Workplace temperature, humidity, and airflow were measured in five locations each, and vapor-phase nicotine levels in 23 different subareas of the main workplace. The sample of potentially exposed workers was 70% female and 25% active smokers; the unexposed group was younger and more predominantly male, but comparable in their active smoking rate (19%). Eleven subjects self-reported as nonsmokers but had salivary cotinine levels greater than 15 ng/mL; these respondents were analyzed separately from those whose smoking histories and biomarkers were concordant. Among validated nonsmokers, there was a positive association between (area) environmental nicotine measurements and both reported symptoms ($r = 0.165$; $p < 0.01$) and saliva cotinine levels ($r = 0.313$; $p < 0.001$). Among the various symptoms reported, eye, nose, and throat irritation were most closely related to environmental nicotine levels.

Experimental Studies
Experimentally, a substantial fraction of published investigations of SHS effects on the upper airway effects have been controlled human exposure studies. Although the earliest of these focused on eye irritation (41,42), the nose itself has subsequently become a major focus of studies.

Bascom et al. (43) and Willes et al. (44) identified a subgroup of research subjects who reported a variety of nasal symptoms (congestion, rhinorrhea, sneezing, and postnasal drip) upon prior exposure to SHS. This group comprised approximately one-third of their study population and was labeled the "historically SHS-sensitive" subgroup in the authors' subsequent provocative testing protocol. Using a climate-controlled exposure chamber, the investigators conducted sidestream tobacco smoke (STS) challenge testing, examining a variety of endpoints. As a group, historically SHS sensitive, but not SHS nonsensitive, subjects showed significant increases in nasal airway resistance (NAR) by rhinomanometry after 15-minute exposures to STS at levels chosen to simulate a smoking lounge. These changes in objectively measured NAR paralleled the onset of symptoms of nasal stuffiness and rhinorrhea. Although the symptoms described above resemble those of allergic rhinitis, the authors noted that only a small proportion of historically SHS-sensitive subjects have positive skin test reactivity to tobacco-leaf extract (see review of tobacco allergy in Ref. 45).

To investigate the mechanism(s) underlying the responses they observed, Bascom et al. (43) performed nasal lavage pre- and post-STS exposure. Although allergy-like nasal symptoms were provoked acutely, traditional markers of allergic nasal response (including histamine, various kinins, and albumin) were not found to be increased post-exposure. These findings were taken as evidence that acute nasal responses to STS/SHS may occur via nonallergic, irritative mechanisms (see discussion of neurogenic reflexes, above). Despite a lack of evidence for direct allergic mechanisms, individuals who display both subjective and objective SHS sensitivity are more likely than nonresponders to have documented nontobacco allergies, implying a modulatory effect of allergy upon the irritant chemoreceptive system (43,46,47).

Bascom et al. (48) studied nasal mucociliary clearance (NMC) in 12 healthy adults, half of whom had a history of ETS sensitivity and an objective, congestive response to a controlled challenge to ETS (ETS-S) and half nonsensitive (ETS-NS). Investigators exposed subjects to either air or sidestream tobacco smoke (SS) on two separate days, at least a week apart, in a climate-controlled chamber. Exposures lasted 60-minutes and the level of SS was regulated to a carbon monoxide concentration of 15 ppm. Roughly an hour after the exposure, 99 mTc-sulfur colloid aerosol was introduced nasally and serial counts were measured with a scintillation detector over the following hour. As a group, ETS-NS subjects showed more rapid clearance of the radiolabeled tracer than did ETS-S subjects. This group difference was based on half (three of six) ETS-S subjects, who showed marked inhibition of NMC. This subgroup did not differ significantly from the other ETS-S subjects with regard to age, gender, or allergy status. The authors acknowledged a marked heterogeneity in response of NMC to SS exposure, and the fact that multiple factors may govern the response. If present, slowed NMC could predispose individuals to respiratory tract infections.

Bascom et al. (49) and Kesavanathan et al. (50) studied 13 ETS-S and 16 ETS-NS subjects exposed to "low-to-moderate" SS levels (1, 5, and 15 ppm CO times 2 hours). A high proportion of subjects in both groups (69% of ETS-S and 50% of ETS-NS) had skin test reactivity to one or more aeroallergens. Objective endpoints included both nasal airway resistance (NAR) measured by posterior rhinomanometry, and nasal cross-sectional area/volume by acoustic rhinometry (AR). In general, postexposure symptoms increased monotonically with exposure level, with eye irritation and odor reaching significance at a lower exposure level (1 ppm CO) than nasal congestion, rhinorrhea, or cough (15 ppm). Differential responses by historical sensitivity status were evident for NAR at 1 and 5 ppm—but not at 15 ppm. The pattern of differences was complex, in that the ETS-NS group showed more objective nasal congestion at 1 ppm and the ETS-S group showed more congestion at 5 ppm. The pattern of differences for AR was even more complex, depending upon the portion of the tracing targeted (anterior, mid-, or posterior nasal cavity). In ETS-S subjects, nasal volume decreased in a dose-dependent manner. ETS-NS showed a qualitatively complex response pattern, with significant dimensional reductions in mid- and posterior nasal at 1 ppm CO but not at 5 ppm CO, and reductions in posterior nasal volume at 15 ppm CO.

Kesavanathan et al. (50) formally compared the endpoints of NAR and AR from this dataset in terms of coefficient of variation and correlation between symptoms and instrumental findings. In this latter regard, baseline subjective congestion correlated with NAR in ETS-S subjects, but with AR in ETS-NS subjects.

Nowak et al. (51) exposed 10 mild asthmatics to sidestream smoke at 22 ppm CO-equivalents for three hours, with control (clean air) exposure on separate days. Although the emphasis of this study was the lower airway, nasal lavage (NL) fluid was also obtained 30 minutes before and 30 minutes after smoke exposure. NL fluid was analyzed for histamine, albumin, eosinophilic cationic protein, myeloperoxidase, hyaluronic acid, and tryptase. Sidestream smoke exposure resulted in significantly greater increases in self-reported eye, nose, and throat irritation compared with clean air exposure ($p < 0.05$). NL mediators post-SS exposure were not significantly different from pre-challenge or post-sham values, however. The authors concluded that a three-hour ETS exposure was not a significant proinflammatory stimulus in the upper airway.

Walker et al. (52) exposed 17 nonsmoking, nonallergic white male subjects to clean air and five different experimentally generated ETS levels between 58 and 765 $\mu g/m^3$ total respirable particles (0.25–3 ppm CO over background). Sessions lasted 90 minutes with a 50-minute "plateau" period. Endpoints included symptom reporting, respiratory behavior, eye blink rate, cognitive performance, and mood state. Subjective eye irritation, eye dryness, odor, annoyance, and lack of air quality acceptability all rose significantly at the lowest ETS level employed, and increased monotonically with concentration thereafter. Nose and throat irritation were significantly elevated at or above the second ETS exposure level (0.5 ppm CO over background). Respiratory changes consisted of decreased respiratory rate and increased tidal volume, with minute ventilation staying relatively constant. Ventilatory changes occurred at all ETS exposure levels, without evidence of a dose–response relationship. Significant increases in eye blink rate occurred at the highest exposure level only. There were no significant exposure-related changes in cognitive performance, but a trend toward increased anxiety and anger—and decreased curiosity—which was significant at the highest exposure level. The authors argued that even the lowest ETS exposure level employed in this experiment was higher than real-life ETS exposures, and that 80% of individuals would be expected to find air containing ETS at 63 $\mu g/m^3$ total respirable particles unacceptable.

Willes et al. (53) studied 23 subjects, 14 ETS-S and 9 ETS-NS, with controlled exposures on two separate days to clean air or SS (15 ppm CO equivalent times two hours). Eight of fourteen ETS-S subjects (57%) were judged to be atopic by skin testing, and an even greater proportion of the ETS-NS subjects (78%) had evidence of allergies. In terms of upper airway endpoints, subjects rated symptoms and had nasal airway resistance (NAR) measured by posterior rhinomanometry both pre- and post-exposure. Nasal lavage (NL), on the other hand, was limited to post-exposure. Urinary cotinine levels were used to validate exposure. Following SS exposure, nasal symptoms increased and NAR rose significantly. Although seven of the eight subjects with the greatest ETS-related increases in NAR were in the ETS-S group, the two groups did not differ significantly in their mean response to ETS challenge. Nasal lavage markers, on the other hand, including total cell counts, neutrophils, and albumin, were unaffected by ETS exposure.

Junker et al. (54) conducted three separate substudies relating to ETS. The first was an emissions study, in which they found that machine-smoked cigarettes yielded significantly more VOCs and CO, but lower particulate mass, than had previously been documented. The second was an "odor threshold" study using an olfactometer, in which 18 female nonallergic, nonsmoking subjects detected SS odor in an ascending series, method of limits paradigm. The mean odor threshold corresponded to fresh air dilution volume of >19,000 m^3 per cigarette, over 100 times more than had previously been suggested for acceptable indoor air conditions. The third substudy was a whole-body ("chamber") study, in which 24 female subjects breathed SS over a wide concentration range (4.4–431 μg/m^3 PM2.25), the lowest of which corresponded to the level yielding odor detection in 95% of the threshold trials. Eye, throat, and nasal irritation, arousal, and annoyance were significantly elevated at the lowest SS exposure level, corresponding to a fresh air dilution volume of >3000 m^3 per cigarette. The authors pointed out that odor threshold concentrations for SS are three and more orders of magnitude lower than typical ETS concentrations measured in field settings, and that symptoms appeared at one order of magnitude lower SS concentrations than previously reported. They concluded that acceptable air quality for nonsmokers in smoking-permitted buildings may only be achievable with complete physical separation of smokers and nonsmokers.

Alteration of Sensory Thresholds

Cigarette smoking has the ability to change apparent chemosensory sensitivity to airborne odorants and irritants; in at least one case, these observations extend to passive smokers. Ahlstrom et al. (55) tested smokers, nonsmokers, and passive smokers for odor acuity to n-butanol and pyridine (the latter being a constituent of tobacco smoke). Both active and passive smokers were less sensitive than nonsmokers. Cometto-Muñiz and Cain (56) and Dunn et al. (57) examined the endpoint of altered respiration (reflex transitory inspiratory pause) as a measure of nasal irritant sensitivity. Both studies reported higher irritation thresholds (i.e., lower sensitivity) among smokers versus nonsmokers exposed to a nonodorant stimulus (high-level carbon dioxide). This finding was recently replicated using a CO_2-detection threshold as the endpoint of interest (58). On the other hand, Kjaergaard et al. (59,60) exposed smokers and nonsmokers to carbon dioxide by mask to determine eye irritation thresholds, and found no appreciable difference in sensitivity between the two groups. No published studies were identified that examined trigeminal irritant thresholds among passive smokers.

A number of mechanisms could explain observed sensory shifts in active and passive smokers. Decreased odor acuity among smoke-exposed individuals could result from increased nasal secretions, which in turn would pose an increased diffusion barrier to odorant molecules. Alternatively, habituation (in effect, ignoring the stimulus) may explain the odor perception findings; Ahlstrom et al. (55) emphasized the latter possibility because passive smokers did not differ from nonsmokers in the number of "zero intensity" responses given.

Eustachian Tube Dysfunction and Otitis Media in Children

Numerous studies have linked exposure to SHS in children to the development of otitis media (OM). These studies are reviewed in detail in two reviews published by the State of California, Office of Environmental Health Hazard Assessment (61,62). Among the biological mechanisms that likely produce this endpoint is irritant-induced Eustachian tube dysfunction. Lending credibility to this mechanism is a study by Dubin et al. (63) in which passive opening and closing pressures for the Eustachian tubes were elevated in rats exposed to sidestream smoke, as opposed to those exposed to clean air (control) conditions.

Dose–Response Considerations
Cain et al. (64), using a climate-controlled exposure chamber, found that 10% of nonsmoking subjects complained of unacceptable air quality (either due to eye irritation or odor annoyance) when SHS raised carbon monoxide (CO) levels by 2 ppm over background, and over 20% expressed dissatisfaction at 5 ppm over background. Muramatsu et al. (42) reported that nearly 30% of experimental subjects had complaints of moderate-to-severe eye irritation with SHS-derived CO levels 2.5 ppm over background. By comparison, CO levels can reach up to 10 ppm over background in smoking-permitted offices (average 2.5–2.8 ppm), and as high as 29 ppm (average 4.8–17 ppm) in taverns (19). Although early experimental work on sensory annoyance has been performed using CO as an index of SHS exposure, most current investigators believe that CO is an insensitive and unreliable surrogate measure for irritant and odorant exposure (65). Alternative exposure metrics include total respirable particles and airborne nicotine levels.

Effects of SHS Constituents
Selected investigators have examined the human sensory and reflex respiratory response to specific SHS constituents. Kendal-Reed et al. (66) demonstrated reflex changes in respiration (decreased tidal volume) among human volunteers exposed briefly (15 seconds) to propionic acid vapor at concentrations of 0.12 to 85 ppm. Walker et al. (67) examined the olfactory and irritant (trigeminal) properties of nicotine in both humans and experimental animals. The investigators pointed out that, on a part-per-million basis, nicotine was more potent than acetic or propionic acids, or amyl acetate, in eliciting: (*i*) olfactory sensation (in subjects with a normal sense of smell) and (*ii*) subjective nasal irritation (in subjects lacking the sense of smell). The investigators were able to corroborate their estimates of the relative stimulatory potencies of these compounds by obtaining electrophysiologic recordings from the trigeminal nerve in rats, finding a 15- to 60-fold lower response threshold for nicotine versus the other study compounds.

SUMMARY
SHS exposure produces a variety of irritative symptoms involving the upper respiratory tract and increasingly, these endpoints are being objectively documented and quantified. In addition to sensory irritation (and related inflammatory conditions), odor annoyance may detract significantly from subjective well-being and productivity among building occupants. Experimental studies conducted by investigators familiar with building ventilation practices suggest that it is impossible to sufficiently dilute airflow or segregate air handling systems to prevent annoyance, sensory irritation, or more serious health effects from affecting resident nonsmokers. Not only have smoking bans been enacted in public buildings on an ever-widening scale, but other traditional bastions of smoking behavior, such as multiunit (apartment) dwellings, are also increasingly being targeted for smoke-free policies to protect the health and well-being of nonsmokers.

REFERENCES

1. McKay J, Eriksen M. The tobacco atlas. Brighton, UK: Myriad Editions Ltd, 2002.
2. Chapman S. Other people's smoke: What's in a name? Tob Control 2003; 12(2): 113–114.
3. US Department of Health and Human Services. The health consequences of involuntary smoking: A report of the surgeon general: US Dept of Health and Human Services, Public Health Service, Centers for Disease Control; 1986. Report No.: 87–8398.
4. Ingebrethsen BJ, Heavner DL, Angel AL, et al. A comparative study of environmental tobacco smoke particulate mass measurements in an environmental chamber. JAPCA 1988; 38(4):413–417.
5. Singer BC, Hodgson AT, Guevarra KS, et al. Gas-phase organics in environmental tobacco smoke. 1. Effects of smoking rate, ventilation, and furnishing level on emission factors. Environ Sci Technol 2002; 36(5):846–853.
6. Kaegler M. Contract Research Center. Project no.: B1006 p 0500/3149 biological activity of fresh and ets-like sidestream smoke of standard reference cigarette 2r1 21-day inhalation study on rats. 2 February, 1995. Philip Morris. http://legacy.library.ucsf.edu/tid/ria32d00. Accessed April 2005.
7. Reininghaus W, Stinn W, Voncken P, et al. INBIFO. Documentation p 0500/5224 determination of the mass concentration ratio of particle/nicotine in room-aged sidestream smoke. 15 August, 1995. Philip Morris. http://legacy.library.ucsf.edu/tid/vhd55c00. Accessed April 2005.
8. Pandian MD, Ott WR, Behar JV. Residential air exchange rates for use in indoor air and exposure modeling studies. J Expo Anal Environ Epidemiol 1993; 3(4):407–416.
9. Singer BC, Hodgson AT, Nazaroff WW. Gas-phase organics in environmental tobacco smoke: 2. Exposure-relevant emission factors and indirect exposures from habitual smoking. Atmos Environ 2003; 37(39–40):5551–5561.
10. Schick SF, Glantz S. Concentrations of the carcinogen 4-(methylnitrosamino)-1-(3-pyridyl)-1-butanone in sidestream cigarette smoke increase after release into indoor air: Results from unpublished tobacco industry research. Cancer Epidemiol Biomarkers Prev 2007; 16(8):1547–1553.
11. Destaillats H, Singer BC, Lee SK, et al. Effect of ozone on nicotine desorption from model surfaces: Evidence for heterogeneous chemistry. Environ Sci Technol 2006; 40(6):1799–1805.
12. Schick S, Glantz S. Philip Morris toxicological experiments with fresh sidestream smoke: More toxic than mainstream smoke. Tob Control 2005; 14(6):396–404.
13. Haussmann HJ, Anskeit E, Becker D, et al. Comparison of fresh and room-aged cigarette sidestream smoke in a subchronic inhalation study on rats. Toxicol Sci 1998; 41(1):100–116.
14. Schick S, Glantz SA. Sidestream cigarette smoke toxicity increases with aging and exposure duration. Tob Control 2006; 15(6):424–429.
15. Hugod C. Indoor air pollution with smoke constituents-an experimental investigation. Prev Med 1984; 13:582–588.
16. Samet J, Cain W, Leaderer B. Environmental tobacco smoke. In: Samet JM, Spengler JD, eds. Indoor Air Pollution. Baltimore, MD: Johns Hopkins University Press, 1991:131–169.
17. Weber A. Acute effects of environmental tobacco smoke. Eur J Respir Dis 1984; 65(suppl 133):98–108.
18. Ayer H, Yeage RD. Irritants in cigarette smoke plumes. Am J Publ Health 1982; 72:1283–1285.
19. Triebig G, Zober M. Indoor air pollution by smoke constituents – a survey. Prev Med 1984; 13:570–581.
20. Cain WS. Contribution of the trigeminal nerve to perceived odor magnitude. Ann N Y Acad Sci 1974; 237(0):28–34.
21. Cain WS, Murphy CL. Interaction between chemoreceptive modalities of odour and irritation. Nature 1980; 284(5753):255–257.

22. Frank ME, Rabin MD. Chemosensory neuroanatomy and physiology. Ear Nose Throat J 1989; 68(4):291–296.

23. Pevsner J, Reed RR, Feinstein PG, et al. Molecular cloning of odorant-binding protein: Member of a ligand carrier family. Science 1988; 241(4863):336–339.

24. Lundberg J, Alving K, Karlsson J, et al. Sensory neuropeptide involvement in animal models of airway irritation and of allergen-evoked asthma. Am Rev Respir Dis 1991; 143(6):1429–1431.

25. Lundberg J, Martling C, Lundblad L. Cigarette smoke-induced irritation in the airways in relation to peptide-containing, capsaicin-sensitive sensory neurons. Klin Wochenschr 1988; 66(suppl 11):151–160.

26. Silver WL. Neural and pharmacological basis for nasal irritation. Ann N Y Acad Sci 1992; 641:152–163.

27. Hummel T, Livermore A, Hummel C, et al. Chemosensory event-related potentials in man: Relation to olfactory and painful sensations elicited by nicotine. Electroencephalogr Clin Neurophysiol 1992; 84(2):192–195.

28. Raphael GD, Baraniuk JN, Kaliner MA. How and why the nose runs. J Allergy Clin Immunol 1991; 87(2):457–467.

29. Baraniuk JN, Kaliner MA. Neuropeptides and nasal secretion. J Allergy Clin Immunol 1990; 86(4 Pt 2):620–627.

30. Widdicombe J. Nasal pathophysiology. Respir Med 1990; 84:3–10.

31. Alarie Y. Sensory irritation by airborne chemicals. CRC Crit Rev Toxicol 1973; 2(3):299–363.

32. Cain W, Tosun T, See L-C, et al. Environmental tobacco smoke: Sensory reactions of occupants. Atmos Environ 1987; 21:347–353.

33. National Research Council, ed. Environmental tobacco smoke: Measuring exposures and assessing health effects. Washington, D. C.: National Academy Press, 1986.

34. Knasko S. Ambient odor's effect on creativity, mood and perceived health. Chem Senses 1992; 17:27–35.

35. Rotton J. Affective and cognitive consequences of malodorous pollution. Basic Appl Soc Psych 1983; 4(2):171–191.

36. Mizoue T, Reijula K, Andersson K. Environmental tobacco smoke exposure and overtime work as risk factors for sick building syndrome in Japan. Am J Epidemiol 2001; 154(9):803–808.

37. Jones S, Love C, Thomson G, et al. Second-hand smoke at work: The exposure, perceptions and attitudes of bar and restaurant workers to environmental tobacco smoke. Aust N Z J Public Health 2001; 25(1):90–93.

38. Wieslander G, Lindgren T, Norback D, et al. Changes in the ocular and nasal signs and symptoms of aircrews in relation to the ban on smoking on intercontinental flights. Scand J Work Environ Health 2000; 26(6):514–522.

39. Eisner MD, Smith AK, Blanc PD. Bartenders' respiratory health after establishment of smoke-free bars and taverns. JAMA 1998; 280(22):1909–1914.

40. Raynal A, Burge PS, Roberston A, et al. How much does environmental tobacco smoke contribute to the building symptom index? Indoor Air 1995; 5(1):22–28.

41. Weber A, Grandjean E. Acute effects of environmental tobacco smoke. In: O'Neill IK, Brunnemann KD, Dodet B, Hoffmann D, eds. Environmental carcinogend: Methods of analysis and exposure measurement. Lyon, France: International Agency for Research on Cancer, 1987:59–68.

42. Muramatsu T, Weber A, Muramatsu S, et al. An experimental study on irritation and annoyance due to passive smoking. Int Arch Occup Environ Health 1983; 51(4):305–317.

43. Bascom R, Kulle T, Kagey-Sobotka A, et al. Upper respiratory tract environmental tobacco smoke sensitivity. Am Rev Respir Dis 1991; 143(6):1304–1311.

44. Willes S, Fitzgerald T, Bascom R. Nasal inhalation challenge studies with sidestream tobacco smoke. Arch Environ Health 1992; 47:223–230.

45. Stankus RP, Menon PK, Rando RJ, et al. Cigarette smoke-sensitive asthma: Challenge studies. J Allergy Clin Immunol 1988; 82(3 Pt 1):331–338.

46. Bascom R. Differential responsiveness to irritant mixtures. Possible mechanisms. Ann N Y Acad Sci 1992; 641:225–247.
47. Cummings K, Zaki A, Markello S. Variation in sensitivity to environmental tobacco smoke among adult non-smokers. Int J Epidemiol 1991; 20(1):121–125.
48. Bascom R, Kesavanathan J, Fitzgerald TK, et al. Sidestream tobacco smoke exposure acutely alters human nasal mucociliary clearance. Environ Health Perspect 1995; 103(11):1026–1030.
49. Bascom R, Kesavanathan J, Permutt T, et al. Tobacco smoke upper respiratory response relationships in healthy nonsmokers. Fundam Appl Toxicol 1996; 29(1): 86–93.
50. Kesavanathan J, Swift DL, Fitzgerald TK, et al. Evaluation of acoustic rhinometry and posterior rhinomanometry as tools for inhalation challenge studies. J Toxicol Environ Health 1996; 48(3):295–307.
51. Nowak D, Jorres R, Martinez-Muller L, et al. Effect of 3 hours of passive smoke exposure in the evening on inflammatory markers in bronchoalveolar and nasal lavage fluid in subjects with mild asthma. Int Arch Occup Environ Health 1997; 70(2):85–93.
52. Walker J, Nelson P, Cain WS, et al. Perceptual and psychophysiological responses of non-smokers to a range of environmental tobacco smoke concentrations. Indoor Air 1997; 7(3):173–188.
53. Willes SR, Fitzgerald TK, Permutt T, et al. Acute respiratory response to prolonged, moderate levels of sidestream tobacco smoke. J Toxicol Environ Health A 1998; 53(3):193–209.
54. Junker MH, Danuser B, Monn C, et al. Acute sensory responses of nonsmokers at very low environmental tobacco smoke concentrations in controlled laboratory settings. Environ Health Perspect. 2001; 109(10):1045–1052.
55. Ahlstrom R, Berglund B, Berglund U, et al. A comparison of odor perception in smokers, nonsmokers, and passive smokers. Am J Otolaryngol 1987; 8(1):1–6.
56. Cometto-Muniz J, Cain W. Perception of nasal pungency in smokers and nonsmokers. Physiol Behav 1982; 29(4):727–731.
57. Dunn J, Cometto-Muniz J, Cain W. Nasal reflexes: Reduced sensitivity to co2 irritation in cigarette smokers. J Appl Toxicol 1982; 2:176–296.
58. Shusterman D, Balmes J. Measurement of nasal irritant sensitivity to pulsed carbon dioxide: A pilot study. Arch Environ Health 1997; 52(5):334–340.
59. Kjaergaard S, Pedersen OF, Molhave L. Common chemical sense of the eyes – influence of smoking, age, and sex. Indoor Air '90: Proceedings of the 5th International Conference on Indoor Air Quality and Climate; Toronto: International Conference on Indoor Air Quality and Climate, Inc.; 1990:257–262.
60. Kjaergaard S, Pedersen O, Molhave L. Sensitivity of the eyes to airborne irritant stimuli: Influence of individual characteristics. Arch Environ Health 1992; 47: 45–50.
61. California Environmental Protection Agency. Health effects of exposure to environmental tobacco smoke: The report of the california environmental protection agency. Smoking and tobacco control monograph number 10. Bethesda: US Department of Health and Human Services, National Institutes of Health, National Cancer Institute; August 1999.
62. California Environmental Protection Agency, Office of Environmental Health Hazard Assessment, Office of Air Resources Board. Proposed identification of environmental tobacco smoke as a toxic air contaminant. Oakland California: California Environmental Protection Agency; June 24, 2005.
63. Dubin MG, Pollock HW, Ebert CS, et al. Eustachian tube dysfunction after tobacco smoke exposure. Otolaryngol Head Neck Surg 2002; 126(1):14–19.
64. Cain WS. A functional index of human sensory irritation. International Conference on Indoor Air Quality and Climate; Berlin: International Conference on Indoor Air Quality and Climate Inc.; 1987:661–665.
65. Chappell S, Parker R. Smoking and carbon monoxide levels in enclosed plublic places in new brunswick. Can J Publ Health 1977; 68:159–161.

66. Kendal-Reed M, Walker JC, Morgan WT. Human responses to odorants studied with precision olfactometry. In: Yoshizawa S, ed. The 7th international conference on indoor air quality and climate. Tokyo: Indoor Air, 1996:1007–1012.
67. Walker J, Kendal-Reed M, Bencherif M, et al. Olfactory and trigeminal responses to nicotine. Drug Dev Res 1996; 38:160–168.

Chlorine Exposure in Humans and Experimental Animals

Dennis J. Shusterman

*Division of Occupational and Environmental Medicine,
University of California, San Francisco, California, U.S.A.*

John B. Morris

*Department of Pharmaceutical Sciences, School of Pharmacy,
University of Connecticut, Storrs, Connecticut, U.S.A.*

INTRODUCTION

Chlorine (CAS No. 7782–50-5) was the seventh most heavily produced industrial chemical in the United States in 2001, with some 24 billion pounds output that year (1). Chlorine is used in a wide variety of industrial processes, including production of chlorinated solvents and plastics, water purification, battery manufacturing, and food processing (2). Chlorine gas is frequently involved in transportation accidents and so-called "fugitive" industrial emissions, with some 16 million pounds of airborne releases being reported to the EPA in 2002 (3). Chlorine can also be generated by the mixing of common household bleach with acidic cleaning products, leading to potentially serious irritant exposures in the home (4). In sum, chlorine is a widely used industrial chemical whose combined occupational and environmental exposure potential is considerable.

Because of its high chemical reactivity and substantial water solubility of 4.7 g/L at 37°C (5), chlorine can be expected to have considerable impact on the upper airway. In aqueous solution, chlorine is rapidly hydrolyzed to hydrochloric and hypochlorous acids. In single-breath studies in humans, it has been estimated that 90% of an inhaled bolus (of up to 3 ppm concentration) is deposited in the upper airway in humans (6). Upper respiratory tract deposition efficiencies in excess of 95% have also been observed in the mouse (7). Occupational chlorine exposures have been associated with, in addition to acute lung injury, both irritant-induced asthma and irritant rhinitis (8). Thus, chlorine is an important agent to consider in the context of upper airway toxicology.

STUDIES IN EXPERIMENTAL ANIMALS

There have been a variety of experimental animal studies on the pathological lesions resulting from acute high concentration and chronic low concentration exposures to chlorine (9–13). The focus here, however, will be on the immediate, presumably reflex responses that occur during short-term exposure to chlorine in animal models. Chlorine has long been known to induce the sensory irritation response in rats (14) and mice (15). This response is mediated by trigeminal nerve stimulation (16) and is characterized in animal models by a decreased respiratory frequency caused by a pause at the start of expiration (17). This response can be quantitated by measuring breathing frequency and/or by measuring the duration of the pause during early expiration, termed the duration of braking. Since breathing frequency can be reduced by a variety of mechanisms, the latter represents a more specific measure.

Barrow et al. examined the breathing frequency during exposure to chlorine in the Swiss-Webster mouse throughout a 10-minute exposure (18). The observed RD_{50} (the concentration required to cause a 50% reduction in breathing frequency) was 9.3 ppm. These authors also determined the RD_{50} for hydrochloric acid (a hydrolysis product of chlorine in solution) and obtained a value of 309 ppm, indicating that the acid hydrolysis product is roughly 30 times less potent than chlorine itself. In subsequent studies (14), it was determined that chlorine was also a sensory irritant in the rat, and that tolerance developed with repeated daily exposures to this irritant. The mechanism of the tolerance is unknown. Gagnaire et al. examined sensory irritation in OF1 mice during a 60-minute exposure to chlorine (15). Breathing frequencies steadily decreased throughout the exposure. An RD_{50} value of 3.5 ppm was obtained based on the maximal decrease observed during the exposure. These studies indicate that the sensory irritation response may be slow to develop and also that it is plastic, in that tolerance can develop depending on the exposure history.

Morris et al. examined mechanistic aspects of the immediate nasal response to chlorine in the female C57Bl/6 J mouse (7). To match the human studies of Shusterman (see below), mice were exposed for 15 minutes to chlorine over a range of concentrations that included those used in the human. The study focused on two responses: sensory irritation, as measured by breathing frequency and duration of braking, and airway obstruction as measured by specific airways resistance (sRaw) in intact mice, and by measurement of nasal airflow resistance in the surgically isolated upper respiratory tract of anesthetized mice.

Chlorine acted as a sensory irritant as indicated by concentration-dependent decreases in breathing frequency and increases in duration of braking. In intact animals, a concentration-dependent increase in sRaw (as measured by plethysmography) was observed. Both the sensory irritation and obstructive responses developed slowly, the maximal response being observed at the end of the 15-minute exposure. Nasal obstruction and the perception of irritation occur in humans exposed to chlorine (see below), thus there appears to be some correlation between animals (mice) and man.

Chlorine is hydrolyzed to hypochlorite and hydrochloric acid in solution. Both could be acting to cause nasal responses; however, the lack of potency of hydrochloric acid compared to chlorine itself suggests that hypochlorite is the critical agent (18). To confirm this supposition, mice were exposed to sodium hypochlorite aerosol in roughly the same concentration as chlorine gas. The hypochlorite produced slowly developing sensory irritation (decreased breathing frequency and increased duration of braking) and obstructive (increased sRaw) responses during the 15-minute exposure. The magnitudes of the responses to hypochlorite were equal to or greater than the response to equimolar chlorine (7). These results confirm that chlorine likely initiates immediate nasal responses via hypochlorite-dependent mechanisms. The molecular studies of Jordt's group—see below—support this concept (19).

The study of Morris et al. also included examination of physiological mechanisms of the response to chlorine in the mouse. The concentration-dependent increases in sRaw in intact animals may be reflective of upper or lower airway obstruction. To assess the contribution of the upper airways, the effect of chlorine in the surgically isolated upper respiratory tract of anesthetized mice was examined. In this model, chlorine induced a progressive increase in nasal flow resistance. The maximal value (observed at the end of the 15-minute exposure) was sufficient to account for the entire sRaw response observed in intact animals. This observation, coupled with the high deposition efficiency (>95%) of chlorine in the mouse nose, strongly suggests that the sRaw response observed in intact mice was reflective of upper airway obstruction (7).

Long-term degeneration of sensory nerves can be induced in laboratory animals by large doses of capsaicin. Capsaicin acts through the TRPV1 receptor to cause axonal degeneration of TRPV1-expressing C fibers, presumably through an excitotoxic mechanism (20,21). Both the sensory irritation and airway obstructive (sRaw) responses to chlorine were virtually absent in mice treated one week earlier with capsaicin (7), suggesting that both responses are critically dependent on sensory C fiber stimulation. Capsaicin pretreatment also results in virtual ablation of the sensory irritation and obstructive responses to acrolein, another potent irritant (22), thus chlorine is not unique in this regard. The mediators responsible for the obstructive response are not known and may include neuropeptides (via the axonal reflex) or acetylcholine (via reflex stimulation of parasympathetic nerves). Atropine was without effect on the chlorine-induced obstructive response suggesting the latter pathway was not significantly involved. ±

In unpublished studies, Morris examined the sensory irritation and sRaw responses to 1.6 ppm chlorine in wild-type and TRPV1-/- (knockout) mice. The sensory irritation response, as measured by duration of braking was similar: 260 ± 42 ms versus 316 ± 44 ms in wild-type and knockout mice, respectively (mean ± SEM). The maximal sRaw (as percent of baseline) was also similar averaging $222 \pm 10\%$ and $199 \pm 11\%$, in these strains, respectively. These results indicate that the TRPV1 receptor is not critical for initiation of chlorine-induced responses. Recent studies by Jordt et al. (19) have revealed a critical role of the TRPA1 receptor (see below).

Induction of allergic airway disease by ovalbumin sensitization and aerosol challenge modulates chlorine responsiveness. Shown in Figure 1 are the sensory irritation response (as measured by expiratory pause duration) and maximal sRaw response in control and ovalbumin-allergic airway diseased (OVA-AAD) mice exposed to 0.8 ppm chlorine (this experiment followed the identical paradigm as in Ref. 22). As can be seen, in allergic airway diseased mice, there was an enhanced sensory irritation response, but the obstructive response was not altered. An identical pattern (enhanced sensory irritation but obstruction) was observed in OVA-AAD mice exposed to acrolein (22)

Bessac et al. (19) examined molecular aspects of trigeminal nerve activation by hypochlorite, the active agent formed by hydrolysis of chlorine in aqueous solution. These authors provide persuasive evidence that sensory nerve stimulation by hypochlorite is mediated via the transient receptor potential A1 (TRPA1) receptor. The TRPA1 receptor is expressed in a subset of TRPV1 (capsaicin) receptor expressing sensory nerves. In vitro, hypochlorite selectively stimulated a subset of nerves sensitive to the TRPA1 agonist mustard oil. Hypochlorite activated cloned human and mouse TRPA1 receptors. Neurons from TRPA1-/- (knockout)

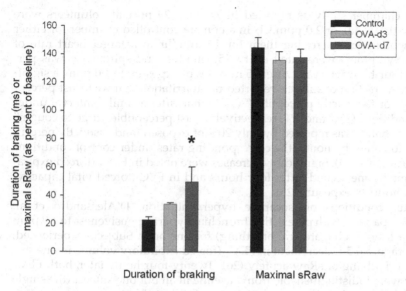

FIGURE 1 Chlorine-induced responses in OVA-AAD model. Responses to 0.8 ppm chlorine (15-minute exposure) in control mice and OVA-AAD mice after three or seven daily ovalbumin aerosol challenges (Ref. 22 for details of the model). ANOVA revealed a significant difference in the duration of braking response ($p < 0.05$), with the response in the day 7, but not day 3 mice, being greater than control. No difference was observed in the sRaw response (ANOVA; $p > 0.05$). *Source*: John Morris, unpublished data, 2004.

mice failed to respond to hypochlorite. Moreover, the sensory irritation response to hypochlorite aerosol was absent in vivo in TRPA1-/- mice. The TRPA1 receptor is also activated by other oxidants (e.g., H_2O_2), by electrophiles, including acrolein and crotonaldehyde and by cigarette smoke extract (23). Elucidating the receptor basis for chlorine (hypochlorite)-stimulation of sensory nerves represents a significant advance in our understanding of sensory nerve-irritant interactions (see chap. 12). There are many similarities in the response of the mouse to chlorine and acrolein: (*i*) both induce sensory irritation and an immediate nasal obstructive response, (*ii*) both responses are virtually abolished by capsaicin-induced nerve degeneration, and (*iii*) the sensory irritation, but not the obstructive response, is enhanced in allergic airway disease. Perhaps these represent general trends for all TRPA1 acting irritants.

CONTROLLED HUMAN EXPOSURE STUDIES
Controlled human exposure studies utilizing chlorine gas have involved a range of exposure apparatus (mask; whole-body chamber), durations (ranging from 15 minutes to 8 hours; single and multiday), and biological endpoints (subjective symptoms, pulmonary function tests, measures of nasal patency, indices of nasal inflammation). Because of chlorine's predilection for the upper airway, some studies have emphasized upper over lower airway endpoints, whereas others have utilized either exercise or hyperventilation to increase Cl_2 delivery to the lower airways. Several of these studies are reviewed below.

In a chamber study as reported in Anglen, 29 normal volunteers were exposed to 0, 0.5, 1.0, or 2.0 ppm Cl_2 in a climate-controlled chamber for either four or eight hours, exercising them for 15 min/hr to a target heart rate of 100 BPM. Symptoms were rated every 15 minutes, and spirometry was performed at four-hour intervals. After 15 minutes of exposure to 1.0 ppm, a significantly larger fraction of subjects reported nasal irritation that was "just perceptible" (70%) or "distinctly perceptible" (35%) than after a similar interval under control conditions (35% and 5%, respectively). Just perceptible "urge to cough," on the other hand, was reported by only 20% of exposed (and distinctly perceptible urge to cough by none), the corresponding rates under control conditions being 15% and 4% (24). Significant decreases were noted in FEV1 (forced expiratory volume in one second) after four hours and in FVC (forced vital capacity) after eight hours of exposure (25).

Under conditions of isocapnic hyperventilation, D'Alessandro et al. exposed five patients with preexisting bronchial hyperresponsiveness to 1.0 ppm Cl_2 via mask (with obligate oral breathing) for one hour. Subjects experienced, on the average, a 0.5-L decrement in FEV1 (forced expiratory volume in one second) and a doubling of sRaw acutely (26). Twenty-four hours later, both FEV1 and sRaw were indistinguishable from baseline in all but one subject (the single patient who reported chest symptoms during the procedure).

Schins et al. exposed eight male volunteers at rest in a chamber for 6 hr/day on three successive days to 0.1, 0.3, or 0.5 ppm Cl_2, or to filtered air. Spirometry and nasal lavage samples were obtained pre- and postexposure. Neither spirometric values nor nasal lavage analytes (cell counts, interleukin-8 levels) changed significantly in response to Cl_2 exposures at these levels (27). Although the authors reported minimal exposure-related symptom reporting, symptoms were ascertained neither systematically nor at baseline, and were subject to interpretation by the experimenters, thus tempering the generalizability of the finding.

Shusterman et al. utilized a nasal mask to expose 16 nonsmoking, nonasthmatic subjects [eight with seasonal allergic rhinitis (SAR) and eight nonrhinitics (NR)] to either Cl_2 gas (0.5 ppm × 15 minutes) or filtered air for a similar period. Exposures occurred a week apart and were in counter-balanced order. Subjects rated nasal symptoms and had their nasal airway resistance measured (by active posterior rhinomanometry) before, immediately after, and 15 minutes postexposure. SAR—but not NR—subjects showed significant increases in nasal airway resistance postexposure ($p < 0.05$; Fig. 2). SAR subjects also reported more significant exposure-related odor, nasal irritation, and nasal congestion (blockage) than did NR subjects (28).

Utilizing an similar experimental protocol in a larger cohort stratified by age, gender, and allergic rhinitis status ($n = 52$), Shusterman et al. (29) again found that seasonal allergic rhinitis predicted greater objective obstruction post-Cl_2 exposure ($p < 0.01$ at 15 minutes postexposure to 1.0 ppm Cl_2). Further, advancing age—but not gender—also predicted a greater obstructive response to Cl_2 ($p < 0.01$ immediately postexposure).

In an attempt to elucidate the pathophysiologic mechanism(s) of the Cl_2-induced obstructive nasal response, Shusterman et al. pretreated 24 subjects (12 with SAR and 12 NR) with either ipratropium bromide topical spray (a cholinergic blocker) or with placebo on a double-blinded basis prior to exposure to either Cl_2 (1.0 ppm × 15 minutes) or filtered air × 15 minutes. The remainder

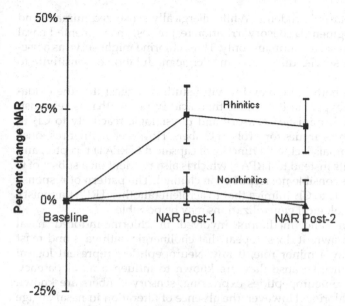

FIGURE 2 Chlorine-induced response in humans. Response to 0.5 ppm chlorine (15-minute exposure) in subjects with and without rhinitis. Shown is net percent change in nasal airway resistance (from baseline; chlorine minus air day) immediately postexposure (time 1) and 15 minutes postexposure (time 2) by rhinitis status ($p < 0.05$). *Source*: From Ref. 28.

of the protocol was as above (30). Cholinergic blockade prior to provocation did not significantly alter the obstructive effect of Cl_2 on SAR subjects, indicating that cholinergic parasympathetic reflexes were not likely to be responsible for the nasal obstructive response to Cl_2.

Also along mechanistic lines, Shusterman et al. analyzed nasal lavage fluid for tryptase (31) and various neuropeptides (32) pre- and postexposure in a subgroup of eight SAR and eight NAR subjects among whom differential physiologic reactivity to 1.0 ppm Cl_2 had previously been established (29). Neither tryptase (which is coreleased with histamine in immediate hypersensitivity reactions) nor various neuropeptides [substance P (SP), calcitonin gene-related protein (CGRP), vasoactive intestinal peptide (VIP), or neuropeptide Y (NPY)] were elevated in nasal lavage fluid post-Cl_2 challenge. Thus, neither mast cell degranulation (tryptase) nor peptidergic neural reflexes (neuropeptides) could be shown to be responsible for the nasal obstructive response to Cl_2.

DISCUSSION/CONCLUSIONS

Chlorine is an important toxic air contaminant, with potential for producing serious and persistent adverse effects in both the upper and lower respiratory tracts. Acute, high-level exposures may produce chemical pneumonitis, irritant-induced asthma, or chronic rhinitis (the latter two endpoints also being observed after multiple exposures at intermediate levels). At lower levels of exposure, sensory irritation, increased airflow resistance, and increased secretion in the upper airway are the predominant responses. Sensory irritation is documented experimentally as subjective nasal irritation in humans, or as respiratory slowing/expiratory pause in rodents. While allergically sensitized humans and rodents both show augmented sensory irritation responses, an augmented nasal obstructive response is seen in humans only. Thus, chlorine might serve as a useful irritant to examine species differences in allergic modulation of sensitivity to irritants.

Mechanistically, both in vivo and in vitro studies suggest it is the oxidative function of Cl_2 (hypochlorite) rather than acidity per se that is responsible for Cl_2's irritancy. Diminished sensory and physiologic reactivity to Cl_2 in capsaicin-treated animals argues for a role of C fibers. However, neither response was diminished in animals lacking a functional capsaicin (TRPA1) receptor, and in vitro evidence points instead to TRPA1, which is also resident on a subset of C fibers, as being the responsible nociceptive ion channel. The pattern of response of experimental animals to Cl_2 mimics that to acrolein (another TRPA1 agonist), suggesting some toxicological generalizations may be possible.

The mechanisms and mediators involved in chlorine-induced nasal obstruction are not known. It does appear that cholinergic pathways and mast cell degranulation play a minor role, if any. Neuropeptides[1] represent logical mediators of importance because they are known to influence nasal patency, and capsaicin-sensitive neuropeptides expressing, sensory C fibers are known to be stimulated by chlorine. However, the absence of alteration in nasal lavage neuropeptides levels by chlorine remains enigmatic. Perhaps alternate methodologies might reveal a role for neuropeptides and/or other C fiber–derived mediators are involved. Alternatively, nonneurogenic mechanisms, such as epithelial cell activation, could participate in the congestive response, as appears to be the case when humans are challenged with selected noxious hypertonic stimuli (Refs. 33 vs. 34). Nevertheless, chlorine is one of the few irritants that has been examined mechanistically in both laboratory animals and humans and, therefore, provides a useful model system to examine irritancy in multiple mammalian species.

REFERENCES

1. Anon. Production: Down but not out. Chem Engr News 2002; 80(15):60–65. http://pubs.acs.org/cen/coverstory/8025/pdf/8025production.pdf. Accessed September 17, 2008.
2. Lewis RJ. Hawley's Condensed Chemical Dictionary, 15th ed. Hoboken, NJ: John Wiley and Sons, 2007.
3. USEPA. Toxics Release Inventory Program. Washington: US Environmental Protection Agency, 2008. http://www.epa.gov/tri/. Accessed September 17, 2008.
4. Centers for Disease Control and Prevention. Chlorine gas toxicity from mixture of bleach with other cleaning products – California. MMWR Morb Mortal Wkly Rep 1991; 40:619–629.
5. Anon. Solubility of gases in water. 2008. www.engineeringtoolbox.com.
6. Nodelman V, Ultman JS. Longitudinal distribution of chlorine absorption in human airways: comparison of nasal and oral quiet breathing. J Appl Physiol 1999; 86: 1984–1993.

---

[1] Potential neuropeptide influences on nasal patency would include the *axon response* (with release of the SP and CGRP from afferent nerves) and *efferent neural reflexes* (with release of VIP from parasympathetic or NPY from sympathetic nerves). All of the above are vasodilators with the exception of NPY, which is a vasoconstrictor.

7. Morris JB, Wilkie WS, Shusterman DJ. Acute respiratory responses of the mouse to chlorine. Toxicol Sci 2005; 83:380–387.
8. Leroyer C, Malo JL, Girard D, et al. Chronic rhinitis in workers at risk of reactive airways dysfunction syndrome due to exposure to chlorine. Occup Environ Med 1999; 56:334–338
9. Barrow CS, Kociba RJ, Rampy LW, et al. An inhalation toxicity study of chlorine in Fischer 344 rats following 30 days of exposure. Toxicol Appl Pharmacol 1979; 49(1): 77–88.
10. Jiang XZ, Buckley LA, Morgan KT. Pathology of toxic responses to the RD_{50} concentration of chlorine gas in the nasal passages of rats and mice. Toxicol Appl Pharmacol 1983; 71:225–236.
11. Klonne DR, Ulrich CE, Riley MG, et al. One-year inhalation toxicity study of chlorine in rhesus monkeys (Macaca mulatta). Fundam Appl Toxicol 1987; 9(3):557–572.
12. McNulty MJ, Chang JC, Barrow CS, et al. Sulfhydryl oxidation in rat nasal mucosal tissues after chlorine inhalation. Toxicol Lett 1983; 17(3–4):241–246.
13. Wolf DC, Morgan KT, Gross EA, et al. Two-year inhalation exposure of female and male B6C3F1 mice and F344 rats to chlorine gas induces lesions confined to the nose. Fundam Appl Toxicol. 1995; 24:111–1131.
14. Chang JCF, Barrow, CS. Sensory irritation tolerance and cross-tolerance n F-344 rats exposed to chlorine or formaldehyde gas. Toxicol Appl Pharmacol 1984; 76:319–327.
15. Gagnaire F, Azim S, Bonnet P, et al. Comparison of the sensory irritation response in mice to chlorine and nitrogen trichloride. J Appl Toxicol 1994; 14:405–409.
16. Alarie Y. Sensory irritation by airborne chemicals. CRC Crit Rev Toxicol 1973; 3: 299–363.
17. Vijayaraghavan R, Schaper M, Thompson R, et al. Characteristic modification of the breathing pattern of mice to evaluate the effects of airborne chemicals on the respiratory tract. Arch Toxicol 1993; 67:478–490.
18. Barrow CS, Alarie Y, Warrick JC, et al. Comparison of the sensory irritation response in mice to chlorine and hydrogen chloride. Arch Environ Health 1977; 32:68–76.
19. Bessac BF, Sivula M, von Hehn CA, et al. TRPA1 is a major oxidant sensor in murine airway sensory neurons. J Clin Invest 2008; 118:1899–1910.
20. Holzer P. Capsaicin: cellular targets, mechanisms of action, and selectivity for thin sensory neurons. Pharmacol Rev 1991; 43:143–201.
21. Szallasi A, Blumberg PM. Vanilloid (Capsaicin) receptors and mechanisms. Pharmacol Rev 1999; 51:159–212.
22. Morris JB, Symanowicz PT, Olsen JE, et al. Immediate sensory nerve mediated respiratory responses to irritants in healthy and allergic airway diseased mice. J Appl Physiol 2003; 94:1563–1571.
23. Andre E, Campi B, Saterazzi S, et al. Cigarette smoke-induced neurogenic inflammation is mediated by a,b-unsaturated aldehydes and the TRPA1 receptor in rodent. J Clin Invest 2008; 118:2574–2582.
24. Anglen DM. Sensory Response of Human Subjects to Chlorine in Air [dissertation]. Ann Arbor, MI: University of Michigan, 1981.
25. Rotman HH, Fliegelman MJ, Moore T, et al. Effects of low concentrations of chlorine on pulmonary function in humans. J Appl Physiol 1983; 54:1120–1124.
26. D'Alessandro A, Kuschner W, Wong H, et al. Exaggerated responses to chlorine inhalation among persons with nonspecific airway hyperreactivity. Chest 1996; 109:331–337.
27. Schins RP, Emmen H, Hoogendijk L, et al. Nasal inflammatory and respiratory parameters in human volunteers during and after repeated exposure to chlorine. Eur Respir J 2000; 16:626–632.
28. Shusterman DJ, Murphy MA, Balmes JR. Subjects with seasonal allergic rhinitis and nonrhinitic subjects react differentially to nasal provocation with chlorine gas. J Allergy Clin Immunol 1998; 101(6 Pt 1):732–740.
29. Shusterman D, Murphy MA, Balmes J. Influence of age, gender, and allergy status on nasal reactivity to inhaled chlorine. Inhal Toxicol 2003; 15:1179–1189.

30. Shusterman D, Murphy MA, Walsh P, et al. 2002. Cholinergic blockade does not alter the nasal congestive response to irritant provocation. Rhinology 40:141–146.
31. Shusterman D, Balmes J, Avila PC, et al. Chlorine inhalation produces nasal congestion in allergic rhinitics without mast cell degranulation. Eur Respir J 2003b; 21: 652–657.
32. Shusterman D, Balmes J, Murphy MA, et al. Chlorine inhalation produces nasal airflow limitation in allergic rhinitic subjects without evidence of neuropeptide release. Neuropeptides 2004; 38:351–358.
33. Koskela H, Di Sciascio MB, Anderson SD, et al. Nasal hyperosmolar challenge with a dry powder of mannitol in patients with allergic rhinitis. Evidence for epithelial cell involvement. Clin Exp Allergy 2000; 30:1627–1636.
34. Baraniuk JN, Ali M, Yuta A, et al. Hypertonic saline nasal provocation stimulates nociceptive nerves, substance P release, and glandular mucous exocytosis in normal humans. Am J Respir Crit Care Med 1999; 160:655–662.

David C. Dorman and Melanie L. Foster
College of Veterinary Medicine, North Carolina State University, Raleigh, North Carolina, U.S.A.

Hydrogen sulfide (H_2S) is a well-recognized toxic gas and environmental hazard. This chapter provides an update of our current understanding of H_2S-induced toxicity with a special emphasis on research of interest to inhalation toxicologists. We also discuss the role H_2S has as a gaseous biological molecule and putative neurotransmitter. No attempt has been made to provide an exhaustive review, and wherever possible references to recent research and review articles are provided.

PHYSICAL AND CHEMICAL PROPERTIES

Hydrogen sulfide (molecular weight $= 34.08$; 1 ppm $= 1.39$ mg/m^3) is a colorless flammable gas with a characteristic odor of rotten eggs. Hydrogen sulfide is heavier than air (specific gravity $= 1.19$ vs. air $= 1.00$), is relatively water soluble (0.398 g/100 g water at 20°C), and exists as a gas at ambient conditions (vapor pressure $= 15,600$ mm Hg at 25°C). Aqueous solutions of H_2S are considered unstable as a result of sulfide reaction with molecular oxygen.

SOURCES

Hydrogen sulfide is produced in large quantities when sulfur-containing proteinaceous materials undergo putrefaction and is therefore also known as swamp or sewer gas. The ambient air concentration resulting from natural sources is estimated as 0.11 to 0.33 ppb (1). Industrial sources of H_2S include pulp and paper operations, tanneries, mining, and petroleum refineries. Hydrogen sulfide is a byproduct of many industrial processes and is used to make inorganic sulfides found in certain dyes, pesticides, polymers, and pharmaceuticals.

Sulfide is present as an endogenous substance in normal mammalian tissues (2–5). The production of sulfide is closely associated with the catabolism of cysteine and methionine as well as with gluthathione metabolism (6,7). Hydrogen sulfide is synthesized endogenously in mammalian tissues by cystathionine beta-synthase and cystathionine gamma-lyase, two pyridoxal-5'-phosphate–dependent enzymes responsible for L-cysteine metabolism. In tissue homogenates, rates of sulfide production are in the range of 1 to 10 pmoles/sec/mg protein (8). This results in low micromolar extracellular concentrations of sulfide.

Literature values for endogenous levels of sulfide are variable and depend on the procedures used to extract the sulfide from the tissue and the analytical chemical methods used to quantify this metabolite (4,9,10). Special care must be taken to minimize the loss of free sulfide from the tissue sample due to the volatility of this gas. Normal tissues contain relatively high (ppm) concentrations of endogenous sulfide ion (HS^-), and sulfide has been measured (mean $\pm$ SEM) in rat hindbrain, lung, nasal respiratory epithelium, and olfactory epithelium at 1.21 ± 0.05 $\mu g/g$, 0.54 ± 0.03 $\mu g/g$, 1.73 ± 0.17 $\mu g/g$, and 1.42 ± 0.11 $\mu g/g$, respectively (11).

Endogenous sulfide also arises from bacterial activity in the lower bowel. Volatile sulfur compounds are produced by oral bacteria (12), with H_2S and methyl mercaptan as the main components, found in mouth air. Humans with severe halitosis may have oral cavity H_2S concentrations that exceed 0.7 ppm (13).

Toxicokinetics

Sulfide is rapidly used and/or metabolized by mammalian tissues. The major metabolic and excretory pathway of H_2S involves oxidation of sulfide to sulfate with subsequent urinary excretion of free and conjugated sulfates (14). The exact mechanism of the oxidation is unknown; however, both enzymatic (sulfide oxidase) and nonenzymatic catalytic systems have been proposed. Data suggest that thiol S-methyltransferase also catalyzes the methylation of H_2S to yield less toxic methanethiol and dimethylsulfide (14).

Several investigators have examined the toxicokinetics of H_2S following inhalation. Kage et al. (1992) (3) reported elevated blood and urinary thiosulfate concentrations in rabbits exposed to 100 to 200 ppm H_2S for 60 minutes. Kangas and Savolainen (1987) (15) likewise reported elevated urinary thiosulfate levels in human volunteers exposed to 8, 18, or 30 ppm H_2S for 30 to 45 minutes. Nasal extraction of H_2S was measured in the isolated upper respiratory tracts of male rats (16). Extraction was measured for constant unidirectional inspiratory flow at 75, 150, and 300 mL/min, which correspond to 50%, 100%, and 200% of the predicted minute volume of the adult male CD rat. Nominal H_2S exposure concentrations of 10, 80, and 200 ppm were used. Nasal extraction was dependent on the concentration of inspired H_2S and the rate of airflow through the nasal cavity and ranged from 32% for a 10 ppm exposure at 75 mL/min to 7% for a 200 ppm exposure at 300 mL/min. Dorman and coworkers (2002) (11) determined nasal and lung sulfide and sulfide metabolite concentrations immediately after the end of a three-hour H_2S exposure to 30 to 400 ppm in adult rats (Fig. 1). They found that nasal and lung sulfide concentrations increased following H_2S exposure in a dose-dependent manner. They also examined lung sulfide concentrations for up to seven hours after the end of a three-hour exposure to 400 ppm H_2S. Lung sulfide concentrations rapidly returned to preexposure levels within minutes after the end of a three-hour exposure, suggesting that rapid pulmonary elimination or metabolism of sulfide occurs. An accumulation of sulfide metabolites was not observed in the lung during the three-hour H_2S exposure. Transient increases in lung sulfite, sulfate, and thiosulfate concentrations were observed, however, immediately after the end of the 400-ppm H_2S exposure. This increase in sulfide metabolite concentrations occurred coincidentally with the rapid decrease in lung sulfide concentration. This observation suggests that the detoxification of sulfide to sulfate may become less effective as the concentration of sulfide increases in blood and other tissues due to H_2S exposure.

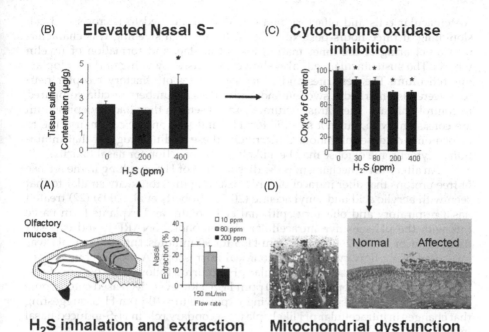

(B) **Elevated Nasal S⁻**

(C) **Cytochrome oxidase inhibition⁻**

(A)

Olfactory mucosa

(D)

Normal   Affected

H₂S inhalation and extraction by the nasal epithelium

Mitochondrial dysfunction and olfactory neuronal loss

FIGURE 1 Hypothesized steps in the pathogenesis of hydrogen sulfide (H₂S)–induced olfactory neuronal loss in rats. Panel (**A**) shows nasal uptake data from Schroeter et al., (16) demonstrating dose-dependent nasal uptake. Extraction of H₂S results in elevated concentrations (**B**) and decreased cytochrome oxidase activity (**C**) in the olfactory epithelium [data from Dorman et al. (11)]. Ultrastructural changes include mitochondrial swelling in the olfactory epithelium (**D**, *left*) and olfactory neuronal loss (**D**, *right*). Data from Brenneman et al. (52). Focal olfactory neuronal loss can occur (**D**) and may be in part due to regional changes in airflow or inherent tissue sensitivity.

Mode of Action

Higher (millimolar) tissue sulfide concentrations tend to be cytotoxic to cells, via free radical and oxidant generation, calcium mobilization, and induction of mitochondrial cell death pathways. Under physiologic conditions, H₂S acts to block the respiratory chain primarily by inhibiting cytochrome c oxidase, and the undissociated species (H₂S) is a more potent inhibitor than the anionic species (HS⁻) (17,18). Tissues with high oxygen demand (e.g., cardiac muscle, brain) are particularly sensitive to sulfide inhibition of electron transport. Cytochrome oxidase inhibition by H₂S may indirectly cause hyperpnea through stimulation of the carotid and aortic body chemosensors by blocking the availability of oxygen (19). The clinical picture resulting from acute lethal exposure to H₂S (500–1000 ppm) is almost identical with hydrogen cyanide poisoning.

The mechanism by which H₂S inhalation damages the nasal epithelium and results in adverse clinical signs is poorly understood. Decreased cytochrome oxidase activity was observed in the rat olfactory and respiratory epithelium following a three-hour exposure to ≥30 ppm H₂S (Fig. 1) (11). Rats exposed once to 400 ppm for three-hour had severe swelling of mitochondria in both sustentacular cells and olfactory neurons that progressed to necrosis and cell sloughing at three hours postexposure. Mitochondrial swelling was characterized by clearing of the inner matrix, loss of cristae, and formation of myelin whorls. The sustentacular cells also showed extensive swelling of the endoplasmic reticulum. The dendrites and olfactory vesicles of olfactory receptor neurons were also markedly swollen and had reduced numbers of cilia compared to control animals. The ultrastructural changes seen in the olfactory epithelium are consistent with, but not specific for, H_2S-induced anoxic cell injury due to cytochrome oxidase inhibition. Collectively, these findings suggest that inhibition of cytochrome oxidase may be a likely mode of action for nasal toxicity.

An alternative mechanism is the dissociation of H_2S resulting in the release of free protons that alter intracellular pH resulting in cytotoxicity similar to that seen with acrylic acid and vinyl acetate (20,21). Roberts et al. (2006) (22) treated nasal respiratory and olfactory epithelial cell isolates and explants from naïve rats with the pH-sensitive intracellular chromophore, SNARF-1, and exposed them to air or 10, 80, 200, or 400 ppm H_2S for 90 minutes. Intracellular pH was measured using flow cytometry or confocal microscopy. A modest, but statistically significant decrease in intracellular pH occurred following exposure of respiratory and olfactory epithelia to 400 ppm H_2S. However, decreased cytochrome oxidase activity was observed following exposure to >10 ppm H_2S, suggesting that changes in intracellular pH likely play a secondary role in H_2S-induced nasal injury.

PHYSIOLOGICAL FUNCTIONS

Like nitric oxide (NO) and carbon monoxide (CO), H_2S has been identified as a putative gaseous biological molecule and neurotransmitter (7,23). At physiological levels, H_2S can exert cytotoxic or cytoprotective effects (23). Hydrogen sulfide can scavenge certain oxyradicals (24), peroxynitrite (25), hypochlorous acid (26), and homocysteine (27). Low levels of H_2S can upregulate endogenous antioxidant systems (28). There is some evidence that H_2S also participates in normal nerve transmission (2,29), although any role in olfaction is currently unknown.

Hydrogen sulfide also upregulates production of heme oxygenase in pulmonary smooth muscle cells (30) and rat nasal tissues (31). Upregulation of heme oxygenase-1 can secondarily induce carbon monoxide production resulting in additional cytoprotective and anti-inflammatory effects (32). Sulfide relaxes smooth muscle and has vasodilatory and cardiac protective effects. These responses can be prevented by pretreatment with the potassium ATP (K_{ATP}) channel inhibitor glibenclamide (7).

TOXICITY

Concentration-dependent toxicity in the respiratory, cardiovascular, and nervous systems can occur following H_2S inhalation. Concentrations that produce effects in these systems are similar (i.e., less than an order of magnitude) in rats, mice, and humans, and species differences in toxicity are not prominent (1).

Low-Dose (<50 ppm) Effects

Occupational exposure to H_2S occurs in the agricultural, gas, oil refining, and certain other industries. Workers often notice the characteristic "rotten-egg" odor associated with H_2S exposure. Most people readily perceive this gas since the olfactory detection limit for H_2S is ≥ 5 ppb (33). Hydrogen sulfide is often considered a "nuisance odor" and community concerns arise at relatively low exposure levels. Areas of the human brain that demonstrate activation by H_2S inhalation (as revealed by functional brain MRI studies) include the medial frontal gyrus, insular gyrus, putamen, and hippocampus, suggesting that this odor is detected by the olfactory system (34). This pattern of brain activation is distinct from that seen with carbon dioxide, a gas that primarily activates the trigeminal nerve (34).

Individuals exposed to low levels of H_2S and other sulfur gases often report headaches, nausea, wheezing, shortness of breath, eye irritation/conjunctivitis, and other symptoms (35–38). Information obtained from questionnaires in a study, in which measured exposure levels were used, indicated that average 24-hour exposures of <0.007 ppm (peak four-hour exposure was 0.095 ppm) H_2S did not cause significant respiratory or ocular effects, but above that level respiratory symptoms were observed (39).

Hydrogen sulfide is a contact irritant causing inflammatory and irritant effects on the moist membranes of the eyes and respiratory tract (14). Ocular effects include intense photophobia, eyelid spasm, excessive tearing, intense congestion, pain, blurred vision, and sluggish pupil reaction (40). Vanhoorne and coworkers (1995) (41) reported that viscose rayon workers exposed chronically to H_2S concentrations of ≤ 3.6 ppm did not have significantly more ocular complaints when compared to a group of unexposed workers. In contrast, workers exposed to ≥ 6.4 ppm H_2S had significantly more eye complaints including pain, burning, irritation, hazy sight, and photophobia. An H_2S concentration of ~ 20 ppm appears to be a threshold for obvious acute impacts on the eye (40).

A series of experiments performed by Bhambhani and coworkers (42–45) examined the dose-related effects of low-dose exposures of H_2S on exercising volunteers. The results indicated that there were no significant changes in arterial blood gases, hemoglobin saturation levels, or cardiovascular and metabolic responses at any exposure level and at any exercise level. Likewise, heart rate, ventilation rate, oxygen rate, and carbon dioxide production were generally unaffected. Other responses demonstrated equivocal results. Citrate synthetase, a marker of aerobic metabolism, was also significantly decreased in activity in one experiment (43). In some experiments, muscle or blood lactate increased significantly during exposure to 5.0 ppm H_2S (45). These results suggest that anaerobic metabolism is increased following H_2S inhalation. Other studies by this group, however, showed that muscle lactate dehydrogenase and cytochrome oxidase were not altered or minimally affected by H_2S inhalation (43). Animal studies performed at much higher exposure concentrations suggest that blood lactate concentrations do not change significantly in piglets exposed to H_2S at 20 to 80 ppm for 1.5 hours (46).

Several studies have evaluated human airway changes following H_2S inhalation. Jappinen et al. (1990) (47) exposed mild asthmatics to 2 ppm H_2S for 30 minutes in a closed chamber. There were no changes in forced vital capacity (FVC), forced expiratory volume in one second (FEV_1), and forced expiratory flow (FEF) as a result of exposure. Jappinen et al. (1990) (47) also studied a cohort of 26 male pulp mill workers to assess the possible effects of daily H_2S exposure (≤ 10 ppm) on respiratory function. Bronchial responsiveness, FVC, and FEV_1 were measured after the end of a work day following at least one day off work. No significant changes in respiratory function or bronchial responsiveness were observed following H_2S exposure compared to the control values. In another study, 21 swine confinement facility owner/operators were tested by spirometry immediately before and after a four-hour work period (48). The confinement workers had statistically ($p < 0.05$) significant reductions in pulmonary flow rates ranging from 3.3% to 11.9% mean FEF (forced expiratory flow) after the four-hour work period. The work environment was sampled for particulates and gases during the exposure period and there was suggestive evidence for a concentration–response relationship between carbon dioxide and H_2S exposure concentrations and lung function decrements. However, these monitoring data were not presented in the study report.

Adverse neurological effects are also of concern in people. Fiedler and colleagues (2008) (49) exposed 74 healthy subjects to H_2S at 0.05, 0.5, and 5 ppm. A dose–response reduction in air quality and increases in ratings of odor intensity, irritation, and unpleasantness were observed. Total symptom severity, including eye irritation, was not significantly elevated across any exposure condition, but anxiety symptoms were significantly greater in people exposed to 5 ppm H_2S when compared to responses seen following exposure to 0.05 ppm. No dose–response effect was observed for sensory or cognitive measures. Verbal learning was compromised during each exposure condition. The authors concluded that although some symptoms increased with exposure, the magnitude of these changes was relatively minor. Increased anxiety was significantly related to ratings of irritation due to odor. Whether the effect on verbal learning represents a threshold effect of H_2S or an effect due to fatigue across exposure could not be determined.

Curtis et al. (1975) (50) exposed three immature pigs continuously (24 hr/day) to either 0 or 8.5 ppm H_2S for 17 days. Pigs were subjected to a complete gross examination at necropsy and histological examination of tissues from the respiratory tract, eye, and viscera. The pigs weighed an average of 13.2 kg at the beginning of the study and gained an average of 0.53 kg/day. The results indicate that the exposure level was a NOAEL (no-observed-adverse-effect-level). Unfortunately the nasal cavity was not examined.

Intermediate-Dose (50–100 ppm) Effects

Studies conducted in rodents have confirmed that the nasal epithelium is especially sensitive to H_2S inhalation (Fig. 2). The most prominent nasal lesion is olfactory mucosal necrosis (51–53). Multifocal, bilaterally symmetrical olfactory neuron loss and basal cell hyperplasia, limited to the olfactory mucosa, are often observed in rodents exposed to H_2S. Lesions occur most commonly in the dorsal medial meatus and dorsal and medial areas of the ethmoid recess. Within the olfactory epithelium, the sites commonly affected by H_2S are likely exposed to a higher local tissue dose of H_2S because they border high velocity air streams emanating from the dorsal medial meatus.

In rodents, olfactory neuronal loss becomes more prevalent and extensive (i.e., a greater percentage of the olfactory epithelium is affected) as the exposure duration and/or exposure concentration increases. Brenneman and coworkers (2000) (54) used a unilateral nasal inhalation model and showed that olfactory neuronal loss in rats is the result of a direct (portal of entry) effect rather than systemic delivery of the gas. Observed NOAEL for olfactory neuronal loss have

Rhinitis

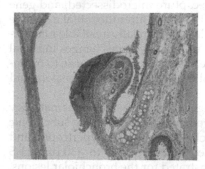

Bronchiolar epithelial hypertrophy and hyperplasia

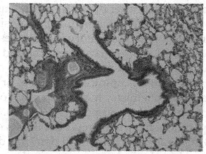

Olfactory neuronal loss

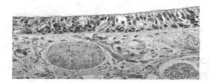

Nasal respiratory metaplasia

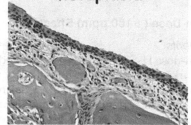

FIGURE 2 Nasal and lower airway changes seen in rats following acute (respiratory epithelial metaplasia) or repeated exposure to moderate concentratons of hydrogen sulfide.

been reported to be 30 and 10 ppm for acute (4-day) and subchronic (90-day) exposure, respectively.

The observed olfactory neuronal loss is partially repaired following exposure. Replacement of olfactory neurons is dependent upon a functional basal layer in the olfactory epithelium. Basal cell hypertrophy and hyperplasia followed by the complete restoration of the olfactory epithelium by six weeks following the end of a five-day H_2S exposure regimen demonstrated that this tissue retained its full regenerative capacity following acute injury (52). However, subchronic exposure to 80 ppm H_2S resulted in severe atrophy of the olfactory mucosa with a minimal hyperplastic response in the basal layer. This finding suggests that subchronic exposure to relatively low levels of H_2S inhibited regeneration of the olfactory epithelium (51).

Mice exposed subchronically to 80 ppm H_2S for 6 hr/day, 5 day/wk for 90 days develop rhinitis (100% incidence) (53). Brenneman et al. also reported transient metaplasia in the nasal respiratory mucosa one day after a three-hour exposure to ≥80 ppm H_2S (52). This lesion is not seen following repeated exposures (5–65 exposure days), suggesting that the regenerated respiratory epithelium undergoes an adaptive response and becomes resistant to further injury (51). Roberts et al. (2008) (31) exposed rats nose-only to either air or 200 ppm

H_2S for 3 hr/day for 1 or 5 consecutive days. Nasal respiratory epithelial cells at the site of injury and regeneration were laser-capture microdissected, and gene expression profiles were generated at 3, 6, and 24 hours after the initial exposure and at 24 hours after the fifth exposure using the commercially available microarrays. Gene ontology enrichment analysis identified early gene changes in signal transduction, inflammatory/defense response, cell cycle, and response to oxidative stress. Later gene changes occurred in cell cycle, DNA repair, transport, and microtubule-based movement.

In general, pulmonary edema, or changes in routine clinical hematology and serum chemistry measures are not observed in rodents following subchronic exposure to up to 80 ppm H_2S (52,53). Bronchiolar epithelial hypertrophy and hyperplasia have been reported in rats exposed subchronically to $\geq$30 ppm H_2S (53). Overall, a NOAEL of 10 ppm was demonstrated for the bronchiolar lesions.

Neurological responses are also seen in neonatal and adult rats exposed to moderate H_2S concentrations. Observed changes have included altered brain neurotransmitter concentrations, increased Purkinje cell dendrite length, and altered electroencephalogram (EEG) activity in the hippocampus (55–58).

High-Dose (>100 ppm) Effects

Rodents

High-dose H_2S inhalation is associated with nasal toxicity in rodents. Lopez and coworkers (1988) (59) found olfactory mucosal necrosis and necrosis with ulceration and inflammation in the nasal respiratory epithelium of rats exposed for 4 hours to 400 ppm H_2S. Numerous studies suggest that acute as well as repeated high dose (80–400 ppm) H_2S exposure can affect brain neurochemistry, physiology, and behavior in rodents (60,61).

The available data demonstrates that Haber's rule ($C \times T = k$) does not appear to apply to H_2S-induced lethality for short-time intervals. LC_{50} values for 10 (835 ppm), 30 (726 ppm), and 50 minutes (683 ppm) have been reported for Wistar rats (62,63). Zwart et al. (1990) (63) reported H_2S inhalation LC_{50}s of 683 and 676 ppm, respectively, for rats and mice following a 50-minute exposure. MacEwen and Vernot (1972) (64) reported LC_{50}s for rats and mice of 712 and 634 ppm after one hour of exposure. Prior et al. (1988) (65) reported that rat LC_{50}s for 2-, 4-, and 6-hour exposures were 587, 501, and 335 ppm, respectively. In this study deaths were attributable to severe pulmonary edema. Lopez et al. (1989) (66) showed that the gross and histologic evidence of pulmonary edema caused by five minutes of exposure to lethal concentrations (1667 ppm) of H_2S was a direct effect, since intraperitoneal injections of 30 mg/kg sodium hydrosulfide did not affect the airways or lungs.

Humans

With exposure to 100 ppm H_2S, humans report a sudden loss of olfaction (so-called olfactory paralysis) that is typically reversible with removal from contaminated air (1). This response does not appear to represent accommodation typical of most odorants. Persistent anosmia and hyposmia with an immediate onset have been reported in workers following acute, near-fatal episodes of H_2S poisoning (67,68). Olfactory deficits (hyposmia) with both a delayed and immediate onset have also been documented following repeated exposures to sublethal doses of H_2S in workers at a refinery construction site (69). The ability of H_2S to compromise the sense of smell creates an even greater hazard in the workplace because exposed workers lose their means of detecting the presence of H_2S or other noxious gases.

Symptoms observed following acute exposure to ~100 to 500 ppm H_2S include ocular and respiratory tract irritation, nausea, vomiting, diarrhea, headaches, loss of equilibrium, memory loss, olfactory paralysis, loss of consciousness, tremors, and convulsions (1). Respiratory symptoms become more severe with longer exposure resulting in pulmonary edema. Hydrogen sulfide–induced ocular effects include tearing, burning, and irritation of the cornea and conjunctivae (40).

Many individuals exposed to ≥500 ppm H_2S develop a rapid loss of consciousness followed by apparent recovery. This syndrome can occur after only one or two breaths of H_2S and is often referred to as "knockdown." The observed comatose state is often rapidly reversed when the victim is evacuated to fresh air and given artificial respiration and/or oxygen. However, long-term neurological deficits including incoordination, memory and motor dysfunction, personality changes, hallucinations, and anosmia can occur. Many of these clinical effects are consistent with organic brain disease resulting from hypoxia and may occasionally persist for several years after the initial H_2S exposure (68,70–73).

Hydrogen sulfide is also associated with fatal exposures in the workplace (74). Exposure concentrations and durations in these accidents are generally poorly defined. Vapor concentrations on the order of 500 to 1000 ppm or more are often fatal within minutes (72). Most fatalities occur in confined spaces (e.g., sewers, animal processing plants, manure tanks) and result from respiratory failure, noncardiogenic pulmonary edema, coma, and cyanosis (70). Pulmonary edema, however, is not uniformly seen in fatal H_2S poisoning. Increased blood sulfide concentrations are occasionally detected following high-dose H_2S exposure. Some individuals exposed to ~1000 ppm develop vagal-mediated apnea and H_2S-induced central respiratory arrest.

RISK ASSESSMENT CONSIDERATIONS

Olfactory mucosal lesions from the aforementioned H_2S subchronic inhalation study (51) have been used as the critical effect by the U.S. Environmental Protection Agency (US EPA) to calculate an inhalation reference concentration (RfC). The RfC is an estimate of a continuous inhalation exposure to humans that is likely to be without appreciable risk of adverse effects over a person's lifetime. As of 2006, the RfC for H_2S is 2×10^{-3} mg/m^3 based on a NOAEL of 10 ppm (13.9 mg/m^3). The RfC calculation relies on dosimetric adjustment factors that are used to account for species differences in delivered dose, to continuous exposure conditions (24 hr/day for 70 years), and the regional gas dose ratio (RGDR). The RGDR is a ratio of the relative minute volume to upper respiratory tract surface area ratios between the rat and the human and assumes complete extraction of the gas with a uniform dose throughout the upper respiratory tract. Uncertainty factors of 300 (a factor of 3 for interspecies extrapolation, 10 for sensitive populations, and 10 for subchronic exposure) were also used by the US EPA in their derivation of the RfC.

An alternative approach using computational fluid dynamics (CFD) models has been used to describe the uptake of inhaled H_2S in rat nasal passages

(75,76). These simulations showed a good correlation between areas of high flux on airway walls with olfactory epithelial responses in the rat nasal cavity. Nasal extraction data (16) was used to estimate kinetic parameters governing systemic and metabolic clearance in nasal tissue with a pharmacokinetic model. The estimated kinetic parameters were then implemented in a boundary condition for the CFD model that incorporated first-order and saturable tissue reaction kinetics in nasal tissue to govern mass flux at the air–tissue interface. Extraction of H_2S in rat nasal passages predicted with this pharmacokinetic-driven CFD model compared well with the rodent experimental data generated by Brenneman et al. (2000) (51). Olfactory flux levels predicted from the human CFD model were used to derive a no-observed-adverse-effect-level human equivalent concentration (NOAEL[HEC]) for H_2S. This NOAEL[HEC] value (5 ppm) can be used as a guide in establishing an inhalation RfC.

REFERENCES

1. ATSDR (Agency for Toxic Substances and Disease Registry). Toxicological Profile for Hydrogen Sulfide. Atlanta, GA: U.S. Department of Health and Human Services, Agency for Toxic Substances and Disease Registry, 2006.
2. Boehning D, Snyder SH. Novel neural modulators. Ann Rev Neuro Sci 2003; 26: 105–131.
3. Kage S, Nagata T, Takekawa K, et al. The usefulness of thiosulfate as an indicator of hydrogen sulfide poisoning in forensic toxicological examination: A study with animal experiments. Jap J Forensic Toxicol 1992; 10:223–227.
4. Mitchell TW, Savage JC, Gould DH. High-performance liquid chromatography detection of sulfide in tissues from sulfide-treated mice. J Appl Toxicol 1993; 13: 389–394.
5. Warenycia MW, Goodwin LR, Francom DM, et al. Dithiothreitol liberates non-acid labile sulfide from brain tissue of H_2S-poisoned animals. Arch Toxicol 1990; 64: 650–655.
6. Fiorucci S, Distrutti E, Cirino G, et al. The emerging roles of hydrogen sulfide in the gastrointestinal tract and liver. Gastroenterology 2006; 131:259–271.
7. Wang R. Two's company, three's a crowd: can H_2S be the third endogenous gaseous transmitter? FASEB J 2002; 16:1792–1798.
8. Doeller JE, Isbell TS, Benavides G, et al. Polarographic measurement of hydrogen sulfide production and consumption by mammalian tissues. Anal Biochem 2005; 341: 40–51.
9. Goodwin LR, Francom D, Dieken FP, et al. Determination of sulfide in brain tissue by gas dialysis/ion chromatography postmortem studies and two case reports. J Anal Toxicol 1989; 13:105–109.
10. Kage S, Nagata T, Kimura K, et al. Extractive alkylation and gas chromatographic analysis of sulfide. J Forensic Sci 1988; 33:217–222.
11. Dorman DC, Moulin FJM, McManus BE, et al. Cytochrome oxidase inhibition induced by acute hydrogen sulfide inhalation: correlations with tissue sulfide concentrations in the rat brain, liver, lung and nasal epithelium. Toxicol Sci 2002; 65:18–25.
12. Persson S, Edlund MB, Claesson R, et al. The formation of hydrogen sulphide and methyl mercaptan by oral bacteria. Oral Microbiol Immunol 1990; 5:195–201.
13. Tsai CC, Chou HH, Wu TL, et al. The levels of volatile sulfur compounds in mouth air from patients with chronic periodontitis. J Periodontal Res 2008; 43(2): 186–193.
14. Beauchamp RO, Bus JS, Popp JA, et al. A critical review of the literature on hydrogen sulfide toxicity. CRC Crit Rev Toxicol 1984; 13:25–97.
15. Kangas J, Savolainen H. Urinary thiosulphate as an indicator of exposure to hydrogen sulphide vapour. Clin Chim Acta 1987; 164:7–10.

16. Schroeter JD, Kimbell JS, Bonner AM, et al. Incorporation of tissue reaction kinetics in a computational fluid dynamics model for nasal extraction of inhaled hydrogen sulfide in rats. Toxicol Sci 2006(b); 90(1):198–207.
17. Hill BC, Woon TC, Nicholls P, et al. Interactions of sulphide and other ligands with cytochrome *c* oxidase. An electron-paramagnetic-resonance study. Biochem J 1984; 224:591–600.
18. Khan AA, Schuler MM, Prior MG, et al. Effects of hydrogen sulfide exposure on lung mitochondrial respiratory chain enzymes in rats. Toxicol Appl Pharmacol 1990; 103:482–490.
19. Ammann HM. A new look at physiologic respiratory response to H_2S poisoning. J Hazard Mater 1986; 13:369–374.
20. Frederick CB, Gentry PR, Bush ML, et al. A hybrid computational fluid dynamics and physiologically based pharmacokinetic model for comparison of predicted tissue concentrations of acrylic acid and other vapors in the rat and human nasal cavities following inhalation exposure. Inhal Toxicol 2001; 13:359–376.
21. Lantz RC, Orozco J, Bogdanffy MS. Vinyl acetate decreases intracellular pH in rat nasal epithelial cells. Toxicol Sci 2003; 75:423–431.
22. Roberts ES, Wong VA, McManus BE, et al. Changes in intracellular pH play a secondary role in hydrogen sulfide-induced nasal cytotoxicity. Inhal Toxicol 2006; 18(3):159–167.
23. Szabó C. Hydrogen sulphide and its therapeutic potential. Nat Rev Drug Discov 2007; 6(11):917–935.
24. Geng B, Chang L, Pan C, et al. Hydrogen sulfide regulation of myocardial injury induced by isoproterenol. Biochem Biophys Res Commun 2004; 318:756–763.
25. Whiteman M, Armstrong JS, Chu SH, et al. The novel neuromodulator hydrogen sulfide: an endogenous peroxynitrite 'scavenger'? J Neurochem 2004; 90: 765–768.
26. Whiteman M, Cheung NS, Zhu YZ, et al. Hydrogen sulphide: a novel inhibitor of hypochlorous acid-mediated oxidative damage in the brain? Biochem Biophys Res Commun 2005; 326:794–798.
27. Yan SK, Chang T, Wang H, et al. Effects of hydrogen sulfide on homocysteine-induced oxidative stress in vascular smooth muscle cells. Biochem Biophys Res Commun 2006; 351(2):485–491.
28. Kimura Y, Kimura H. Hydrogen sulfide protects neurons from oxidative stress. FASEB J 2004; 18:1165–1167.
29. Kimura H. Hydrogen sulfide induces cyclic AMP and modulates NMDA receptor. Biochem Biophys Res Commun 2000; 267:129–133.
30. Qingyou Z, Junbao D, Weijin Z, et al. Impact of hydrogen sulfide on carbon monoxide/heme oxygenase pathway in the pathogenesis of hypoxic pulmonary hypertension. Biochem Biophys Res Commun 2004; 317:30–37.
31. Roberts ES, Thomas RS, Dorman DC. Gene expression changes following acute hydrogen sulfide (H_2S)-induced nasal respiratory epithelial injury. Toxicol Pathol 2008; 6(4):560–756.
32. Ryter SW, Alam J, Choi AM. Heme oxygenase-1/carbon monoxide: from basic science to therapeutic applications. Physiol Rev 2006; 86:583–650.
33. Hoshika Y, Imamura T, Muto G, et al. International comparison of odor threshold values of several odorants in Japan and in the Netherlands. Environ Res 1993; 61(1): 78–83.
34. Bensafi M, Iannilli E, Gerber J, et al. Neural coding of stimulus concentration in the human olfactory and intranasal trigeminal systems. Neuroscience 2008; 154(2):832–838.
35. Glass DC. A review of the health effects of hydrogen sulphide exposure. Ann Occup Hyg 1990; 34:323–327.
36. Jaakkola JJ, Vilkka V, Marttila O, et al. The South Karelia Air Pollution Study. The effects of malodorous sulfur compounds from pulp mills on respiratory and other symptoms. Am Rev Respir Dis 1990; 142:1344–1350.

37. Marttila OT, Haahtela H, Vaittinen I, et al. The South Karelia air pollution study: relationship of outdoor and indoor concentrations of malodorous sulfur compounds released by pulp mills. J Air Waste Manage Assoc 1994; 44:1093–1096.

38. Shusterman D. Community health and odor pollution regulation. Am J Public Health 1992; 82(11):1566–1567.

39. Marttila O, Jaakkola JJK, Partti-Pellinen K, et al. South Karelia air pollution study: daily symptom intensity in relation to exposure levels of malodorous sulfur compounds from pulp mills. Environ Res 1995; 71:122–127.

40. Lambert TW, Goodwin VM, Stefani D, et al. Hydrogen sulfide (H_2S) and sour gas effects on the eye. A historical perspective. Sci Total Environ 2006; 367(1):1–22.

41. Vanhoorne M, de Rouck A, de Bacquer D. Epidemiological study of eye irritation by hydrogen sulfide and/or carbon disulphide exposure in viscose rayon workers. Ann Occup Hyg 1995; 39:307–315.

42. Bhambhani Y, Burnham R, Snydmiller G, et al. Comparative physiological responses of exercising men and women to 5 ppm hydrogen sulfide exposure. Am Ind Hyg Assoc J 1994; 55:1030–1035.

43. Bhambhani Y, Burnham R, Snydmiller G, et al. Effects of 5 ppm hydrogen sulfide inhalation on biochemical properties of skeletal muscle in exercising men and women. Am Ind Hyg Assoc J 1996(a); 57:464–468.

44. Bhambhani Y, Burnham R, Snydmiller G, et al. Effects of 10 ppm hydrogen sulfide inhalation on pulmonary function in healthy men and women. J Occup Environ Med 1996(b); 38:1012–1017.

45. Bhambhani Y, Singh M. Physiological effects of hydrogen sulfide inhalation during exercise in healthy men. J Appl Physiol 1991; 71:1872–1877.

46. Li J, Zhang G, Cai S, et al. Effect of inhaled hydrogen sulfide on metabolic responses in anesthetized, paralyzed, and mechanically ventilated piglets. Pediatr Crit Care Med 2008; 9(1):110–112.

47. Jappinen P, Vilkka V, Marttila O, et al. Exposure to hydrogen sulfide and respiratory function. Br J Ind Med 1990; 47:824–828.

48. Donham KJ, Zavala DC, Merchant J. Acute effects of the work environment on pulmonary functions of swine confinement workers. Am J Ind Med 1984; 5:367–376.

49. Fiedler N, Kipen H, Ohman-Strickland P, et al. Sensory and cognitive effects of acute exposure to hydrogen sulfide. Environ Health Perspect 2008; 116(1):78–85.

50. Curtis SE, Anderson CR, Simon J, et al. Effects of aerial ammonia, hydrogen sulfide and swine-house dust on rate of gain and respiratory-tract structure in swine. J Animal Sci 1975; 41:735–739.

51. Brenneman KA, James JA, Gross EA, et al. Olfactory neuron loss in adult male CD rats following subchronic inhalation exposure to hydrogen sulfide. Toxicol Pathol 2000; 28:326–333.

52. Brenneman KA, Meleason DF, Sar M, et al. Olfactory mucosal necrosis in male CD rats following acute inhalation exposure to hydrogen sulfide: reversibility and possible role of regional metabolism. Toxicol Pathol 2002; 30:200–208.

53. Dorman DC, Struve MF, Gross EA, et al. Respiratory tract toxicity of inhaled hydrogen sulfide in Fisher-344 rats, Sprague-Dawley rats, and $B_6C_3F_1$ mice following subchronic (90 day) exposure. Toxicol Appl Pharmacol 2004; 198:29–39.

54. Brenneman KA, Wong BA, Buccellato MA, et al. Direct olfactory transport of inhaled manganese ($^{54}MnCl_2$) to the rat brain: toxicokinetic investigations in a unilateral nasal occlusion model. Toxicol Appl Pharmacol 2000; 169(3):238–248.

55. Hannah RS, Hayden LJ, Roth SH. Hydrogen sulfide exposure alters amino acid content in developing rat CNS. Neurosci Lett 1989; 99:323–327.

56. Hannah RS, Roth SH. Chronic exposure to low concentrations of hydrogen sulfide produces abnormal growth in developing cerebellar Purkinje cells. Neurosci Lett 1991; 122:225–228.

57. Skrajny B, Hannah RS, Roth SH. Low concentrations of hydrogen sulfide alter monoamine levels in the developing rat central nervous system. Can J Physiol Pharmacol 1992; 70:1515–1518.

58. Skrajny B, Reiffenstein RJ, Sainsbury RS, et al. Effects of repeated exposures of hydrogen sulfide on rat hippocampal EEG. Toxicol Lett 1996; 84:43–53.
59. Lopez A, Prior M, Yong S, et al. Nasal lesions in rats exposed to hydrogen sulfide for four hours. Am J Vet Res 1988; 49:1107–1111.
60. Partlo LA, Sainsbury RS, Roth SH. Effects of repeated hydrogen sulphide exposure on learning and memory in the adult rat. Neurotoxicol 2001; 22:177–189.
61. Struve MF, Brisbois JN, James RA, et al. Neurotoxicological effects associated with short-term exposure of Sprague-Dawley rats to hydrogen sulfide. Neurotoxicol 2001; 22:375–385.
62. Arts JHE, Zwart A, Schoen ED, et al. Determination of concentration–time–mortality relationships versus LC50s according to OECD guideline 403. Exp Pathol 1989; 37: 62–66.
63. Zwart A, Arts JHE, Klokman-Houweling JM, et al. Determination of concentration—time–mortality relationships to replace LC50 values. Inhal Toxicol 1990; 2:105–117.
64. MacEwen JD, Vernot EH. Toxic Hazards Research Unit Annual Report. Ohio: Aerospace Medical Research Laboratory, Air Force Systems Command, Wright-Patterson Air Force Base; 1972. Report No. ARML-TR-72–62. pp. 66–69.
65. Prior MG, Sharma AK, Yong S, et al. Concentration–time interactions in hydrogen sulfide toxicity in rats. Can J Vet Res 1988; 52:375–379.
66. Lopez A, Prior M, Reiffenstein RJ, et al. Peracute toxic effects of inhaled hydrogen sulfide and injected sodium hydrosulfide on the lungs of rats. Fund Appl Toxicol 1989; 12:367–373.
67. Schneider JS, Tobe EH, Mozley PD Jr, et al. Persistent cognitive and motor deficits following acute hydrogen sulfide poisoning. Occup Med (Oxf) 1998; 48:255–260.
68. Tvedt B, Skyberg K, Aaserud O, et al. Brain damage caused by hydrogen sulfide: a follow-up study of six patients. Am J Ind Med 1991; 20:91–101.
69. Hirsch AR, Zavala G. Long term effects on the olfactory system of exposure to hydrogen sulfide. Occup Environ Med 1999; 56:284–287.
70. Arnold IM, Dufresne RM, Alleyne BC, et al. Health implication of occupational exposures to hydrogen sulfide. J Occup Med 1985; 27(5):373–376.
71. Kilburn KH, Warshaw RH. Hydrogen sulfide and reduced-sulfur gases adversely affect neurophysiological functions. Toxicol Ind Health 1995; 11:185–197.
72. Reiffenstein RJ, Hulbert WC, Roth SH. Toxicology of hydrogen sulfide. Ann Rev Pharmacol Toxicol 1992; 32:109–134.
73. Tvedt B, Edlund A, Skyberg K, et al. Delayed neuropsychiatric sequelae after acute hydrogen sulfide poisoning: affection of motor function, memory, vision, and hearing. Acta Neurol Scand 1991; 84:348–351.
74. Fuller DC, Suruda AJ. Occupationally related hydrogen sulfide deaths in the United States from 1984 to 1994. J Occup Med 2000; 42:939–942.
75. Moulin FJM, Brenneman KA, Kimbell JS, et al. Predicted regional flux of hydrogen sulfide correlates with distribution of nasal olfactory lesions in rats. Toxicol Sci 2002; 66:7–15.
76. Schroeter JD, Kimbell JS, Andersen ME, et al. Use of a pharmcokinetic driven computational fluid dynamics model to predict nasal extraction of hydrogen sulfide in rats and humans. Toxicol Sci 2006; 94(2):359–367.

22 Sulfur Dioxide Exposure in Humans

Jane Q. Koenig

Department of Environmental and Occupational Health Sciences,
University of Washington, Seattle, Washington, U.S.A.

INTRODUCTION AND BACKGROUND

Sulfur dioxide (SO_2) is one of the six air contaminants regulated as "criteria " pollutants by the Environmental Protection Agency. The criteria pollutants are ones that are widespread and relatively toxic. SO_2 is emitted from both natural sources, such as volcanoes, and man-made sources. The most common source of SO_2 emissions in the United States is coal-fired power plants. Other stationery sources of SO_2 are refineries, smelters, paper and pulp mills, and food processing plants. A small amount of SO_2 is emitted from the mobile fleet of vehicles. SO_2 has been an air pollutant of concern for human populations for centuries. It was one of the air pollutants strongly associated with mortality during the London Killer Fog in December 1952. Approximately 20,000 premature deaths were associated with SO_2 and black smoke during and in the month after that episode. In addition being a health threat itself, SO_2 is the precursor of sulfuric acid (H_2SO_4), a pollutant prominent in acid rain and a fine particulate matter.

EPA sets National Ambient Air Quality Standards (NAAQS) for the criteria pollutants. The current primary (health-related) standards for SO_2 are 0.14 ppm for a 24-hour average and 0.03 ppm for an annual average. There is a secondary (welfare-related) standard of 0.5 ppm for a three-hour average. There has been an ongoing discussion within EPA over the last decade regarding the setting of a short-term (for example, one hour) primary standard for SO_2. This would appear to be sensible, as SO_2, being a very soluble gas, does not remain long in the atmosphere. Moreover, physiologic effects occur rapidly (within minutes) upon exposure to SO_2. Several states do have short-term standards for SO_2. For instance, in Washington State there is a three-hour average (0.50 ppm), a one-hour average not to be exceeded more than twice in seven days (0.25 ppm), and a one-hour average not to be exceeded more than once a year (0.40 ppm).

This chapter attempts to summarize (*i*) the general human health effects associated with SO_2; and (*ii*) the association between SO_2 exposure and biological responses in the human nose and the upper airways. SO_2 was originally considered the most harmful criteria pollutant due to its properties of uptake, efficient respiratory tract deposition, and its wide spread presence in the United States. There is a wealth of data showing associations between SO_2 exposure and respiratory morbidity and mortality. A number of reviews summarizing the health effects of inhaled SO_2 have been published (1–4). However, as control technologies (SO_2 scrubbers, etc.) have been developed for sources, SO_2 concentrations have decreased substantially. As part of the response to acid rain, a cap-and-trade system was imposed on SO_2 emissions in the United States in the mid-1990s. As a consequence, SO_2 emissions rapidly dropped to around 41% of 1980 levels by 2002 (http://www.grinningplanet.com/2004/02–12/cap-and-trade-pollution). However, many experts feel SO_2 emissions should be ratcheted down further as provided by the Clean Air Act. There also is concern that the United States may increase its reliance on coal combustion as means to energy independence. The SO_2 cap-and-trade program has been so successful that it is offered as a model for a cap-and–trade effort to remove carbon from the atmosphere.

EFFECTS OF SO_2 ON THE LOWER AIRWAY

The history of the search for human health effects of SO_2 is of relevance to this review. In the late-1970s and 1980s, there was increased interest in the health effects of inhaled SO_2 due to the oil embargo by Middle Eastern countries. The U.S. government backed a move toward energy independence based on large increases in the use of coal for electricity. The public health community cringed because coal combustion is associated with significant SO_2 (as well as heavy metal) emissions. EPA funded several research projects investigating the respiratory health effects of SO_2 exposure. One of these projects, directed by N. Robert Frank, was at the School of Public Health at University of Washington (UW) in Seattle. Since the Clean Air Act of 1970 specifies that the NAAQS be set to protect the most sensitive groups, investigators began studies of the effects of SO_2 in populations with asthma (a notably sensitive population). Asthma is the most common chronic disease of childhood, thus knowledge of SO_2 effects in that population was needed. A series of controlled human studies were conducted at UW using adolescents with asthma. Many of these subjects were adolescents from a nearby asthma and allergy clinic who had well-documented diagnoses of asthma and/or allergy. In general, these studies found that a 10-minute exposure during moderate exercise to 1.0 or even 0.5 ppm SO_2 was associated with large decrements in lung functions (1). Some of the subjects complained of shortness-of-breath and occasionally needed to use their rescue inhalers after exposure. Other symptoms associated with short-term exposure to SO_2 are cough and wheeze. The effects from SO_2 exposure in children were subsequently replicated in adults (5). Across studies, the primary physiologic effect of SO_2 exposure was bronchoconstriction and pretreatment with albuterol (a beta$_2$-agonist causing bronchodilation) abolished the SO_2-induced effect (6). However, there are also suggestions that SO_2 exposure is associated with some degree of airway inflammation. For instance, Sandstrom et al. (7) detected increased number of macrophages, lymphocytes, and mast cells in bronchioalveolar lavage fluid up to 72 hours post-SO_2 exposure in healthy nonsmoking subjects. Pretreatment with other medications used in asthma therapy (cromolyn sodium, theophylline, and ipratropium bromide) all mitigate the SO_2 effect (see Ref. 1 for details). These medications have varied mechanisms of action. As a result of this research, we now know that individuals with asthma are much more sensitive to the effects of inhaled SO_2 than those without asthma (1).

While studies discussed above pertain to acute short-term SO_2 exposures, there also is literature pertaining to chronic exposures. In the United States, investigators conducted studies in a large epidemiologic endeavor tracking populations exposed to sulfur oxide air pollution. Unfortunately, the largest study, the so-called Community Health Environmental Surveillance System (CHESS) study sponsored by EPA, was criticized for inaccurate statistical analyses, and the data from that study were ignored. (The CHESS study found significant associations between SO_2 exposure and mortality in susceptible populations.) These findings were published years after the CHESS research program. An unfortunate outcome of this event was that EPA deemphasized the value of epidemiological air pollution studies for a decade or more. The effects attributed to SO_2 were often confounded by concomitant exposure to sulfuric acid aerosols. Thus, the EPA entitled its 1982 summary document "Air Quality Criteria for Particulate Matter and Sulfur Oxides."

A summary of the epidemiologic evidence of an association between chronic respiratory health effects and SO_2 exposure appears in a review by Dockery and Speizer (8). In addition to data from the London Killer Fog episode mentioned earlier, these authors noted data from an industrial exposure incident in Japan. Japanese researchers found a mortality rate of $4.4/10^5$ in SO_2-polluted areas compared to $0.4/10^5$ in the nonpolluted area. Japan eventually compensated persons who lived in sulfur-polluted areas and developed lung disease.

EFFECTS OF SO_2 ON THE UPPER AIRWAY

Dosimetry

Although the effects of SO_2 on asthmatic airways are well-known, upper airway effects also have been documented. SO_2 is a highly water-soluble gas, thus breathing through the nose scrubs much of the SO_2 from the inhaled air (9,10). Early studies showed that individuals quietly breathing SO_2 in concentrations from 1 to 25 ppm by nose are likely to absorb over 95% of the inhaled gas in the airways above the larynx (11). Since SO_2 is highly water soluble, the gas will be rapidly transformed when it deposits on aqueous surfaces in the airways. In terms of chemical reactivity, one study showed, using nasal lavage fluid, that S-sulfonate levels were statistically elevated after exposure to 1 ppm (in children with asthma) or 7 ppm (in healthy adult subjects) (12).

Although this defense mechanism (absorption) protects the lung from inhaled SO_2, the nasal passages themselves are prone to adverse effects as we will see. Further the defensive role of the upper airway can be overwhelmed by a variety of factors. For example, increased ventilation rate, such as seen during exercise, can effect oral–nasal partitioning of airflow and decrease residence time of SO_2 sufficiently to allow more of the gas to penetrate into the lungs. Similarly, chronic nasal obstruction due to inflammatory disease (rhinosinusitis) can predispose to mouth breathing, even at rest, and thereby decrease the upper airway efficiency at scrubbing SO_2.

The effect of SO_2 on the upper airways has been evaluated in controlled human studies similar to those researching its effects on the lower airway. The studies showing that subjects with asthma were much more susceptible to the effects of inhaled SO_2 than those without (1) were carried out with subjects breathing on a mouthpiece with nose clips in place. The question arose whether the same adverse effects would be seen if subjects breathed in a natural manner, that is, through the nose or through the nose and mouth. Most individuals breathe through both the nose and mouth when exercising. There is a switching point from nose to mouth breathing at a ventilatory rate of around 35 L/min (13). Since most exposure to SO_2 by subjects with asthma is likely to occur outdoors where they are relatively active, investigators decided to investigate the importance of the route of inhalation.

During the 1980s, a group of studies were conducted comparing the effects of SO_2 inhaled by nose versus by mouth. It was found that SO_2 exposure via a mouthpiece with nose clips in place resulted in a greater lung response than exposure during unencumbered breathing during which some of the gas was inhaled through the nasal passages (14–17). Thus, normal oronasal breathing apparently provides some protective scrubbing of SO_2. Consequently, modeling SO_2 lung dosimetry, in addition to concentration, duration, and ventilatory rate, requires consideration of the water solubility of the gas and the state of oronasal partitioning.

Nasal Patency

Although it appears that inhalation of SO_2 through the nose decreases to some extent the magnitude of the lung response, another question arises, namely, what is the effect on the nasal passages themselves of exposure to SO_2. There is a paucity of short-term health data on the effects of SO_2 inhalation on the upper airways. However, techniques are available to measure selected nasal physiologic parameters (see chap. 9). One such parameter is posterior rhinomanometry. The investigators at UW used such a system to quantify the effect of nasal inhalation of SO_2 on work of breathing (17). In this system, nasal work of breathing was measured using a modified skin diving mask and two differential pressure transducers. Pressure in the oropharynx is detected using a tube extending to the back of the mouth and connected to one side of a differential pressure transducer. The opposite side of the transducer monitors pressure within the mask. The energy expended during nasal breathing is calculated as a means of characterizing nasal flow energy expenditure during expiration. Nasal work of breathing can be calculated as the rate of energy required in joules per second or as nasal power in milliwatts.

Using this apparatus, a study was carried out to investigate the effect of SO_2 exposure on the nasal work of breathing (17). Ten adolescent subjects with extrinsic asthma breathed either clean filtered air or 0.5 ppm SO_2 (on separate days) via a face mask. For comparison, they also breathed filtered air or 0.5 ppm SO_2 through a mouthpiece with nose clips in place. The exposures were for 30 minutes at rest followed by 20 minutes during moderate exercise on a treadmill. Nasal work of breathing varied greatly within and between the subjects. As a group, the average increase in nasal power after exposure to SO_2 was 30% after exposure through the face mask and 32% after exposure through the mouthpiece. These changes represented statistically significant increases from baseline nasal power values.

As a follow-up to this study, Koenig et al. (18) conducted a controlled SO_2 exposure of 13 adolescent subjects with exercise-induced bronchoconstriction. The objective of this second study was to determine whether an antihistamine (chlorpheniramine) pretreatment of the nasal passages would inhibit the increased work of nasal breathing. Subjects were treated with 4 mg or 12 mg chlorpheniramine or placebo on the evening before the SO_2 exposure. Nasal work of breathing was measured as described above. The SO_2 exposure increased work of breathing approximately 35%. There were no significant effects of chlorpheniramine treatment at a dose of 4 mg. However, pretreatment with

12 mg was associated with a mitigation of SO_2-induced work of breathing (1% increase compared to the 35% increase with placebo).

In another study of nasal breathing, Tam et al. (19) studied 19 adult subjects with chronic rhinitis (and asthma) to determine whether a relatively high concentration of SO_2 (4 ppm) would cause an increase in nasal symptoms or nasal resistance to air flow. Unlike the above-referenced study in children, they found no difference in nasal effects between exposure to SO_2 or to room air. Eight of these subjects participated in a study to measure specific airway resistance. Again they saw no difference in airway resistance between exposures to 4 ppm SO_2 or room air.

Other Physiologic Endpoints

Nasal Mucociliary Clearance

As the nasal passages serve as the primary defense of the lung, objective measures of the effectiveness of the nasal defense mechanism are important. In addition to the nose's scrubbing function for water-soluble gases, it also filters larger particles and disposes them via the mucociliary blanket. Thus, alterations in nasal mucociliary function would be of interest to researchers studying effects of air pollutants. One method used to assess mucociliary function uses a tagged, nondispersible, insoluble particle whose movement can be monitored (20). A second method is the time required for a grain of artificial sweetener (saccharine) to travel from the placement point in the nose to the pharynx (20; as well as chapter 9 of this book).

It is known that cigarette smoke (21) and sulfuric acid aerosol (22) alter mucociliary transport. Along these lines, there have been a few studies of the relationship between SO_2 exposure and mucociliary transport. Kienast et al. (23) conducted an in vitro study investigating the effects of SO_2 and NO_2 inhalation on ciliary beat frequency in cells taken from 12 healthy adult volunteers. In an exposure chamber, the ciliated cells were exposed for 30 minutes to either 2.50–12.5 ppm SO_2 or 3.0–15.0 ppm NO_2 or a mixture of both pollutants. Ciliary beat frequency (CBF) was measured by video-interference microscopy (24). After the SO_2 exposures, the beat frequency was decreased from 7.3 to 6.7 Hz ($p = 0.0005$). Another study of the relationship between SO_2 exposure and ciliary activity in human cells retrieved respiratory cells from the nasal cavities of 25 healthy human subjects (25). As in the previous study, CBF was measured using video-interference microscopy. A concentration-dependent decrease in ciliary beat was seen following exposure to 2.5 and 12.5 ppm SO_2. In this study, exposure to NO_2 was not associated with a decrease in ciliary beat frequency.

McManus et al. (26) obtained human nasal epithelial cells from donors undergoing elective surgery. The cells were grown to confluence until they formed a homogeneous monolayer. They then were exposed to 1 to 5 ppm SO_2 or filtered air. Cells exposed to SO_2 demonstrated a significantly decreased [3H] leucine incorporation compared to cells exposed to air. The effect was both concentration and time-dependent.

Another study of cilia in the nasal mucosa following acute exposure to SO_2 was conducted by Carson et al. (27). Electron microscopic examination of ultra-thin sections of ciliated nasal epithelium from seven volunteers was conducted following a two-hour exposure to 0.75 ppm SO_2. The investigators found an apparent increase in the prevalence of compound cilia in four of the seven subjects.

Resistance to Viral Infection

Another area of inquiry concerns the potential effect of SO_2 exposure on resistance to viral upper respiratory tract infections. One study was conducted to evaluate the interaction between short-term SO_2 exposure and experimentally induced rhinovirus infection (28). Thirty-two volunteers were divided into two groups; one group was exposed to 5 ppm SO_2 and the other group served as a control. Infection rate was measured by viral shedding. The SO_2 exposure caused a decrease in nasal mucous flow, a known risk factor for infection. However, there was not a clear difference between the two groups in terms of signs of infection. The group exposed to SO_2 had fewer symptoms, and a trend toward a shorter incubation period could be observed. No differences in antibody response were shown.

COMBINED STUDIES OF UPPER AND LOWER AIRWAY EFFECTS

Experimental

Tunnicliffe et al. (29) explored the effects of exposure to 200 ppb SO_2 in adult subjects with and without asthma. Subjects were exposed for one hour at rest. The health endpoints measured included nasal lavage as well as spirometry, symptom ratings, exhaled nitric oxide, and heart rate variability. Each subject underwent nasal lavage prior to the exposure and at approximately six hours postexposure. The supernatant was assayed for ascorbic acid and uric acid. In the subjects with asthma, postexposure ascorbic acid values were significantly lower after SO_2 than after air. There were no similar changes in uric acid values. Nevertheless, the nasal lavage changes may have been related to preexposure lavage, as that has been shown to disrupt values for up to 72 hours. Postexposure uric acid concentrations in nasal lavage were significantly increased for other pollutants studied (ammonium be sulfate or sulfuric acid). Although there were no changes in lung function after exposure to 200 ppb SO_2, there was a significant increase in respiratory rate as compared with a clean air exposure.

Epidemiologic

The effects of air pollution on upper and lower respiratory symptoms and peak expiratory flow in children were studied in the Netherlands (30). The children kept a daily diary for three months. The air pollutants considered were particulate matter, black smoke, NO_2, and SO_2. None of the upper respiratory symptoms were associated with SO_2 exposure, although they were weakly associated with black smoke. Lower respiratory symptoms were associated with all the pollutants.

One study of chronic exposure to SO_2 was conducted with workers in a broom manufacturing factory in Yugoslavia (31). SO_2 is used to bleach the broomcorn. SO_2 concentrations ranged from 17.1 to 149.4 mg/m^3 in winter and from 0 to 0.75 in summer. The workers, many who had been employed for up to 34 years, complained of coughing, dyspnea, burning in nose, eyes, and throat, sore throat, and tearing compared to a nonexposed control group. The greatest number of workers complained of coughing (94.2%) and difficulty in breathing (83.7%). Others reported burning sensations in the eye and throat (80%) and in the nose (74.7%). Sulfate concentrations were higher in urine of exposed workers than in controls.

INTERNATIONAL PERSPECTIVE

Exposures to SO_2 within the United States have decreased significantly since 1970. During the past several years, no community in the United States was categorized as being out of compliance for the SO_2 NAAQS. However, we must remember that there is not a short-term (1–2 hours) primary standard for SO_2 that is considered by many as necessary to protect public health. On the other hand, the situation regarding SO_2 exposure in China and other Asian countries is much different. Coal combustion is the major source of energy necessary for the rapid development occurring in China with attendant health effects. One study estimates that outdoor air pollution is associated with 300,000 deaths and 20 million cases of respiratory illness annually (32). It can be expected that much of the air pollution contains SO_2. Young children with developing lungs are likely to suffer most from coal combustion (33). It is expected that upper airway effects are among the adverse effects. Hedley et al. (34) found a decrease in cardiorespiratory and all cause deaths in Hong Kong following restrictions on sulfur content in fuel. A benefit–cost analysis of SO_2 exposures in Hong Kong was conducted recently (35). In this population, achieving lower air pollutant levels was associated with an estimated 1335 fewer deaths, 60,587 fewer hospital bed days, and 6.7 million fewer doctor visits for yearly respiratory complaints.

It was mentioned earlier that the major source of SO_2 is volcanoes. In most cases, the volcanoes are not near large population sites; however, in Hawaii, the Kilauea Volcano has been emitting sulfurous air pollution into nearby communities since 1983 (36). Investigators there have found associations between SO_2 and acute bronchitis. In one study, the cumulative incidence rate for bronchitis was 117.74 cases per 1000 persons (36).

CONCLUSION

In conclusion, this review demonstrates the close link between SO_2 exposure and adverse health effects. Although lower airway effects and mortality attract the most scientific and regulatory attention, effects on the upper airways are of substantial concern for quality of life. Of the six criteria pollutants, SO_2, due to its high water solubility, is clearly the primary pollutant of greatest concern for effects upon the nose and upper respiratory system. Against this backdrop, an estimated one-fifth of the U.S. population suffers from either nonallergic or mixed (allergic + nonallergic) rhinitis (37). As a group, these individuals may be more susceptible to the upper airway effects of air pollutants (38). Therefore, regardless of progress achieved in controlling the pollutant in so-called developed countries, SO_2 should remain a pollutant of concern, both as a primary pollutant and as a contributor to particulate matter, acid aerosols, and acid deposition.

REFRENCES

1. Koenig JQ. Health effects of ambient air pollution: How safe is the air we breathe? Boston: Kluwer Academic Publishers, 1999.
2. Holgate ST. Air pollution and health. San Diego, CA: Academic Press, 1999.

3. Schwartz J, Hayward SF. Air quality in America: a dose of reality on air pollution levels, trends, and health risks. Washington, D.C.: AEI Press, 2007.
4. United States Environmental Protection Agency. Environmental Criteria and Assessment Office. Supplement to the second addendum (1986) to air quality criteria for particulate matter and sulfur oxides (1982) assessment of new findings on sulfur dioxide, acute exposure health effects in asthmatic individuals. NC: Research Triangle Park, 1994.
5. Sheppard D, Wong WS, Uchara CF, et al. Lower threshold and greater bronchoalveolar responsiveness of asthma subjects to sulfur dioxide. Am Rev Respir Dis 1980; 122: 873–878.
6. Koenig JQ, Marshall SG, Horike M, et al. The effects of albuterol on SO_2-induced bronchoconstriction in allergic adolescents. J Allergy Clin Immunol 1987; 79: 54–58.
7. Sandstrom T, Stjernberg N, Andersson M-C, et al. Cell response in bronchoalveolar lavage fluid after exposure to sulfur dioxide: A time-response study. Am Rev Respir Dis 1989; 140:1828–1831.
8. Dockery DW, Speizer FE. Epidemiological evidence for aggravation and promotion of COPD by acid air pollution. Lung Biol Health Dis 1989; 43:201–225.
9. Frank NR, Amdur MO, Worcester H, et al. Effects of acute controlled exposure to SO_2 on respiratory mechanics in healthy male adults. J Appl Physiol 1962; 17: 252–258.
10. Brain JD. The uptake of inhaled gases by the nose. Ann Otol Rhinol Laryngol 1970; 79:529–539.
11. Speizer FE, Frank NR. The uptake and release of SO_2 by the human nose. Arch Environ Health 1966;12:725–728.
12. Bechtold WE, Waide JJ, Sandstrom T, et al. Biological markers of exposure to SO_2: S-Sulfonates in nasal lavage. J Expo Anal Environ Epidemiol 1993;3:371–382.
13. Niinimaa V, Cole P, Mintz S, et al. The switching point from nasal to oronasal breathing. Respir Physiol 1980; 42:61–71.
14. Kirkpatrick MB, Sheppard D, Nadel JA, et al. Effect of oronasal breathing route on sulfur-dioxide bronchoconstriction in exercising asthmatic subjects. Am Rev Respir Dis 1982; 124:627–630.
15. Linn WS, Shamoo DA, Spier CE, et al. Respiratory effects of 0.75 ppm sulfur dioxide in exercising asthmatics: Influenced of upper respiratory defenses. Environ Res 1983; 30:340–348.
16. Bedi JF, Horvath SM. Inhalation route effects on exposure to 2.0 parts per million sulfur dioxide in normal subjects. JAPCA 1989; 39:1448–1452.
17. Koenig JQ, Morgan MS, Pierson WE. The effects of sulfur oxides on nasal and lung function in adolescents with extrinsic asthma. J Allergy Clin Immunol 1985; 76: 813–818.
18. Koenig JQ, McManus MS, Bierman CW, et al. Chlorpheniramine–sulfur dioxide interactions on lung and nasal function in allergic adolescents. Pediatr Asthma Allergy Immunol 1998; 2:199–205.
19. Tam EK, Liu J, Bigby BG, et al. Sulfur dioxide does not acutely increase nasal symptoms of nasal resistance in subjects with rhinitis or in subjects with bronchial hyperresponsiveness to sulfur dioxide. Am Rev Respir Dis 1988; 138:1559–1564.
20. Ballenger JJ. Some effects of the respired environment on the nose. Laryngoscope 1981; 91:1622–1626.
21. Bascom R, Kesavanathan J, Fitzgerald TK, et al. Sidestream tobacco smoke exposure acutely alters human nasal mucociliary clearance. Environ Health Perspect 1995; 103:1026–1030.
22. Grahame T, Schlesinger R. Evaluating the health risk from secondary sulfates in eastern North America. Inhal Toxicol 2005; 17:15–27.
23. Kienast K, Riechlmann H, Knorst M, et al. An experimental model for the exposure of human ciliated cells to sulfur dioxide at different concentrations. Clin Invest 1994; 72: 215–219.

24. Kienast K, Riechlmann H, Knorst M, et al. Combined exposures of human ciliated cells to different concentrations of sulfur dioxide and nitrogen dioxide. Eur J Med Res 1996; 1:533–536.
25. Riechlmann H, Kienast K, Schellenberg J, et al. An in vitro model to study effects of airborne pollutants on human ciliary activity. Rhinology 1994; 32:105–108.
26. McManus MS, Altman LC, Koenig JQ, et al. Human nasal epithelium: characterization and effects of in vitro exposure to sulfur dioxide. Exp Lung Res 1989; 15:849–865.
27. Carson JL, Collier AM, Hu S, et al. The appearance of compound cilia in the nasal mucosa of normal human subjects following acute, in vivo exposure to sulfur dioxide. Environ Res 1987; 42:155–165.
28. Andersen IB, Jensen L, Reed SE, et al. Induced rhinovirus infection under controlled exposure to sulfur dioxide. Arch Environ Health 1977; 32:120–125.
29. Tunnicliffe WS, Harrison RM, Kelly FJ, et al. The effect of sulphurous air pollutant exposures on symptoms, lung function, exhaled nitric oxide, and nasal epithelial lining fluid antioxidant concentrations in normal and asthmatic adults. Occup Environ Med 2003; 60:1–7.
30. Boezen HM, van der Zee SC, Postma DS, et al. Effects of ambient air pollution on upper and lower respiratory symptoms and peak expiratory flow in children. Lancet 1999; 353:874–878.
31. Savic M, Siriski-Sasic J, Djulizibaric D. Discomforts and laboratory findings in workers exposed to sulfur dioxide. Int Arch Occup Environ Health 1987; 59:513–518.
32. Millman A, Tang D, Perera FP. Air pollution threatens the health of children in China. Pediatrics 2008; 122: 620–628.
33. Perera FP. Children are likely to suffer most from our fossil fuel addiction. Environ Health Perspect 2008; 116:987–990.
34. Hedley AJ, Wong C-M, Thach TQ, et al. Cardiorespiratory and all-cause mortality after restriction on sulphur content of fuel in Hong Kong: an intervention study. Lancet 2002; 360:1646–1652.
35. Hedley AJ, McGhee SM, Barron B, et al. Air pollution costs and paths to a solution in Hong Kong—Understanding the connections among visibility, air pollution, and health costs in pursuit of accountability, environmental justice, and health protection. J Toxicol Environ Health A 2008; 71: 544–554.
36. Longo BM, Yang W. Acute bronchitis and volcanic air pollution: a community-based cohort study at Kilauea Volcano, Hawaii, USA. J Toxicol Environ Health A 2008; 71:1565–1571.
37. Settipane RA, Charnock DR. Epidemiology of rhinitis: allergic and nonallergic. In: Baranuik JN, Shusterman D, eds. Nonallergic Rhinitis. New York: Informa Healthcare, 2007:23–334.
38. Shusterman D, Murphy MA. Nasal hyperreactivity in allergic and non-allergic rhinitis: a potential risk factor for non-specific building-related illness. Indoor Air 2007; 17:328–333.

23 Exposure to Volatile Organic Compounds in Humans

Christoph van Thriel

IfADo—Leibniz Research Centre for Working Environment and Human Factors, Dortmund, Germany

INTRODUCTION

Within chemistry, organic chemistry is a discipline investigating compounds consisting primarily of carbon and hydrogen, and thus, volatile organic compounds (VOCs) are molecules assembled around carbon and hydrogen atoms. With increasing number of carbon atoms, these compounds become solid matter but many of these organic compounds are small and volatile molecules. Chemically identical VOCs can be found both as environmental air pollutants and indoor air contaminants. VOCs are emitted from different sources such as combustion by-products, cooking, construction material, office equipment, consumer products, and industrial facilities. So-called microbial VOCs (mVOCs) can also be produced by the growth of molds and bacteria. In general, this class of chemicals consists of gases or vapors, but a generally accepted definition of the term VOCs is not available in the scientific literature.

According to the World Health Organization (WHO; World Health Organization, 1989), VOCs are defined as all organic compounds having boiling point in the range of 50°C to 260°C (1). In addition, compounds having boiling points in the range from ≤0°C to 50°C were classified as very volatile organic compounds (vVOCs), and compounds having boiling points above 260°C were described as semivolatile organic compounds (sVOCs) (2). Regardless of this subclassification, the volatility of a compound is important and under normal climatic conditions humans are exposed to VOCs in the form of true gases or vapors. Within the class of VOCs, the most important subgroups are aliphatic and cyclic hydrocarbons, aromatic hydrocarbons, aldehydes, terpenes, alcohols, esters, ethers, glycols/glycolesters/glycolethers, halocarbons, ketones, alkenes, and organic acids (1). Important representatives of VOCs are formaldehyde or acetaldehyde, styrene, toluene, benzene, methylene chloride, and vinyl chloride.

CURRENT DEFINITIONS OF VOCs

Within the EU legislation, volatility, indicated by the boiling point, is also the primary characteristic of VOCs. In their official documents, the EU defined VOCs as carbon-based chemical compounds emitted into the atmosphere from natural sources or as a result of human activities (e.g., the use of solvents, paints, and varnishes; the storage of motor fuel and the use of motor fuel in filling stations; and vehicle exhaust gases) (3). The second descriptive aspect of the term VOC, organic, is mainly represented by the fact that carbon atoms in various structures compose these substances. In contrast, the United States Environmental

Protection Agency (US EPA) recently updated their definition to "Volatile organic compounds (VOC) means any compound of carbon, excluding carbon monoxide, carbon dioxide, carbonic acid, metallic carbides or carbonates, and ammonium carbonate, which participates in atmospheric photochemical reactions" (Definition per 40 CFR Part 51.100(s), as amended through January 18, 2007). By this revision US EPA no longer refers to the vapor pressure (volatility) as primary characteristic of a VOC, but considers the potency of a compound to produce ozone when in the sunlight as more relevant. According to this revision, several "former" VOCs, like acetone or dichloromethane, were excluded from the US EPA list of VOCs due to their negligible photochemical reactivity. This revision was made due to the concern about the significant contribution of environmental VOCs to tropospheric ozone (4) and subsequently to global warming. This revision also reflects that ambient air pollutants are of more significance to the US EPA. Thus, the revised definition of VOCs does not only reflect physicochemical aspects but also an implicit evaluation of their predominant effects, namely, the contribution to environmental air pollution.

Regardless of these two concurrent definitions, both the European legislation and the US EPA are concerned about health risk associated with human exposures to VOCs. Since ecological effects currently focused by the definition of US EPA will, at least in the long run, affect human health, these effects of VOCs will be described in more detail.

HEALTH EFFECTS OF VOCs

VOCs are potential health hazards and due to the predominately inhalational route of exposure, the respiratory tract is their major target site (5–8). Based on the water solubility of VOCs, different compartments of the respiratory tract can be affected (8). This fractionation reflects the physiology of the respiratory tract, namely, the lining of the surface of the respiratory tract with ciliated epithelium. The major constituent of the epithelial lining fluid is water, and the affinity of volatile chemical toward water (hydrophilicity) determines the compartment of the respiratory tract where the VOCs will predominantly unfold their effects. Compounds with higher hydrophilicity, like ammonia or aldehydes, will affect the eyes, the nasal cavity, the pharynx, and larynx. Chemicals of medium water solubility, for example, ozone, will affect the trachea and bronchi. Finally, compounds of low water solubility, like nitrogen dioxide, will target the bronchioles and alveoli (8).

In general, the elicited health effects range from sensory-mediated symptoms, like odor annoyance, burning eyes, and intranasal irritations (9), to severe diseases, like asthma and chemical pneumonitis (6). Moreover, several of the aforementioned compounds like formaldehyde and acetaldehyde are probably carcinogenic to humans (10) and in rats these substances are known to induce nasal cancer (11). The carcinogenic effect of these compounds has been hypothesized by some to be strictly a high-dose effect based upon the observation that carcinogenesis appears to occur only in the presence of chronic inflammation (12). This "threshold concept" has been applied to several chemicals (e.g., naphthalene) based on so-called "mechanism of action," but only some of them (e.g., vinyl acetate) fit well into a nonlinear, dose-effect model (13). Such threshold models of carcinogenesis are extremely controversial and thought by many scientists, including those at the United States and California Environmental

Protection Agencies, to represent a statistical artifact due to the practical necessity to limit the numbers of animals in cancer bioassays. These agencies instead establish "no significant risk levels" (NSRLs) based upon "de minimus" risk levels (e.g., inducing no more than one cancer per 100,000 individuals exposed over the course of their lifetime) (14). For a more in-depth discussion of issues related to the toxicokinetics and mechanism-of-action of nasal carcinogens, refer to chapters 5 and 18.

While the significance of carcinogenic and pulmonary toxic effects is undoubted, the evaluation of predominantly sensory-mediated effects, especially subjectively reported irritation, is debated (Ref. 15 vs. Ref. 16 and 17). Therefore, the sensory systems involved in the perception of VOCs and the subsequent generation of symptoms and measurable health effects at the upper respiratory tract and the eyes will be described.

Sensory Perception of VOCs
The epithelium of the upper respiratory tract as well as the cornea of the eye is innervated by various sensory nerves that inform the organism of the presence of chemicals in its breathable air. The most prominent sense within the nasal cavity is the olfactory nerve mediating thousands of distinguishable perceptions (18), which can be elicited by very low airborne concentration of a particular odorant (e.g., musky odor). In addition to this very sensitive sentinel, the trigeminal nerve carries information about chemicals in the inhaled air (19,20). In the context of VOCs, it seems noteworthy to mention that volatility is not the only primary characteristic of this class of chemicals. Volatility does act as a precondition for environmental chemicals to reach the nasal cavity, and thus, to become perceivable by the chemical senses. It has also been shown that volatility, expressed by the saturated vapor pressure ($P°$), is a major predictor of the nasal pungency threshold (NPT) measured in anosmics (21). Even though there are more sophisticated quantitative structure–activity relationships (QSAR; see also chap. 25) predicting NPT by various physicochemical properties (22), volatility of a compound is important to reach the free nerve endings of the trigeminal nerve as well as to cause irritation.

The fifth cranial nerve also innervates the cornea, and VOCs might stimulate this chemosensory pathway even without inhalation (23). Nonnasal compartments of the upper respiratory tract are also innervated by the glossopharyngeal and vagus nerves, which also respond to VOC irritancy.

Even though trigeminal perceptions (e.g., stinging, piquancy, burning, tingling, freshness, prickling, itching, or cooling) are not as fine-tuned as olfactory perceptions, and higher VOC concentrations are required to elicit these perceptions (24), they can be very detailed. This perceptional differentiation of the intranasal trigeminal system was confirmed in normosmics and anosmics. Both groups were able to discriminate between different intensity-matched odorants (e.g., menthol and cineole), presumably based on trigeminal perception alone (25). However, the trigeminal nerve is not only capable of describing the perceptions of odorants with strong trigeminal component like the aforementioned compounds. Intranasal branches of the trigeminal nerve are also part of the "pain system," and a recent fMRI study (26) revealed that the activation patterns of intranasal simulation with CO_2, a pure trigeminal irritant, were, at least to some extent, comparable to the painful stimulation provoked by weak cutaneous, electrical, or mechanical pain stimuli. Since pain is an important warning signal informing the organism about the risk of injury and severe tissue damage, there are several sensory-mediated defense mechanisms protecting the respiratory system (see chap. 14). Such trigeminal nerve-mediated reflexes are sneezing and rhinorrhea (runny nose) (27), and at lower compartments of the respiratory system—mainly mediated by the glossopharyngeal and vagus nerves—cough, bronchospasm, and other pulmonary effects that can occur in response to inhaled irritants (28). In contrast, the olfactory system is less implicated in such protective reflexes (29) and therefore the pure olfactory perception of VOCs seems to be of less health significance than the aforementioned effects. Cometto-Muniz and coworkers accurately described this distinction between olfactory and trigeminal effects in the sentence ". . . whereas odor may annoy, irritation may become a symptom and may in some cases reach the point of a clinical sign" (30). Shusterman (17), on the other hand, argued that some purely odor-related health complaints (e.g., headaches, nausea), if predictable and reproducible in populations, should be treated as health *effects*.

The underlying problem of this "evaluation dilemma" is the fact that VOCs usually stimulate the two aforementioned chemosensory systems dose-dependently, olfaction in low concentrations and trigeminal chemoreception in higher concentrations. In more psychological terms, the olfactory pathway is capable of informing the organism about the presence of a VOC while the trigeminal pathway helps inform the organism about the risk of health hazards and injury. The dilemma in humans exposed to VOCs is that the cut-off point for the transition from the feeling of *being exposed* to the feeling of *being at risk* is subjected to a huge number of confounding factors. Such factors are demographic factors like age and gender, personality traits (31), cognitive biasing by attribution styles (32), adaptation due to prolonged exposure periods (33), occupational exposures (34), and olfactory functioning (35). Many of these factors are uncontrollable in epidemiological and experimental studies, investigating the effects of VOCs either as environmental or as indoor air pollutants and therefore, sensitive biological effect measures are needed to determine the onset of physiological dysregulation. Such measures will now be described in more details.

Biological Effect Measures of VOC Effects

Eye and intranasal irritations are prime symptoms of acute exposures to VOCs, and several methods for their "objective" assessment in humans are available (36). As a rule in toxicology, the prevention of acute effects usually reduces the likelihood of chronic effects (see earlier section on formaldehyde) leading into irreversible pathophysiological changes (37). Doty et al. (36) describe a broad array of physiological correlates of acute sensory upper airway irritation, such as nasal airflow changes, changes of nasal volume, breathing patterns change, changes in secretion, changes in ciliary beat frequency (nasal mucociliary clearance), changes in blood flow, or the release of biochemical mediators associated with pathophysiological processes (8,36). Moreover, another paper described several biological effect markers of eye irritation (23). These measures, mainly related to tear film stability, are especially helpful for the evaluation of indoor air VOC pollution in office environments. In contrast to traditional pulmonary measures, often used to evaluate the effect of criteria air pollutants like sulfur dioxide (38), no normative data is available for most of the biological markers of intranasal and eye irritations (8). Relying on normative data for pulmonary measures like the forced expiratory volume in one second (FEV1) might be insufficient for preventional purposes, and, currently, noninvasive and more experimental methods (e.g., induced sputum and exhaled nitric oxide) have also been developed to detect early lung effects due to chemical exposures in the working environment, and to prevent occupational asthma more effectively (39,40). The application of such "experimental methods" is valuable in particular study designs, namely, experimental within-subject designs or cross-shift analysis in epidemiological research (41). Because of the large proportion of VOCs with high-hydrophilicity biological methods, assessing intranasal effects are more relevant and some of them have been used both in controlled challenge studies using environmental chambers (42) and, at least to some extent, in cross-sectional epidemiological studies investigating indoor or outdoor pollution by VOCs (43).

In general, functional, morphological, and biochemical indicators of nasal irritation can be distinguished. Two methods broadly used in otorhinolaryngology, rhinomanometry and acoustic rhinometry have been applied to study nasal congestion due to chemical exposures (44). While rhinomanometry is a functional marker measuring the nasal airflow, acoustic rhinometry studies the geometry of the nasal cavity by evaluating the cross-sectional area of the nose and the volume of the nasal cavity by analysis of incident and reflected sound. Both techniques are capable of measuring swelling or other alternations of the nasal mucosa in response to chemical exposures (8). Biochemical indicators are used to search for and to verify the underlying mechanisms of such functional changes. Inflammation, either neurogenic (e.g., axon reflex of trigeminal C-fibers) or immunologic (e.g., neutrophil activation), is thought to contribute to such dysfunctional effects of VOCs on the upper respiratory tract. Allergic reactions mediated by mast cells seem to be unlikely in the context of VOC exposures, but for some environmental compounds like environmental tobacco smoke such mechanisms are conceivable (45).

Since more details about mechanisms and measures of adverse effects within the nasal cavity of humans have been given throughout this book, the three final sections will now focus on VOC effects in the context of (a) environmental air pollution, (b) indoor air quality, and (c) working environment.

ENVIRONMENTAL AIR POLLUTION

When talking about environmental air pollution one has to bear in mind that there is a complex mixture of particulate matter of different sizes, ozone, and other criteria air pollutants, such as nitrogen oxides or sulfur dioxide, and VOCs. VOCs can be important due to their particular mode of action (e.g., formaldehyde) or due to the fact that many VOCs are important for the formation of ozone due to their aforementioned photoreactivity. Accordingly, governmental agencies like the US EPA (46) and the European Commission (47) developed strategies to reduce VOCs in the environment. The Thematic Strategy on Air Pollution of the EU, for instance, claimed that, in combination with the reduction of other forms of air pollution, a reduction of VOCs by 51% from year 2000 to 2020 is needed to achieve levels of air quality that do not give rise to significant negative impacts on, and risks to, human health. The US EPA initiated similar programs and monitoring programs are currently running (48).

The dramatic impact of environmental air pollution on the nasal epithelium or the respiratory tract in general has been described for highly polluted cities like Mexico City (49) or Bombay (50). In the study by Calderon-Garciduenas et al. (49), regarding single VOCs, formaldehyde and acetaldehyde were measured in high concentrations (CH_3CHO: 5.9–110 ppbv; CH_2O: 2–66.7 ppbv) in some areas of Mexico City and these concentrations were considered among the highest reported in urban air around the world (51). The investigated inhabitants of these highly polluted areas with daily exposures of more than 10 hours had severe nasal lesions with nasal crustings, episodes of epistaxis, and nasal squamous metaplasia replacing the normal ciliated respiratory epithelium. Because of the complex mixture of compounds to which this population was exposed to, the causative role of VOCs, especially formaldehyde and acetaldehyde, on nasal pathophysiology cannot be estimated precisely. Experimental animal studies on the carcinogenic effect of formaldehyde and acetaldehyde (52) suggest at least evidence that the extremely high air pollution with these particular pollutants might be jointly responsible for the metaplasia. Since the outdoor environment is less controllable, several authors try to estimate the role of VOCs by extrapolation from indoor-air scenarios to environmental air pollution (7). This review found that indoor-air formaldehyde concentration was associated with asthma and bronchitis in children. On the other hand, an experimental study recently showed that low levels of a mixture of VOCs (25 mg/m^3) and their ozone oxidation products (40 ppb) were not capable of inducing nasal symptoms or biochemical effects during short-term exposures of two hours (53). Even when using such extrapolations, the real contribution of VOCs in general or of single VOCs (e.g., aldehydes) to nasal toxicity within the context of environmental air pollution needs further investigation. Nevertheless, at very high doses like in Mexico City or other megacities, nasal toxicity caused by VOCs, as part of general air pollution, is a severe public health problem that needs to be addressed with programs like the aforementioned ones in the United States or the EU.

Indoor Air Quality

This section discusses the effects of VOCs in nonindustrial working environments like office rooms, hospitals, schools, public transportation, or other public sector buildings. According to a recent review, there are more than 45 chemicals or classes of chemicals (e.g., PCBs, PAHs) emitted from various sources (e.g., building materials, maintenance and cleaning products, and electronic equipment) contaminating the ambient air of nearly all buildings (1). In this review, the authors stated that most occurring VOCs in indoor environment include ethanol, limonene, acetone, toluene, and methylene chloride. However, under regular indoor-air conditions these VOCs will always appear as mixtures, and therefore the exposure is often described as TVOC (total VOC concentration). Research in the area of indoor air pollution by VOCs was mainly triggered by the occurrence of the sick building syndrome (SBS) in the 1980s (5). Symptoms of the upper airways and eye irritation are the most prominent features of SBS (54). This syndrome is controversially discussed since the reported complaints are vague, with no clear causal agent (55) and very low exposures (e.g., average exposure 0.1–1.2 mg/m^3; up to 32 mg/m^3 in one building, Danish Town Hall Study) that often do not fall into the dose-range of conventional toxicological concern (56). However, the SBS exposure scenarios are different from "classic," single-component toxicology, and unknown interactions among the components of these mixtures might lead to additive or supraadditive effects causing low-dose toxicity (57) in "polluted" indoor air environments.

Biological effect measures, as previously described in this chapter, are rarely used in cross-sectional studies on SBS. Thus, most of the existing epidemiological studies, summarized in a review of a scientific workshop in Stockholm 1996 (5), only used self-reports of health symptoms and their relationship to various building characteristics, sometimes including measures of TVOCs. The results of the summarized studies are highly inconsistent and the scientific quality of the studies is sometimes equivocal due to uncertain exposure assessment. However, a well-documented longitudinal study in 129 employees in primary schools (58) showed a significant association of the total-indoor concentration of volatile hydrocarbons with the prevalence of health symptoms related to irritation of the upper airways or the eyes (e.g., eye irritation, swollen eyelids, nasal catarrh, and blocked-up nose). Moreover, certain groups of the measured VOCs like aromatics, n-alkanes, terpenes, and butanols showed significant associations while others, like 2-ethylhexanol or other "unidentified low-boiling hydrocarbons" showed no such association. Another large epidemiological study, the aforementioned Danish Town Hall Study, with more than 4000 participants showed that certain tasks like photoprinting, working at video display terminals, and handling carbonless paper are correlated with the frequency of mucosal irritation and of general symptoms (59). These might be indirect exposure indices for the contributions of various VOCs to upper airway health effects, since VOCs are generated for instance by photocopy machines (60) or personal computers (61). In contrast, in one of the investigated building (No. 60) of the Danish Town Hall study, a TVOC concentration of 32 mg/m^3 was measured, almost 30 times higher than the concentrations in the other buildings. Despite this extreme difference, the prevalence rate of irritation of the mucous membranes was not elevated within this particular building.

Accordingly, many epidemiological studies revealed that SBS, as a syndrome based on subjective reports of health symptoms, is a multicausal syndrome and many nonchemical factors, such as atopy, gender, and psychosocial factors, are known to modulate eye irritation (23) as well as nasal irritation (59). Moreover, other environmental factors such as ventilation, humidity, building age, molds, bacteria, and dampness might cause effects on the upper respiratory tract (62). This study, for instance, showed that building dampness per se, dampness-related alkaline degradation of di-(2-ethylhexyl) phthalate (DEHP) in building material (indicated by increased 2-ethyl-1-hexanol in the air) and ammonia under the floor carpet were associated with the occurrence of ocular and nasal symptoms. Moreover, one of the investigated biological effect markers, the concentration of lysozyme (indicator of an innate immune system response) in nasal lavage, was elevated in those participants from "dampness-affected" buildings. The TVOC concentration, comprehensively measured in that study, was reported to explain neither the symptoms nor the biological measures of upper airway or eye irritation.

In general, based on the limited number of well-conducted epidemiological studies, the evaluation of the role of VOCs within the context of indoor air pollution is almost impossible. For example, Andersson et al. concluded in their review more than 10 years ago, "... the group cannot recommend the present use of TVOC as a risk indicator for health effects and discomfort problems in buildings" (5). Thus, the contribution of VOC mixtures to adverse effects on the upper airways is far from being conclusive.

In addition to epidemiological studies, there are some experimental studies investigating primary and secondary reactions of human volunteers exposed to standard VOC mixtures ("M22"; "Molhave cocktail") or *n*-dexane in environmental chambers (2,63). The author summarized five experimental studies and distinguished between primary reactions, related to the recognition of the exposure by chemosensory perceptions, and secondary reactions, referring to biological effects occurring after prolonged exposures. The studies investigated VOC concentrations ranging from 0 to 25 mg/m^3, but only one study investigated a broader range of VOCs capable to establish a dose–response relationship (64). However, in his summary of these experimental studies, Molhave concluded that VOC concentration below 0.2 mg/m^3 were within the comfort range, while concentration between 0.2 and 3 mg/m^3 might be capable of provoking, possibly in interaction with other exposures, irritation and discomfort (2). In his summary, he also concluded that concentrations above 25 mg/m^3 were within the toxic range, eliciting even neurotoxic effects. Such neurotoxic effects could not be confirmed by neurobehavioral tests, but reports about fatigue and mental confusion (65) were considered to support such CNS effects.

More recently, Wolkoff and coworkers published an update for the classification of VOCs in the indoor environment. They proposed four categories (*i*) chemically nonreactive, (*ii*) chemically "reactive," (*iii*) biologically reactive (i.e., from chemical bonds to receptor sites in mucous membranes), and (*iv*) toxic compounds, of which the first class, nonreactive VOCs, are considered nonirritants at typical indoor-air level (54). This reclassification is partly related to the QSAR modeling of the sensory effects of VOCs as described in chapter 25 and provides promising approaches to elucidate the role of VOCs within the etiology for upper airway effects and eye irritation caused by indoor-air exposures more systematically.

This new approach, also the loose ends of the empirical studies on VOCs in the indoor environment, as well as the growing population spending their work time indoors, together indicate that more research on this topic is needed.

Working Environment
This last section addresses the regulation of VOCs in the industrial work environment, where occupational exposure limits (OELs) are recommended by scientific committees (e.g., American Conference of Governmental Industrial Hygienists, ACGIH; German Commission for the Investigation of Health Hazards of Chemical Compounds in the Work Area, DFG MAK Commission) to avoid adverse health effects due to occupational exposures. Two premises have to be mentioned before going into more details.

   (i)   In the working environment only single compounds are regulated by OELs and mixtures of chemicals are handled by an additive mixture formula (66) and coexposures are usually neglected.
   (ii) VOCs are not considered as a homogenous class of chemicals since their adverse or critical effect might differ according to their particular toxicity.

The critical effects falling into the scope of this chapter are the avoidance of (*i*) eye irritation, (*ii*) upper airway irritation, and (*iii*) lower airway irritation. Additionally, distal pulmonary effects are considered, but since they are of minor relevance they will not be addressed here.

In Europe and in the United States about 40% of the OELs are set to avoid sensory irritation (67,68). Under physiochemical considerations, many of these compounds, like ethyl acetate, acetone, 2-ethylhexanol, or the carboxylic acids, fall into the class of VOCs. During the last decade there was a scientific debate about the methods that are appropriate to derive a NOAEL (no-observable-adverse-effect level) from experimental human exposure studies (69). This debate was driven by the growing knowledge about nonsensory factors affecting reports of irritation (9,70,71) and the lack of adequate data to derive appropriated OELs for local irritants. The OELs of several compounds were based on very old studies (72,73) using unknown or unstandardized tools for the assessment of (a) the exposure, and (b) the elicited irritation. Meanwhile, several irritants could be reevaluated due to new experimental exposure studies using the aforementioned biological measures of intranasal and eye irritations. Muttray and coworkers, for instance, investigated acute effects of 1,1,1-trichloroethane at 20 and 200 ppm for four hours, assessing nasal mucociliary transport time (by saccharine instillation) and sampling of nasal fluid (74). After the 200-ppm exposure, several interleukins (IL-1β, IL-6, IL-8) were significantly elevated and the authors concluded that subclinical inflammation of nasal mucosa might occur when workers are exposed to these concentrations. Despite this study results, the German MAK-value of 200 ppm was not changed, although it is more protective than the ACGIH TLV of 350 ppm. This compound is a good example showing that an OEL initially set to prevent acute neurotoxic effects needs to be reevaluated if adverse chemosensory and/or inflammatory effects occur at lower concentrations. The most sensitive endpoint or effect should be considered in order to ensure the best possible prevention of the exposed workers.

An overview of our own experiments (42) showed that biological effects after experimental exposures could only be verified for a limited number of compounds. In our experiments, ethyl benzene, 2-butanone (methyl ethyl ketone; MEK), isopropyl alcohol, 1-octanol, and 2-ethylhexanol were investigated. Anterior active rhinomanometry (AAR) and biochemical parameters from nasal secretion (nasal lavage fluid; NLF) were used to measure intranasal irritation physiologically. The functional measurement of the nasal airflow (AAR) revealed a dose-dependency only for 1-octanol. The nasal airflow decreased by 5% when 24 volunteers were exposed to 0.1 ppm 1-octanol for four hours, while this decrease was around 16% when the exposure was as high as 6.4 ppm. The biochemical analyses revealed that the neuropeptide substance P was significantly elevated after exposures to 20 ppm of 2-ethylhexanol for four hours. A neurogenic axon reflex of the trigeminal C-fibers might be responsible for the release of substance P into the nasal fluid (75). Thus, a sensory-mediated defense mechanism might have been activated in the exposed volunteers, reflecting their elevated reports of intranasal irritation (76), which were accompanied by physiological responses to the chemical. Moreover, another paper of our group gave further support that at acute exposures of 20 ppm, biological indicators of trigeminal-mediated responses to 2-ethylhexanol were affected (77). This paper showed that the blinking frequency was dose-dependently elevated in two experiments investigating independent samples of volunteers either exposed to constant concentrations of 2-ethylhexanol (1.5, 10, and 20 ppm throughout the exposure period) or variable concentrations investigating the role of exposure peaks of 20 and 40 ppm that were embedded into the 10- and 20-ppm C_{TWA} exposure scenarios. Compared to the 1.5-ppm exposure period preceding, during the 40-ppm exposure peaks, the blinking frequency was threefold higher. As mentioned previously, this VOC is also important within the context of SBS and very recently, an in vitro study using spleen cells revealed that 2-ethylhexanol might function as a modulator of immune response (78). 2-Ethylhexanol is one of the rare VOCs where convincing evidence exists that biological responses are triggered by acute exposure of approximately 20 ppm. Accordingly, the German MAK-value was lowered from 50 to 20 ppm.

There are several other irritants like propylene glycol monomethyl ether (PGME) (79), isopropyl alcohol (34), acetic acid (80,81), and acetaldehyde (82) that have been investigated in human volunteer studies. The results, as summarized in Table 1, revealed some reports of sensory irritation that could only be confirmed physiologically among vulnerable subjects [e.g., seasonal allergic rhinitis (SAR)] (81) or with only one of serveral physiological measures (34).

This brief overview of the results revealed that exposures as high as the existing OELs of the investigated irritants did only cause selective effects on the applied physiological measures. When evaluating these studies with respect to the overall effect of the irritants on the upper respiratory tract and the eyes, no consistent effect on the biological effect markers could be confirmed. Thus, the existing OELs of the compounds could be confirmed by these experimental exposure studies.

A still unsolved problem of VOC exposures in the working environment is the evaluation of mixtures. The premise of pure additivity might underestimate the real health hazards of multiple exposures. The only mixtures that have been investigated in chamber studies with respect to their component chemosensory effects are standard and dearomatized white spirits (83). During and after acute exposures of these two types of white spirits (WS) in concentrations of up to 300 mg/m^3 for two hours, no biological signs of inflammation or sensory irritation (e.g., increased blinking frequency) could be revealed. Only the smell of the standard WS, containing up to 19% aromatics, was stronger than that of the dearomatized WS. Psychophysical studies (see chaps. 9 and 13) might be useful to evaluate the adequacy of the additive mixture formula for the estimation of health effects of local irritants.

In conclusion, the true contribution of VOCs to toxicology of the nose and upper airways is far from being conclusive, especially within the framework of indoor and environmental air pollution. Multifold interactions among airborne chemicals, particles, and other factors hamper the evaluation of the causative effects of individual components. More epidemiological studies with comprehensive exposure assessments are needed to help prevent adverse effects of the upper respiratory tract due to chemical exposures in the environments in which humans are living and working.

TABLE 1 Brief Summary of the Results of Volunteer Studies Investigating Propylene Glycol Monomethyl Ether (PGME), Isopropyl Alcohol (IPA), Acetic Acid, and Acetaldehyde

Compound	Study	Concentrations and duration	Subjects	Measures	Results
PGME	(79)	0, 100, and 150 ppm 2.5 hr	12 healthy male volunteers	Eye redness, corneal thickness, tear film break-up time, conjunctival epithelial damage, blinking frequency, subjective ratings	Mild subjective eye irritation at 150 ppm; no other measure affected
Isopropyl alcohol (IPA)	(34)	0 and 400 ppm IPA, PEA (odorous control condition) 4 hr	12 Naïve controls 12 phlebotomists (occupationally exposed to IPA) 7 females and 5 males each	Ocular hyperemia, nasal congestion, nasal secretion, respiration, subjective ratings	Increase in respiration frequency during 400 ppm IPA
Acetic acid	(81)	0 and 15 ppm 15 min	8 Healthy volunteers 8 SAR subjects 4 Females and 4 males each	Nasal airways resistance (NAR)	Increased NAR only among SAR subjects (one with an extreme increase)
	(80)	0, 5, and 10 ppm 2 hr	12 Healthy volunteers: 6 females and 6 males	Pulmonary function, nasal swelling, nasal airway resistance, blinking frequency, plasma inflammatory markers, subjective ratings	Mild subjective nasal irritation at 10 ppm; no other measure affected
Acetaldehyde	(82)	0 and 50 ppm 4 hr	20 healthy male volunteers	Odor threshold for *n*-butanol, mucociliary transport time, nasal secretions, nasal biopsies, subjective ratings	No subjective reports of irritation; no other measure affected

REFERENCES

1. Wang S, Ang HM, Tade MO. Volatile organic compounds in indoor environment and photocatalytic oxidation: state of the art. Environ Int 2007; 33(5):694–705.
2. Molhave L. Indoor climate, air pollution, and human comfort. J Expo Anal Environ Epidemiol 1991; 1(1):63–81.
3. EU. 2004. http://eur-lex.europa.eu/LexUriServ/LexUriServ.do?uri = OJ:L:2004: 143:0087:0096:EN:PDF. Accessed February 23, 2009.
4. Mohamed MF, Kang D, Aneja VP. Volatile organic compounds in some urban locations in United States. Chemosphere 2002; 47(8):863–882.
5. Andersson K, Bakke JV, Bjorseth O, et al. TVOC and health in non-industrial indoor environments. Indoor Air 1997; 7(2):78–91.
6. Bernstein JA, Alexis N, Barnes C, et al. Health effects of air pollution. J Allergy Clin Immunol 2004; 114(5):1116–1123.
7. Delfino RJ. Epidemiologic evidence for asthma and exposure to air toxics: linkages between occupational, indoor, and community air pollution research. Environ Health Perspect 2002; 110(suppl 4):573–589.
8. Shusterman D. Environmental nonallergic rhinitis. Clin Allergy Immunol 2007; 19:249–266.
9. Dalton P. Upper airway irritation, odor perception and health risk due to airborne chemicals. Toxicol Lett 2003; 140–141(1–3):239–248.
10. Feron VJ, Til HP, Vrijer Fd, et al. Review: toxicology of volatile organic compounds in indoor air and strategy for further research. Indoor Built Environ 1992; 1(2): 69–82.
11. Arts JH, Rennen MA, de Heer C. Inhaled formaldehyde: evaluation of sensory irritation in relation to carcinogenicity. Regul Toxicol Pharmacol 2006; 44(2):144–160.
12. Bolt HM. Genotoxicity—threshold or not? Introduction of cases of industrial chemicals. Toxicol Lett 2003; 140–141(1–3):43–51.
13. Hengstler JG, Bogdanffy MS, Bolt HM, et al. Challenging dogma: thresholds for genotoxic carcinogens? The case of vinyl acetate. Annu Rev Pharmacol Toxicol 2003; 43:485–520.
14. CEPA, 2001. http://www.oehha.ca.gov/prop65/policy_procedure/pdf_zip/Safe HarborProcess.pdf. Accessed February 23, 2009.
15. Paustenbach DJ, Gaffney SH. The role of odor and irritation, as well as risk perception, in the setting of occupational exposure limits. Int Arch Occup Environ Health 2006; 79(4):339–342.
16. Cometto-Muniz JE, Cain WS. Sensory irritation. Relation to indoor air pollution. Ann N Y Acad Sci 1992; 641:137–151.
17. Shusterman D. Critical review: the health significance of environmental odor pollution. Arch Environ Health 1992; 47(1):76–87.
18. Hatt H. Molecular and cellular basis of human olfaction. Chem Biodivers 2004; 1(12):1857–1869.
19. Doty RL. Intranasal trigeminal detection of chemical vapors by humans. Physiol Behav 1975; 14(6):855–859.
20. Livermore A, Hummel T, Pauli E, et al. Perception of olfactory and intranasal trigeminal stimuli following cutaneous electrical stimulation. Experientia 1993; 49(10): 840–842.
21. Abraham MH, Kumarsingh R, Cometto-Muniz JE, et al. An algorithm for nasal pungency thresholds in man. Arch Toxicol 1998; 72(4):227–232.
22. Abraham MH, Sanchez-Moreno R, Cometto-Muniz JE, et al. A quantitative structure-activity analysis on the relative sensitivity of the olfactory and the nasal trigeminal chemosensory systems. Chem Senses 2007; 32(7):711–719.
23. Wolkoff P, Skov P, Franck C, et al. Eye irritation and environmental factors in the office environment—hypotheses, causes and a physiological model. Scand J Work Environ Health 2003; 29(6):411–430.
24. Cometto-Muniz JE, Cain WS. Thresholds for odor and nasal pungency. Physiol Behav 1990; 48(5):719–725.

25. Laska M, Distel H, Hudson R. Trigeminal perception of odorant quality in congenitally anosmic subjects. Chem Senses 1997; 22(4):447–456.
26. Iannilli E, Del Gratta C, Gerber JC, et al. Trigeminal activation using chemical, electrical, and mechanical stimuli. Pain 2008; 139(2):376–388.
27. Shusterman D. Toxicology of nasal irritants. Curr Allergy Asthma Rep 2003; 3(3):258–265.
28. Arts JH, de Heer C, Woutersen RA. Local effects in the respiratory tract: relevance of subjectively measured irritation for setting occupational exposure limits. Int Arch Occup Environ Health 2006; 79(4):283–298.
29. Brand G. Olfactory/trigeminal interactions in nasal chemoreception. Neurosci Biobehav Rev 2006; 30(7):908–917.
30. Cometto-Muniz JE, Cain WS, Abraham MH. Nasal pungency and odor of homologous aldehydes and carboxylic acids. Exp Brain Res 1998; 118(2):180–188.
31. Shusterman D. Individual factors in nasal chemesthesis. Chem Senses 2002; 27(6):551–564.
32. Dalton P. Cognitive influences on health symptoms from acute chemical exposure. Health Psychol 1999; 18(6):579–590.
33. Dalton P, Dilks D, Hummel T. Effects of long-term exposure to volatile irritants on sensory thresholds, negative mucosal potentials, and event-related potentials. Behav Neurosci 2006; 120(1):180–187.
34. Smeets MA, Maute C, Dalton PH. Acute sensory irritation from exposure to isopropanol (2-propanol) at TLV in workers and controls: objective versus subjective effects. Ann Occup Hyg 2002; 46(4):359–373.
35. van Thriel C, Kiesswetter E, Schaper M, et al. Odor annoyance of environmental chemicals: sensory and cognitive influences. J Toxicol Environ Health A 2008; 71(11–12):776–785.
36. Doty RL, Cometto-Muniz JE, Jalowayski AA, et al. Assessment of upper respiratory tract and ocular irritative effects of volatile chemicals in humans. Crit Rev Toxicol 2004; 34(2):85–142.
37. Bender J. The use of noncancer endpoints as a basis for establishing a reference concentration for formaldehyde. Regul Toxicol Pharmacol 2002; 35(1):23–31.
38. Nowak D, Jorres R, Berger J, et al. Airway responsiveness to sulfur dioxide in an adult population sample. Am J Respir Crit Care Med 1997; 156(4, Pt 1):1151–1156.
39. Hoffmeyer F, Harth V, Merget R, et al. Exhaled breath condensate analysis: evaluation of a methodological setting for epidemiological field studies. J Physiol Pharmacol 2007; 58(suppl 5, Pt 1):289–298.
40. Lemiere C. Non-invasive assessment of airway inflammation in occupational lung diseases. Curr Opin Allergy Clin Immunol 2002; 2(2):109–114.
41. Raulf-Heimsoth M, Pesch B, Schott K, et al. Irritative effects of fumes and aerosols of bitumen on the airways: results of a cross-shift study. Arch Toxicol 2007; 81(1):35–44.
42. van Thriel C, Seeber A, Kiesswetter E, et al. Physiological and psychological approaches to chemosensory effects of solvents. Toxicol Lett 2003; 140–141(1–3): 261–271.
43. Norback D, Wieslander G. Biomarkers and chemosensory irritations. Int Arch Occup Environ Health 2002; 75(5):298–304.
44. Kesavanathan J, Swift DL, Fitzgerald TK, et al. Evaluation of acoustic rhinometry and posterior rhinomanometry as tools for inhalation challenge studies. J Toxicol Environ Health 1996; 48(3):295–307.
45. Vinke JG, KleinJan A, Severijnen LW, et al. Passive smoking causes an 'allergic' cell infiltrate in the nasal mucosa of non-atopic children. Int J Pediatr Otorhinolaryngol 1999; 51(2):73–81.
46. Wilson JH Jr, Mullen MA, Bollman AD, et al. Emission projections for the U.S. Environmental Protection Agency Section 812 second prospective Clean Air Act cost/benefit analysis. J Air Waste Manag Assoc 2008; 58(5):657–672.
47. EU, 2005. http://eur-lex.europa.eu/LexUriServ/site/en/com/2005/com2005_0446 en01.pdf. Accessed February 23, 2009.

48. Woo JH, He S, Tagaris E, et al. Development of North American emission inventories for air quality modeling under climate change. J Air Waste Manag Assoc 2008; 58(11):1483–1494.
49. Calderon-Garciduenas L, Rodriguez-Alcaraz A, Villarreal-Calderon A, et al. Nasal epithelium as a sentinel for airborne environmental pollution. Toxicol Sci 1998; 46(2):352–364.
50. Kamat SR, Patil JD, Gregart J, et al. Air pollution related respiratory morbidity in central and north-eastern Bombay. J Assoc Physicians India 1992; 40(9):588–593.
51. Baez AP, Belmont R, Padilla H. Measurements of formaldehyde and acetaldehyde in the atmosphere of Mexico City. Environ Pollut 1995; 89(2):163–167.
52. Feron VJ, Arts JH, Kuper CF, et al. Health risks associated with inhaled nasal toxicants. Crit Rev Toxicol 2001; 31(3):313–347.
53. Laumbach RJ, Fiedler N, Gardner CR, et al. Nasal effects of a mixture of volatile organic compounds and their ozone oxidation products. J Occup Environ Med 2005; 47(11):1182–1189.
54. Wolkoff P, Wilkins CK, Clausen PA, et al. Organic compounds in office environments—sensory irritation, odor, measurements and the role of reactive chemistry. Indoor Air 2006; 16(1):7–19.
55. Tsai YJ, Gershwin ME. The sick building syndrome: what is it when it is? Compr Ther 2002; 28(2):140–144.
56. Montgomery MR, Reasor MJ. A toxicologic approach for evaluating cases of sick building syndrome or multiple chemical sensitivity. J Allergy Clin Immunol 1994; 94(2, Pt 2):371–375.
57. Rothman AL, Weintraub MI. The sick building syndrome and mass hysteria. Neurol Clin 1995; 13(2):405–412.
58. Norback D, Torgen M, Edling C. Volatile organic compounds, respirable dust, and personal factors related to prevalence and incidence of sick building syndrome in primary schools. Br J Ind Med 1990; 47(11):733–741.
59. Skov P, Valbjorn O; The Danish Indoor Air Study Group. The "sick" building syndrome in the office environment: The Danish town hall study. Environ Int 1987; 13(4–5):339–349.
60. Stefaniak AB, Breysse PN, Murray MP, et al. An evaluation of employee exposure to volatile organic compounds in three photocopy centers. Environ Res 2000; 83(2):162–173.
61. Bako-Biro Z, Wargocki P, Weschler CJ, et al. Effects of pollution from personal computers on perceived air quality, SBS symptoms and productivity in offices. Indoor Air 2004; 14(3):178–187.
62. Wieslander G, Norback D, Nordstrom K, et al. Nasal and ocular symptoms, tear film stability and biomarkers in nasal lavage, in relation to building-dampness and building design in hospitals. Int Arch Occup Environ Health 1999; 72(7):451–461.
63. Kjaergaard S, Molhave L, Pedersen OF. Human reactions to indoor air pollutants: n-decane. Environ Int 1989; 15(1–6):473–482.
64. Molhave L. Indoor air quality in relation to sensory irritation due to volatile organic compounds. ASHRAE Transactions 1986; 92(Pt 1) (Paper 2954).
65. Otto DA, Hudnell HK, House DE, et al. Exposure of humans to a volatile organic mixture. I. Behavioral assessment. Arch Environ Health 1992; 47(1):23–30.
66. ACGIH. 2008 TLVs® and BEIs®. Cincinnati, OH: ACGIH; 2008.
67. Dick RB, Ahlers H. Chemicals in the workplace: incorporating human neurobehavioral testing into the regulatory process. Am J Ind Med 1998; 33(5):439–453.
68. Edling C, Lundberg P. The significance of neurobehavioral tests for occupational exposure limits: an example from Sweden. Neurotoxicology 2000; 21(5):653–658.
69. van Thriel C, Triebig G, Bolt HM. Editorial: Evaluation of chemosensory effects due to occupational exposures. Int Arch Occup Environ Health 2006; 79(4):265–267.
70. Dalton P. Odor perception and beliefs about risk. Chem Senses 1996; 21(4):447–458.
71. Dalton P. Odor, irritation and perception of health risk. Int Arch Occup Environ Health 2002; 75(5):283–290.

72. Nelson KW, Ege JF, Ross M, et al. Sensory response to certain industrial solvent vapors. J Ind Hyg Toxicol 1943; 25:282–285.
73. Silverman L, Schulte HF, First MW. Further studies on sensory response to certain industrial solvent vapors. J Ind Hyg Toxicol 1946; 28:262–266.
74. Muttray A, Klimek L, Faas M, et al. The exposure of healthy volunteers to 200 ppm 1,1,1-trichloroethane increases the concentration of proinflammatory cytokines in nasal secretions. Int Arch Occup Environ Health 1999; 72(7):485–488.
75. Bascom R, Meggs WJ, Frampton M, et al. Neurogenic inflammation: with additional discussion of central and perceptual integration of nonneurogenic inflammation. Environ Health Perspect 1997; 105(suppl 2):531–537.
76. van Thriel C, Kiesswetter E, Schaper M, et al. An integrative approach considering acute symptoms and intensity ratings of chemosensory sensations during experimental exposures. Environ Toxicol Pharmacol 2005; 19(3):589–598.
77. Kiesswetter E, van Thriel C, Schaper M, et al. Eye blinks as indicator for sensory irritation during constant and peak exposures to 2-ethylhexanol. Environ Toxicol Pharmacol 2005; 19(3):531–541.
78. Yoshida Y, Liu J, Sugiura T, et al. The indoor air pollutant 2-ethyl-hexanol activates CD4 cells. Chem Biol Interact 2009; 177(2):137–141.
79. Emmen HH, Muijser H, Arts JH, et al. Human volunteer study with PGME: eye irritation during vapour exposure. Toxicol Lett 2003; 140–141(1–3):249–259.
80. Ernstgard L, Iregren A, Sjogren B, et al. Acute effects of exposure to vapours of acetic acid in humans. Toxicol Lett 2006; 165(1):22–30.
81. Shusterman D, Tarun A, Murphy MA, et al. Seasonal allergic rhinitic and normal subjects respond differentially to nasal provocation with acetic acid vapor. Inhal Toxicol 2005; 17(3):147–152.
82. Muttray A, Gosepath J, Brieger J, et al. No acute effects of an exposure to 50 ppm acetaldehyde on the upper airways. Int Arch Occup Environ Health 2009; 82(4):481–488.
83. Ernstgard L, Iregren A, Juran S, et al. Acute effects of exposure to vapours of standard and dearomatized white spirits in humans. 2. Irritation and inflammation. J Appl Toxicol 2009; 29(3):263–274.

24 Benchmark Dose and Noncancer Risk Assessment for the Upper Airways

Bruce S. Winder, Karen Riveles, and Andrew G. Salmon

Office of Environmental Health Hazard Assessment, California Environmental Protection Agency, Oakland, California, U.S.A.

INTRODUCTION

Inclusion of a risk assessment chapter in an edited work on nasal toxicology is appropriate for a variety of reasons. First, whether we speak of human or experimental animal data, the upper respiratory tract is often the direct target of toxic action. Second, biological effects observed in the upper airway may serve as a surrogate for potential airway or parenchymal injury in the more distal respiratory tract. The latter is particularly relevant for inhalation toxicity studies involving rodents, since the anatomy of their upper airways provides a more efficient barrier to lung exposure than is the case in humans. Whether the task at hand is interpreting experimental data from laboratory animals or humans, or the interpretation of epidemiologic data, the role of risk assessment is to extrapolate across dose rates, durations, and often species to arrive at a quantitative estimate of the risk of exposure-related harm to humans. This chapter concentrates on risk assessment for noncancer endpoints in the upper airway, beginning with a review of generic risk assessment tools. The reader who is interested in assessing cancer risk related to inhalational exposure is referred to Ref. 1.

BACKGROUND: DOSE–RESPONSE MODELS FOR NONCANCER ENDPOINTS

Modeling of noncancer health endpoints may include either discrete or continuous biological processes. The simplest example of a discrete noncancer endpoint would be the presence or absence of a pathological finding. On an aggregate level, the presence or absence of a disease (as defined by combined clinical, biochemical, radiologic, and pathologic criteria) can be modeled. For example, chronic obstructive lung disease (COPD) has been defined by consensus criteria and can be treated as an all-or-none endpoint in epidemiologic studies. On the other hand, a physiologic measure that is a component of this diagnosis, such as the forced expiratory volume in one second (FEV_1), can be treated as a continuous variable, sometimes using data from the same study from which incidence or prevalence data are derived. Similarly, a score representing the degree or severity of a particular response (e.g., a histopathological lesion) may

The views expressed are those of the authors and do not necessarily represent those of the Office of Environmental Health Hazard Assessment, the California Environmental Protection Agency, or the State of California.

be treated as a continuous variable provided the multilevel score for each case is based on a progressive and standardized evaluation protocol. Extension of these analogies to the human upper airway is subject to limitations in case definition (e.g., for diseases such as rhinitis and sinusitis), as well as difficulties with standardization and interpretation of subjective and objective measures (chaps. 9 and 13). At times, interpretation of the human health significance of noncancer pathologic lesions observed in experimental animals (e.g., squamous metaplasia of the turbinates in rodents) can be challenging because of interspecies differences in anatomy and regional dosimetry, as well as high background prevalence rates in control animals.

DOSE THRESHOLDS FOR APPEARANCE OF TOXIC RESPONSES

Noncancer health risk assessment has been based on the concept that a threshold concentration or dose exists below which no adverse effects occur. While such thresholds are observed among individuals, the existence and magnitude of a population threshold below which no members of the population experience adverse effects cannot be demonstrated. In any study, the entire population of concern is not examined; rather a sample of the population from which inferences are drawn is studied. Therefore, it is not possible to distinguish whether a concentration is truly below a population threshold level for an adverse effect or is rather a level associated with a relatively low incidence of adverse effects. The latter cannot be distinguished from background rates in the population.

There may also be cases where no threshold exists in the general population for a particular effect. This situation may occur for responses for which there is no theoretical threshold due to the mechanism of toxicity. The most accepted example of this is chemical carcinogenesis, particularly for genotoxic carcinogens. However, there may, at least in principle, be other types of toxicity that do not show a threshold at any dose level.

Even where a true threshold exists in the dose–response of a particular individual to a chemical exposure, there may in fact be no identifiable threshold in the response of the general population. This may occur in the case where some individuals in a diverse population show a threshold whereas others do not, which is at least theoretically possible if genetic polymorphisms exist that inactivate a protective mechanism. However, the most likely case is where a true threshold in the response occurs in all individuals at low doses, but the background rate or extent of that toxic response in the population is already above zero due to population-wide exposure to that pollutant or another causative factor that produces the same endpoint or disease. In this case, any increment in exposure to the pollutant of concern will cause an increase in the prevalence or severity of the disease, despite the existence of a threshold in the individual dose–response relationship. A probable example of this is seen in the neurodevelopmental effects of lead exposure in children, which recent risk assessments have described using linear or other continuous dose–response functions (2). The data available for criteria pollutants such as ozone or particulate matter are consistent with linear no-threshold dose–response curves for cardiovascular mortality (3,4).

Where these special cases (of continuous dose–response) are demonstrated to exist on the basis of population health data, or appear likely based on mechanistic studies, it will be appropriate to use these data to develop risk-based or continuous-response models to describe the population impacts of exposure to these pollutants, rather than relying on the threshold dose–response description to identify a "safe" exposure level. It should be noted that lack of a true threshold does not necessarily imply linearity of response at all doses. Conversely, the observation of a nonlinear dose–response curve does not necessarily imply the existence of a threshold. However, in the majority of cases for noncancer effects the existence of a threshold in the dose-response is both plausible, and often, within the acknowledged limitations, demonstrable. Therefore, the threshold assumption is regarded as the default for noncancer risk assessment and is most often used.

USE OF THRESHOLD DOSE–RESPONSE DATA IN HEALTH RISK ASSESSMENT

The basic methodology for the development of health protective levels by agencies undertaking public health risk assessment has, until recently, remained largely unchanged. This consists of:

1. Identification of critical studies and toxicological endpoints
2. Determination of a point of departure:
   a. An exposure level in an animal experiment or an epidemiological study at which no adverse effects (or at least minimal adverse effects) are observed (NOAEL)
   b. A benchmark dose (a statistical estimate of a low response rate, typically 5%, in the dose–response curve for the chemical of concern)
3. Extrapolation from this point of departure to a health protective level for the target human population is by means of:
   c. Explicit models where possible
   d. Uncertainty factors.

Uncertainty factors (UFs) are used to derive health-protective exposure guidelines for humans based on the point of departure: no-observed-adverse-effect levels (NOAELs) or the lower limits of the benchmark dose (BMCLs) from animal experiments. The application of uncertainty factors to the NOAEL or BMCL for a critical endpoint is similar.

Uncertainty factors are applied to take into consideration:

- Humans may be more sensitive than experimental animals
- Certain individuals may be more sensitive than the general population
- Uncertainties in the toxicological database
- Issues such as duration of exposure, effect severity, and route-to-route extrapolation.

Model-based extrapolation procedures or, where these are unavailable, uncertainty factors are used by the Office of Health Hazard Assessment (OEHHA) in deriving RELs to account for:

1. the magnitude of effect observed at a lowest observed-adverse-effect level (LOAEL) compared with a NOAEL (5,6);
2. for chronic RELs, the potentially greater effects from a continuous lifetime exposure compared to a subchronic exposure (5,7,8);

3. the potentially greater sensitivity of humans relative to experimental animals not accounted for by differences in relative inhalation exposure (5,9);
4. the potentially increased susceptibility of sensitive individuals, for example, due to interindividual variability in response (9–13); and
5. other deficiencies in the study design (5,7,8,14,15).

The use of UFs for determining "safe" or "acceptable" levels has been discussed extensively in the toxicological literature (5,9,16–20).

As noted above, UFs are used when insufficient data are available to support the use of chemical-specific and species-specific extrapolation factors. Some examples of UFs include:

1. LOAEL uncertainty factor—UF_L
2. subchronic uncertainty factor—UF_S
3. interspecies uncertainty factor—UF_A
4. intraspecies uncertainty factor—UF_H
5. database deficiency factor—UF_D

Historically, UFs have most often been order-of-magnitude factors, indicating the broad level of uncertainty in addressing the area of concern (5). More recently, OEHHA and the U.S. Environmental Protection Agency (U.S. EPA) have used intermediate UFs, usually having a value of 3 (the rounded square root of 10) in areas estimated to have less residual uncertainty (20,21). In special cases, other UF values may be considered appropriate. While the actual value of $\sqrt{10}$ is 3.16, in practice, a single intermediate UF is calculated as 3 rather than 3.16, while two such intermediate UFs cumulate to 10. Thus, cumulative UFs could equal 1, 3, 10, 30, 100, 300, 1000, or 3000 (21).

IDENTIFICATION OF A POINT OF DEPARTURE: NO-EFFECT LEVELS VS. BENCHMARKS

Two major strategies are used for dose–response assessment methods to estimate "thresholds" of responses from study data. These are the benchmark dose (BMD) or benchmark concentration (BMC) approach and the no-observed-adverse-effect level (NOAEL) or NOAEL/LOAEL approach. In both approaches, uncertainty factors (UFs) are applied to account for various uncertainties in extrapolating from the study results to the general population to ensure the protection of public health. In recent years, the benchmark dose method has taken advantage of available data to address uncertainties quantitatively in the traditional NOAEL/LOAEL approach. OEHHA (21) recommends the use of the benchmark dose modeling (BMD) method wherever possible when data are adequate in order to address quantitatively the adequacy of reference exposure levels to protect the health of both children and adults.

The traditional NOAEL/LOAEL approach for safety assessments of chemicals for noncancer health effects was based upon dividing a NOAEL by uncertainty or safety factors to provide an acceptable daily intake (ADI), reference dose (RfD), or reference exposure level (REL). The NOAEL is the highest dose or exposure level with no biological or statistical increase in the frequency or severity of adverse effects among the exposed group relative to a control group (20). The NOAEL is based on the premise that an exposure threshold exists below which exposures have no effect. A lowest observed-adverse-effect level (LOAEL) is defined as the lowest exposure level in a study or series of studies with a biologically and/or statistically significant increase in the frequency or severity of adverse effects among an exposed population relative to a control group.

The NOAEL/LOAEL method is a simple approach for the setting of exposure guidelines for potentially toxic substances for noncancer health risks. However, this approach has several limitations and must be tempered by appropriate statistical interpretation. The NOAEL and LOAEL are highly dependent on the test doses chosen by the original investigators and are restricted to the experimental dose groups. Other limitations are that the NOAEL approach does not explicitly incorporate information on the shape of the dose–response curve or account for the study size appropriately.

Both Gaylor (22) and Leisenring and Ryan (23) demonstrated the sensitivity of the NOAEL to sample size and its high variability from experiment to experiment. A NOAEL could be associated with a substantial (1–20%) but undetected incidence of adverse effects among the exposed experimental group or population. This may be because only a subset of individuals from the population was observed, and the experiment may not have been designed to observe all adverse effects associated with the substance. Therefore, one may not safely conclude that the study concentration or dose is not associated with any adverse effects (20). Alternatively, a NOAEL could be many-fold lower than a true population threshold due to study design and dose spacing (22,23).

The U.S. EPA (20) determined that a NOAEL not associated with an identified biological effect (a "no-observed-effect-level" or NOEL) identified from a study with only one dose level is unsuitable for derivation of a reference concentration (RfC) for chronic exposure. However, there is a limited availability of multidose studies for a variety of chemicals. Therefore, other agencies such as OEHHA, may use a NOAEL without an associated LOAEL identified in the same study (termed a free-standing NOAEL) if no other suitable studies are available, and the NOAEL is consistent with the overall health hazard database for that chemical (including any case reports or studies with shorter durations) (21).

Traditionally, if not identifiable empirically, a NOAEL was estimated from the lowest exposure concentration reported to produce the adverse effect (LOAEL). An UF is applied to the LOAEL to estimate the NOAEL. If there exist multiple, non-identical NOAELs and LOAELs for the same compound and critical effect, the study of the best quality reporting the highest value for a NOAEL (preferred) or the lowest value for the LOAEL is used for the development of RELs.

A newer approach to identify an appropriate dose for noncancer toxicity is the benchmark dose or, for inhalation, benchmark concentration (BMC) approach. This was developed from the concept that a concentration estimated to be associated with a predefined low risk could provide an alternative to the NOAEL (24–30). Crump (26) defined the benchmark dose as a lower confidence limit corresponding to a moderate increase in risk (1–10%) above the background risk. The BMC method allows a mathematical and statistical approach to the calculation of reference concentrations or exposure levels (26,31–36).

In both approaches, the literature is examined to identify relevant endpoints. Toxicological endpoints are evaluated to determine the most sensitive effect (occurring at the lowest exposure level), and a dose–response relationship is determined. The most sensitive adverse effect of relevance to human health (termed the "critical effect") is selected and is usually a mild adverse effect.

BENCHMARK DOSE METHODOLOGY

This chapter describes the use of benchmark dose analysis as an alternative to the traditional NOAEL/LOAEL approach in setting exposure guidelines based on effects in the upper airways. For this purpose, the upper airways comprise the nasal passages and pharynx.

To estimate levels using the BMC method, quantal or continuous dose–response data for a toxicant (measured for at least two dose levels and a control) are required. Supporting toxicological data are not always sufficient to permit this level of quantification. In most cases, the method will allow determination of a benchmark concentration even with relatively sparse data, but with less confidence in the result. The alternative NOAEL method may give the appearance of providing a result more easily with poor data, but in fact, the uncertainty in such a result can be extremely large, and the situation is not improved by the inability to quantify this uncertainty.

In defining a point of departure for the derivation of health protective levels, the usual criterion chosen is the BMCL, defined as the 95% lower confidence limit of the concentration expected to produce toxic responses in a chosen percentage of subjects (the benchmark response rate) exposed at this dose. A suitable mathematical function is fitted to the concentration versus response relationship using likelihood methodology. The function used is selected according to defined quality-of-fit criteria. The concentration expected to produce the benchmark response rate (BMC) and the lower confidence bound on that concentration (BMCL) are identified from the fitted curve. In the case of quantal data in an animal toxicity experiment, the benchmark response rate is usually selected at 5%. Other types of data, including continuous measures of toxic response, and data from epidemiological studies, require an appropriate benchmark response rate identified on a case-by-case basis.

Despite its advantages, there are sources of uncertainty in the experimentally derived BMC value. For example, the studies used to estimate the BMCL have usually been performed with animals rather than humans, for example Ref. 37. Also, the experimental duration of exposure may differ from that which is of interest for the establishment of exposure guidelines. Additionally, the dose of toxicant delivered to the target tissue may differ between species and among humans and may depend on the type of activity in which the individual is engaged. Another area of uncertainty is that there can be a large degree of variability in the number of people who respond at any exposure level. For example, there may be over a 10-fold variability in the irritation threshold (the concentration of a substance at which irritation of the eyes, nose, and/or throat is first detectable) for chlorine (38). To estimate a health protective level such as a REL for the population of concern, the BMCL is therefore modified by UFs, except where explicit extrapolation models are available to allow for these differences.

BMCL/UFs = REL

Most frequently, the characteristics of the BMCL are chosen so that its properties are similar to that of the NOAEL described below. Thus, similar UFs are applied with both approaches. Specific data sets may, however, result in the use of UFs different from what would be used with a standard NOAEL, determined on a case-by-case basis; the rationale would be described in each toxicity summary for the individual chemicals.

SELECTION OF APPROPRIATE BENCHMARK CONCENTRATION RESPONSE RATE

A response range of 1% to 5% approximates the lower limit of adverse effect detection likely to occur in typical human epidemiological studies. In large laboratory animal studies, the detectable response rate is typically in the range of 5% to 10% (22). In 1995, using animal developmental toxicity data, the U.S. EPA concluded that a 1% response rate was likely to be too low to be detected and therefore too uncertain to use as a point of departure, while either 5% (BMC$_{05}$) or 10% (BMC$_{10}$) response rates were adequate for the purposes of estimating a benchmark concentration (34). One reason for this conclusion was the large difference (29-fold) between observed NOAELs and the 1% benchmark using developmental toxicity data. Subsequently, the U.S. EPA (39) used a 10% response rate for benchmark concentrations when deriving chronic inhalation reference concentrations (RfCs). More recently, RfC determinations for various endpoints by the U.S. EPA have used either 5% or 10% as the benchmark response rate, depending on the statistical uncertainty in the data (40,41). OEHHA has used the 5% response rate in several chronic RELs and showed that the lower 95% confidence bound on the BMCL$_{05}$ typically appears equivalent for risk assessment purposes to a NOAEL in well designed and conducted animal studies, where a quantal measure of toxic response is reported (31–36,42–44). Therefore, OEHHA typically uses a 5% response rate as the default for determination of the BMCL from quantal data (i.e., the effect is either present or it is not) in animals (45).

Other response rates may be selected if the data indicate that this is appropriate. For instance, large epidemiological studies examining a relatively severe endpoint such as clinical disease may support the use of a 1% response criterion, as in the case of the chronic REL recently developed for respirable crystalline silica (35). In that case, the size of the epidemiological database was large and as a result there was high confidence in the response at low exposures. In the case of a steep dose–response relationship, the selection of response rate is less influential on the final value. For acute lethality studies, the 1% and 5% response rate benchmark concentrations differed, on average, by less than twofold from the respective NOAEL (45).

Various criteria have been proposed for selecting an appropriate benchmark response rate for continuous data such as body weight, blood cell numbers, and levels of enzyme activity (30,46–48). One criterion is statistical confidence, for example, criteria based on some multiple (1.0–3.0) of the standard deviation of the reported measurements, either above or below the mean, particularly in controls or low-dose groups. A standard deviation of 2.33 from the mean identifies values at the first and 99th percentiles, extreme values even if not adverse. If values greater than the 98th to 99th percentile are abnormal, then a concentration that changes the mean by one standard deviation yields roughly 10% excess risk in subjects in the abnormal range (49). A second criterion is scientific judgment as to what constitutes a biologically relevant perturbation in a measured parameter, such as one that exceeds the likely range of physiological compensation. Some clinical guidelines are generally accepted as cutoff points, although they are not necessarily thresholds. These might include:

1. reduction in lung function [<80% of expected forced expiratory volume (FEV_1)],
2. a carboxyhemoglobin level of 1.1% to 1.3%, and
3. a pesticide worker's blood cholinesterase level less than 80% of the individual's baseline level.

The choice of an appropriate benchmark criterion for continuous data is currently based on the particular nature of those data, including supporting information on severity of the effect and possible mechanisms of repair or compensation, rather than on any overall policy-based guidance. In the development of the chronic REL for carbon disulfide, OEHHA used as the benchmark response rate a 5% reduction in peroneal motor conduction velocity ($BMCL_{05}$), a mild effect and definitely within the range of normal variation. In some cases, population shifts in a continuous variable such as FEV_1, blood pressure, birth weight, thyroid hormone levels (50), or IQ [e.g., effects of lead as reported by Lanphear et al. (51)] may result in pushing more individuals into a high-risk category, and thus small shifts can be considered adverse.

SELECTION OF CONFIDENCE LIMITS
The benchmark dose or concentration is selected by fitting an assumed dose–response curve to the observed response data. Mathematical curve fitting of this type necessarily involves recognition of uncertainty and variability in the input data. Fitted curves or interpolated values are generally described in terms of both maximum likelihood estimates (MLE) and confidence bounds on these estimates. Variation around the predicted values is generally assumed to follow a χ^2 (Chi squared) distribution, and the χ^2 statistic is used as a criterion of fit quality and in deriving "p" values and confidence limits on estimates. The 95% lower confidence limit (LCL) of the concentration at the chosen benchmark response rate or level is generally used as the BMCL, rather than the MLE. This is preferred since it takes into account sources of uncertainty intrinsic to the source data, including the variability of the test population and the number of subjects in the study. This provides an incentive for the generation and use of higher quality data, unlike the NOAEL/LOAEL methodology, which makes no explicit quantitative allowance for uncertainty in the underlying data. Use of the 95% LCL in a benchmark calculation also takes into account the quality of fit for the dose–response curve. The Benchmark Dose Workshop in 1985 (34) recommended using the 95% LCL in benchmark dose calculations. With robust data sets, the 90%, 95%, and 99% LCLs are close to each other and to the MLE (52).

SELECTION OF MODELS TO FIT THE DOSE–RESPONSE CURVE
It is important to select an appropriate mathematical model for the type of data used for benchmark concentration calculations (53). The U.S. EPA's Benchmark Dose Software (BMDS) contains a variety of models (54).

For dichotomous data, the models include the following:

1. gamma distribution,
2. logistic,

3. multistage,
4. probit,
5. quantal linear,
6. quantal quadratic, and
7. Weibull.

The quantal linear and the quantal quadratic are special cases of the Weibull model in which the exponents are one and two, respectively. The probit and logistic models can be run using either the dose or the logarithm of the dose. These models are useful for data where the subjects at each level of exposure did or did not experience a specific adverse effect such as eye irritation, liver enlargement, or an impaired nervous system (based on passing or failing a specific test). For nested dichotomous data, such as found in animal developmental data in which individual offspring are nested in litters, the models available are:

1. NLogistic (logistic nested),
2. NCTR (National Center for Toxicological Research), and
3. Rai & Van Ryzin (after the authors who described the model).

While quantal data show a dichotomous response reported as either the presence or absence of an effect, continuous data represent actual measurements, some examples of which are body weight, enzyme activity level, blood cell counts, IQ, and nerve conduction velocity. In addition to measurements of true continuous variables, a scoring system for severity or multiplicity of effects may be used to develop values for individual experimental subjects and for groups. Such scores may be regarded as pseudocontinuous variables and analyzed similarly to true continuous measures of response, although it should be recognized that there are limitations to this assumption. In particular, the individual scores are effectively step functions, and not all steps may represent similar proportional increments in response. The description for acetaldehyde given below shows an example of continuous benchmark dose modeling using a pseudocontinuous data set with the endpoint of severity rating of nasal olfactory degeneration.

For continuous data, the available models are:

1. linear,
2. polynomial,
3. power, and
4. Hill.

To date, the models most used by OEHHA are those for dichotomous data. Usually each model is fit to a dose–response data set of the most sensitive endpoint available, and both the MLE and the lower 95% confidence bound benchmark confidence level ($BMCL_{05}$) of the effective dose (ED_{05}) are derived from each model. When the number of subjects is very large as in the case for some occupational and epidemiological exposures such as respirable, crystalline silica, the MLE_{01} and $BMCL_{01}$ (a 1% benchmark) can be determined (35). The models that give an acceptable fit ($p \geq 0.10$ by χ^2) using the Chi square test (χ^2) are further examined. Some models may fit the entire range of the data equally well by the χ^2 test, but one may be better than another in describing the shape of the dose–response curve at the lower end of the dose range, which is critical in defining a benchmark such as $BMCL_{05}$. If more than one model gives an acceptable fit to the data, then some judgment is used in balancing a model's goodness of fit (as possibly indicated by a much higher p-value or as determined visually from the plotted curve) versus the level of health protection provided by the $BMCL_{05}$ derived using that model. From the perspective of protecting public health, the lowest value of the $BMCL_{05}$ from a model having an acceptable fit might be taken. However, with certain data sets, some models (including the often used log-probit model) may indicate an MLE, which is very far from the BMCL value (55). For well-fitting models, the BMCL is seldom less than one-third of the corresponding MLE, unless the overall precision of the data is poor. The analyst should also beware of attempts to fit complex models to data sets with insufficient precision to specify all the model parameters accurately (54). Thus, there must be allowance for professional judgment by toxicologists and statisticians.

EXAMPLES OF THE USE OF BENCHMARK DOSE METHODS FOR RESPIRATORY ENDPOINTS

Upon acute exposure of the upper airways to reactive substances, the most commonly observed endpoint is irritation. While some manifestations of irritation may be amenable to objective evaluation (e.g., nasal rhinometry, pulmonary function, eye redness), at some level, irritation is an inherently subjective experience reflecting an individual's perceptions. As a result, quantifying the response to irritants generally involves subjective evaluation of the experience by test subjects using a severity scale. In some instances, these data may be used as a continuous endpoint in a benchmark analysis. However, variation in the sensitivity of individuals to irritation often tends to be large, and it may therefore be more appropriate to reduce the data to a measure of the concentration-dependent presence or absence of subjective irritation.

RD$_{50}$: A UNIQUE SPECIAL CASE

One of the best-known examples of a response of the respiratory system to toxic chemical exposures is the slowing of breathing rate in rodents exposed to sensory irritants (56). This is a reflex response to stimulation of chemoreceptors associated with the trigeminal nerve and located in the cornea and nasal mucosa (57). The response in rodents is a steady decrease in respiratory rate with increasing exposure concentration. Related, but more complex, reflex responses are noted in humans, who typically report burning sensations and choking when exposed to these chemicals. Alarie (56–58) described a systematic procedure to evaluate and compare the sensory irritant effects of a wide range of chemicals using the RD_{50}, that is, the concentration of the chemical that produces a 50% depression in breathing rate in the standardized mouse bioassay. This parameter, with an appropriate safety/uncertainty factor, has also been used as a point of departure for setting ceiling levels or threshold limit values (TLVs) for occupational health protection (59,60) and has also been suggested as a starting point in determining reference exposure levels (RELs) for public health protection (43,61).

The procedure for determining the RD_{50} is in fact a type of benchmark dose analysis, but with important simplifying characteristics. Values of the respiratory rate are fit to a model which is linear with the log of exposure concentration, and the dose for 50% effect is determined by interpolation. The requirement to identify a large benchmark response ratio such as 50%, and the fact that the data follow a simple log-linear relationship with concentration over the range of interest simplifies the fit procedure needed. In this case the error function may be assumed to be normally distributed, and an exact least-squares linear regression procedure used to determine the parameters of the fit line. This contrasts with the more general case where it is necessary to use iterative fitting and more complex error function assumptions since the dose-response function may be non-linear and the target response rate is at the low end of the range of observed data (where departure from linearity may be more notable).

Obviously, the choice of a 50% response rate as the point of departure implies a different approach to extrapolation if this is used to determine a tolerable level for human exposure. This extrapolation may be made by a simple factor or a more complex function, but in any case it will not bear much relationship to the procedure used to extrapolate to a protective level from a NOAEL. Use of high benchmark response rates in this way is not unprecedented. However, in the absence of more informative data it has been suggested that a tentative protective level may be estimated by applying a large uncertainty factor (10^3 or 10^4) to the dose causing lethality in half the cases (LD_{50}).

An Example of the Use of Benchmark Dose Modeling (BMC) with a Quantal Endpoint: Sensory Irritation Due to Ammonia

Ammonia vapors are an example of a reactive compound, acute exposures to which manifest as irritation of the nose and throat, and may include coughing and/or lacrimation. As levels increase, minute ventilation volumes decrease, chest discomfort may be experienced, and ultimately pulmonary edema can occur. Studies of ammonia inhalation have found large variations in individual sensitivities that are in part dependent on exposure history (62–64). Tolerance to elevated ammonia levels (100 ppm) has been seen to develop in a matter of weeks following exposure to lower concentrations (64). Comparisons of naïve and previously exposed individuals have found substantial differences in reported symptoms (smell, eye irritation, cough, general discomfort, headache, and chest irritation) after exposure to 110 ppm ammonia and above for one hour (63). These data are thus amenable to a quantal approach where subjects at each exposure level do or do not experience a specific adverse effect.

For the purposes of a benchmark analysis, the results from four human studies of experimental ammonia inhalation were combined (62,63,65,66). Exposure times varied among studies from 5 to 120 min with exposure levels ranging from 30 to 500 ppm. To facilitate comparison, the concentrations in the studies used for the analysis were time-adjusted to one hour using the formula $C^n \times T = K$ where $n = 4.6$ (Table 1). The value of 'n' was determined empirically from the data sets by sequentially varying its value in the log-normal probit analysis. The value of 4.6 provided the best fit to the data by Chi square analysis.

For the quantal analysis, the response data were reduced to the presence or absence of irritation symptoms. In Figure 1, generated with U.S. EPA's BMD software version 1.4.1b, the fraction of individuals affected on a probit scale is plotted against the log of the dose in ppm. A 5% response rate was selected and the resulting benchmark concentration (BMC) is 20 ppm, while the lower bound of the 95% confidence interval ($BMCL_{05}$) is 14 ppm. Thus 14 ppm represents the point of departure. For comparison, using the same data from the

TABLE 1 Ammonia Irritation: Adjusted 60-Minute Exposures

Concentration (ppm)	30	50	50	72	50	80	134	110	140	500
Exposure time (min)	10	5	10	5	120	120	5	60	60	30
Adjusted 1 hr concentration	20	29	34	42	43	69	78	95	120	430
Response	0/15	0/10	4/6	3/10	7/16	9/16	8/10	12/16	15/16	7/7

combined studies in a LOAEL/NOAEL analysis, the LOAEL is 34 ppm, and the point of departure would be the NOAEL of 29 ppm. In this example, the benchmark approach yields a point of departure half of that obtained from the LOAEL/NOAEL approach, resulting in a more health-protective value.

An Example of the Use of Benchmark Dose Modeling (BMC) with a Continuous Endpoint: Olfactory Pathology Due to Acetaldehyde

A pseudo-continuous data set was available on severity of hyperplastic changes in olfactory nasal epithelium of rats exposed to acetaldehyde (67,68) to use as a model for continuous benchmark dose analysis. The subchronic studies by Appelman exposed rats to acetaldehyde vapor (6 hr/day, 5 days/week for four weeks to 150, 400, 500, 1000, 2200, or 5000 ppm) and collected incidence and severity data. Two qualities that made this data set acceptable for use in

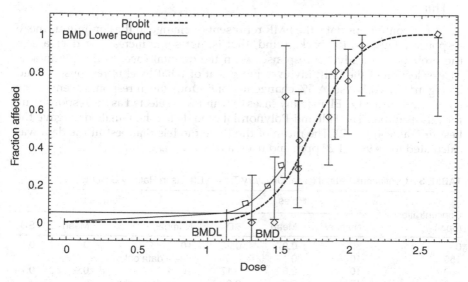

Probit model with 0.95 confidence level

FIGURE 1 Benchmark analysis using quantal data. Data from four human studies measuring sensory irritation associated with ammonia exposure were pooled and analyzed with benchmark dose software. A probit model provided the best fit to the data. A 5% response rate was selected and gave a benchmark concentration (BMD) of 20 ppm. The 95% lower confidence level (BMDL) was 14 ppm.

TABLE 2   Severity Coding Values for Pseudocontinuous Dataset

No effect = 0	Marked = 4
Minimal = 1	Moderate with hyperplasia = 5
Slight = 2	Severe with hyperplasia = 6
Moderate = 3	Very severe with hyperplasia = 7

benchmark dose modeling were adequate dose spacing and a large enough number of animals per dose group (n = 10 per group). In addition, the data for both males and females allowed for separate and combined analysis by gender.

For use in the benchmark dose analysis, the data for acetaldehyde were converted to a pseudo-continuous data set using a method of severity coding. The severity ratings and assigned numerical values are in Table 2. The original data set also indicated how many animals had a particular severity of olfactory degeneration within a dose group. For each dose group, the number of animals per severity group was counted and their corresponding severity code values were added (Table 2). The mean and standard deviaton of the severity code values were calculated for all animals in a particular dose group. The means and standard deviations for all dose groups (Table 3) were entered into the BMC program. The data for males and females were analyzed separately since females showed an increase in severity ratings at several doses.

For the acetaldehyde data, the following BMC models were run:

- Linear
- Polynomial (2nd & 3rd degree)
- Power
- Hill

For continuous data, the BMR represents a change in the continuous mean response relative to the background, that is, not some increase or decrease in the probability of adverse response, as in the quantal case. In the continuous case, selection of the BMR involves judgment of what level of response should be regarded as adverse. A 5% change in continuous mean response seems to be comparable to a NOAEL in many cases (53) and was selected as the response rate for this example. The Hill and Polynomial models for the female data gave the best fit (Table 4; Fig. 2). The mean of the three models that best fit the data was calculated to be 99 ± 1.20 ppm and used as the $BMCL_{05}$.

TABLE 3   Continuous Data of Nasal Olfactory Tissue Effects in Rats by Severity

Concentration (ppm)	Males			Females		
	Number	Mean	SD	Number	Mean	SD
0	30	0.07	0.25	10	0	0
150	10	0	0	(Male data only)		
400	10	2.6	1.17	10	0.9	0.74
500	10	2.5	0.97	(Male data only)		
1000	10	2.8	0.63	10	3.6	0.70
2200	10	5.3	2.21	10	5.1	1.91
5000	10	6.7	0.67	10	6.9	0.32

Source: From Refs. 67, 68.

TABLE 4 BMC Modeling of Continuous Data in the Female Rat (Appelman Data)

Method	BMCL	BMC	p-Value	AIC
Hill	100	205	0.07	55.96
Polynomial (2nd degree)	101	126	0.02	56.18
Polynomial (3rd degree)	97	165	0.03	55.95

In the case of acetaldehyde, this analysis is a departure from previous analyses that used quantal data of incidence of degeneration of olfactory epithelium and a NOAEL/LOAEL approach. The point of departure for a NOAEL/LOAEL approach would be 150 ppm, whereas the benchmark dose approach identified 99 ppm as the point of departure. It should be noted also that the severity data from the Appelman study provide a measure of the intensity of the response in one specific region of the upper respiratory tract. This is a valid basis for treating the score as a pseudo-continuous variable, since there is an underlying assumption that the value of an individual score is proportional to the amount of tissue damage caused at that site. Some other data have been reported where the increasing value of the score developed might represent increasing area of damage, but without necessarily increasing severity at the target site(s). It is less easy to defend the score as a proportionate measure of response in this case, especially if the increasing area of impact extends to different regions of the respiratory tract. In such cases, it may be better to determine severity-based dose-response relationships for the different areas. It may also be informative to compare these

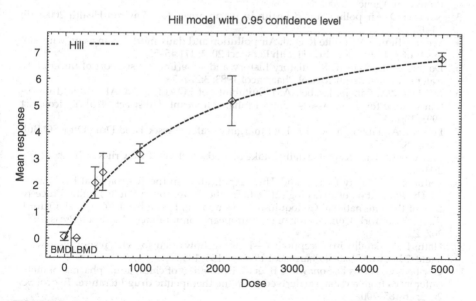

FIGURE 2 Benchmark analysis using CONTINUOUS data. Benchmark dose software was used to analyze the data from two animal studies measuring severity of degeneration of olfactory epithelium in rats after acetaldehyde exposure. The Hill model provided one of the best fits to the data. A 5% response rate was selected and gave a benchmark concentration (BMD) of 205 ppm. The 95% lower confidence level (BMDL) was 100 ppm.

in light of aerodynamic deposition models, which are now available for the respiratory tracts of humans and several experimental species.

CONCLUSION
Benchmark dose methodology is useful in evaluating the dose-response characteristics for various types of response to toxic chemicals in the upper respiratory tract. This approach is preferable to the more traditional NOAEL/LOAEL approach since it takes advantage of the entire range of the data, rather than a single exposure group. It is also less susceptible to influence by the idiosyncrasies of individual study designs and choices of exposure concentrations, and provides a statistically based measure of the uncertainty in the estimate of the benchmark concentration. This benchmark may be used as a point of departure from which to extrapolate to health protective levels appropriate for occupational or community exposures.

REFERENCES
1. Salmon AG. Risk assessment for chemical carcinogens. In: Landolph J, Warshawsky D, eds. Molecular Carcinogenesis and the Molecular Biology of Human Cancer. Boca Raton, FL: CRC Press/Taylor and Francis, 2006; 501–545.
2. Carlisle J, Kaley K, Dowling K. Development of Health Criteria for School Site Risk Assessment Pursuant to Health and Safety Code Section 901(g); Proposed Child-specific Benchmark Change in Blood Lead Concentration for School Site Risk Assessment. Office of Environmental Health Hazard Assessment, California Environmental Protection Agency; 2006.
3. Schwela D. Air pollution and health in urban areas. Rev Environ Health 2000; 15: 13–42.
4. Vedal S, Brauer M, White R, et al. Air pollution and daily mortality in a city with low levels of pollution. Environ Health Perspect 2003; 111:45–52.
5. Dourson ML, Stara JF. Regulatory history and experimental support of uncertainty (safety) factors. Regul Toxicol Pharmacol 1983; 3:224–38.
6. Mitchell WA, Gift JS, Jarabek AM. Suitability of LOAEL to NOAEL 10-fold uncertainty factor for health assessments of inhaled toxicants [Abstract #475]. Toxicologist 1993; 13:140.
7. Lehman AJ, Fitzhugh OG. 100-Fold margin of safety. Assoc Food Drug Off USQ Bull 1954; 18:33–35.
8. Bigwood EJ. The acceptable daily intake of food additives. CRC Crit Rev Toxicol 1973; 2:41–93.
9. Vettorazzi G. Safety Factors and Their Application in the Toxicological Evaluation. In: The Evaluation of Toxicological Data for the Protection of Public Health: Proceedings of the International Colloquium, Luxembourg, December 1976. 1st ed. Oxford, [Eng.]; New York: Commission of the European Communities, Pergamon Press; 1977: 207–223.
10. Hattis D. Variability in susceptibility—how big, how often, for what responses to what agents? Environ Toxicol Pharmacol 1996a; 2:135–145.
11. Ginsberg G, Hattis D, Sonawane B, et al. Evaluation of child/adult pharmacokinetic differences from a database derived from the therapeutic drug literature. Toxicol Sci 2002; 66:185–200.
12. Miller MD, Marty MA, Arcus A, et al. Differences between children and adults: implications for risk assessment at California EPA. Int J Toxicol 2002; 21:403–418.
13. Dorne JL, Renwick AG. The refinement of uncertainty/safety factors in risk assessment by the incorporation of data on toxicokinetic variability in humans. Toxicol Sci 2005; 86:20–26.

14. NRC. Guidelines for Developing Community Emergency Exposure Levels for Hazardous Substances. Washington, D.C.: Committee on Toxicology, National Research Council, National Academy Press, 1993.
15. U.S. EPA. Reference Dose (RfD): Description and Use in Health Risk Assessments. Background Document1A. March 15. Washington, D.C.: United States Environmental Protection Agency, 1993.
16. NRC. Drinking Water and Health. In: Safe Drinking Water Committee CC, Safety and Risk Assessment, National Academy of Sciences, ed. Washington, D.C.: National Research Council, National Academy Press, 1977–1987.
17. Alexeeff GV, Lipsett MJ, Kizer KW. Problems associated with the use of immediately dangerous to life and health (IDLH) values for estimating the hazard of accidental chemical releases. Am Ind Hyg Assoc J 1989a; 50:598–605.
18. Alexeeff GV, Lewis DC. Factors influencing quantitative risk analysis of noncancer health effects from air pollutants. Pittsburgh, PA: Air & Waste Management Association; 1989b:1–14.
19. Dourson ML, Felter SP, Robinson D. Evolution of science-based uncertainty factors in noncancer risk assessment. Regul Toxicol Pharmacol 1996; 24:108–120.
20. U.S. EPA. Methods for Derivation of Inhalation Reference Concentrations and Application of Inhalation Dosimetry. Cincinnati, OH: Office of Research and Development, United States Environmental Protection Agency, 1994.
21. OEHHA. The Air Toxics Hot Spots Program Risk Assessment Guidelines. Technical Support Document for the Derivation of Reference Exposure Levels. California: Air Toxicology and Epidemiology Section, Office of Environmental Health Hazard Assessment, California Environmental Protection Agency, 2008.
22. Gaylor DW. Incidence of developmental defects at the no observed adverse effect level (NOAEL). Regul Toxicol Pharmacol 1992; 15:151–160.
23. Leisenring W, Ryan L. Statistical properties of the NOAEL. Regul Toxicol Pharmacol 1992; 15:161–71.
24. Mantel N, Bryan WR. "Safety" testing of carcinogenic agents. J Natl Cancer Inst 1961; 27:455–470.
25. Mantel N, Bohidar NR, Brown CC, et al. An improved Mantel-Bryan procedure for "safety" testing of carcinogens. Cancer Res 1975; 35:865–872.
26. Crump KS. A new method for determining allowable daily intakes. Fundam Appl Toxicol 1984; 4:854–871.
27. Dourson ML. New approaches in the derivation of acceptable daily intake (ADI). Comm Toxicol 1986; 1:35–48.
28. Hartung R. Dose–response relationships. In: Tardiff RG, Rodricks JV, eds. Toxic Substances and Human Risk. New York, NY: Plenum Press; 1987:29–46.
29. Gaylor DW. Applicability of cancer risk assessment techniques to other toxic effects. Toxicol Ind Health 1988; 4:453–459.
30. Gaylor D, Ryan L, Krewski D, et al. Procedures for calculating benchmark doses for health risk assessment. Regul Toxicol Pharmacol 1998; 28:150–164.
31. Lewis DC, Alexeeff GV. Quantitative Risk Assessment of Noncancer Health Effects for Acute Exposure to Air Pollutants. In: 82nd Annual Meeting of the Air and Waste Management Association. Anaheim, CA: Air and Waste Management Association, 1989:89–91.
32. Alexeeff GV, Lewis DC, Lipsett MJ. Use of toxicity information in risk assessment for accidental release of toxic gases. J Hazard Mater 1992; 29:387–403.
33. Alexeeff GV, Lewis DC, Ragle NL. Estimation of potential health effects from acute exposure to hydrogen fluoride using a "benchmark dose" approach. Risk Anal 1993; 13:63–69.
34. Barnes DG, Daston GP, Evans JS, et al. Benchmark dose workshop: criteria for use of a benchmark dose to estimate a reference dose. Regul Toxicol Pharmacol 1995; 21:296–306.
35. Collins JF, Salmon AG, Brown JP, et al. Development of a chronic inhalation reference level for respirable crystalline silica. Regul Toxicol Pharmacol 2005; 43:292–300.

36. Starr TB, Goodman JI, Hoel DG. Uses of benchmark dose methodology in quantitative risk assessment. Regul Toxicol Pharmacol 2005; 42:1–2.
37. Kuwabara Y, Alexeeff G, Broadwin R, et al. Evaluation of the RD_{50} for determining acceptable levels of exposure to airborne sensory irritants. [Abstract #882]. Toxicologist 2006; 90:179.
38. Anglen DM. Sensory Response of Human Subjects to Chlorine in Air [Dissertation]. Ann Arbor, MI: University of Michigan, 1981.
39. U.S. EPA. Integrated Risk Information System (IRIS) Database. California: United States Environmental Protection Agency; 2007.
40. U.S. EPA. A Review of the Reference Dose and Reference Concentration Process. California: Risk Assessment Forum, United States Environmental Protection Agency; 2002.
41. U.S. EPA. An Examination of EPA Risk Assessment Principles and Practices. Washington, D.C.: Office of the Science Advisor, United States Environmental Protection Agency, 2004.
42. Collins JF, Alexeeff GV, Lewis DC, et al. Development of acute inhalation reference exposure levels (RELs) to protect the public from predictable excursions of airborne toxicants. J Appl Toxicol 2004; 24:155–166.
43. Alexeeff GV, Deng KK, Broadwin RL, et al. Benchmark dose evaluation for human irritation [Abstract # 901]. Toxicologist 2006; 90:183.
44. Brown JP, Collins JF, Salmon AG, et al. Use of benchmark dose methodology on human non-cancer data to develop protective criteria for child exposures to arsenic [Abstract # 2185]. Toxicologist 2006; 90:448.
45. Fowles JR, Alexeeff GV, Dodge D. The use of benchmark dose methodology with acute inhalation lethality data. Regul Toxicol Pharmacol 1999; 29:262–278.
46. U.S. EPA. The Use of the Benchmark Dose Approach in Health Risk Assessment. California: Risk Assessment Forum, United States Environmental Protection Agency; 1995.
47. Crump KS. Critical issues in benchmark calculations from continuous data. Crit Rev Toxicol 2002; 32:133–153.
48. Sand SJ, von Rosen D, Filipsson AF. Benchmark calculations in risk assessment using continuous dose–response information: the influence of variance and the determination of a cut-off value. Risk Anal 2003; 23:1059–1068.
49. Crump KS. Calculation of benchmark doses from continuous data. Risk Anal 1995; 15:78–89.
50. Ting D, Howd RA, Fan AM, et al. Development of a health-protective drinking water level for perchlorate. Environ Health Perspect 2006; 114:881–886.
51. Lanphear BP, Hornung R, Khoury J, et al. Low-level environmental lead exposure and children's intellectual function: an international pooled analysis. Environ Health Perspect 2005; 113:894–899.
52. Sand S, Filipsson AF, Victorin K. Evaluation of the benchmark dose method for dichotomous data: model dependence and model selection. Regul Toxicol Pharmacol 2002; 36:184–197.
53. Filipsson AF, Sand S, Nilsson J, et al. The benchmark dose method—review of available models, and recommendations for application in health risk assessment. Crit Rev Toxicol 2003; 33:505–542.
54. U.S. EPA. Benchmark Dose Software (BMDS) Tutorial. California: United States Environmental Protection Agency, 2006.
55. Murrell JA, Portier CJ, Morris RW. Characterizing dose–response I: critical assessment of the benchmark dose concept. Risk Anal 1998; 18:13–26.
56. Alarie Y. Irritating properties of airborne materials to the upper respiratory tract. Arch Environ Health 1966; 13:433–449.
57. Alarie Y. Sensory irritation of the upper airways by airborne chemicals. Toxicol Appl Pharmacol 1973; 24:279–297.
58. Alarie Y. Bioassay for evaluating the potency of airborne sensory irritants and predicting acceptable levels of exposure in man. Food Cosmet Toxicol 1981; 19:623–626.

59. Sangha GK, Alarie Y. Sensory irritation by toluene diisocyanate in single and repeated exposures. Toxicol Appl Pharmacol 1979; 50:533–547.
60. Barrow CS, Alarie Y, Warrick JC, et al. Comparison of the sensory irritation response in mice to chlorine and hydrogen chloride. Arch Environ Health 1977; 32:68–76.
61. Kuwabara Y, Alexeeff GV, Broadwin R, et al. Evaluation and application of the RD50 for determining acceptable exposure levels of airborne sensory irritants for the general public. Environ Health Perspect 2007; 115:1609–1616.
62. Silverman L, Whittenberger JL, Muller J. Physiological response of man to ammonia in low concentrations. J Ind Hyg Toxicol 1949; 31:74–78.
63. Verberk MM. Effects of ammonia in volunteers. Int Arch Occup Environ Health 1977; 39:73–81.
64. Ferguson WS, Koch WC, Webster LB, et al. Human physiological response and adaption to ammonia. J Occup Med 1977; 19:319–326.
65. Industrial Bio-Test Laboratories I. Report to International Institute of Ammonia Refrigeration: Irritation threshold evaluation study with ammonia; March 23, 1973.
66. MacEwan J, RTheodore J, Vernot E. Human exposure to EEL, concentration of monomethylhydrazine. In: AMRL-TR-1970;70–102, ed. Wright-Patterson Air Force Base (OH): SysteMed Corp; 1970.
67. Appelman LM, Woutersen RA, Feron VJ. Inhalation toxicity of acetaldehyde in rats. I. Acute and subacute studies. Toxicology 1982; 23:293–307.
68. Appelman LM, Woutersen RA, Feron VJ, et al. Effect of variable versus fixed exposure levels on the toxicity of acetaldehyde in rats. J Appl Toxicol 1986; 6:331–336.

25 Physicochemical Modeling of Sensory Irritation in Humans and Experimental Animals

Michael H. Abraham, Ricardo Sánchez-Moreno, and Javier Gil-Lostes
Department of Chemistry, University College London, London, U.K.

J. Enrique Cometto-Muñiz and William S. Cain
Chemosensory Perception Laboratory, Department of Surgery (Otolaryngology), University of California, San Diego, La Jolla, California, U.S.A.

INTRODUCTION

Chemosensory detection of volatile organic compounds (VOCs) by humans rests on the senses of smell and sensory irritation, both being part of the body warning system. Several terms have been employed that subsume the sensations evoked by chemicals that are typically viewed as irritative. For example, in 1912, Parker introduced the concept of the "common chemical sense" (CCS) to describe general mucosal sensitivity to chemicals (1,2). More recently, the terms "chemesthesis" and "pungency" have been used to describe sensations evoked by chemicals that are not properly odors. Such pungency includes piquancy, tingling, prickling, irritation, stinging, burning, and freshness, among others.

The sense of smell gives rise to the perception of odors, whereas chemesthesis gives rise to the perception of pungency. Whereas odors are detected in the olfactory mucosa that covers the upper back portion of the nasal cavity via the olfactory nerve (cranial nerve I), chemesthetic sensations are mainly detected in all three mucosa: ocular, nasal, and oral, mainly via the trigeminal nerve (cranial nerve V) (3). The pharmacological and toxicological characterization of the senses of smell and chemesthesis includes the study of the breadth and sensitivity of responses toward the spectrum of chemicals. Recent studies on the molecular biology of smell (4) have provided additional support to the long-held view of the existence of a large number of different odorant receptors, probably in the order of 800 in humans, of which about half are functional (5). Since a given odorant can activate more than one type of receptor, and since a given type of receptor can be activated by several odorants, the number of possible odorants that can be recognized may reach several million (6,7). Of course, odor detection is another matter altogether, and within the context of odor detection, a systematic strategy to study the breadth of chemical tuning and sensitivity in olfaction and chemesthesis has considerable merit. One strategy to uncover the physicochemical basis for odor detection thresholds and pungency thresholds of VOCs consists of measuring chemosensory thresholds for homologous series of compounds. Over the past 20 years, Cometto-Muñiz et al. have carried out a systematic investigation into thresholds for sensory irritation and odor, using panels of human subjects under carefully controlled conditions and homologous series of compounds, including alcohols, esters (acetates), ketones, aldehydes, carboxylic acids, aromatic hydrocarbons, and terpenes.

Previous research has shown that olfaction is more sensitive than chemesthesis in the detection of VOCs (8). One approach to avoid the influence of smell is to probe nasal chemosensory detection in participants lacking olfaction, that is anosmics. Another approach consists in testing participants with an intact olfaction, that is, normosmics, but in terms of nasal localization or lateralization rather than detection. This technique tests the ability of subjects to identify the nostril receiving the VOC when the contralateral nostril simultaneously receives plain air (9), and it has been shown that such localization is mediated by trigeminal, not olfactory input (10).

There are more than 100,000 industrial chemicals, and even if only a third could be classed as VOCs or semivolatile organic compounds, it is obvious that experimental determination of potency toward humans cannot possibly be extended to more than a very small proportion. The use of animal experiments allows the study of VOCs that are too toxic to be tested on humans, but, again, the number of VOCs that can be tested on animals is but a small fraction of the number of VOCs that are actual or potential irritants.

There is therefore a very definite need for some type of prediction of the potency of VOCs toward humans. Even if restricted to VOCs that act through "physical" or "nonreactive" mechanisms, rather than through "chemical" or "reactive" mechanisms, such predictions would considerably help to fill the gap between the relatively small number of VOCs studied to date, and the very large number of chemicals that could be encountered. We define and explain the terms "nonreactive" and "reactive" in the section "Upper respiratory tract irritation in mice."

As mentioned, irritant VOCs comprise a very diverse group of compounds and the structural basis of their activity is not well understood. Nerve recordings (11) and human psychophysical studies (12,13) indicate that response thresholds are strongly related to hydrophobicity within homologous series. The high correlation between hydrophobicity and pungency has led to the suggestion that irritants interact with a hydrophobic biophase (14), either the epithelium through which irritants must diffuse or the nerve ending itself.

The trigeminal nerve fibers reaching the nasal mucosa ramify repeatedly, terminating in free nerve endings (15). Trigeminal nerve fibers that respond to irritating compounds, contain the neuropeptide substance P (SP) and calcitonin gene-related peptide (CGRP) and extend close to the nasal epithelia surface (16). Electron microscopic studies suggest that the vast majority of the CN V free nerve endings terminate within the lamina propria. Nevertheless, a few trigeminal fibers terminate at the line of tight junctions only a few micrometers from the surface (16). For volatile chemicals to stimulate these nerve endings, they must (*i*) pass into the nasal cavity, (*ii*) partition into and diffuse through the mucus, and (*iii*) cross the epithelial membranes and/or intercellular tight junctions.

Several mechanisms have been proposed to explain how irritative chemicals initiate transduction at the surface of cell membranes (17) although the nature of these processes is still poorly understood. Compounds that are chemically reactive, that is their potency is induced by a chemical mechanism (18), can produce irritation directly by reacting with a receptor or indirectly by mucosal tissue damage via chemical reaction without the need to interact with any particular receptor (17). In the latter case, damaged cells would release endogenous chemicals such as ATP, H^+, and bradykinin (a nonapeptide messenger), which, in turn, could act specifically upon ion channels to produce the neural response (19,20). On the other hand, there are other compounds that are likely to act on specific receptors such as menthol (21), capsaicin (22), and nicotine (23).

Studies of homologous series of irritant VOCs shed light on the contribution of the chemical structure in their interactions with receptors. Inoue and Bryant (24) found that (*i*) longer chain aldehydes and alcohols activate neurons at lower concentration than their shorter homologous, while chain length does not affect enaldehyde responses; (*ii*) aldehydes are more active, stimulating more neurons than the same concentration of corresponding alcohols; (*iii*) the presence of a double bond in enaldehydes increases the number of neurons that respond to lower concentration than corresponding aldehydes; (*iv*) the three homologous series activate a subpopulation of capsaicin-sensitive nociceptors (25), and partially activate a TRPA1-bearing subpopulation of nociceptors, enough to cause pungency, as mustard oil activation of TRPA1 does (26).

Some of these findings are in very good agreement with those by Cometto-Muñiz et al. For instance, odor detection thresholds (ODTs) and nasal pungency thresholds (NPTs) decrease when the chain length increases (12). Cometto-Muñiz et al. also reported lower threshold values for aldehydes than alcohols (13). These findings are also consistent with those of Alarie et al. (18), since they demonstrate that enaldehydes have lower irritation thresholds in mice than the corresponding saturated aldehydes.

Methods for the determination of NPTs in humans have been described in Chapter 13 ("Nasal chemosensory irritation in humans"), but there has also been considerable work on the effect of VOCs on upper respiratory tract irritation in mice, which requires a different experimental methodology. The most sophisticated method is that due to Alarie et al., who have described it in some detail (27,28). An all-glass exposure chamber holds four mice, the bodies of which are held in plethysmographs connected to a pressure transducer. The head of each mouse projects into the chamber, and a constant flow of air containing a known concentration of a chemical is passed through the chamber. The plethysmograph/pressure transducer records the respiratory rate of the mouse, and the biological endpoint is taken as the concentration of chemical that leads to a 50% decrease in respiratory rate, denoted as RD_{50}. The first quantitative structure–activity relationships for irritation of vapors were constructed using the RD_{50} endpoint in mice, and it is this work that we first consider, before the work of Cometto-Muñiz and Cain on nasal pungency thresholds in humans.

UPPER RESPIRATORY TRACT IRRITATION IN MICE

A great deal of work on upper respiratory tract irritation in mice has been carried out by Alarie et al., who not only developed a rigorous experimental procedure to obtain values of RD_{50} that is an ASTM standard (29), but who also showed that the mouse bioassay could be used to estimate acceptable exposure levels in man (30–33). Kuwabara et al. (34) have also reviewed the use of RD_{50} values to determine acceptable exposure levels in man. Not surprisingly, there have been

TABLE 1 The FBP Rule for Some Alkylbenzenes (32)

VOC	RD_{50} in ppm	P^o in ppm at 37°C	RD_{50}/P^o
Toluene	5300	73,000	0.073
Ethylbenzene	4060	29,400	0.138
n-Propylbenzene	1530	11,100	0.138
Isopropylbenzene	2490	12,700	0.196
n-Butylbenzene	710	5600	0.126
tert-Butylbenzene	760	8600	0.088
n-Pentylbenzene	230	1020	0.226
tert-Butyltoluene	360	2050	0.176
n-Hexylbenzene	125	359	0.349

numerous attempts to correlate RD_{50} values with various physicochemical properties of VOCs.

Much of the early work revolved around the suggestion of Ferguson (35) that for VOCs that acted by a "physical" mechanism, there was a correlation between the gaseous toxic concentration and solubility or vapor pressure. Brink and Pasternak (36) considered the ratio P^{nar}/P^o, where P^{nar} is the gaseous narcotic partial pressure of a VOC in some particular bioassay and P^o is the saturated vapor pressure of the VOC, and referred the ratio as the "thermodynamic activity." They put forward the rule of equal narcotic activity at equal thermodynamic activity; in other words, the ratio P^{nar}/P^o is constant. The Ferguson–Brink–Pasternak (FBP) rule was extended to cover not just gaseous narcosis, but gaseous bioassays in general, as in Eq. [1].

$$P^{bioassay}/P^o = c \qquad (1)$$

It was never very clear how the FBP rule, Eq. [1], could be regarded as a thermodynamic rule, because in practice it was not actually constant. For example, Nielsen and Alarie (32) obtained the data shown in Table 1 for a series of alkylbenzenes. The ratio RD_{50}/P^o is only approximately constant. Abraham et al. (37) then showed that there was no thermodynamic basis for the FBP rule at all, but that it was just a useful empirical observation.

Nevertheless, several analyses of RD_{50} values have used the VOC vapor pressure or normal boiling point. Muller and Gref (38) analyzed four series of VOCs: ketones, alcohols, acetates, and various benzene derivatives. For each individual series, they showed correlations between $\log(1/RD_{50})$ and various physicochemical properties such as the water to octanol partition coefficient, and the normal boiling point, T_B. However, since each series was analyzed separately, it is doubtful if the various correlations have any mechanistic significance. The importance of the work is only to show that it is possible to connect values of RD_{50} (or $\log 1/RD_{50}$) to physicochemical properties.

Roberts (39) also analyzed the four series of compounds separately, but was able to generate a general QSAR covering 42 saturated ketones, saturated alcohols, and aromatic compounds, Eq. [2]

$$\log(M/RD_{50}) = 0.0173 T_B' - 4.090 \qquad (2)$$

$$N = 42, SD = 0.119, R^2 = 0.974.$$

In Eq. [2], M is the molecular weight of the VOC and T_B' is an adjusted normal boiling point taken as $(26.5 T_B / 22) - 8$ for the alcohols, and $(24.0 T_B / 22) - 4$ for phenol; all the other compounds have $T'_B = T_B$. N is the number of data points, that is, the number of VOCs, SD is the standard deviation, and R is the correlation coefficient. Five compounds were omitted from Eq. [2]: allyl alcohol, methyl vinyl ketone, crotyl alcohol, mesityl oxide, and vinyl toluene. Roberts pointed out that the excess potency of the first four compounds could be due to their reactive electrophilic character, and subsequent work seems to confirm this. Also omitted from Eq. [2] was a series of acetate esters. Roberts suggested that these esters were hydrolyzed to the corresponding alcohol and acetic acid, the latter being a much more potent irritant.

Even for nonreactive compounds, the problem with Eq. [2] concerns the adjusted normal boiling point, which is a purely empirical correction. It is impossible to use Eq. [2] to predict further values of RD_{50}, outside the series of saturated ketones and saturated alcohols, because the adjustment itself cannot be predicted.

Nielsen and Alarie (32) showed that for a series of alkylbenzenes and a series of aliphatic alcohols, the FBP ratio was roughly constant (Table 1).

The publication of an important database of RD_{50} values by Schaper (40) led to the analysis of a much larger number of RD_{50} values than had hitherto been possible. Alarie et al. (18) used the FBP rule to divide a database of 145 VOCs into those that acted by a "physical" mechanism (nonreactive VOCs) and for which $P^{bioassay} / P^o > 0.1$, and those that acted by a "chemical" mechanism (reactive VOCs) and for which $P^{bioassay} / P^o < 0.1$ (the bioassay being the RD_{50} value). The division into nonreactive and reactive VOCs based on $P^{bioassay} / P^o > 0.1$ or < 0.1 is arbitrary, but there will always be an arbitrary element in such a division—how much more reactive than expected does a VOC have to be before being assigned to the 'reactive' class? They later showed (41) that for 50 nonreactive VOCs, a reasonable correlation of $\log RD_{50}$ against $\log P^o$ was obtained:

$$\log RD_{50} = 0.844 \log P^o + 2.634. \tag{3}$$

$$N = 50, SD = 0.257, R^2 = 0.89.$$

In their analysis of the FBP equation, Abraham et al. (37) set out various models for the biological activity of gaseous compounds, and showed that experimental observations could be analyzed through the simple model illustrated in Figure 1. The VOC is transferred from the gas phase to a receptor phase or receptor area in step 1, and then the VOC interacts with the receptor in step 2. For nonreactive compounds, step 1 was the most important in upper respiratory tract irritation in mice (37), and subsequently was shown also to be the main step

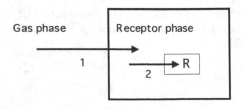

FIGURE 1 The two-step mechanism proposed by Abraham et al. (37).

in inhalation anesthesia (42). Since the main step involves the transfer of a VOC from the gas phase to a receptor phase, Abraham et al. (37) argued that it should be possible to discover a gas phase to solvent process that was a good model for stage 1. The appropriate physicochemical factor is K (sometimes denoted as L), the gas to solvent partition coefficient defined through Eq. [4]; K is then dimensionless.

$$K = \frac{\text{Concentration in solvent at equilibrium (mol/dm}^3)}{\text{Concentration in the gas phase at equilibrium (mol/dm}^3)} \tag{4}$$

It was then shown that the solvents tri(2-ethylhexyl)phosphate and N-formylmorpholine were good models for upper respiratory tract irritation in male Swiss OF_1 mice, wet octanol was not quite such a good model, and water was a very poor model (37,43).

Nielsen et al. (44) correlated log RD_{50} values against log K for solubility of gaseous compounds in wet octanol, but only for a restricted set of VOCs, and also used the water to octanol partition coefficient, as log P_{oct}, as a descriptor; other workers did likewise (45), again for a very restricted set of VOCs.

The first really general QSAR for RD_{50} values used a multiple linear regression analysis, MLRA to correlate $\log(1/RD_{50})$ against a number of independent variables, or descriptors (46), leading to Eq. [5]. In MLRA, a dependent variable such as $\log(1/RD_{50})$, in Eq. [5], is linearly correlated against two or more independent variables. The procedure is essentially the same as simple correlation of a dependent variable against one independent variable.

$$\log(1/RD_{50}) = 0.60 + 1.35S + 3.19A + 0.77L \tag{5}$$

$$N = 39, SD = 0.10, R^2 = 0.98.$$

The independent variables in Eq. [5] were S the VOC dipolarity/polarizability, A the VOC overall hydrogen bond acidity, and L the logarithm of the gas to hexadecane partition coefficient at 25°C. Note that in Eq. [5] the units of RD_{50} were mol/dm^3, so that the constant term is quite different to that in equations where RD_{50} is in ppm. The independent variables have more recently been discussed in depth (47). Eq. [5] is consistent with the analysis of Abraham et al. (37) of the biological activity of gases and their suggestion that solubility of gases in solvents could be a possible model, because equations on the lines of Eq. [5] have been shown to account for the solubility of gases in a variety of solvents (47).

Alarie et al. (41) expanded Eq. [5] to include 50 nonreactive VOCs as in Eq. [6], and also constructed a number of other, slightly less successful QSARs.

$$\log(1/RD_{50}) = -6.834 + 1.280S + 2.230A + 0.764L \tag{6}$$

$$N = 50, SD = 0.244, R^2 = 0.91$$

A slightly larger database led to Eq. [7] with somewhat less satisfactory statistics.

$$\log(1/RD_{50}) = -7.049 + 1.437S + 2.316A + 0.774L \tag{7}$$

$$N = 58, SD = 0.354, R^2 = 0.84$$

The most recent analysis of RD_{50} values is by Luan et al. (48), who used a number of methods to classify VOCs into reactive and nonreactive groups, and to

TABLE 2   The Independent Variables (Descriptors) Used by Luan et al.

Descriptor	Chemical meaning (48)
Re	Refractivity
RPCG	RPCG relative positive charge
IC_{ave}	Average information content (order 2)
CH_{donor}	Count of H-donor sites
RNSB	Relative number of single bonds
ZX	ZX shadow

correlate $\log(RD_{50})$ for 59 nonreactive VOCs. They start by calculating about 612 theoretical descriptors for each VOC, and reduce them to a pool of 166 descriptors. The "best" set of descriptors from the pool of 166 were chosen for a linear correlation of 47 of the nonreactive VOCs:

$$\log(RD_{50}) = 5.50 - 0.043Re - 6.329RPCG - 0.377IC_{ave}$$
$$-0.049CH_{donor} + 3.826RNSB - 0.047ZX \tag{8}$$

$$N = 47, SD = 0.362, R^2 = 0.844$$

The meaning of the six descriptors is given in Table 2. Although Eq. [8] is more complicated, and much more difficult to interpret in a chemical way, than Eqs. [6] and [7], it is statistically no more successful. Luan et al. (48) then used Eq. [8] to predict values for 12 nonreactive VOCs that had not been used to set up Eq. [8]. Table 3 gives the statistics for the predictions in terms of AE the average error, AAE the absolute average error, RMSE the root mean square error, and SD the standard deviation; this is the first time (48) that the predictive ability of any equation for RD_{50} has been assessed. In Table 3, this linear model is denoted as LM.

Luan et al. (48) also used two nonlinear methods to analyze 47 of the nonreactive compounds, one method is based on a support vector machine (SVM) and the other method is a radial basis function neural network (RBFNN). Details of the predictions for the 12 nonreactive VOCs used as a test set are in Table 3. Results are somewhat better than those from Eq. [8]. Unfortunately, it is not possible to compare the SVM and RBFNN methods with the linear Eq. [6] or [7] in any exact way, but the nonlinear methods do not seem to be very much better than the linear method in the correlation of $\log(RD_{50})$ values for nonreactive VOCs.

However, for the classification of 142 VOCs into a set of reactive and a set of nonreactive compounds, the nonlinear SVM method was much superior to linear methods. For 28 test compounds, the linear method correctly classified

TABLE 3   Statistics for the Prediction of $\log(RD_{50})$ for 12 Nonreactive VOCs (48)

Method	AE	AAE	RMSE	SD
LM	−0.148	0.358	0.507	0.529
SVM	−0.084	0.348	0.452	0.472
RBFNN	−0.081	0.346	0.480	0.501

75% of the VOCs, whereas the SVM method correctly classified 85.7%. This is the first time that VOCs have been classed as nonreactive or reactive without use of the FBP rule.

NASAL PUNGENCY THRESHOLDS IN MAN

Nasal pungency involves the transfer of a compound, for example a VOC, from an air stream through a mucus layer into a receptor or receptor area, and interaction of the VOC with the receptor. The model set out by Abraham et al. (37), Figure 1, thus seems appropriate for the analysis of NPTs. Abraham et al. (47) have devised a linear free energy relationship (LFER) and have applied it to the correlation of the solubility of gases in various solvents and biological phases. Since the first step of the model (Fig. 1) involves the solubility of VOCs in the receptor area, it seems logical to apply the LFER to values of NPTs. The LFER can be stated as in Eq. [9]:

$$SP = c + e \cdot E + s \cdot S + a \cdot A + b \cdot B + l \cdot L \tag{9}$$

In Eq. [9], E, S, A, B, and L are independent variables that are properties, or descriptors, of the VOC, and c, e, s, a, b, and l are regression coefficients determined by multiple linear regression. The equation has been described in detail previously (47). Briefly, E is the excess molar refraction, S is the dipolarity/polarizability, A and B are the overall or effective hydrogen bond acidity and basicity, respectively, of the VOC, and L is defined as the logarithm of the VOC gas–hexadecane partition coefficient at 298 K, which is a measure of the lipophilicity of the VOC. The regression coefficients in Eq. [9] are not merely fitted coefficients since they define the complementary physicochemical properties that characterize the receptor environment or biophase most receptive to the VOC (47). The dependent variable, SP, is either a physicochemical property of a VOC, such as log K where K is the gas to solvent partition coefficient for a series of VOCs into a given solvent or condensed phase; or a biological property of a VOC, such as an NPT for a series of VOCs, in the form of log(1/NPT). The reciprocal is used so that the larger is the log(1/NPT), the more potent is the VOC.

The values of NPT, or log(1/NPT), for a variety of VOCs have been determined by Cometto-Muñiz and Cain as reported in Chapter 13. When Eq. [9] was applied to NPTs as log(1/NPT), a very good correlation that accounted for 95% of the total effect was obtained (49) see Eq. [10]. The term in $e \cdot E$ was not significant and was left out. Eq. [10] strongly suggests that the factors that influence NPTs are those that influence the transfer of VOCs from the gas phase to condensed phases, that is from the gas phase to the receptor phase, and that VOC–receptor interactions are of secondary importance:

$$\log(1/NPT) = -8.562 + 2.209S + 3.417A + 1.535B + 0.865L \tag{10}$$

$$N = 34, R^2 = 0.953, SD = 0.27, F = 144$$

A recent equation, based on 47 VOCs, rather than the 34 VOCs in Eq. [10] is shown as Eq. [11]. Values of the descriptors and of the log(1/NPT) are in Table 4:

$$\log(1/NPT) = -7.770 + 1.543S + 3.296A + 0.876B + 0.816L \tag{11}$$

$$N = 47, R^2 = 0.901, SD = 0.312, F = 95, Q^2 = 0.874,$$
$$PRESS = 5.1901, PSD = 0.351.$$

TABLE 4 VOC Descriptors Used in Eq. [11]

VOC	E	S	A	B	L	SP[a]
Methanol	0.278	0.44	0.43	0.47	0.970	−4.54
Ethanol	0.246	0.42	0.37	0.48	1.485	−3.95
Propan-1-ol	0.236	0.42	0.37	0.48	2.031	−3.40
Propan-2-ol	0.212	0.36	0.33	0.56	1.764	−4.26
Butan-1-ol	0.224	0.42	0.37	0.48	2.601	−3.04
Butan-2-ol	0.217	0.36	0.33	0.56	2.338	−3.76
tert-Butyl alcohol	0.180	0.30	0.31	0.60	1.963	−4.52
Pentan-1-ol	0.219	0.42	0.37	0.48	3.106	−3.23
Hexan-1-ol	0.210	0.42	0.37	0.48	3.610	−2.60
Heptan-1-ol	0.211	0.42	0.37	0.48	4.115	−2.32
Heptan-4-ol	0.180	0.36	0.33	0.56	3.850	−2.53
Octan-1-ol	0.199	0.42	0.37	0.48	4.619	−1.85
Methyl acetate	0.142	0.64	0.00	0.45	1.911	−5.05
Ethyl acetate	0.106	0.62	0.00	0.45	2.314	−4.83
Propyl acetate	0.092	0.60	0.00	0.45	2.819	−4.24
Butyl acetate	0.071	0.60	0.00	0.45	3.353	−3.56
sec-Butyl acetate	0.044	0.57	0.00	0.47	3.054	−3.50
tert-Butyl acetate	0.025	0.54	0.00	0.47	2.802	−3.98
Pentyl acetate	0.067	0.60	0.00	0.45	3.844	−3.22
Hexyl acetate	0.056	0.60	0.00	0.45	4.290	−2.80
Heptyl acetate	0.050	0.60	0.00	0.45	4.796	−2.49
Octyl acetate	0.046	0.60	0.00	0.45	5.270	−1.95
Propanone	0.179	0.70	0.04	0.49	1.696	−5.12
Pentan-2-one	0.143	0.68	0.00	0.51	2.755	−3.47
Heptan-2-one	0.123	0.68	0.00	0.51	3.760	−2.91
Nonan-2-one	0.113	0.68	0.00	0.51	4.735	−2.53
Toluene	0.601	0.52	0.00	0.14	3.325	−4.47
Ethyl benzene	0.613	0.51	0.00	0.15	3.778	−4.00
Propyl benzene	0.604	0.50	0.00	0.15	4.230	−3.17
Butanal	0.187	0.65	0.00	0.45	2.270	−4.77
Pentanal	0.163	0.65	0.00	0.45	2.851	−4.57
Hexanal	0.146	0.65	0.00	0.45	3.357	−3.70
Heptanal	0.140	0.65	0.00	0.45	3.865	−3.13
Octanal	0.160	0.65	0.00	0.45	4.361	−3.24
Formic acid	0.343	0.75	0.76	0.33	1.545	−2.50
Butanoic acid	0.210	0.64	0.61	0.45	2.750	−1.79
Hexanoic acid	0.174	0.63	0.62	0.44	3.697	−1.30
Cumene	0.602	0.49	0.00	0.16	4.084	−3.22
p-Cymene	0.607	0.49	0.00	0.19	4.590	−3.05
Δ-3-Carene	0.511	0.22	0.00	0.10	4.649	−3.21
Linalool	0.398	0.55	0.20	0.67	4.794	−2.55
1,8-Cineole	0.383	0.33	0.00	0.76	4.688	−2.37
α-Terpinene	0.526	0.25	0.00	0.15	4.715	−3.30
Pyridine	0.631	0.84	0.00	0.52	3.022	−3.11
Menthol	0.400	0.50	0.23	0.58	5.177	−1.71
Oct-1-yne	0.155	0.22	0.09	0.10	3.521	−4.49
Chlorobenzene	0.718	0.65	0.00	0.07	3.657	−4.02

[a]log(1/NPT).

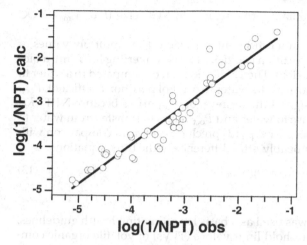

FIGURE 2   A plot of calculated values of log(1/NPT) on Eq. [11] against the observed values of log(1/NPT).

There are not enough data points in Table 4 to divide them into a training set and a test set. Hawkins (50) suggests that assessment of predictive ability using a test set are unreliable unless the set contains at least 50 data points—compare the test set of only 12 compounds used by Luan et al. (48) above.

It has been shown that statistics based on the leave-one-out method yield a reasonable assessment of predictive ability (51), even though based on the total data set. The leave-one-out statistics are Q^2, PRESS, and PSD the "predictive" standard deviation as explained in detail (51). In Eq. [11], PSD = 0.35 log units, and this can be taken as an assessment of the predictive capability of Eq. [11]. A plot of calculated values of log(1/NPT) on Eq. [11] against the observed values is shown in Figure 2; there is random scatter about the line of best fit.

A remarkable feature of the equations for log(1/NPT) is that they include carboxylic acids, aliphatic aldehydes, and the lower alkyl acetates, all of which are regarded as reactive compounds in the mouse bio-assay (18,41,43). The reasons why these VOCs behave as nonreactive compounds in the sensory irritation assay in man and as reactive compounds in upper respiratory tract irritation in mice are not clear.

There have been a few other QSARs constructed for the correlation of NPTs. Famini et al. (52) obtained the linear relationship, Eq. [12] for 42 VOCs, with a statistical fit about the same as Eq. [11]:

$$\log(1/NPT) = -16.40 + 2.258 V_{mc} + 3.471 q^- + 50.56 \varepsilon_\alpha + 13.63 q^+ \qquad (12)$$

$$N = 42,\ R^2 = 0.923,\ SD = 0.357,\ F = 110,\ Q^2 = 0.885$$

In Eq. [12], V_{mc} is a molecular volume, q^- is the absolute value of the most negative formal charge on the VOC (equivalent to the electrostatic basicity), ε_α is a covalent acidity term, and q^+ is the partial charge of the most positive hydrogen in the VOC (equivalent to the electrostatic acidity). Although the formalism differs from that used in Eq. [11], the terms in Eq. [12] are comparable to those in Eq. [11]; both equations contain terms related to VOC size (L or V_{mc}), VOC basicity and VOC acidity.

Hau et al. (53) adopt a rather different approach. They multiply values of NPT by K_w, the gas to water partition coefficient thus converting NPT into a bio-phase to water partition coefficient. Then NPT$\times K_w$ can be compared to a water to solvent partition coefficient such as the water to octanol partition coefficient, P_{oct}, as in Eq. [13]. The negative sign of the term in log P_{oct} arises because NPT$\times K_w$ refers to transfer from biophase to water and P_{oct} refers to transfer from water to octanol. The paucity of statistics for Eq. [13] precludes any firm comparison with Eq. [11] or [12], but there is probably little difference in the three equations:

$$\log(\text{NPT} \times K_w) = 7.69 - 1.16 \log P_{oct} \tag{13}$$

$$N = 33, R^2 = 0.943$$

Subsequently, Eq. [13] was used as a basis for estimating health guidelines, and it was suggested that threshold limit values (TLVs) of volatile organic compounds should not be more than 20% of a predicted NPT value (54).

CONCLUSIONS

The prediction of further values of sensory irritation from already known experimental data is an on-going and very important area of research. One major problem is that nearly all predictive methods are restricted to irritants that act through "physical" or "nonreactive" mechanisms, rather than through "chemical" or "reactive" mechanisms. A useful prerequisite to the setting up of an equation or algorithm for the correlation and then prediction of sensory irritation is therefore the ability to distinguish between the two types of irritants. Alarie et al. used the function $P^{bioassay}/P^o > 0.1$ to define nonreactive irritants in upper respiratory irritation in mice, and then used a linear method to correlate the bioassay end-point, as log RD_{50} against various properties or descriptors of the nonreactive irritants. Although various other workers have proposed equations for upper respiratory tract irritation in mice by nonreactive irritants, the equations set out by Alarie et al. are the most general. Eq. [6] or [7] with 50 and 58 irritants respectively seem to be the most useful equations yet proposed. Very little work has been carried out on predictions for reactive irritants, although Luan et al. use the definition of Alarie et al. to construct a method for the classification of irritants as nonreactive and reactive. Whether this is more useful than the criterion of Alarie et al. remains to be seen.

There have been a number of studies leading to equations for the correlation and prediction of NPTs in man. There is little difference in the statistics for goodness-of-fit, but Eq. [11], with 47 irritants, is the most general. It is of interest that in all of the studies on NPTs in man, no preformulated criterion for nonreactive or reactive irritants was set out. Reactive irritants are identified as outliers to the general equations for log(1/NPT). Thus, if an irritant is much more potent than expected from the general equation, say Eq. [11], it will be regarded as a reactive irritant.

The two methods of defining nonreactive and reactive irritants both involve assignment of some more-or-less arbitrary criterion. In the procedure of Alarie et al., the function $P^{bioassay}/P^o > 0.1$ is used to define nonreactive irritants, but 0.1 is a rather arbitrary criterion derived by inspection of the chemical nature of the total set of compounds. The identification of reactive compounds as outliers to the general equation also involves an arbitrary assignment based on the extent of deviation. For example, reactive irritants might be those that are more irritant than calculated by three standard deviations.

Although there is clear scope for the construction of equations for sensory irritation that are more general than those proposed to date, the restrictive factor is the lack of extra experimental data. The present equations, as set out here, are probably statistically as good as can be obtained, and only by incorporating further experimental results are they likely to be improved significantly.

ACKNOWLEDGMENTS

Preparation of this chapter was supported in part by grants number R01 DC 002741 and R01 DC 005003 from the National Institute on Deafness and Other Communication Disorders (NIDCD), National Institutes of Health (NIH).

REFERENCES

1. Parker GH. The relation of smell, taste and the common chemical sense in vertebrates. J Acad Nat Sci Phila 1912; 15: 221–234.
2. Keele CA. The common chemical sense and its receptors. Arch Int Pharmacodyn Ther 1962; 139: 547–557.
3. Bryant B, Silver WL. Chemesthesis: the common chemical senses. In Finger TE, Silver WL, Restropo D, eds The neurobiology of taste and smell, 2nd ed., New York, NY: Wiley-Liss, 2000:73–100.
4. Ressler KJ, Sullivan SL, Buck LB. A molecular dissection of special patterning in the olfactory system. Curr Opin Neurobiol 1994: 4:588–596.
5. Niimura Y, Nei M. Evolutionary dynamics of olfactory and other chemosensory receptor genes in vertebrates. J Hum Genet 2006; 51:505–517.
6. Firestein S. Olfaction: scents and sensibility. Curr Biol 1996; 6:666–667.
7. Mori K, Yoshihara Y. Molecular recognition and olfactory processing in the mammalian olfactory system. Prog Neurobiol 1995; 45:585–619.
8. Cometto-Muñiz JE, Cain WS. Thresholds for odor and nasal pungency. Physiol Behav 1990; 48:719–725.
9. Cometto-Muñiz JE, Cain WS. Trigeminal and olfactory sensitivity: comparison of modalities and methods of measurement. Int Arch Occup Envirion Health 1998; 71:105–110.
10. Kobal G, Van Toller S, Hummel T. Is there directional smelling? Experimentia 1989; 45:130–132.
11. Silver WL, Mason JR, Marshall DA, et al. Rat trigeminal olfactory and taste responses after capsaicin desensitization. Brain Res 1985; 333:45–54.
12. Cometto-Muñiz JE, Cain WS. Perception of odor and nasal pungency from a homologous series of volatile organic compounds. Indoor Air 1994; 4: 140–145.
13. Cometto-Muñiz JE, Cain WS, Abraham MH. Nasal pungency and odor of homologous aldehydes and carboxylic acids. Exp Brain Res 1998; 118:180–188.
14. Abraham MH, Kumarsingh R, Cometto-Muñiz JE, et al. Draize eye scores and eye irritation thresholds in man can be combined into one quantitative structure-activity relationship. Toxicol In Vitro 1998; 12: 403–408.
15. Cauna NK, Hinderer K, Wentges RT. Sensory receptor organs of the human nasal respiratory mucosa. Am J Anat 1969; 124:187–210.
16. Finger TE, St Jeor V, Kinnamon JC, et al. Ultrastructure of substance P- and CGRP-immunoreactive nerve fibres in the nasal epithelium of rodents. J Comp Neurol 1990; 294: 293–305.
17. Nielsen GD. Mechanisms of activation of the sensory irritant receptor by airborne chemicals. Crit Rev Toxicol 1991; 21:183–208.

18. Alarie Y, Schaper M, Nielsen GD, et al. Structure-activity relationships of volatile organic chemicals as sensory irritants. Arch Toxicol 1998; 72:125–140.
19. Cesare P, McNaughton P. Peripheral pain mechanisms. Curr Opin Neurobiol 1997; 7:493–499.
20. McCleskey EW, Gold MS. Ion channels of nociception. Ann Rev Physiol 1999; 61: 835–856.
21. Eccles R, Jawad MS, Morris S. The effects of oral administration of (-) menthol on nasal resistance to airflow and nasal sensation of airflow in subjects suffering from nasal congestion associated with the common cold. J Pharm Pharmacol 1990; 42: 652–654.
22. Jancso N, Jancso-Gabor A, Szolcsanyi J. Direct evidence for neurogenic inflammation and its prevention by denervation and by pre-treatment with capsaicin. Br J Pharmacol Chemother 1967; 31:138–151.
23. Alimohammadi H, Silver WL. Evidence for nicotinic acetylcholine receptors on nasal trigeminal nerve endings of the rat. Chem Senses 2000; 25:61–66.
24. Inoue T, Bryant BP. Multiple types of sensory neurons respond to irritating volatile organic compounds (VOCs): calcium fluorimetry of trigeminal ganglion neurons. Pain 2005; 117:193–203.
25. Schmelz M, Schmidt R, Bickel A, et al. Specific C-receptors for itch in human skin. J Neurosci 1997; 17:8003–8008.
26. Jord SE, Bautista DM, Chuang HH, et al. Mustard oils and cannabinoids excite sensory nerve fibres through the TRP channel ANKTM1. Nature 2004; 427:260–265.
27. Barrow C, Alarie Y, Warrick J, et al. Comparison of the sensory irritation response in mice to chlorine and hydrogen chloride. Arch Environ Health 1977; 32:68–76.
28. Alarie Y, Kane L, Barrow C. Sensory irritation: the use of an animal model to establish acceptable exposure to airborne chemical irritants. In: Reeves AL, ed. Toxicology: principles and practice, Vol 1. New York, NY: John Wiley and Sons, 1980: 48–92.
29. American Society for testing and materials (ASTM). Standard test method for estimating sensory irritancy of airborne chemicals. Designation E 981–84. Philadelphia, PA: ASTM, 1984.
30. Alarie Y. Bioassay for evaluating the potency of airborne sensory irritants and predicting acceptable levels of exposure in man. Food Cosmet Toxicol 1981; 19: 623–626.
31. Alarie Y. Dose-response analysis in animal studies: prediction of human responses. Environ Health Perspect 1981; 42:9–13.
32. Nielsen GD, Alarie Y. Sensory irritation, pulmonary irritation, and respiratory stimulation by airborne benzene and alkyl benzenes: prediction of safe industrial exposure levels and correlation with their thermodynamic properties. Toxicol Appl Pharmacol 1982; 65:459–477.
33. Nielsen GD, Wolkoff P, Alarie Y. Sensory irritation: risk assessment approaches. Regul Toxicol Pharmacol 2007; 48:6–18.
34. Kuwabara Y, Alexeeff GV, Broadwin R, et al. Evaluation and application of the RD_{50} for determining acceptable exposure levels of airborne sensory irritants for the general public. Environ Health Perspect 2007; 115:1609–1616.
35. Ferguson J. Use of chemical potentials as indexes of toxicity. Proc R Soc (London) 1939; B127:387–404.
36. Brink F, Pasternak JM. Thermodynamic analysis of the effectiveness of narcotics. J Cell Comp Physiol 1948; 32: 211–233.
37. Abraham MH, Nielsen GD, Alarie Y. The Ferguson principle and an analysis of biological activity of gases and vapors. J Pharm Sci 1994; 83:680–688.
38. Muller J, Gref G. Recherche de relations entre toxicite de molecules d'interet industriel et proprietes physico-chimiques: test d'irritation des voies aeriennes superieures appliqué a quatre familles chimiques. Food Chem Toxicol 1984; 22: 661–664.
39. Roberts DW. QSAR for upper-respiratory tract irritation. Chem Biol Interact 1986; 57:325–345.

40. Schaper M. Development of a data base for sensory irritants and its use in establishing occupational exposure limits. Am Ind Hyg Assoc J 1993; 54:488–544.
41. Alarie Y, Schaper M, Nielsen GD, Abraham MH. Estimating the sensory irritating potency of airborne nonreactive volatile organic chemicals and their mixtures. SAR QSAR Environ Res 1996; 5:151–165.
42. Abraham MH, Acree WE Jr., Mintz C, Payne S. Effect of anesthetic structure on inhalation anesthesia: implications for the mechanism. J Pharm Sci 2008; 97:2373–2384.
43. Alarie Y, Nielsen GD, Andonian-Haftvan J, Abraham MH. Physicochemical properties of nonreactive volatile organic compounds to estimate RD_{50}: alternatives to animal studies. Toxicol Appl Pharmacol 1995; 134:92–99.
44. Nielsen GD, Thomsen ES, Alarie Y. Sensory irritation receptor compartment properties. Acta Pharm Nord 1990; 1:31–44.
45. Gagnaire F, Azim S, Simon P, et al. Sensory and pulmomary irritation of aliphatic amines in mice: a structure-activity relationship. J Appl Toxicol 1993; 13:129–135.
46. Abraham MH, Whiting GS, Alarie Y, et al. Hydrogen bonding.12. A new QSAR for upper respiratory tract irritation by airborne chemicals in mice. Quant Struct Act Relat 1990; 9:6–10.
47. Abraham MH, Ibrahim A, Zissimos AM. The determination of sets of solute descriptors from chromatographic measurements. J Chromatogr A 2004; 1037:29–47.
48. Luan F, Ma W, Zhang X, et al. Quantitative structure-activity relationship models for prediction of sensory irritants (log RD_{50}) of volatile organic compounds. Chemosphere 2006; 63:1142–1153.
49. Abraham MH, Andonian-Haftvan J, Cometto-Muñiz JE, Cain WS. An analysis of nasal irritation thresholds using a new solvation equation. Fundam App Toxicol 1996; 31:71–76
50. Hawkins DM. The problem of overfitting. J Chem Inf Comput Sci 2004; 44:1–12.
51. Abraham MH, Acree WE Jr., Leo AJ, et al. Partition from water and from air into wet and dry ketones. New J Chem. 2009; 33:568–573
52. Famini GR, Aguiar D, Payne MA, et al. Using the theoretical linear solvation energy relationship to correlate and predict nasal pungency thresholds. J Mol Graphics Model 2002; 20:277–280.
53. Hau KM, Connell DW, Richardson BJ. Quantitative structure–activity relationship for nasal pungency thresholds of volatile organic compounds. Toxicol Sci 1999; 47:91–98.
54. Hau KM, Connell DW, Richardson BJ. Use of partition models in setting health guidelines for volatile organic compounds. Regul Toxicol Pharmacol 2000; 31: 22–29.

26 Effect of Allergic Inflammation on Irritant Responsiveness in the Upper Airways

Thomas E. Taylor-Clark* and Bradley J. Undem

Division of Allergy & Clinical Immunology, Johns Hopkins School of Medicine, Baltimore, Maryland, U.S.A.

INTRODUCTION

For those suffering with allergy to aeroallergens, it is self-evidently clear that the inhalation of allergen (pollen, cat dander, etc.) directly leads to symptoms that are best explained by nerve activation (itchy sensations, sneezing, coughing increased secretions, etc.). The simplest explanation for this is the allergen activation of tissue mast cells resulting in the release of neuroactive mediators. It is also evident that a subset of allergic subjects becomes hypersensitive to irritants other than the allergens that may include particulate matter, excessive temperature, anosmotic solutions, acid, smoke, volatile gases, perfumes, pollutants, etc.

This notion of a "neurally hypersensitive state" is not only based on anecdotal information commonly obtained from patient complaints, but has also found support in several clinical research studies. For example, a seasonally allergic subject has been shown to have an exaggerated neuronal response (sneezing, parasympathetic reflex secretion) to a given amount of sensory stimulus when studied "in season" as compared to "out of season." Such sensory hyperresponsiveness with allergic rhinitis has been confirmed in controlled human trials using various irritants including carbon dioxide, capsaicin, bradykinin, hypertonic saline, histamine, n-propanol, and Cl_2 (1–7). In addition, epidemiological studies have also shown that nasal allergies correlate with high incidence of irritant hyperrresponsiveness (8,9) and chronic cough (10,11). The data therefore indicate that for many, allergen provocation not only leads to the production of neuroactive mediators, but also over time leads to a state of nonspecific neuronal hypersensitivity. In this review, we attempt to summarize some of the basic physiological information that may subserve this allergen-induced neuronal activation and hypersensitivity.

The airways can be divided into three sections: the nasal cavity (a highly vascularized tissue lacking contractile smooth muscle), the large airways including trachea and bronchi (containing contractile smooth muscle), and the parenchymal compartment including the alveoli and respiratory bronchioles (the site of gas exchange, but lacking contractile smooth muscle). The present discussion focuses on the *upper airways* defined here as the nasal cavity, trachea, and main bronchi (airways that are not directly involved with gas exchange).

Current affiliation: Department of Molecular Pharmacology and Physiology, University of South Florida, Tampa, Florida, U.S.A.

SENSORY INNERVATION OF UPPER AIRWAYS

Airway defensive responses to noxious or potential noxious stimuli are mediated by a large range of peripheral cell types and tissues (including smooth muscle, goblet cells, skeletal muscle, endothelial cells, and the autonomic nervous system). Their initiation, however, is largely dependent on the function of one cell type—afferent sensory nerves innervating the airways.

In relative terms, sensory nerves are very long cells that are able to conduct information, encoded as all-or-nothing action potentials, between organs and the central nervous system. Sensory (afferent) nerves, including those that innervate the airways, consist of a peripheral terminal (which can have a wide range of arborization) that feeds into a single axon that extends to the neuronal cell body (housed in discrete sensory ganglia) and beyond to the central terminal in the CNS. Sensory nerves are commonly defined by the ganglion in which the cell body is located and by the degree of myelination/conduction velocity of the projected fiber. Myelinated afferent nerves conduct action potentials at relatively high velocities (5–50 m/s) and are referred to as A-fibers. Unmyelinated afferent nerves conduct action potentials slowly (<2 m/s) and are referred to as C-fibers. In airways, the A-fibers are often "physiological sensors" in that they respond to changes in the physiological environment (e.g., low threshold mechanosensors). The C-fiber populations respond vigorously to "irritants" and potentially noxious stimuli. As such they are referred to as nociceptors. Activation of nociceptors leads to pain sensations in the somatosensory system. In visceral tissues, activation of nociceptors may or may not lead to pain, but often leads to reflexes aimed at limiting the potential tissue damage. In the respiratory tract, these "defensive" reflexes might include sneezing, coughing, reflex-mediated airway narrowing, and increases in secretions.

Anatomy

The nasal mucosa is innervated by olfactory afferent nerves (detection of odors) and trigeminal afferent nerves (detection of airflow, temperature, touch, osmolarity, pH, and chemicals). Afferent nerves, derived from the maxillial and ophthalmic branches of the trigeminal ganglion, project nerve fibers that terminate in the nasal epithelium, blood vessels, and submucosal glands. There are only a handful of studies that have attempted a characterization of nasal afferent nerves based on conduction velocities, but it is probably that both A-fiber and C-fiber populations are present and that the majority are nociceptive C-fibers (12–14). Trigeminal afferent nerves project centrally to the subnucleus caudalis and subnucleus interpolaris of the medulla in the brain stem.

The pharyngeal mucous membrane beneath the nasopharynx is innervated by the glossopharyngeal nerve. The sensory innervation of the pharynx is relatively sketchy, but low threshold rapidly and slowly adapting nerves have been identified (15). It is also quite likely that this tissue is also innervated by glossopharyngeal nociceptive C-fibers, that can transmit signals leading to pain, that is, "sore throat" (16).

The major sensory afferent innervation of the larynx, trachea, and main bronchi is derived from the vagus nerves, which have their cell bodies situated in either the vagal nodose ganglion or the vagal jugular ganglion. Bronchopulmonary tissue is also innervated to a lesser extent by afferent nerves derived from spinal dorsal root ganglia (sometimes referred to as sympathetic afferents because they follow nerve tracts of the sympathetic autonomic system). Vagal afferent nerves innervating the airways can be broadly classified into three groups based on conduction velocity (17). First, fast conducting myelinated A-fibers typically have a low threshold for mechanical stimuli (e.g., stretch or punctate) and are relatively unresponsive to most inflammatory mediators (although if these substances cause mechanical changes through vasodilation or bronchoconstriction, this can indirectly activate low threshold mechanosensitive A-fibers). These low threshold mechanosensitive nerves have been further subdivided into rapidly and slowly adapting receptors (SARs and RARs) based on the rate of adaptation to prolonged lung distension. Second, slow conducting unmyelinated C-fibers, making up 80% of the vagal afferent supply, typically respond to a variety of mechanical, osmotic, temperature, and chemical (including inflammatory mediators such as bradykinin) stimuli (18). Structures throughout the lung are innervated by these afferent nerves, such as the bronchial and pulmonary vasculature, the airway epithelium, the airway smooth muscle, the submucosal glands, and the intrinsic parasympathetic ganglia. Lastly, at least in guinea pigs, there is strong evidence that a relatively slow conducting myelinated Aδ-fiber ($\sim$2–5 m/s) innervates the larynx, trachea, and main bronchi (19). These fibers are insensitive to stretch but are exquisitely sensitive to punctate mechanical stimulation. They are also sensitive to anosmotic solution and to rapid reductions in pH, but are insensitive to temperature and inflammatory mediators. This particular fiber-type leads to coughing when stimulated (20). All vagal afferent nerves project centrally to overlapping areas in the nucleus tractus solitarious (nTS).

Most airway afferent nerves contain the excitatory neurotransmitter glutamate and synaptic transmission between airway afferent fibers, and the second-order neurons in the brain stem is largely dependent on this transmitter. Activation of brain stem neurons ultimately leads to the stimulation or inhibition of preganglionic parasympathetic nerves (central reflex regulation), as well as the activation of thalamic and higher centers connected with sensation, emotion, and behavior. In many mammalian species, a substantial population of trigeminal and vagal C-fibers innervating the airways contain tachykinins, such as substance P and neurokinin A, which are synthesized in the cell body and transported along the axon to the peripheral and central terminals (18,21–23). Studies have shown that centrally acting inhibitors of neurokinin receptors (NK1, NK2, and NK3) reduce central reflexes evoked by the activation of airway C-fibers. These reports suggest that in the brain stem, tachykinins can play an important excitatory role in synaptic transmission, although this has not yet been demonstrated electrophysiologically (24).

Activation of airway tachykinergic C-fibers can also result in the local peripheral release of tachykinins, a process termed the "local axon reflex." Depending on the species, tachykinin release onto smooth muscle, bronchial and pulmonary circulation, submucosal glands, and airway epithelium occurs through afferent nerve collaterals. The net effect of locally released neuropeptides is plasma extravasation and an inflammatory response that collectively are referred to as "neurogenic inflammation." Neurogenic inflammatory responses have been shown to occur in human nasal mucosa, especially in allergic subjects (25). In addition, there is substantial collateral tachykinergic innervation of parasympathetic ganglionic neurons. Stimulation of these nerves leads to neurokinin-dependent depolarization and activation of bronchial parasympathetic neurons (26). This is sometimes referred to as a "peripheral reflex" (i.e., an afferent-parasympathetic reflex that is not dependent on transmission through the CNS). Other vasoactive peptides have been demonstrated in subsets of airway C-fibers, such as calcitonin gene–related peptide, vasoactive intestinal peptide, and neuropeptide Y, but little is known about their contribution to neurogenic inflammatory responses.

Activation Mechanisms of Sensory Nerves

Afferent nerve fibers innervating the airways project terminals to specific tissue compartments depending on their modality. This allows for the efficient detection of specific factors within the local environment (such as stretch, pH, temperature, inflammatory mediators). Regardless of the stimuli, afferent nerve activation at the peripheral terminal occurs following the opening of specific ion channels, which causes ion fluxes at the nerve membrane leading to a graded membrane depolarization referred to as a "generator potential." If the generator potential is great enough, it surpasses the threshold of voltage-gated sodium channels leading to the initiation of action potentials that then travel in an all-or-nothing fashion along the nerve fiber to the terminals in the CNS (and in some cases antidromically back along collateral branches in an axon reflex as discussed above).

The gating of peripheral terminal ion channels that is required for afferent nerve activation can occur through ionotropic and metabotropic mechanisms (27). The ionotropic mechanism refers to an ion channel that has a self-contained activation/binding site for a specific stimulus. For example, the TRPV1 channel is an ionotropic receptor for capsaicin, the pungent ingredient in chili peppers, and has a separate site that is activated by heat ($>43°C$). Certain ion channels can also be gated indirectly through the actions of second messenger systems (metabotropic), typically following the activation of G-protein–coupled receptors (GPCRs).

Mechanical Activation

Most airway sensory nerves are reproducibly responsive to some sort of mechanical stimulation. For example, stretch, which occurs during eupneic respiration, activates SAR and RAR Aβ-fibers in the lower airways, whereas punctate perturbation activates Aδ-fibers in the trachea and Aβ- and C-fibers throughout the airways. The molecular identity(s) of the ion channel(s) responsible for mechanical-induced nerve activation are not known. Indeed, there is some debate whether mechanical stimulation directly gates the unknown ion channel (physical hypothesis), or that nonneuronal mechanosensitive cells release paracrine mediators in response to mechanical stimulation that activate an ion channel on the nerve terminal (chemical hypothesis). Studies of mechanotransduction in nonmammalian species have demonstrated mechanosensitive ion channels that have high sequence homology with mammalian ion channel superfamilies such as the epithelial sodium channels (ENaCs) and the transient receptor potential (TRP) channels. So far, however, no definitive candidates have been identified. The "chemical hypothesis," alternatively, could help indicate a possible function of neuroepithelial bodies, structures that have been shown in immunohistochemical studies to surround populations of lower airway afferent terminals (28,29). Mechanical perturbation of neuroepithelial bodies in the airways and similar cell types in other visceral tissues has been shown to release autacoids such as 5-HT and ATP (30,31), which can directly gate specific ion channels on airway afferents (see below). However, many airway nerves are not associated with neuroepithelial bodies, and some mechanosensitive afferents, for example the Aδ-fibers in the trachea, are insensitive to these autacoids (19,20).

Chemical Activation

Many chemical mediators delivered to the airways can induce airway afferent action potential discharge. However, this can frequently be due to indirect actions of the mediators. For example, mediators that induce bronchospasm, edema, and changes in lung compliance can result in the activation of airway afferents due to mechanical changes in the airway tissue. In addition, stimuli that cause the release of neuroactive mediators (e.g., bradykinin, ATP) may induce nerve activation indirectly through these autacoids. The following list summarizes those ion channels expressed on airway afferent nerves that are known to play a direct role in the chemical initiation of action potentials.

TRPV1

A member of the TRP channel superfamily, TRPV1 has been shown on almost all C-fibers innervating the upper and lower airways (14,32). TRPV1 is a nonselective ion channel that is activated directly by capsaicin and protons, leading to robust action potential discharge in these neurons (33,34). TRPV1 is also activated by certain arachidonic acid metabolites (35) and as such may play a critical role in the initiation of action potentials downstream of various GPCRs, including bradykinin B2 receptor (34), PAR2 (36), and histamine H1 receptor (37).

TRPA1

Like TRPV1, TRPA1 is a nonselective ion channel that is expressed on airway C-fibers (38). TRPA1 channels are not activated by capsaicin, but are strongly activated by allyl isothiocyanate, the active ingredient in mustard oil, wasabi, and horseradish (39). Tasting a small sample of these substances leaves little doubt that they are effective activator of upper airway nociceptors. TRPA1 channels are directly activated by a large range of irritants (40,41), due to their sensitivity to reactive electrophilic moieties (42). Thus, TRPA1 channels are crucial to the activation of airway C-fibers by exogenous reactive compounds such as acrolein, industrial diisocyanates, and formaldehyde, as well as by endogenously produced inflammatory mediators including reactive oxygen species (including hydrogen peroxide and hypochlorite) and unsaturated aldehydes (e.g., 4-hydroxynonenal and 4-oxononenal). TRPA1 has also been implicated in activation of nociceptors by cigarette smoke extracts (43).

TRPM8

Although there is little evidence to suggest that TRPM8 channels are expressed on lower airway afferents, TRPM8 is expressed on a subset of nasal trigeminal neurons that also express TRPV1 (C-fibers) (44). TRPM8 can be gated by cold temperature, and chemically by menthol (45). It is therefore possible that TRPM8 may be involved in the cooling sensation felt in the upper airway upon inhalation of menthol and related chemicals.

Purinergic Receptors

ATP can act on purinergic ionotropic receptors referred to as P2X receptors (P2Y receptors are the purinergic metabotropic GPCRs). P2X channels display a wide variety of biophysical and pharmacological properties based on their subunit composition (46). In a subset of airway nerves, $P2X_2$ and $P2X_3$ subunits are coexpressed to form the $P2X_{2/3}$ heteromer (47). Activation of these receptors leads to large depolarizing currents and action potential discharge in both C-fibers and A-fibers in the airways (18).

5-HT Receptors

5-HT effectively stimulates action potential discharge in subsets of bronchopulmonary C-fibers (32,48). The receptor most often involved in this is the 5-HT3 receptor. With the exception of 5-HT3, all other 5-HT receptors are GPCRs. The 5-HT3 receptor is an ionotropic receptor, with the ion channel serving as a nonselective cation channel.

Acetylcholine Nicotinic Receptors

Inhalation of cigarette smoke, in naïve smokers, causes intense airway irritation and cough. Experimental studies support the hypothesis that these sensations are dependent on activation of ionotropic nicotinic cholinergic receptors on vagal nociceptors in the airway epithelium (49).

Un-identified Receptors

The use of selective antagonists and receptor knock out mice has identified the molecular mechanisms by which many compounds activate sensory nerves. However, this is not the case for all irritants, particularly with respect to airway physiology. A wide array of alcohols, ketones, aldehydes, and other hydrocarbons [termed loosely as volatile organic compounds (VOCs)] cause sensory irritation upon exposure to the human airways. These compounds likely directly modulate sensory nerves, as they cause an influx of calcium in dissociated trigeminal neurons (50). Those VOCs that are reactive (electrophilic) activate airway sensory afferents through TRPA1 (41), but the majority of VOCs activate neurons through unknown mechanisms. Indeed, given that VOC sensitivity seems to be distributed across different modalities of nerve subtypes, it is plausible that some VOC-induced activation is through nonspecific mechanisms.

G-Protein–Coupled Receptors (GPCRs)

In general, agonists of GPCRs linked to Gs, Gi, or Gq lead to phosphorylation of various ion channels that lead to changes in the excitability of afferent nerve membranes, rather than overt activation. A couple of notable exceptions to this are adenosine and bradykinin. Adenosine, a mediator known to cause dyspneic sensations in humans (51), strongly activates pulmonary C-fibers in various species (52,53). This involves activation of adenosine A1 and in some cases also A2A receptors subtypes. Likewise bradykinin, another autacoid that has at least circumstantially been implicated in dyspneic sensations (54), overtly activates bronchopulmonary C-fibers (32,55). This in effect is mediated through the Gq-coupled B2 receptor subtype. The ion channels involved in the generator potential initiated by signaling through GPCRs have not been worked out in all cases. Evidence is accumulating however, that certain TRP channels including TRPV1 and TRPA1 and chloride channels may be involved in the transduction process (34,35,56,57).

Temperature Activation

The effect of temperature on airway afferents has not been as extensively studied as have cutaneous temperature sensors. Cutaneous afferents are activated by absolute temperatures, not by the rate of temperature change, and some of the channels identified as temperature-sensitive in somatosensory systems are present on airway afferents. For example, subsets of airway C-fibers express the TRPV1, TRPM8, and TRPA1 receptors (38,44), which have been shown to be directly activated by noxious heat (>43°C), cold (<23°C), and noxious cold (<15°C), respectively (33,39,45). It is likely that these channels play crucial roles in temperature-induced activation of airway afferents.

Osmolarity Changes

Airway afferents can be activated by both hypo- and hyperosmotic solutions (58). A proportion of C-fibers and tracheal Aδ-fibers (cough fibers) are activated by a specific decrease in Cl-ions (59,60), but other afferents seem to respond to overall changes in tonicity (61). Hyperosmolar solutions, however, appear to stimulate capsaicin-sensitive nociceptors more than tracheal Aδ-fibers (62). It is uncertain whether changes in osmolarity induce afferent activation through specific osmolarity sensing ion channels or through mechanical forces caused by swelling or shrinking of airway tissue. Recent work has shown that TRPV4 and a variant of TRPV1 are possible sensors of osmolarity (63,64); whether these channels contribute to airway afferent osmolarity detection is unknown.

AUTONOMIC INNERVATION

Nasal Mucosa

In general, the effectors of the central and peripheral reflexes are autonomic nerves: there is substantial parasympathetic and sympathetic innervation of the entire airways. Again, like the afferent innervation, the efferent innervation of the nasal mucosa is derived from a separate source to the rest of the airways. With respect to the nasal mucosa, parasympathetic preganglionic nerve fibers, originating in the lacrimo-nasopalatine nucleus in the brain stem, synapse onto postganglionic fibers in the spenopalatine ganglion. The postganglionic fibers are carried through the vidian nerve to the blood vessels and submucosal glands in the nasal mucosa. These fibers contain acetylcholine, nitric oxide, and vasoactive intestinal peptide, and activation causes glandular hypersecretion and increased vascular permeability (combining to produce rhinorrhea) and, arguably, minor vasodilation of sinusoids (nasal congestion) (65–68).

Sympathetic preganglionic nerve fibers, originating in the thoracolumbar region of the spinal cord, synapse onto postganglionic neurons in the superior cervical ganglion. These adrenergic nerve fibers, some of which also contain neuropeptide Y, are carried by the vidian nerve to innervate the nasal mucosal blood vessels (including sinusoids), where their activation causes vasoconstriction (nasal decongestion). Nasal blood vessels are under basal sympathetic tone, which can be altered by inflammatory mediators (69,70). Surgical section of sympathetic nerves and α-adrenoceptor antagonists cause nasal congestion indicating that basally released noradrenaline maintains a degree of vasoconstriction (71,72).

Trachea/Bronchi

With respect to the conducting airways, parasympathetic preganglionic nerve fibers, originating in the area of the nucleus ambiguous and the dorsal motor nucleus in the brain stem, are carried by the vagus nerve into the trachea, main bronchi, and lungs to discrete parasympathetic ganglia, where they synapse with intrinsic postganglionic neurons. The intrinsic ganglia filter much of the preganglionic fiber activity, and this filtering can be inhibited by inflammatory mediators within the airways (73). Postganglionic parasympathetic neurons can be subcategorized by their neurotransmitter content: cholinergic fibers that release acetylcholine and noncholinergic nonadrenergic (NANC) fibers that release nitric oxide (NO) and vasoactive intestinal peptide (VIP). Both cholinergic and NANC parasympathetic fibers innervate airway smooth muscle, bronchial and pulmonary blood vessels, and submucosal glands (74–77). Activation of cholinergic nerves causes bronchoconstriction, glandular hypersecretion, and bronchial and pulmonary vasodilation. Activation of NANC parasympathetic nerves causes bronchodilation (i.e., only observable under conditions of atropine cholinergic blockade) and mucus secretion.

Sympathetic preganglionic nerve fibers, originating in the spinal cord, synapse onto postganglionic neurons in the stellate and superior cervical ganglion. The postganglionic sympathetic neurons frequently contain noradrenaline and neuropeptide Y as their primary transmitter. These nerves innervate bronchial and pulmonary blood vessels and submucosal glands, and their activation causes bronchial and pulmonary vasoconstriction and glandular hypersecretion (74,75,77). In man there is very little sympathetic innervation of airway smooth muscle. Although bronchial smooth muscle expresses β-adrenoceptors, these receptors are activated primarily by circulating epinephrine (or therapeutic bronchodilators) (76).

FUNCTIONAL CONSEQUENCE OF NOCICEPTOR ACTIVATION

The activation of nociceptive nerves innervating the airways leads to aversive sensations (e.g., itch, dyspnea, and urge to cough) and the initiation of defensive reflexes (cough, sneeze, secretion, bronchoconstriction, and vasodilation) (78). Defensive reflexes include those initiated locally through neuropeptide release from afferent peripheral terminals (local axon reflex); those mediated through the direct activation of peripheral parasympathetic ganglionic nerves (peripheral reflex); and those mediated through the activation of central synapses in the brain stem (central reflex) leading to the stimulation of efferent autonomic nerves (see above). For many reflexes, the contribution of each of these pathways depends on the nociceptive sensory nerve activated, the tissue the nerve innervates, and the species. Indeed even a single effector pathway can have opposing afferent inputs. An example of this complexity is the nociceptor-central reflex modulation of respiratory rate in anesthetized guinea pigs: the TRPV1 agonist capsaicin (C-fiber stimulant) causes apnea and a decrease in breathing rate when applied to the nasal mucosa, very little effect when applied to the trachea, an increase in breathing rate when capsaicin is inhaled, and a multiphasic response consisting of an abrupt drop in breathing rate followed by tachypnea followed by apnea followed by a slow return to basal with i.v. injection (79).

Sneeze

A sneeze is an expulsive reflex that serves to remove unwanted substances (debris and chemical irritants dissolved in secretions) from the nasal cavity. It requires a complex control of lower airway smooth muscle tone, glottis closure, and breathing pattern. Sneezing can occasionally be induced by mechanical stimulation of the nasal mucosa and by chemicals such as ammonia, capsaicin, 5-HT, and bradykinin (80–83). Histamine seems to consistently evoke sneeze (66,84). It is not known what specific trigeminal afferent nerves are responsible for the initiation of sneeze, but it is likely that C-fibers play a predominant role (85).

Cough

The cough reflex involves a complex coordination of breathing pattern and smooth muscle tone. Tussive stimuli include inhalation of particulates, hypo- and hypertonic solutions, sulfur dioxide, acid, capsaicin, and various autacoids in addition to mechanical punctate stimulation of the bronchi and trachea (86,87). These chemicals can initiate cough directly, by activating lung afferents, or indirectly, following induced mechanical changes such as mucus secretion or edema. In experimental animals, cough can be directly induced through two distinct afferent pathways: through the punctate-mechanical and pH activation of the vagal nodose Aδ-fibers, as well as through the polymodal activation of vagal C-fibers (20). Cough is universally abolished by vagotomy and vagal cooling, whereas anesthesia only reduces cough initiated by lung C-fibers.

Experimental evidence suggests that the cough reflex can also be augmented (likely through modulation of central synaptic transmission) by the stimulation of C-fibers innervating both the lower airway and the nasal mucosa. For example, bradykinin infused into the lungs of cannulated anesthetized guinea pigs augmented the cough elicited from citric acid–induced activation of tracheal Aδ-fibers (88), and capsaicin instillation into the nasal mucosa of anesthetized guinea pigs and cats augmented the cough elicited from mechanical stimulation of tracheal Aδ-fibers (89).

Dyspnea

Several lines of evidence support the hypothesis that dyspnea may be brought about or at least influenced by perturbations in vagal sensory nerve activity. First, electrical stimulation of the vagus nerves, using stimulation paradigms that do not cause bronchoconstriction or changes in heart rate, can lead to dyspnea in humans (90); and both bilateral block of action potential conduction in the cervical vagus and lidocaine inhalation effectively reduces dyspnea in patients with various airway diseases including asthma (91,92). Second, inhalation of known direct C-fiber activators such as histamine, adenosine, and ATP lead to a more profound dyspnea than methacholine inhalation despite similar or even less degrees of bronchoconstriction (93,94). In fact in some studies adenosine has been found to evoke dyspneic sensations even in the absence of increases in airway resistance (51,95). In addition, inhalation of prostaglandin E2, a bronchodilating mediator that is commonly associated with increasing sensory nerve excitability, exacerbates the dyspnea associated with exercise, in the absence of increases in bronchoconstriction (96). Finally, evidence for a role of vagal afferent nerves in dyspnea can be found in the results from studies with inhaled furosemide. Furosemide reduces dyspneic sensations leading to a significant increase in the duration of a voluntary breath hold (97,98). The mechanism for this effect is not clear, but it likely involves the modulation by furosemide of vagal afferent nerve activity (59). Confusing the issue are findings that furosemide can inhibit vagal C-fiber activity, but may also increase the activity of low threshold mechanosensitive SAR fibers (99).

As mentioned above, the process of smooth muscle contraction and airway narrowing will lead to changes in activity in the low threshold mechanosensors in the airways. It is not known whether the amount of dyspnea per increase in airway resistance would be decreased or increased if the bronchoconstriction would occur independently of activity in the vagal mechanosensors.

NEUROMODULATION BY ALLERGIC INFLAMMATION

In atopic individuals that have been sensitized to aeroallergens, exposure to the specific allergen leads to the cross-linking of the IgE, followed by mast cell activation and degranulation. This results in the release of a wide range of autacoids and cytokines that can then modulate the function of many cell types in the upper airways including secretory cells, vascular endothelium, and bronchial smooth muscle. In addition, many of these mediators can directly interact with nerves within the upper airways inducing action potential discharge (*activation*), by augmenting nerve excitability (*modulation*) and by inducing changes in protein synthesis (*plasticity*).

Afferent Neuromodulation

Activation

Based on the evidence present above regarding the pharmacology and physiology of sensory afferent nerves innervating the airways, one would predict that allergen would induce the activation of afferent airway nerves. This has been shown repeatedly for nasal trigeminal afferents, with allergen inducing sneeze and nasal itch, and central reflex-dependent increases in parasympathetic-mediated hypersecretion (100). This response is largely blocked by histamine H1 receptor antagonists and is mimicked by histamine nasal challenge (84). Electrophysiological studies on guinea pig nasal trigeminal neurons demonstrate that histamine directly activates nasal C-fibers through H1 receptor activation (14,22), and, based on studies in other tissues, this is likely mediated in part by the gating of TRPV1 channels (37). Evidence that allergen-induced mediator release causes the direct activation of afferent nerves in the large airways and lung parenchyma is harder to interpret, given that substantial mechanical changes (e.g., bronchoconstriction) may occur that can indirectly cause afferent activation. Nevertheless, allergen directly evokes central-reflex bronchoconstriction (101,102), as well as evoking cough (103) that is largely resistant to bronchodilation, thus it is likely that a direct activation of airway afferents occurs during allergen challenge. In general, the electrophysiological data support the reflex data: allergen activates rat bronchopulmonary C-fibers (104); and rabbit RAR fibers (105).

Excitability Changes

The threshold of airway afferent nerves to various stimuli is not static, and allergen (through the release of inflammatory mediators) can, independent of overtly initiating action potential discharge, increase the excitability of certain airways nerves to certain stimuli. There are multiple mechanisms by which nerve excitability can be modulated (27,106), but all result in greater afferent action potential discharge and induction of reflexes for a given degree of activating stimulus. For example, cysteinyl leukotrienes (e.g., LTD4) do not overtly activate guinea pig nasal nociceptors, but substantially augment their response to other activating stimuli including histamine (107). Similar increases in excitability of uncharacterized trigeminal nociceptive neurons have been observed with 5-HT, IL-1β, and PGE2 (108–110).

In clinical studies, challenge of the nasal mucosa with nociceptive stimuli such as capsaicin, histamine, bradykinin, and hypertonic saline produces greater reflex hypersecretion in allergic rhinitis subjects, although methacholine, which directly stimulates the nasal glands, produces the same response in normal subjects and rhinitics (2–4). In addition, the sneeze response to a given dose of histamine is increased in rhinitics.

The potential for allergen-induced excitability changes in vagal afferent nerves innervating the trachea/bronchi has been demonstrated with increased cough responses to capsaicin (111) and acid (112), and increased reflex-bronchoconstriction to bradykinin (113). Data from electrophysiological studies support these observations. Guinea pig tracheal Aδ fibers are sensitive to mechanical punctate stimuli and initiate cough once activated (see above). Exposure of sensitized trachea to allergen dramatically decreases the Aδ-fiber threshold for punctate stimuli (114) (Fig. 1). This increase in neuronal excitability was not prevented by pyrilamine (H1 antagonist) or indomethacin (COX inhibitor) or zileuton (LOX inhibitor). In vivo recordings of rat pulmonary C-fibers have shown that exposure of sensitized rats increases the action potential discharge to a given dose of right atrial injections of capsaicin (104). These allergen-induced augmentations of nerve excitability are likely mediated by direct and acute interactions of mast cell–derived mediators and the sensory nerves (115). Similar hyperexcitable states of vagal airway C-fiber activity have been demonstrated following challenge with cysteinyl leukotrienes (116), adenosine (117), and PGE2 (118,119).

Plasticity

The activation and augmented excitability of airway sensory nerves described above focuses on acute interactions of allergic mediators with neuronal ion channels and nerve terminal proteins, and as such these mechanisms would not be expected to long outlast the presence of the inflammatory stimulus. Another important type of allergen-induced neuromodulation is where the allergic inflammation leads to long-term changes in nerve function; changes that may long outlive exposure. These long-lasting events are referred to as neuroplastic changes or simply "neuroplasticity." The bulk of the evidence thus far obtained demonstrates changes in transcription/translation of genes and the resulting changes of ion channel or neurotransmitter expression, but it is conceivable that physical changes in dendrite and axonal structure may also occur resulting in altered synaptic transmission. Large families of neurotrophin molecules, the

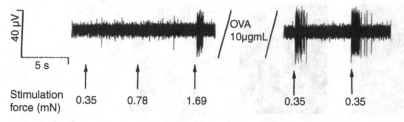

FIGURE 1 Representative trace of the effects of antigen challenge (10 μg/mL ovalbumin) on the mechanical threshold of a single afferent nerve ending from sensitized guinea-pig large airways. Large airways and extrinsic innervation (including vagal ganglia) were dissected out and placed in a custom-made dish perfused with Kreb's solution (37°C). A glass extracellular recording electrode was micropositioned within the vagal nodose ganglion such that mechanical or electrical stimulation of the large airways elicited action potentials from a single afferent fiber. Mechanical punctate force was applied using von Frey fibers. Action potential responses were recorded to increasing force of mechanical stimulation prior to and following ovalbumin was added to the Kreb's perfusion. *Source*: Adapted from Ref. 114.

prototypical example being nerve growth factor, are common mediators of sensory and autonomic neuroplastic changes. That allergic inflammation can lead to neuroplasticity might be expected given the observation allergic inflammation is commonly associated with elevation of neurotrophic molecules including nerve growth factor (NGF) (3,120–124).

Perhaps one of the more dramatic changes that occurs during inflammation is the induction of neuropeptide content in vagal and trigeminal nerves that typically do not contain these mediators. Inferior turbinate biopsies from rhinitics indicate higher substance P content in trigeminal nasal nerves compared to normals subjects (125). Although the identity of these nerves has yet to be described, exposure of mice to NGF causes an increase in substance P expression in large diameter, neurofilament positive neurons (which correspond to the mechanosensitive A-fibers) (126). Both allergen and NGF induce similar changes in vagal neurons innervating the airways: substance P expression in large diameter neurofilament positive neurons increased from virtually 0 to 32% within a day of allergen challenge (127–130) (Fig. 2). This phenotypic switch of mechanosensitive A-fibers innervating the airways into neuropeptide expressing nerves could have profound effects on neurogenic inflammation in the periphery as well as synaptic transmission in the brain stem.

Allergic inflammation may also change the activation profile of sensory nerves. For example, in the rat airways allergic inflammation was associated with TRPV1 expression on non–C-fiber neurons (131). This may explain why in the inflamed rat airways, the low threshold stretch receptors became sensitive to capsaicin.

Central Sensitization
The mechanisms leading to exaggerated and inappropriate sensations and reflexes in airway inflammatory diseases have received relatively little attention from the pulmonary research community. Studies on the somatosensory system, however, may prove instructive in this regard. In the study of pain, it has long been recognized that inflammation can lead to a state of hyperalgesia such that

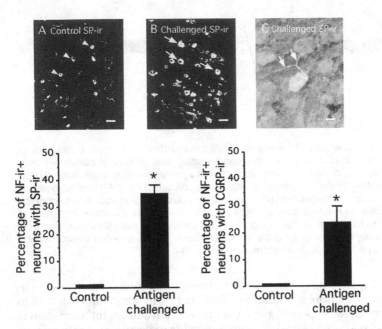

FIGURE 2 Aerosol challenge with ovalbumin induces the expression of neuropeptides in large-diameter nodose ganglion neurons. (**A**): Section of control nodose ganglion showing that only small, <25-μm-diameter neurons are immunoreactive for substance P (SP) (*arrowheads*; scale bar = 50 μm). (**B**): 24 hours after a sensitized guinea pig was challenged with ovalbumin aerosol, large-diameter (35–45 μm) neurons are immunoreactive for substance P (*arrows*; scale bar = 50 μm). Similar increases in calcitonin gene–related peptide (CGRP) immunoreactivity were observed following antigen challenge (not shown). (**C**): In a 16-μm-thick section, a large-diameter neuron in a ganglion from a challenged guinea pig is positive for substance P immunoreactivity. The immunostained peptide fills the axon which projects in both the rostral and caudal (*arrows*) directions (scale bar = 20 μm). *Bottom*: histograms showing the percentage of large-diameter, neurofilament (NF)-positive neurons that expressed substance P (*left*) or CGRP (*right*) immunoreactivity in nodose ganglia isolated from nonsensitized (control or passively sensitized and antigen challenged) guinea pigs 24 hours after ovalbumin inhalation. Bars represent means + SE from six animals. The total number of NF-positive neurons counted within each group was similar, averaging 852 ± 95 neurons. *Significant difference between the nonsensitized and sensitized animals ($p < 0.01$). *Source*: Adapted from Ref. 128.

the threshold for painful stimuli is decreased (132). Inflammation can also lead to a state of inappropriate pain, i.e., painful sensations to a nonpainful stimulus such as gentle brushing of the skin or hair. The term given to this inappropriate pain sensation is *allodynia* (133). A similar phenomenon occurs with respect to inappropriate itch sensations, to which the term *alloknesis* is given. By analogy, in asthma and rhinitis, rather than leading to painful sensations, inflammation may lead to "inappropriate" hunger for air, that is, an "allodyspneic" sensation. Also, analogous to alloknesis, asthmatics may experience itch sensations in their airways leading to an urge to cough despite nothing physical in the airway provoking the itchy irritation.

Allodynia (inappropriate pain response) is thought to involve converging interactions that occur in the central nervous system between large-diameter, low-threshold mechanosensing nerves that respond to gentle touch or brushing, and nociceptors that are stimulated by inflammatory mediators (133). The large diameter mechanosensing neurons usually release excitatory amino acids (EAA) as their neurotransmitter, causing fast synaptic transmission in the spinal cord. Nociceptors typically project neuropeptide-containing unmyelinated (C-type) fibers to the spinal cord, and are thought to be involved in transmitting pain sensations. Some of the central projections of nociceptive neurons converge on the same secondary neurons in the spinal cord that are innervated by the aforementioned large-diameter mechanosensing fibers. When nociceptive peptides such as substance P are released into this convergent synapse, they cause slow excitatory potentials and other processes that amplify fast synaptic transmission. This results in tactile stimuli sensed as pain, that is, allodynia. This process of synaptic augmentation by converging presynaptic inputs in the central nervous system is referred to as "central sensitization" (24).

Based on the evidence it is now reasonable to hypothesize that central sensitization within the brainstem, specifically in the nucleus of the solitary tract, can lead to symptoms of allergic airway disease. An important aspect of central sensitization is that, depending on the sphere of influence of transmitters released from nociceptors in the brain stem, the nociceptive terminals involved may be physically separated from the pathways being amplified. For example, studies have shown that stimulation of nociceptors in a region of the lungs devoid of true cough receptors can centrally sensitize the cough pathways that initiates from Aδ fibers in the larynx and trachea (79,88). Likewise stimulating nociceptors in the esophagus or nasal mucosa can centrally sensitize the tracheal cough pathways (89,134). The point here is that nasal inflammation may lead to urge to cough sensations in the absence of a "postnasal drip" of substances onto the cough fibers in the larynx and trachea. This idea is not limited to cough; stimulating laryngeal nociceptors with capsaicin can centrally sensitize the parasympathetic pathways leading to a generalized increased parasympathetic drive to the airways (135).

Allergen-Induced Autonomic Modulation

The neuromodulation associated with airway inflammation is not limited to primary afferent nerve activity and CNS integration. Airway inflammation also directly affects autonomic neuronal activity. Allergen challenge has been shown to increase neuronal excitability in all types of autonomic neurons including those in myenteric, sympathetic, and parasympathetic ganglia (73,136,137). Allergen challenge of a guinea pig isolated bronchus leads to long-lasting increases in the synaptic efficacy in the parasympathetic ganglia. This decreases the filtering function of airway ganglia leading to a generalized increase in parasympathetic tone. Airway inflammation can also lead to an increase in the amount of acetylcholine released per action potential from the postganglionic fibers at the level of the neuro-effector cells. This has been explained by mediators causing a decrease in the presynpatic inhibitory influence of cholinergic muscarinic 2 receptors (138).

As described above for sensory nerves, there may also be neuroplastic changes in autonomic neurons associated with allergic inflammation. Preliminary studies have shown that allergic inflammation is associated with a phenotypic switch of parasympathetic VI/NO neurons into cholinergic neurons (139).

This, in theory, could lead to a loss of inhibitory innervation to the airway smooth muscle, simultaneously with an increase in cholinergic contractile innervation. Cytokines and neurotrophins associated with allergic inflammation have also been shown to induce the expression of tachykinins in parasympathetic neurons within ferret airways (140).

CONCLUSIONS

The idea that allergic inflammation leads to a "neuronal hypersensitivity" is supported by a preponderance of anecdotal observations as well as by clinical laboratory investigation. In susceptible individuals, the allergic reaction somehow sets in motion events that lead to a perverted and/or inappropriate sensitivity to disparate sensory inputs. A stimulus that might normally go unnoticed is recognized as irritation that causes sneezing, urge to cough, and excessive autonomic reflex action. This can in some cases be merely bothersome, but in other cases can become debilitating and can exacerbate underlying disease states. Based largely on work in animal models, the data support the hypothesis that the allergic inflammatory response can lead to acute increases excitability of sensory nerves in the upper airways. It can also lead to long-lived changes in gene expression in these nerves; so-called neuroplastic changes. These changes in nerve excitability and phenotype can result in central sensitization such that the interpretation of the information within the CNS that arises from the afferent nerves is altered. In addition, allergic inflammation can lead to both acute changes in excitability and neuroplastic changes in the autonomic nerves that innervate the glands and smooth muscle of the airways.

As more mechanistic information becomes available, drugs may be developed that directly target these neuronal mechanisms. A therapeutic strategy aimed at normalizing the nervous system would be expected to act in an additive or even synergistic way with available anti-inflammatory treatments.

REFERENCES

1. Stjarne P, Lundblad L, Lundberg JM, et al. Capsaicin and nicotine-sensitive afferent neurones and nasal secretion in healthy human volunteers and in patients with vasomotor rhinitis. Br J Pharmacol 1989; 96:693–701.
2. Baraniuk JN, Silver PB, Kaliner MA, et al. Perennial rhinitis subjects have altered vascular, glandular, and neural responses to bradykinin nasal provocation. Int Arch Allergy Immunol 1994; 103:202–208.
3. Sanico AM, Koliatsos VE, Stanisz AM, et al. Neural hyperresponsiveness and nerve growth factor in allergic rhinitis. Int Arch Allergy Immunol 1999; 118:154–158.
4. Sanico AM, Philip G, Lai GK, et al. Hyperosmolar saline induces reflex nasal secretions, evincing neural hyperresponsiveness in allergic rhinitis. J Appl Physiol 1999; 86:1202–1210.
5. Shusterman D. Toxicology of nasal irritants. Curr Allergy Asthma Rep 2003; 3: 258–265.
6. Doerfler H, Hummel T, Klimek L, et al. Intranasal trigeminal sensitivity in subjects with allergic rhinitis. Eur Arch Otorhinolaryngol 2006; 263:86–90.
7. Shusterman D. Environmental nonallergic rhinitis. Clin Allergy Immunol 2007; 19:249–266.
8. Brasche S, Bullinger M, Morfeld M, et al. Why do women suffer from sick building syndrome more often than men?—Subjective higher sensitivity versus objective causes. Indoor Air 2001; 11:217–222.

9. Shusterman D, Murphy MA. Nasal hyperreactivity in allergic and non-allergic rhinitis: A potential risk factor for non-specific building-related illness. Indoor Air 2007; 17:328–333.

10. Millqvist E, Bende M. Role of the upper airways in patients with chronic cough. Curr Opin Allergy Clin Immunol 2006; 6:7–11.

11. Tatar M, Plevkova J, Brozmanova M, et al. Mechanisms of the cough associated with rhinosinusitis. Pulm Pharmacol Ther 2009; 22(2):121–126.

12. Sekizawa S, Tsubone H. Nasal mechanoreceptors in guinea pigs. Respir Physiol 1996; 106:223–230.

13. Sekizawa S, Tsubone H, Kuwahara M, et al. Nasal receptors responding to cold and l-menthol airflow in the guinea pig. Respir Physiol 1996; 103:211–219.

14. Sekizawa S, Tsubone H, Kuwahara M, et al. Does histamine stimulate trigeminal nasal afferents? Respir Physiol 1998; 112:13–22.

15. Nail BS, Sterling GM, Widdicombe JG. Epipharyngeal receptors responding to mechanical stimulation. J Physiol 1969; 204:91–98.

16. Ninomiya Y, Nakashima K, Fukuda A, et al. Responses to umami substances in taste bud cells innervated by the chorda tympani and glossopharyngeal nerves. J Nutr 2000; 130:950S–953S.

17. Carr MJ, Undem BJ. Bronchopulmonary afferent nerves. Respirology 2003; 8: 291–301.

18. Undem BJ, Chuaychoo B, Lee MG, et al. Subtypes of vagal afferent C-fibres in guinea-pig lungs. J Physiol 2004; 556:905–917.

19. Ricco MM, Kummer W, Biglari B, et al. Interganglionic segregation of distinct vagal afferent fibre phenotypes in guinea-pig airways. J Physiol 1996; 496(Pt 2):521–530.

20. Canning BJ, Mazzone SB, Meeker SN, et al. Identification of the tracheal and laryngeal afferent neurones mediating cough in anaesthetized guinea-pigs. J Physiol 2004; 557:543–558.

21. Grunditz T, Uddman R, Sundler F. Origin and peptide content of nerve fibers in the nasal mucosa of rats. Anat Embryol (Berl) 189:327–337.

22. Taylor-Clark TE, Kollarik M, MacGlashan DW Jr.,et al. Nasal sensory nerve populations responding to histamine and capsaicin. J Allergy Clin Immunol 2005; 116: 1282–1288.

23. Dehkordi O, Rose JE, Balan KV, et al. Neuroanatomical relationships of substance P-immunoreactive intrapulmonary C-fibers and nicotinic cholinergic receptors. J Neurosci Res 2009; 87(7):1670–1678.

24. Mazzone SB, Canning BJ. Central nervous system control of the airways: Pharmacological implications. Curr Opin Pharmacol 2002; 2:220–228.

25. Sanico AM, Atsuta S, Proud D, et al. Dose-dependent effects of capsaicin nasal challenge: In vivo evidence of human airway neurogenic inflammation. J Allergy Clin Immunol 1997; 100:632–641.

26. Myers AC, Undem BJ. Electrophysiological effects of tachykinins and capsaicin on guinea-pig bronchial parasympathetic ganglion neurones. J Physiol 1993; 470: 665–679.

27. Taylor-Clark T, Undem BJ. Transduction mechanisms in airway sensory nerves. J Appl Physiol 2006; 101:950–959.

28. Van Genechten J, Brouns I, Burnstock G, et al. Quantification of neuroepithelial bodies and their innervation in fawn-hooded and Wistar rat lungs. Am J Respir Cell Mol Biol 2004; 30:20–30.

29. Brouns I, Pintelon I, De Proost I, et al. Neurochemical characterisation of sensory receptors in airway smooth muscle: Comparison with pulmonary neuroepithelial bodies. Histochem Cell Biol 2005:1–17.

30. Sauer H, Hescheler J, Wartenberg M. Mechanical strain-induced Ca(2+) waves are propagated via ATP release and purinergic receptor activation. Am J Physiol Cell Physiol 2000; 279:C295–C307.

31. Pan J, Copland I, Post M, et al. Mechanical stretch-induced serotonin release from pulmonary neuroendocrine cells: Implications for lung development. Am J Physiol Lung Cell Mol Physiol 2006; 290:L185–L193.

32. Coleridge JC, Coleridge HM. Afferent vagal C fibre innervation of the lungs and airways and its functional significance. Rev Physiol Biochem Pharmacol 1984; 99: 1–110.

33. Caterina MJ, Schumacher MA, Tominaga M, et al. The capsaicin receptor: A heat-activated ion channel in the pain pathway. Nature 1997; 389:816–824.

34. Kollarik M, Undem BJ. Activation of bronchopulmonary vagal afferent nerves with bradykinin, acid and vanilloid receptor agonists in wild-type and TRPV1-/- mice. J Physiol 2004; 555:115–123.

35. Shin J, Cho H, Hwang SW, et al. Bradykinin-12-lipoxygenase-VR1 signaling pathway for inflammatory hyperalgesia. Proc Natl Acad Sci U S A 2002; 99:10150–10155.

36. Amadesi S, Nie J, Vergnolle N, et al. Protease-activated receptor 2 sensitizes the capsaicin receptor transient receptor potential vanilloid receptor 1 to induce hyperalgesia. J Neurosci 2004; 24:4300–4312.

37. Kim BM, Lee SH, Shim WS, et al. Histamine-induced Ca(2+) influx via the PLA(2)/lipoxygenase/TRPV1 pathway in rat sensory neurons. Neurosci Lett 2004; 361:159–162.

38. Nassenstein C, Kwong K, Taylor-Clark T, et al. Expression and function of the ion channel TRPA1 in vagal afferent nerves innervating mouse lungs. J Physiol 2008; 586:1595–1604.

39. Story GM, Peier AM, Reeve AJ, et al. ANKTM1, a TRP-like channel expressed in nociceptive neurons, is activated by cold temperatures. Cell 2003; 112:819–829.

40. Bessac BF, Jordt SE. Breathtaking TRP channels: TRPA1 and TRPV1 in airway chemosensation and reflex control. Physiology (Bethesda) 2008; 23:360–370.

41. Taylor-Clark TE, Nassenstein C, McAlexander MA, et al. TRPA1: A potential target for anti-tussive therapy. Pulm Pharmacol Ther 2009; 22(2):71–74.

42. Hinman A, Chuang HH, Bautista DM, et al. TRP channel activation by reversible covalent modification. Proc Natl Acad Sci U S A 2006; 103:19564–19568.

43. Andre E, Campi B, Materazzi S, et al. Cigarette smoke-induced neurogenic inflammation is mediated by alpha,beta-unsaturated aldehydes and the TRPA1 receptor in rodents. J Clin Invest 2008; 118:2574–2582.

44. Damann N, Rothermel M, Klupp BG, et al. Chemosensory properties of murine nasal and cutaneous trigeminal neurons identified by viral tracing. BMC Neurosci 2006; 7:46.

45. McKemy DD, Neuhausser WM, Julius D. Identification of a cold receptor reveals a general role for TRP channels in thermosensation. Nature 2002; 416:52–58.

46. Khakh BS, Burnstock G, Kennedy C, et al. International union of pharmacology. XXIV. Current status of the nomenclature and properties of P2X receptors and their subunits. Pharmacol Rev 2001; 53:107–118.

47. Kwong K, Kollarik M, Nassenstein C, et al. P2X$_2$ Receptors Differentiate Placodal vs Neural Crest C-fiber Phenotypes Innervating Guinea Pig Lungs and Esophagus. Am J Physiol Lung Cell Mol Physiol 2008; 295(5):L858–865.

48. Chuaychoo B, Lee MG, Kollarik M, et al. Effect of 5-hydroxytryptamine on vagal C-fiber subtypes in guinea pig lungs. Pulm Pharmacol Ther 2005; 18:269–276.

49. Lee LY, Burki NK, Gerhardstein DC, et al. Airway irritation and cough evoked by inhaled cigarette smoke: Role of neuronal nicotinic acetylcholine receptors. Pulm Pharmacol Ther 2007; 20:355–364.

50. Inoue T, Bryant BP. Multiple types of sensory neurons respond to irritating volatile organic compounds (VOCs): Calcium fluorimetry of trigeminal ganglion neurons. Pain 2005; 117:193–203.

51. Burki NK, Dale WJ, Lee LY. Intravenous adenosine and dyspnea in humans. J Appl Physiol 2005; 98:180–185.

52. Hong JL, Ho CY, Kwong K, et al. Activation of pulmonary C fibres by adenosine in anaesthetized rats: Role of adenosine A1 receptors. J Physiol 1998; 508(Pt 1):109–118.

53. Chuaychoo B, Lee MG, Kollarik M, et al. Evidence for both adenosine A1 and A2A receptors activating single vagal sensory C-fibres in guinea pig lungs. J Physiol 2006; 575:481–490.

54. Auchincloss JH, Streeten DH, Gilbert R, et al. Dyspnea in patients with hyper-bradykininism and excessive venous pooling. Am J Med 1986; 81:260–266.

55. Kajekar R, Proud D, Myers AC, et al. Characterization of vagal afferent subtypes stimulated by bradykinin in guinea pig trachea. J Pharmacol Exp Ther 1999; 289: 682–687.

56. Lee MG, Macglashan DW Jr., Undem BJ. Role of chloride channels in bradykinin-induced guinea pig airway vagal C-fibre activation. J Physiol 2005; 566:205–212.

57. Bautista DM, Jordt SE, Nikai T, et al. TRPA1 Mediates the Inflammatory Actions of Environmental Irritants and Proalgesic Agents. Cell 2006; 124:1269–1282.

58. Pisarri TE, Jonzon A, Coleridge HM, et al. Vagal afferent and reflex responses to changes in surface osmolarity in lower airways of dogs. J Appl Physiol 1992; 73: 2305–2313.

59. Fox AJ, Barnes PJ, Dray A. Stimulation of guinea-pig tracheal afferent fibres by non-isosmotic and low-chloride stimuli and the effect of frusemide. J Physiol 1995; 482 (Pt 1):179–187.

60. Undem BJ, Oh EJ, Lancaster E, et al. Effect of extracellular calcium on excitability of guinea pig airway vagal afferent nerves. J Neurophysiol 2003; 89:1196–1204.

61. Sant'Ambrogio G, Anderson JW, Sant'Ambrogio FB, et al. Response of laryngeal receptors to water solutions of different osmolality and ionic composition. Respir Med 1991; 85(suppl A):57–60.

62. Pedersen KE, Meeker SN, Riccio MM, et al. Selective stimulation of jugular ganglion afferent neurons in guinea pig airways by hypertonic saline. J Appl Physiol 1998; 84:499–506.

63. Alessandri-Haber N, Yeh JJ, Boyd AE, et al. Hypotonicity induces TRPV4-mediated nociception in rat. Neuron 2003; 39:497–511.

64. Naeini RS, Witty MF, Seguela P, et al. An N-terminal variant of Trpv1 channel is required for osmosensory transduction. Nat Neurosci 2006; 9:93–98.

65. Konno A, Togawa K. Role of the vidian nerve in nasal allergy. Ann Otol Rhinol Laryngol 1979; 88:258–266.

66. Doyle WJ, Boehm S, Skoner DP. Physiologic responses to intranasal dose-response challenges with histamine, methacholine, bradykinin, and prostaglandin in adult volunteers with and without nasal allergy. J Allergy Clin Immunol 1990; 86: 924–935.

67. Baraniuk JN, Lundgren JD, Okayama M, et al. Vasoactive intestinal peptide in human nasal mucosa. J Clin Invest 1990; 86:825–831.

68. Georgitis JW. Nasal atropine sulfate: Efficacy and safety of 0.050% and 0.075% solutions for severe rhinorrhea. Arch Otolaryngol Head Neck Surg 1998; 124:916–920.

69. Varty LM, Gustafson E, Laverty M, et al. Activation of histamine H3 receptors in human nasal mucosa inhibits sympathetic vasoconstriction. Eur J Pharmacol 2004; 484:83–89.

70. Taylor-Clark T, Sodha R, Warner B, et al. Histamine receptors that influence blockage of the normal human nasal airway. Br J Pharmacol 2005; 144(6):867–874.

71. Whittet HB, Fisher EW. Nasal obstruction after cervical sympathectomy: Horner's syndrome revisited. ORL J Otorhinolaryngol Relat Spec 1988; 50:246–250.

72. Kirby RS. Clinical pharmacology of alpha1-adrenoceptor antagonists. Eur Urol 1999; 36(suppl 1):48–53; discussion 65.

73. Myers AC, Undem BJ, Weinreich D. Influence of antigen on membrane properties of guinea pig bronchial ganglion neurons. J Appl Physiol 1991; 71:970–976.

74. Downing SE, Lee JC. Nervous control of the pulmonary circulation. Annu Rev Physiol 1980; 42:199–210.

75. Coleridge HM, Coleridge JC. Neural regulation of bronchial blood flow. Respir Physiol 1994; 98:1–13.

76. Canning BJ. Reflex regulation of airway smooth muscle tone. J Appl Physiol 2006; 101:971–985.

77. Wine JJ. Parasympathetic control of airway submucosal glands: Central reflexes and the airway intrinsic nervous system. Auton Neurosci 2007; 133:35–54.

78. Richardson PS, Peatfield AC. Reflexes concerned in the defence of the lungs. Bull Eur Physiopathol Respir 1981; 17:979–1012.
79. Chou YL, Scarupa MD, Mori N, et al. Differential effects of airway afferent nerve subtypes on cough and respiration in anesthetized guinea pigs. Am J Physiol Regul Integr Comp Physiol 2008; 295:R1572–R1584.
80. Tonnesen P, Mygind N. Nasal challenge with serotonin and histamine in normal persons. Allergy 1985; 40:350–353.
81. Lindberg S, Dolata J, Mercke U. Mucociliary activity and the diving reflex. Am J Otolaryngol 1990; 11:182–187.
82. Brunnee T, Nigam S, Kunkel G, et al. Nasal challenge studies with bradykinin: Influence upon mediator generation. Clin Exp Allergy 1991; 21:425–431.
83. Philip G, Baroody FM, Proud D, et al. The human nasal response to capsaicin. J Allergy Clin Immunol 1994; 94:1035–1045.
84. Taylor-Clark T, Foreman J. Histamine-mediated mechanisms in the human nasal airway. Curr Opin Pharmacol 2005; 5:214–220.
85. Konno A, Nagata H, Nomoto M, et al. Role of capsaicin-sensitive trigeminal nerves in development of hyperreactive nasal symptoms in guinea pig model of nasal allergy. Ann Otol Rhinol Laryngol 1995; 104:730–735.
86. Page C, Reynolds SM, Mackenzie AJ, et al. Mechanisms of acute cough. Pulm Pharmacol Ther 2004; 17:389–391.
87. Canning BJ, Mori N, Mazzone SB. Vagal afferent nerves regulating the cough reflex. Respir Physiol Neurobiol 2006; 152:223–242.
88. Mazzone SB, Mori N, Canning BJ. Synergistic interactions between airway afferent nerve subtypes regulating the cough reflex in guinea-pigs. J Physiol 2005; 569: 559–573.
89. Plevkova J, Kollarik M, Brozmanova M, et al. Modulation of experimentally-induced cough by stimulation of nasal mucosa in cats and guinea pigs. Respir Physiol Neurobiol 2004; 142:225–235.
90. Handforth A, DeGiorgio CM, Schachter SC, et al. Vagus nerve stimulation therapy for partial-onset seizures: A randomized active-control trial. Neurology 1998; 51: 48–55.
91. Guz A, Noble M, Eisele J, et al. Experimental results of vagal block in cardiopulmonary disease. In: Porter R, ed. Breathing: Hering-Breuer Centenary Symposium. London: Churchill, 1970:315–329.
92. Taguchi O, Kikuchi Y, Hida W, et al. Effects of bronchoconstriction and external resistive loading on the sensation of dyspnea. J Appl Physiol 1991; 71:2183–2190.
93. Marks GB, Yates DH, Sist M, et al. Respiratory sensation during bronchial challenge testing with methacholine, sodium metabisulphite, and adenosine monophosphate. Thorax 1996; 51:793–798.
94. Tetzlaff K, Leplow B, ten Thoren C, et al. Perception of dyspnea during histamine- and methacholine-induced bronchoconstriction. Respiration 1999; 66:427–433.
95. Burki NK, Alam M, Lee LY. The pulmonary effects of intravenous adenosine in asthmatic subjects. Respir Res 2006; 7:139.
96. Taguchi O, Kikuchi Y, Hida W, et al. Prostaglandin E2 inhalation increases the sensation of dyspnea during exercise. Am Rev Respir Dis 1992; 145:1346–1349.
97. Nishino T, Ide T, Sudo T, et al. Inhaled furosemide greatly alleviates the sensation of experimentally induced dyspnea. Am J Respir Crit Care Med 2000; 161:1963–1967.
98. Moosavi SH, Binks AP, Lansing RW, et al. Effect of inhaled furosemide on air hunger induced in healthy humans. Respir Physiol Neurobiol 2007; 156:1–8.
99. Sudo T, Hayashi F, Nishino T. Responses of tracheobronchial receptors to inhaled furosemide in anesthetized rats. Am J Respir Crit Care Med 2000; 162:971–975.
100. Widdicombe JG. Nasal pathophysiology. Respir Med 1990; 84(suppl A):3–9; discussion–10.
101. Takahashi Y, Mizuno H, Ohno H, et al. Neural reflex-mediated tracheal response during bronchoconstriction induced by ovalbumin antigen in guinea pigs. Gen Pharmacol 1997; 28:399–404.

102. Zimmermann I, Ulmer WT, Weller W. The role of upper airways and of sensoric receptors on reflex bronchoconstriction. Res Exp Med (Berl) 1979; 174:253–265.
103. McLeod RL, Fernandez X, Correll CC, et al. TRPV1 antagonists attenuate antigen-provoked cough in ovalbumin sensitized guinea pigs. Cough 2006; 2:10.
104. Zhang G, Lin RL, Wiggers ME, et al. Sensitizing effects of chronic exposure and acute inhalation of ovalbumin aerosol on pulmonary C fibers in rats. J Appl Physiol 2008; 105:128–138.
105. Mills JE, Sellick H, Widdicombe JG. Activity of lung irritant receptors in pulmonary microembolism, anaphylaxis and drug-induced bronchoconstrictions. J Physiol 1969; 203:337–357.
106. Carr MJ, Undem BJ. Inflammation-induced plasticity of the afferent innervation of the airways. Environ Health Perspect 2001; 109(suppl 4):567–571.
107. Taylor-Clark TE, Nassenstein C, Undem BJ. Leukotriene D(4) increases the excitability of capsaicin-sensitive nasal sensory nerves to electrical and chemical stimuli. Br J Pharmacol 2008; 154:1359–1368.
108. Kadoi J, Takeda M, Matsumoto S. Prostaglandin E2 potentiates the excitability of small diameter trigeminal root ganglion neurons projecting onto the superficial layer of the cervical dorsal horn in rats. Exp Brain Res 2007; 176:227–236.
109. Tsutsui Y, Ikeda M, Takeda M, et al. Excitability of small-diameter trigeminal ganglion neurons by 5-HT is mediated by enhancement of the tetrodotoxin-resistant sodium current due to the activation of 5-HT(4) receptors and/or by the inhibition of the transient potassium current. Neuroscience 2008; 157:683–696.
110. Takeda M, Takahashi M, Matsumoto S. Contribution of activated interleukin receptors in trigeminal ganglion neurons to hyperalgesia via satellite glial interleukin-1beta paracrine mechanism. Brain Behav Immun 2008; 22:1016–1023.
111. Liu Q, Fujimura M, Tachibana H, et al. Characterization of increased cough sensitivity after antigen challenge in guinea pigs. Clin Exp Allergy 2001; 31:474–484.
112. Kamei J, Takahashi Y, Itabashi T, et al. Atopic cough-like cough hypersensitivity caused by active sensitization with protein fraction of Aspergillus restrictus strain A-17. Pulm Pharmacol Ther 2008; 21:356–359.
113. Berman AR, Togias AG, Skloot G, et al. Allergen-induced hyperresponsiveness to bradykinin is more pronounced than that to methacholine. J Appl Physiol 1995; 78:1844–1852.
114. Riccio MM, Myers AC, Undem BJ. Immunomodulation of afferent neurons in guinea-pig isolated airway. J Physiol 1996; 491(Pt 2):499–509.
115. Greene R, Fowler J, MacGlashan D Jr., et al. IgE-challenged human lung mast cells excite vagal sensory neurons in vitro. J Appl Physiol 1988; 64:2249–2253.
116. McAlexander MA, Myers AC, Undem BJ. Inhibition of 5-lipoxygenase diminishes neurally evoked tachykinergic contraction of guinea pig isolated airway. J Pharmacol Exp Ther 1998; 285:602–607.
117. Gu Q, Ruan T, Hong JL, Burki N, et al. Hypersensitivity of pulmonary C fibers induced by adenosine in anesthetized rats. J Appl Physiol 2003; 95:1315–1324; discussion 4.
118. Ho CY, Gu Q, Hong JL, et al. Prostaglandin E(2) enhances chemical and mechanical sensitivities of pulmonary C fibers in the rat. Am J Respir Crit Care Med 2000; 162:528–533.
119. Kwong K, Lee LY. Prostaglandin E2 potentiates a TTX-resistant sodium current in rat capsaicin-sensitive vagal pulmonary sensory neurones. J Physiol 2005; 564:437–450.
120. Bonini S, Lambiase A, Angelucci F, et al. Circulating nerve growth factor levels are increased in humans with allergic diseases and asthma. Proc Natl Acad Sci U S A 1996; 93:10955–10960.
121. de Vries A, Dessing MC, Engels F, et al. Nerve growth factor induces a neurokinin-1 receptor- mediated airway hyperresponsiveness in guinea pigs. Am J Respir Crit Care Med 1999; 159:1541–1544.
122. Braun A, Quarcoo D, Schulte-Herbruggen O, et al. Nerve growth factor induces airway hyperresponsiveness in mice. Int Arch Allergy Immunol 2001; 124:205–207.

123. Farraj AK, Haykal-Coates N, Ledbetter AD, et al. Neurotrophin mediation of allergic airways responses to inhaled diesel particles in mice. Toxicol Sci 2006; 94:183–192.
124. Raap U, Fokkens W, Bruder M, et al. Modulation of neurotrophin and neurotrophin receptor expression in nasal mucosa after nasal allergen provocation in allergic rhinitis. Allergy 2008; 63:468–475.
125. O'Hanlon S, Facer P, Simpson KD, et al. Neuronal markers in allergic rhinitis: Expression and correlation with sensory testing. Laryngoscope 2007; 117:1519–1527.
126. Mingomataj E, Dinh QT, Groneberg D, et al. Trigeminal nasal-specific neurons respond to nerve growth factor with substance-P biosynthesis. Clin Exp Allergy 2008; 38:1203–1211.
127. Hunter DD, Myers AC, Undem BJ. Nerve growth factor-induced phenotypic switch in guinea pig airway sensory neurons. Am J Respir Crit Care Med 2000; 161: 1985–1990.
128. Myers AC, Kajekar R, Undem BJ. Allergic inflammation-induced neuropeptide production in rapidly adapting afferent nerves in guinea pig airways. Am J Physiol Lung Cell Mol Physiol 2002; 282:L775–L781.
129. Chuaychoo B, Hunter DD, Myers AC, et al. Allergen-induced substance P synthesis in large-diameter sensory neurons innervating the lungs. J Allergy Clin Immunol 2005; 116:325–331.
130. Dinh QT, Mingomataj E, Quarcoo D, et al. Allergic airway inflammation induces tachykinin peptides expression in vagal sensory neurons innervating mouse airways. Clin Exp Allergy 2005; 35:820–825.
131. Zhang G, Lin RL, Wiggers M, et al. Altered expression of TRPV1 and sensitivity to capsaicin in pulmonary myelinated afferents following chronic airway inflammation in the rat. J Physiol 2008; 586:5771–5786.
132. Meyer RA, Raja SN, Cambell JN. Neural mechanisms of primary hyperalgesia. In: Belmonte C, Cervero F, eds. Neurobiology of Nociceptors. Oxford University Press, 1996:370–389.
133. LaMotte RH. Secondary cutaneous dysaesthesiae. In: Belmonte C, Cervero F, eds. Neurobiology of Nociceptors. Oxford University Press, 1996:390–417.
134. Kollarik M, Brozmanova M. Cough and gastroesophageal reflux: Insights from animal models. Pulm Pharmacol Ther 2009; 22(2):130–134.
135. Mazzone SB, Canning BJ. Synergistic interactions between airway afferent nerve subtypes mediating reflex bronchospasm in guinea pigs. Am J Physiol Regul Integr Comp Physiol 2002; 283:R86–R98.
136. Undem BJ, Myers AC, Weinreich D. Antigen-induced modulation of autonomic and sensory neurons in vitro. Int Arch Allergy Appl Immunol 1991; 94:319–324.
137. Liu S, Hu HZ, Gao N, et al. Neuroimmune interactions in guinea pig stomach and small intestine. Am J Physiol Gastrointest Liver Physiol 2003; 284:G154–G164.
138. Fryer AD, Wills-Karp M. Dysfunction of M2-muscarinic receptors in pulmonary parasympathetic nerves after antigen challenge. J Appl Physiol 1991; 71:2255–2261.
139. Pan J, Undem BJ, Myers AC. Repeated allergen exposure induces neural plasticity of adult neurons in lower airway parasympathetic ganglia. Faseb J 2007; 21:912–913.
140. Wu ZX, Dey RD. Nerve growth factor-enhanced airway responsiveness involves substance P in ferret intrinsic airway neurons. Am J Physiol Lung Cell Mol Physiol 2006; 291:L111–L118.

The Effect of Air Pollutants on Allergic Upper Airway Disease

Dennis J. Shusterman

Division of Occupational and Environmental Medicine, University of California, San Francisco, California, U.S.A.

INTRODUCTION

Nasal allergy and nasal irritation are two phenomena that are frequently confused by both clinicians and toxicologists alike. Classical "allergy" (as in rhinitis and asthma) is an antibody (IgE)-mediated immediate hypersensitivity response to *allergens*, the vast majority of which are high molecular weight and biological in origin (1). "Irritation," on the other hand, can be defined variously as chemically induced tissue damage, subjective irritation of the airway, skin, or mucous membranes, stimulation of nociceptive nerves, reflex changes triggered by such nerve stimulation, or a combination of these effects (2–4). On an inflammatory level, allergic conditions are generally associated with allergen-induced mast cell activation and T_H2 (T-helper lymphocyte, type 2 differentiation) lymphocyte and eosinophil infiltration, whereas pathologic chemical irritation is generally marked by polymorphonuclear leukocyte influx (similar to bacterial or viral infection). Allergy is further defined clinically by the presence of circulating allergen-specific IgE antibodies and/or a wheal-and-flare reaction after epicutaneous presentation of an allergen during skin testing, the result of in vivo mast cell activation through allergen/IgE interaction (5).

Despite these mechanistic distinctions, the nose, with its limited repertoire of physiologic responses, may respond in overlapping ways to allergens and irritants. Both allergic reactions and irritation can trigger reflex nasal secretion and nasal airflow obstruction. On a sensory level, chemical irritants (e.g., capsaicin, nicotine, allyl isothiocyanate, acrolein) trigger a variety of nociceptive ion channels on airway nerves (generally C or Aδ-fibers) (6–9). Histamine, an endogenous mediator which is one of the hallmarks of the allergic response, stimulates a subset of C fibers, eliciting the sensation of pruritus (itching) and the urge to sneeze (10,11). Thus, the acute subjective response to allergens and irritants can be, to a limited extent, differentiated based upon the relative predominance of itching versus the irritant sensations of stinging, burning, tingling, or aching.

Notwithstanding our modest ability to distinguish between allergic and irritative phenomena based upon clinical history and laboratory criteria, the two processes interact in everyday life, and in fact can be shown to influence one another. As reviewed in this and the preceding chapter, preexisting allergy can influence an organism's response to chemical irritants, and air pollutants—in turn—can influence an individual's response to allergens. This chapter deals with the effect of chemical air pollutants on the allergic response, with an emphasis upon human studies.

DEFINITIONS

To set the stage for reviewing the potential effects of chemical agents on upper airway allergy, it is first important to understand the allergic response, which for convenience will be divided into two stages. In contradistinction to irritation (which can occur on first exposure to an agent), allergy requires *sensitization*, the immunologic process that renders B lymphocytes capable of producing specific antibodies. An agent that accelerates or reinforces the sensitization process is referred to as an *adjuvant* (12). In formulating strategies for immunization against infectious diseases, for example, adjuvants (e.g., aluminum potassium sulfate, or alum) are sometimes included in vaccines to boost their immugenicity. Although allergic sensitization differs from vaccination for infectious diseases in terms of immunologic mechanism, the principle of adjuvant action is analogous, rendering appropriate a discussion of the *adjuvant effects of air pollutants on allergic sensitization.*

Once allergic sensitization to an aeroallergen has occurred, *symptoms are triggered* via mucosal challenge by the allergen in question. Allergen reaches IgE molecules bound to surface ($F_{C\varepsilon}$) receptors on mast cells residing in the mucosa, bridging adjacent receptors and resulting in *mast cell degranulation.* Acutely, histamine, sulfidopeptide leukotrienes, prostaglandins, and other vasoactive mediators are released, resulting in itching, sneezing, nasal secretion, and vascular congestion (in allergic rhinitis), as well as in bronchospasm manifesting with chest tightness and wheezing, and bronchial secretion manifesting with productive cough (in asthma). In addition to the acute response, allergic reactions may give rise to late-phase (or dual-phase) rhinitis/asthma symptoms, manifested over hours after the acute response, as well as tissue inflammation, which may last for days. It is believed that the late phase reaction and the development of inflammation involve a complex interaction between various cell types, including, but not limited to, basophils, eosinophils, and T lymphocytes. If exposure to an agent *after initial sensitization* acts to boost the allergic response, then we speak of a *priming* effect. The term priming is often applied to the boosted response to allergen after prior allergen exposure, but can also be used to describe intensification of an allergic reaction by prior chemical exposure. *Priming of the allergic response by exposure to air pollutants* has been studied in a number of experimental settings, and will be reviewed in this chapter as well.

Somewhat complicating the dichotomy between adjuvant effects and priming is the fact that allergen exposure of an already-sensitized individual can result both in the triggering of an acute allergic reaction and in further allergic sensitization (i.e., increased production of allergen-specific IgE). Intervening (or co-) exposure to a chemical agent may intensify both these processes, resulting in *both* adjuvant *and* priming effects. As a result, it is not practical to consider adjuvant and priming studies as mutually exclusive in the discussion below. To add yet another layer of complexity, a limited number of studies have examined—explicitly or implicitly—the metabolism of chemicals which produce adjuvant/priming effects. These investigations have focused attention on genetic variants in phase II (detoxication) enzymes and on the effect of exogenous antioxidants and/or phase II enzyme-inducers on the adjuvant effects of environmental chemicals (chap. 5).

Finally, in addition to adjuvant effects and priming, we will consider the newly emerging question of *physical and chemical effects of air pollutants on aeroallergens*, including potential effects on pollen allergenicity.

In order to set the stage for the discussion that follows, the basic experimental features of the reviewed studies are listed in Table 1. Studies include human and experimental animal models, in vivo and in vitro studies, and both specific and nonspecific indices of sensitization. Among the methodologic features tabulated is the order of various elements within each study, including sensitization (S), chemical exposure (C), drug administration (D), specific allergen provocation (P), and finally physiologic/biochemical assay (A), as well as an indication as to whether the organism was sensitized naturally or as a result of experimental procedures.

NONSPECIFIC CHEMICAL EFFECTS ON THE ALLERGY PATHWAY

A subset of studies concerns the nonspecific effects of environmental chemicals on immediate hypersensitivity (allergy) pathways. These effects include increased production of mucosal IgE (without regard to antigen specificity), as well as upregulation of signaling molecules, including both T_H2 cytokines and receptors for the mediators of allergy (specifically, histamine), as sampled in a baseline (nonallergen provoked) state.

Diaz-Sanchez et al. (13) studied the effect of diesel exhaust particles (DEP) on local immunoglobulin production in the human nose. They applied three different doses of DEP to the nasal mucosa of 11 asymptomatic individuals, 4 of whom had nasal allergies but were outside of their relevant aeroallergen season. Immunoglobulin levels (including IgA, IgE, IgG, IgG3, IgG4, IgM) and levels of mRNA for various alternative epsilon chains were assayed in nasal lavage fluid four days postchallenge, as well as after challenge with a saline control. The authors found that an intermediate dose of DEP (0.3 mg), but not a low dose (0.15) or high dose (1.0), resulted in a significant rise in total IgE in nasal lavage fluid, with no significant change in other immunoglobulin classes. Further, the relative balance of mRNAs for different epsilon isoforms was altered in the intermediate dose group. The authors concluded that DEP had both a quantitative and a qualitative effect on IgE production in the human nasal mucosa, regardless of allergy status.

In a follow-up study, Diaz-Sanchez et al. (14) examined the effect of DEP nasal challenge on levels of cytokine mRNA in nasal lavage fluid. Fourteen non-smoking volunteers, six with seasonal allergies but asymptomatic at the time of testing, were challenged nasally with either 0.3 mg DEP or with saline (on separate days) and sampled by nasal lavage four days later. Lavage fluid was analyzed for mRNA for IL-2, -4, -5, -6, 10, -13, and IFN-γ, as well as for β-Actin, and CD28. In addition, assays were carried out for total IgE, albumin, and IL-4 protein. As in the study reviewed above, DEP, but not saline, increased total IgE levels. In addition, mRNA was significantly increased for all of the above cytokines post-DEP administration, as was IL-4 protein. Again, these effects occurred without regard to allergy status.

Terada et al. (15) isolated nasal epithelial and microvascular endothelial cells in cell culture and exposed them to DEP in vitro. They then analyzed the growth medium for mRNA for histamine (H1) receptor, and examined histamine-induced release of IL-4 and granulocyte/macrophage colony-stimulating factor (GM-CSF). Compared to control conditions, DEP administration upregulates H1 receptor expression and histamine-induced IL-4 and GM-CSF release in both cell types.

TABLE 1 Effects of Environmental Chemicals on Allergic Upper Airway Disease

Reference No.	Model	Sensitization	Sequence	Agent	Endpoint(s)	Results
Nonspecific effects						
13	Human	Natural (4 SAR / 4 Cntrl)	C-A	DEP	NL	↑ Total IgE, (not IgG, A or M) ↑ ε mRNA
14	Human	Natural (6 SAR / 8 Cntrl)	C-A	DEP	NL	↑ mRNA for cytokines (IL-2, -4, -5, -6, -10, -13, IFN-γ) ↑ IL-4 protein
15	Human (in vitro)	N/A	C-A	DEP	Mediator release in culture	↑ H1Receptor ↑ Hist-induced IL-8 + GM-CSF from nasal epith cells
Specific effects: adjuvant or priming						
16	Human	Natural (13 SAR)	S-C/P-A	DEP	NL	No change total IgE w/ combined DEP +Ag; ↑ Ag-specific IgE " " ↑ epsilon-mRNA ↑ cytokine mRNA (IL-4, -5, -6, -10, -13, IFN-γ)
17	Human	Natural (8 SAR)	S-C/P-A	DEP + Amb a 1	NL	Isotype switching (IgG to IgE)
18	Human	Induced (10 atopic)	C-S-P-A	DEP + KLH	NL	KLH -> IgG + IgA only KLH + DEP -> IgE prod'n ↑ IL-4; n.c. IFN-γ
19	Human	Natural (11 PAR)	S-C-P-A	DEP Der p 1	Symptoms, NL	Decreased Provocat. dose ↑ Symptoms ↑ Histamine release
20	Mouse	Induced	C/S-P-A	Sidestream cigarette smoke + OVA		OVA alone -> No IgE OVA + smoke -> IgE BAL post-OVA challenge had eos. only w/ OVA-smoke
21	Human	Natural (19 SAR)	S-C-P-A	Sidestream cigarette smoke (2 h) + Amb a 1	NL	↑ Histamine release ↑ Ag-specific IgE + IgG4 ↑ IL-4, -5, -13 Decreased IFN-γ

22	Human	Natural (12 SAR)	S-C-P-A	Ozone 0.5 ppm x 4 h	Symptoms, NL	No change Provocat. Dose ↑ PMN + eos w/O3 alone, but no change in Ag response
23	Human	Natural (11 asthmatics sensitive to Der f)	S-C-P-A	Ozone 0.4 ppm x 2 h	Symptoms, NL	Decreased Provocat. Dose ↑ PMNs + eos w/ combined O_3 + Ag
24	Human	Natural (16 SAR)	S-C-P-A	NO_2 400 ppb x 6 h	NL	↑ ECP w/ combined NO_2 + Ag challenge.

Modulation of effect: Genetic or Pharmacol.

25	Human	Natural (18 SAR)	S-C-P-A (x 2)	DEP + Amb a 1	NL	↑ production of IgE, IL-4, and INF-γ was reproducible by individual
26	Human	Natural (19 SAR)	S-C-P-A (by GST genotype)	DEP + Amb a 1	NL	DEP effect on Ag-induced ↑ in IgE & histamine greater w/ GSTM1-null and GSTP1 I105 genotypes
27	Human	Natural (19 SAR)	S-C-P-A (by GST genotype)	Sidestream cigarette smoke (2 h) + Amb a 1	Symptoms, NL	Smoke effect on Ag-induced ↑ in IgE & histamine greater w/ GSTM1-null and GSTP1 I105 genotypes
28	Mouse	Induced	D-S/C-A	DEP + OVA + antioxidants		DEP effect on OVA sensitization reversed by thiol-containing antioxidants
29	Human (in vitro; peripheral B lymphs)	?	S-D-C-P-A	DEP + Sulforaphane	Mediator release in culture	Both DEP and sulforaphane can induce phase-II enzymes. Sulforaphane inhibits DEP-induced increases in IgE.
30	Human (in vitro; bronch epith cells)	?	D-C-A	DEP + Sulforaphane	Mediator release in culture	Sulforaphane inhibited DEP-induced production of IL-8, GM-CSF, and IL-1β

Abbreviations: A, assay (biochemical and/or physiologic); C, chemical exposure (e.g., DEP); D, drug; P, provocation; S, sensitization.

To summarize the studies reviewed in this section, DEP administration to human nasal tissue at the doses cited, either in vivo or in vitro, results in increased IgE and cytokine production, even absent specific allergen challenge.

POTENTIATION OF ALLERGIC SENSITIZATION (ADJUVANT EFFECTS) AND OF THE ACUTE ALLERGIC REACTION (PRIMING) IN THE UPPER AIRWAY

The majority of studies to be considered involve reactions to specific aeroallergens and the effect of preceding (or simultaneous) chemical exposures thereon. As noted in the introduction, the impact of these chemicals may be classified as adjuvant effects, priming, or both.

Diesel Exhaust Particles

Diaz-Sanchez et al. (16) conducted nasal challenges on 13 asymptomatic, nonsmoking, ragweed-sensitive volunteers with either ragweed extract (containing the antigen, *Amb* a 1) alone or—on a separate occasion—with ragweed plus DEP (0.3 mg). IgE (total and Ag-specific), as well as mRNA for epsilon chains and for various cytokines, were assayed in nasal lavage fluid at baseline and postchallenge (days 1, 4, and 8, depending upon the analyte). Postchallenge increases in Ag-specific—but not total—IgE were significantly greater on days 1 and 4 with DEP as opposed to Ag alone. Also increased four days post-Ag challenge was mRNA expression for various epsilon chain isoforms, with DEP altering the effect of Ag challenge on the expression of one of the splice variants studied. Finally, the Ag-induced increase in mRNA expression for the majority of cytokines assayed on post-challenge day 2 (i.e., IL-4, -5, -6, -10, -13, and IFN-γ) was significantly potentiated by DEP.

Fujieda et al. (17) studied DEP-facilitated in vivo Ig isotype switching among eight nonsmoking volunteers with ragweed allergy. Subjects first underwent a titrated nasal allergen challenge with ragweed extract, preceded and followed by nasal lavage. At eight-week intervals, subjects were challenged again with either Ag (at its minimum symptom-provoking dose) + DEP (0.3 mg) or with DEP alone. Total and Ag-specific IgE were assayed in nasal lavage fluid on each occasion. In addition, switch (S) circular DNAs indicative of μ to ϵ isotype switching (i.e., a switch from IgM to IgE production) were assayed. The investigators found comparable postchallenge increases in total IgE after DEP, Ag, and Ag + DEP challenge. On the other hand, both Ag-specific IgE and Sϵ/Sμ circular DNAs, while not materially altered by DEP challenge alone, were increased after Ag challenge, and increased to an even greater degree by combined Ag + DEP challenge. The authors concluded that DEP enhances Ag-driven isotype switching, and speculate that secular trends in allergy prevalence might reflect trends in environmental DEP exposure.

In a ground-breaking study, Diaz-Sanchez et al. (18) studied exposure to a novel allergen ("neoallergen") to determine whether initial IgE-mediated sensitization could be promoted by coexposure to DEP. The neoallergen utilized was keyhole limpet hemocyanin (KLH), a circulating glycoprotein from a marine mollusk (*Megathura cenulata*) for which natural exposure to humans is exceedingly unlikely. Twenty-five asymptomatic, atopic, nonsmoking volunteers were studied outside of their relevant aeroallergen season. KLH-specific IgA, IgE, IgG, and IgG4 were assayed in nasal lavage fluid of 10 subjects before and after intranasal administration of KLH (three doses at two-week intervals). The above regimen was repeated in another 15 subjects, with KLH instillation preceded by 24 hours by administration of DEP (0.3 mg) on each occasion. In addition, IL-4 and IFN-γ proteins were assayed in nasal lavage fluid. KLH administration alone resulted in the production of Ag-specific IgA, IgG, and IgG4—but not IgE—antibodies in the majority of subjects. Preadministration of DEP, on the other hand, resulted in the production of KLH Ag-specific IgE in the majority (9 of 15) of subjects, and also resulted in the production of higher levels of KLH-specific IgG4 antibody. Finally, nasal lavage levels of IL-4—but not IFN-γ— were significantly higher post-KLH immunization after preadministration of DEP than under control conditions. The authors theorized that a combination of facilitating a T_H2 cytokine milieu and enhancing the action of antigen-presenting cells may be responsible for DEPs adjuvant effects in this system.

In a study revealing a distinct *priming* effect of DEP, Diaz-Sanchez et al. (19) performed a titrated dose Ag-challenge of 11 dust mite–sensitive individuals. An initial symptom-provoking dose was first established to achieve a target rating score of 5 (of a possible 12) using Ag (*D. pteronnysinus*) alone. At six-week intervals thereafter, titrated Ag-challenge (beginning with 1/10 the symptom-provoking dose determined in the first session) was repeated *immediately after* instillation (in random order) of DEP, a carbon black suspension, or placebo. Initial Ag-provoked symptoms were unchanged after carbon black or placebo pretreatment, but were increased after DEP administration. Further, the titrated Ag dose necessary to reach the criterion rating of symptoms (5/12) was significantly reduced after DEP administration. When the experiment was repeated with nasal lavage histamine as the primary study endpoint, the findings were similar, with preadministration of DEP being the only intervention that resulted in significant augmentation of Ag-provoked histamine levels.

Sidestream Tobacco Smoke

Rumold et al. (20) studied the effect of side-stream tobacco smoke ("STS") on allergic sensitization to ovalbumin (OVA) in two different strains of mice, C57BL/6 and BALB/c. Animals were exposed to OVA aerosol for 20 minutes on 10 consecutive days after a one-hour exposure to either SHS or filtered air. Peripheral blood was sampled at roughly six-day intervals for 30 days, and analyzed for various Ig classes (including OVA-specific IgE). In addition, animals were challenged with OVA on day 30 and bronchoalveolar lavage (BAL) fluid analyzed for cellularity. Among low IgE-responding mice (C57BL/6), combined SHS/OVA exposure—but not OVA exposure alone—induced allergic sensitization. Among high IgE-responding mice (BALB/c), OVA exposure alone produced only a transient Ag-specific IgE response (day 12); however, this response was both sustained and enhanced if OVA exposure was preceded by SHS. In BALB/c mice, OVA challenge at day 30 produced a cellular infiltrate with eosinophil predominance, but only in the group jointly exposed to SHS/OVA. C57BL/6 mice responded similarly, although the cellular response in that strain included both eosinophils and neutrophils.

Diaz-Sanchez et al. (21) nasally challenged ragweed-sensitive human subjects with ragweed extract after exposure to either STS or filtered air for two hours. Nasal lavage fluid was obtained pre- and post-STS and Ag challenge. Subjects were subjected to the two experimental conditions at least six weeks apart and in counter-balanced order. Compared to the filtered air conditions, STS preexposure was associated with increased Ag-specific IgE, T_H2 cytokines (increased IL-4, -5, and -13; decreased IFN-γ), and histamine levels in lavage fluid. The authors interpreted these results as showing an adjuvant effect of STS.

Ozone

Bascom et al. (22) studied the effect of ozone exposure (0.5 ppm for four hours) on the response to nasal allergen challenge in 12 asymptomatic allergic rhinitics. Endpoints included self-rated symptoms, as well as analysis of nasal lavage fluid for histamine, albumin, TAME-esterase (an enzymatic assay that represents both the activity of plasma and glandular kallikrein and mast cell tryptase), and both total and differential cell counts. Prechallenge exposures alternated between O_3 and clean air on separate days two weeks apart. The authors found that O_3 exposure alone produced nasal symptoms and had proinflammatory effect (increased neutrophils, eosinophils, and mononuclear cells; elevated albumin levels). However, allowing for this effect, they did not find that O_3 alters ("primes") the response to Ag challenge.

In contrast to the above study, Peden et al. (23) found that preexposure to ozone (0.4 ppm for two hours) decreases the titrated Ag provocation dose required to produce a criterion nasal symptom score ($\geq$5 of a possible 9). In addition, preexposure to O_3 increased eosinophil and neutrophil influx—as well as ECP (eosinophil cationic protein) levels—in nasal lavage fluid sampled at 4 and 18 hours. A total of 10 asthmatics with allergy to dust mite (*D. farinae*) were studied on two occasions, separated by at least four weeks, comparing the effects of O_3 with sham (filtered air) exposure. Unique to the study design was the so-called "split nose" method, with bilateral exposure to test atmospheres, but unilateral exposure to Ag. The authors concluded that, in addition to having its own intrinsic proinflammatory effect on the nose, O_3 exposure potentiated ("primed") the late-phase allergic response to Ag in the noses of perennial allergic asthmatics.

Nitrogen Dioxide

Wang et al. (24) studied the effect of nitrogen dioxide preexposure on the nasal response to Ag challenge in 16 asymptomatic seasonal allergic rhinitics. Exposures consisted of 400 ppb NO_2 or filtered air randomized to separate days. Eight of the subjects underwent nasal lavage and measurement of nasal airway resistance (NAR) by active posterior rhinomanometry with no additional procedures. The other eight subjects, immediately after NO_2 or clean air exposure, underwent escalating Ag doses until they exhibited a threefold increase in NAR, followed by nasal lavage. In both arms of the experiment, nasal lavage fluid was analyzed for ECP, MCT (mast cell tryptase), MPO (myeloperoxidase), and IL-8. Of these mediators, only ECP showed an augmented Ag-induced response post-NO_2 exposure. By contrast, Ag-induced MCT increases were similar to post-NO_2 and clean air, and neither MPO nor IL-8 were significantly elevated post Ag-challenge, regardless of preceding exposure conditions. The authors concluded that preexposure to NO_2 increased eosinophil activation in the early phase of Ag response in allergic rhinitics.

MODULATION OF ADJUVANT AND/OR PRIMING EFFECTS BY DETOXICATION PATHWAYS—GENETIC AND PHARMACOLOGIC EFFECTS

Superimposed upon the investigation of adjuvant/priming effects has been increasing attention to biochemical pathways through which such effects are mediated, as well as the potential modulating effects of genetic variation and pharmacology on these pathways. Discussion of these factors follows.

Genetic Effects

Bastain et al. (25) examined individual reproducibility in the adjuvant effect of DEP when coadministered with ragweed extract in seasonal allergic rhinitic patients. At 30-day intervals, subjects underwent nasal lavage on a total of four alternating occasions, after challenge with Ag alone or in combination with DEP, counter-balancing sequences such that half of the subjects started with Ag alone, and half with the combined exposure. There was a high degree of both inter-subject variability and within-subject reproducibility in the extent to which coad-ministration of DEP increased Ag-specific IgE concentrations in nasal lavage fluid post-Ag challenge. On the other hand, the reproducibility of IL-4 and IFN-γ levels was restricted to post-combined challenge only. The authors concluded that susceptibility to the adjuvant effects of DEP was a stable individual trait, and hypothesized that genetic variation in phase-II detoxication enzymes (e.g., glutathione S-transferases or GSTs) might underlie this variability.

Gilliland et al. (26) tested the hypothesis that variations in GST genotype influence the magnitude of DEP's adjuvant effect across individuals. A total of 19 seasonal allergic rhinitic subjects underwent Ag challenge with ragweed extract on two occasions—with Ag alone and with Ag + DEP. The order of exposure was randomized, and nasal lavage fluid was obtained pre- and postexposure. Subjects also had their genomic DNA analyzed for GSTM, GSTP, and GSTT geno-type, acknowledging that the distribution of genotypes among allergic rhinitics may differ from that in the population at large. Judging by both the Ag-specific IgE response and nasal lavage histamine levels, the authors found that sub-jects with GSTM1 null or homozygous GSTP1 Ile105 wild-type genotypes have an enhanced DEP-related adjuvant response. Presence of both GSTM1 null and homozygous GSTP1 Ile105 wild-type genotypes appeared to exert (at least) an additive effect on DEP adjuvancy. By contrast, neither IL-4 nor IFN-γ responses were predicted by GST genotype.

Extending their model to another important environmental pollutant—second-hand tobacco smoke (SHS)—Gilliland et al. (27) again examined the role of GST genotype on adjuvancy. Using an analogous protocol to that in Gilliland et al. (26)—as well as the same panel of subjects as in the DEP/genotyping study—the authors evaluated the effect on Ag-specific IgE and histamine of pre-exposure to two hours of SHS versus preexposure to filtered air before nasal ragweed extract challenge. Genomic DNA obtained from buccal smears had pre-viously been analyzed for GSTM1, GSTP1, and GSTT1 genotype. On an indi-vidual basis, the extent to which SHS enhanced the response to Ag was signif-icantly correlated with each subject's prior response to DEP. Not surprisingly, GSTM1 null genotype and homozygous GSTP1 Ile105 wild-type—both singly and in combination—were associated with an enhanced adjuvant effect of prior SHS exposure on nasal Ag challenge.

Pharmacologic Effects

Whitekus et al. (28) examined the influence of various antioxidants on DEP's adjuvant effect on OVA sensitization in mice. After an in vitro screening procedure that narrowed consideration to the thiol agents N-acetyl cysteine (NAC) and bucillamine (BUC), the authors proceeded to conduct a multiarm experiment in which mice were exposed by inhalation to saline (control), OVA alone, DEP alone, or coexposed to OVA and DEP. In addition, all noncontrol conditions were repeated after intraperitoneal NAC or BUC administration. Both NAC and BUC significantly blunted the OVA-specific peripheral antibody response (including both IgE and IgG1) in DEP-treated animals, without affecting the OVA-alone response. In addition, lung homogenates obtained post-Ag-challenge in OVA/DEP-treated (i.e., sensitized) animals showed significantly lower levels of carbonyl proteins and lipid peroxides after NAC or BUC pretreatment, evidence of a protective effect of these agents against oxidative stress.

Two additional experiments have examined the effect of sulforaphane—an inducer of the phase II (detoxifying) enzymes GSTM1 and NAD(P)H: quinine oxidoreductase (NQO1)—on DEP oxidative stress signaling. Wan and Diaz-Sanchez (29) isolated B lymphocytes (CD19+/CD20+) from peripheral blood and propagated them in culture. They then assayed for GSTM1 and NQO1 mRNA in the presence of increasing concentrations of sulforaphane. For NQO1 in particular, there was a dose-related increase in mRNA with increasing sulforaphane concentrations in growth medium, consistent with enzyme induction. In addition, the authors observed a dose-related inhibition by sulforaphane of DEP-enhanced IgE production in cultured cells, consistent with chemoprotective effect. Ritz et al. (30) studied the effect of sulforaphane on the DEP response in cultured human bronchial epithelial cells. Pretreatment with sulforaphane dampened the cytokine (IL-1β and IL-8) and growth factor (GM-CSF) response when these cells were exposed to DEP extract. Of note, sulforaphane is derived from cruciferous vegetables, and could potentially be administered orally as a chemopreventive agent if shown to be both safe and efficacious.

OTHER AIR POLLUTANT EFFECTS ON UPPER AIRWAY ALLERGIES

Effect of Photochemical Oxidants on Aeroallergens

The final topic of discussion in this chapter focuses "upstream" of allergic sensitization and triggering: specifically on the potential environmental effects of air pollutants on the *quality and quantity of prevalent aeroallergens*.

Masuch et al. (31) published a report comparing antigen levels in rye grass (*Lolium perenne*) pollen in different areas of Germany with different ozone (O_3) levels. In a separate controlled experiment, rye grass was grown in chambers containing either filtered air or O_3 at 130 $\mu g/m^3$. In both cases, the group-5 antigen content per gram of pollen protein was greater in the high- than in the low-O_3 growing condition. Similar findings were reported by Armentia et al. (32), comparing *Lolium perenne* pollen gathered in an urban versus rural environment in Spain.

Ruffin et al. (33) exposed pollen grains from Red Oak (*Quercus rubra*), Meadow Fescue (*Festuca elatior*), and Chinese Elm (*Ulmas pumila*) to carbon monoxide (CO), sulfur dioxide (SO_2), and nitrogen dioxide (NO_2). They observed pollutant-related changes in antigen quality, using two-dimensional thin-layer chromatography. A change in immunologic reactivity was confirmed using rabbit antisera in a double-diffusion chamber. The potential clinical significance of these changes remains unclear, however.

In addition to effects of conventional air pollutants on plant growth and pollen chemistry, increasing atmospheric CO_2 and temperature (i.e., climate change) may lengthen pollen seasons and alter geographic distributions of plant species, as well as change the patterns of the winds distributing aeroallergens. All of these factors may, over time, influence the prevalence and severity of allergic airway disease. The interested reader is referred to several excellent reviews on this topic (34–36).

SUMMARY/CONCLUSIONS

Upon airborne exposure to humans, both allergens and air pollutants have their initial impact on the upper airway. Individuals with allergic sensitization, as compared to those who are not sensitized, respond differently to both classes of agents. Air pollutants can exert both chemically nonspecific (i.e., irritant) and specific effects. Among the latter are the adjuvant and/or priming effects reviewed above. Combustion products such as cigarette smoke and diesel exhaust contain families of compounds, such as polycyclic aromatic hydrocarbons, which may be responsible for their adjuvancy and priming action, in all likelihood mediated by the oxidative stress they produce (37). The studies reviewed in this chapter provide a rationale for aggressive control of exposures to selected pollutants. In addition, together with environmental exposure data, these studies may provide at least a partial explanation of the increasing prevalence of atopic disease in many populations (38).

ACKNOWLEDGMENT

The author wishes to than Dr. Alkis Togias for his generous expenditure of time in reviewing and commenting upon this manuscript.

REFERENCES

1. Leung DY. Allergic immune response. In: Bierman CW, Pearlman DS, Shapiro GG, Busse WW, eds. Allergy, Asthma and Immunology from Infancy to Adulthood. Philadelphia, PA: WB Saunders, 1996:68–78.
2. Alarie Y. Sensory irritation by airborne chemicals. CRC Crit Rev Toxicol 1973; 2: 299–363.
3. Green BG, Mason JR, Kare MR. Preface. In: Green BG, Mason JR, Kare MR, eds. Chemical Senses, Vol 2: Irritation. New York, NY: Marcel Dekker, 1990:v-vii.
4. Widdicombe JG. Nasal pathophysiology. Respir Med 1990; 84(suppl A):3–9; discussion 9–10.
5. Ownby DR. Tests for IgE antibody. In: Bierman CW, Pearlman DS, Shapiro GG, Busse WW, eds. Allergy, Asthma and Immunology from Infancy to Adulthood. Philadelphia, PA: WB Saunders, 1996:144–156.
6. Caterina MJ, Schumacher MA, Tominaga M, et al. The capsaicin receptor: a heat-activated ion channel in the pain pathway. Nature 1997; 389(6653):816–824.
7. Julius D, Basbaum AI. Molecular mechanisms of nociception. Nature 2001; 413(6852):203–210.
8. Jordt SE, Bautista DM, Chuang HH, et al. Mustard oils and cannabinoids excite sensory nerve fibres through the TRP channel ANKTM1. Nature 2004; 427(6971):260–265.

9. Bessac BF, Sivula M, von Hehn CA, et al. TRPA1 is a major oxidant sensor in murine airway sensory neurons. J Clin Invest 2008; 118(5):1899–1910.

10. Tai CF, Baraniuk JN. A tale of two neurons in the upper airways: pain versus itch. Curr Allergy Asthma Rep 2003; 3(3):215–220.

11. Taylor-Clark TE, Kollarik M, MacGlashan DW Jr, et al. Nasal sensory nerve populations responding to histamine and capsaicin. J Allergy Clin Immunol 2005; 116(6):1282–1288.

12. Kuby J. Immunology. New York, NY: Freeman, 1992:91–92.

13. Diaz-Sanchez D, Dotson AR, Takenaka H, et al. Diesel exhaust particles induce local IgE production in vivo and alter the pattern of IgE messenger RNA isoforms. J Clin Invest 1994; 94(4):1417–1425.

14. Diaz-Sanchez D, Tsien A, Casillas A, et al. Enhanced nasal cytokine production in human beings after in vivo challenge with diesel exhaust particles. J Allergy Clin Immunol 1996; 98(1):114–123.

15. Terada N, Jamano N, Maesako K-I, et al. Diesel exhaust particulates upregulate histamine receptor mRNA and increase histamine-induced IL-8 and GM-CSF production in nasal epithelial cells and endothelial cells. Clin Exper Allergy 1999; 29:52–59.

16. Diaz-Sanchez D, Tsien A, Fleming J, et al. Combined diesel exhaust particulate and ragweed allergen challenge markedly enhances human in vivo nasal ragweed-specific IgE and skews cytokine production to a T helper cell 2-type pattern. J Immunol 1997; 158(5):2406–2413.

17. Fujieda S, Diaz-Sanchez D, Saxon A. Combined nasal challenge with diesel exhaust particles and allergen induces In vivo IgE isotype switching. Am J Respir Cell Mol Biol 1998; 19(3):507–512.

18. Diaz-Sanchez D, Garcia MP, Wang M, et al. Nasal challenge with diesel exhaust particles can induce sensitization to a neoallergen in the human mucosa. J Allergy Clin Immunol 1999; 104(6):1183–1188.

19. Diaz-Sanchez D, Penichet-Garcia M, Saxon A. Diesel exhaust particles directly induce activated mast cells to degranulate and increase histamine levels and symptom severity. J Allergy Clin Immunol 2000; 106(6):1140–1146.

20. Rumold R, Jyrala M, Diaz-Sanchez D. Secondhand smoke induces allergic sensitization in mice. J Immunol 2001; 167(8):4765–4770.

21. Diaz-Sanchez D, Rumold R, Gong H Jr. Challenge with environmental tobacco smoke exacerbates allergic airway disease in human beings. J Allergy Clin Immunol 2006; 118(2):441–446.

22. Bascom R, Naclerio RM, Fitzgerald TK, et al. Effect of ozone inhalation on the response to nasal challenge with antigen of allergic subjects. Am Rev Respir Dis 1990; 142(3):594–601.

23. Peden DB, Setzer RW Jr, Devlin RB. Ozone exposure has both a priming effect on allergen-induced responses and an intrinsic inflammatory action in the nasal airways of perennially allergic asthmatics. Am J Respir Crit Care Med 1995; 151(5):1336–1345.

24. Wang JH, Devalia JL, Duddle JM, et al. Effect of six-hour exposure to nitrogen dioxide on early-phase nasal response to allergen challenge in patients with a history of seasonal allergic rhinitis. J Allergy Clin Immunol 1995; 96(5 Pt 1):669–676.

25. Bastain TM, Gilliland FD, Li YF, et al. Intraindividual reproducibility of nasal allergic responses to diesel exhaust particles indicates a susceptible phenotype. Clin Immunol 2003; 109(2):130–136.

26. Gilliland FD, Li YF, Saxon A, et al. Effect of glutathione-S-transferase M1 and P1 genotypes on xenobiotic enhancement of allergic responses: randomised, placebo-controlled crossover study. Lancet 2004; 363(9403):119–125.

27. Gilliland FD, Li YF, Gong H Jr, et al. Glutathione s-transferases M1 and P1 prevent aggravation of allergic responses by secondhand smoke. Am J Respir Crit Care Med 2006; 174(12):1335–1341.

28. Whitekus MJ, Li N, Zhang M, et al. Thiol antioxidants inhibit the adjuvant effects of aerosolized diesel exhaust particles in a murine model for ovalbumin sensitization. J Immunol 2002; 168(5):2560–2567.

29. Wan J, Diaz-Sanchez D. Phase II enzymes induction blocks the enhanced IgE production in B cells by diesel exhaust particles. J Immunol 2006; 177(5):3477–3483.
30. Ritz SA, Wan J, Diaz-Sanchez D. Sulforaphane-stimulated phase II enzyme induction inhibits cytokine production by airway epithelial cells stimulated with diesel extract. Am J Physiol Lung Cell Mol Physiol 2007; 292(1):L33–L39.
31. Masuch GI, Franz JT, Schoene K, et al. Ozone increases group 5 allergen content of Lolium perenne. Allergy 1997; 52(8):874–875.
32. Armentia A, Lombardero M, Callejo A, et al. Is Lolium pollen from an urban environment more allergenic than rural pollen? Allergol Immunopathol (Madr) 2002; 30(4):218–224.
33. Ruffin J, Liu MY, Sessoms R, et al. Effects of certain atmospheric pollutants (SO_2, NO_2 and CO) on the soluble amino acids, molecular weight and antigenicity of some airborne pollen grains. Cytobios 1986; 46(185):119–129.
34. Emberlin J. Interaction between air pollutants and aeroallergens. Clin Exp Allergy 1995; 25(suppl 3):33–39.
35. D'Amato G, Cecchi L. Effects of climate change on environmental factors in respiratory allergic diseases. Clin Exp Allergy 2008; 38:1264–1274.
36. Shea KM, Truckner RT, Weber RW, et al. Climate change and allergic disease. J Allergy Clin Immunol 2008; 122:443–453.
37. Saxon A, Diaz-Sanchez D. Air pollution and allergy: You are what you breathe. Nat Immunol 2005; 6:223–226.
38. Diaz-Sanchez D. Pollution and the immune response: Atopic diseases – are we too dirty or too clean? Immunology 2000; 101:11–18.

Index

Printed in the United States
by Baker & Taylor Publisher Services